Proceedings of the International Symposium on

Exotic Nuclei

Proceedings of the International Symposium on

Exotic Nuclei

Petrozavodsk, Russia 10 – 15 September 2018

Editors

Yu. E. Penionzhkevich

Yu. G. Sobolev

Joint Institute for Nuclear Research, Dubna, Russia

World Scientific

NEW JERSEY • LONDON • SINGAPORE • BEIJING • SHANGHAI • HONG KONG • TAIPEI • CHENNAI • TOKYO

Published by

World Scientific Publishing Co. Pte. Ltd.
5 Toh Tuck Link, Singapore 596224
USA office: 27 Warren Street, Suite 401-402, Hackensack, NJ 07601
UK office: 57 Shelton Street, Covent Garden, London WC2H 9HE

Library of Congress Control Number: 2019030147

British Library Cataloguing-in-Publication Data
A catalogue record for this book is available from the British Library.

Cover design and color photos compilation by Yu. Sobolev
Photos by V. Maslov

EXOTIC NUCLEI
EXON-2018
Proceedings of the International Symposium on Exotic Nuclei

ISBN 978-981-120-944-4

For any available supplementary material, please visit
https://www.worldscientific.com/worldscibooks/10.1142/11530#t=suppl

PREFACE

From 10 to 15 September Petrozavodsk (Russian Federation) hosted the International Symposium on one of the most important and the most rapidly developing areas of nuclear physics — the physics of exotic states of nuclei — "EXON 2018". The organizers of the Symposium were the five largest scientific centers where this area is being successfully developed — the Joint Institute for Nuclear Research in Dubna, the GANIL National Center (France), the RIKEN Research Center (Japan), the GSI Helmholtz Centre for Heavy Ion Research (Germany), and the National Superconducting Cyclotron Laboratory (Michigan, USA). The Symposium was held with the active participation of Petrozavodsk Federal University (PetrGU).

Currently, the most complex physical experiments conducted with the use of the large accelerators facilities and requiring huge financial investments cannot be performed by a single, even highly developed, country. An example of this might be the construction of the Large Hadron Collider at the European Organization for Nuclear Research (CERN), the NICA collider and heavy-ion accelerators at the Joint Institute for Nuclear Research in Dubna, which includes 20 member countries. Therefore, these studies are carried out in close cooperation of research centers of several countries, each of which makes its financial and intellectual contribution to the construction of the largest facilities. These fundamental studies and methods used in them are also of great importance for the related areas of science and technology.

This is the Ninth Symposium on Exotic Nuclei held in Russia. The first symposium took place 27 years ago in 1991 in Foros (Crimea); later symposiums were held on Baikal Lake, in Peterhof, Khanty-Mansiysk, Sochi, Vladivostok, Kaliningrad and Kazan.

Considerable interest in this research and new experimental possibilities have emerged with the launch of radioactive ion beam facilities, and the collaboration has extended to five research centers, that are the co-founders of the EXON workshops. Already in 2004, the co-founders of the symposium, held in Peterhof (near St. Petersburg) were GSI, GANIL, RIKEN, JINR, and later MSU (Michigan, USA).

The Symposium "EXON 2018" was attended by about 140 scientists from 20 countries, most of whom are leading experts in the field of nuclear physics. The most representative delegations were from Japan and China. Research centers of these countries are interested in developing cooperation with JINR and scientific centers of Russia.

The symposium was devoted to the investigation of nuclei in extreme states and, in particular, at the limits of nuclear stability (from very light neutron- and proton-rich up to superheavy nuclei).

The EXON 2018 symposium dealt with the following next topics:

- Properties of light exotic nuclei
- Superheavy elements. Synthesis and properties
- Rare processes and decays
- Radioactive beams. Production and research programs
- Experimental facilities and future projects.

The scientific program included invited talks on important areas of the physics of exotic nuclei and new projects on the largest accelerator complexes and experimental facilities. In addition, discussions with the participation of leading scientists from various research centers in the world were organized. They discussed issues of cooperation in the field of fundamental physics of heavy ions and applied research.

Exoticism in Nuclear Physics

It has been 27 years since the First International Symposium on Exotic Nuclei (EXON) was held. One of the main trends in modern nuclear physics is to obtain and study the properties of nuclear matter (or nuclei) in extreme state. The appearance of powerful accelerators of heavy ions allows the synthesis of such nuclei. Exotic nuclei are basically a nuclear matter in a very unusual state. There are several unusual states that define nuclei as exotic: nuclei of large angular momentum ("rabidly" rotating nuclei), nuclei of high excitation energy ("hot" nuclei), strongly deformed nuclei (super and hyper-deformation, nuclei in unusual configuration form), nuclei with abnormally high number of neutrons or protons (neutron-rich and proton-rich nuclei), and superheavy nuclei up to atomic number of protons $Z > 110$. The study of the properties of nuclear matter in its unusual state provides the important information about the properties of the microcosm and thus allows us to simulate various processes that take place in the Universe. For EXON Symposium we set ourselves several tasks:

1. The symposium should summarize the latest results obtained on the actual subject in collaboration with other research centers.
2. The symposium should take place only in exotic locations of Russia.
3. The symposium results should promote decisions on future programme of the joint experiments.
4. All the symposiums should be based on local universities under active support and immediate participation of the students and teachers of these universities.

We seem to have managed to cope with all the set tasks according to the evidence of geography of the symposium venues. Exotic nuclei show unusual properties in reactions with other nuclei as well. There have recently been a number of interesting results in the study of excitation functions of full cross sections of reactions, fusion-fission reactions and particle evaporation from the excited compound nucleus. These reactions were studied in the beams of radioactive nuclei ^{6}He, ^{8}He, ^{11}Li, and others. This trend of research on reaction with exotic nuclei is one of the major ones in the programme of radioactive beam-factories.

Superheavy Elements

During the symposium, the results of recent experiments on the synthesis and study of properties of nuclei of new superheavy elements were discussed: the discovery of new superheavy elements shows high efficiency of international cooperation. Interesting results were obtained in joint experiments on chemical identification of elements 112 and 114, performed at the Laboratory of Nuclear Reactions (FLNR) JINR (Russia), the GSI (Germany), and the Paul Scherrer Institute (Switzerland). A striking example of cooperation with US scientists is an experiment on the synthesis of element 117, held at the cyclotron of FLNR JINR under the leadership of Yu. Ts. Oganessian. In 2014 Moscow hosted the inauguration of element 114 (Flerovium) and element 116 (Livermorium) discovered in Dubna.

In November 2016, the International Union of Pure and Applied Chemistry (IUPAC) and the International Union of Pure and Applied Physics (IUPAP) approved the discovery of new chemical elements of the periodic table of D. I. Mendeleev with atomic numbers 113, 115, 117, and 118.

For the element with atomic number 113 the discoverers at RIKEN Nishina Center for Accelerator-Based Science (Japan) proposed the name Nihonium and the symbol Nh. Nihon is one of the ways to say "Japan" in Japanese, and literally means "the Land of Rising Sun".

For the element with atomic number 115 the name proposed is Moscovium with the symbol Mc and for the element with atomic number 117, the name proposed is Tennessine with the symbol Ts. These are in line with the tradition of honoring a place or geographical region and are proposed jointly by the discoverers at the Joint Institute for Nuclear Research, Dubna (Russia), Oak Ridge National Laboratory (USA), Vanderbilt University (USA) and Lawrence Livermore National Laboratory (USA). The name Moscovium is in recognition of the city of Dubna, Moscow Region. The name Tennessine is in recognition of the contribution of the Tennessee region, including Oak Ridge National Laboratory, Vanderbilt University, and the University of Tennessee at Knoxville, to superheavy element research.

For the element with atomic number 118 the collaborating teams of discoverers at the Joint Institute for Nuclear Research, Dubna (Russia) and Lawrence Livermore National Laboratory (USA) proposed the name Oganesson and the symbol Og. The proposal is in line with the tradition of honoring a scientist and recognizes Professor Yuri Oganessian for his pioneering contributions to transactinoid elements research. His many achievements include the discovery of superheavy elements and significant advances in the nuclear physics of superheavy nuclei including experimental evidence for the "island of stability".

Thus, the experiments on synthesis of superheavy elements take a major part of the research programme of the leading research centers. In this regard, an important step in further understanding of the properties of superheavy elements were the experiments on research of the structure of transfermium nuclei. The experiments on measurement of gamma transitions in nuclei of 101, 102 and 103 elements have begun. These experiments are currently running at the University of Jyvaskyla (Finland), in GANIL (France) and in FLNR (JINR).

Interesting results were obtained in joint experiments on chemical identification of elements 105, 112 and 114, performed at the Laboratory of Nuclear Reactions (FLNR) JINR (Russia), the GSI (Germany), and the Paul Scherrer Institute (Switzerland). Radiochemists from Dubna, from Technical University of Munich and from GSI, who have performed the first experiments on chemical properties of superheavy elements, have made an important step in chemical identification of superheavy elements and in determination of their place in Mendeleev periodic table. This important trend in research of the properties of superheavy elements is expected to be continued in a broad collaboration of physicists and radiochemists of the leading centers.

New Facilities — New Projects

A special day of the Symposium was devoted to the present and future accelerator complexes for heavy ions and radioactive nuclei in the leading scientific centers of the world. Five laboratories, which are the co-founders of the symposium, are now creating a new generation of accelerators, which will enable significant progress towards the synthesis and study of properties of new exotic nuclei. Projects SPIRAL2, RIKEN RI Beam Factory, FAIR, DRIBs, NICA, FRIB were presented respectively by the heads of these projects — H. En'yo, S. Dmitriev, D. Ackermann, H. Savajols, J. J. Lawri, F. Ibrahim *et al.*

A new accelerator complex NUSTAR in GSI, which includes pilot plants for precision studies of nuclear structure, is now under creation on international-collaboration-basis. At present, a new accelerator complex is in RIKEN; it is intended to produce intense beams of stable and radioactive nuclei, which are going to be used to research the structure of exotic nuclei. Now, the GANIL accelerator complex for production of radioactive beams is under development. The complex SPIRAL in

GANIL has started its first experiments. There began the development of the SPIIRAL-2 project, which is scheduled to receive the first beams of heavy radioactive nuclei in 2018. Accelerator complex DRIBs3, new fragment separator AKKULINNA-2 and SHNF in Dubna will begin. The ALTO Project (Orsay, France) is based on acceleration of photo-fission fragments as well and in other major research centers. Thus, in 2018–2020 there will be launched a number of new accelerator complexes equipped with the unique experimental facilities, which will let us make a new step in synthesis and study of properties of new exotic states of nuclear matter.

One of the organizers and the basis of the Symposium is Petrozavodsk Federal University, which contributed to the preparation of highly qualified personnel and joint research. For the university, the contacts with leading scientists and heads of worldwide major research centers in the area of nuclear physics are the basis for future cooperation in the field of fundamental physics and applied research in related fields of science and technology, in particular, medicine, ecology, geology, information technology, etc.

Before the beginning of the symposium, a satellite school “Contemporary Physics and Nuclear Medicine” was held for two days. At the school, the leading JINR professors delivered lectures to students, post-graduates, and professors of PetrGU on modern problems of nuclear physics and applied physics.

Yu. Penionzhkevich
Chairman EXON 2018

Co-chairmen:

Yu. Oganessian (JINR, Dubna)
H. En'yo (RIKEN, Japan)
A. Navin (GANIL, France)
Ch. Scheidenberger (GSI, Germany)
G. Bollen (MSU, USA)

International Organizing Committee:

V. Matveev (Russia)
G. Trubnikov (Russia)
T. Uesaka (Japan)
M. Lewitowicz (France)
B. Sherrill (USA)
S. Yennello (USA)
A. Maj (Poland)
A. Jokinen (Finland)
Z. Vilakazi (South Africa)

Local Organizing Committee:

Yu. Penionzhkevich (Chairman)
Yu. Sobolev (Scientific secretary)
S. Dmitriev (JINR, Dubna)
D. Kamanin (JINR, Dubna)
E. Kuznetsova (JINR, Dubna)
S. Pakuliak (JINR, Dubna)
V. Ivanshin (JINR, Dubna)
A. Voronov (PetrSU, Petrozavodsk)

Organized by:

Joint Institute for Nuclear Research
(Dubna, RUSSIA)

RIKEN Nishina Center
for Accelerator-Based Science
(Wako-shi, JAPAN)

GSI Helmholtzzentrum für Schwerionenforschung
(Darmstadt, GERMANY)

Grand Accélérateur National d'Ions Lourds
(Caen, FRANCE)

Facility for Rare Isotope Beams
Michigan State University
(Michigan, USA)

Petrozavodsk State University
(Petrozavodsk, Russia)

CONTENTS

1

Light exotic nuclei and study of their properties

^{11}Li + ^{9}Be Reaction in the Framework of the Time-dependent Schrödinger Equation

A. K. Azhibekov[1,2,*] and K. A. Kuterbekov[1]

[1]Faculty of Physics and Technical Sciences,
L. N. Gumilyov Eurasian National University,
Astana, Kazakhstan
[*]E-mail: azhibekoaidos@mail.ru
www.enu.kz

V. V. Samarin[2]

[2]Flerov Laboratory of Nuclear Reactions,
Joint Institute for Nuclear Research Dubna,
Moscow Region, Russia
E-mail: samarin@jinr.ru

Neutron transfer and nucleus breakup cross sections in ^{11}Li+^{9}Be reaction are calculated at energy range up to 32 MeV/nucleon. The evolution of probability density of external weakly bound neutrons of ^{11}Li and the probabilities of neutron transfer and nucleus breakup are determined on the basis of a numerical solution of the time-dependent Schrödinger equation. Our calculation results are agree with the experiment.

Keywords: Neutron transfer; nucleus breakup; time-dependent Schrödinger equation; shell model.

1. Introduction

The article studies processes of neutron transfer and nucleus breakup in reaction (^{11}Li,^{9}Li) at ^{9}Be-target. Recently, study of the mechanisms of interaction of weakly bound light exotic nuclei arouse great interest. It is due to effects of a sharp increase of the total cross section in reactions with these nuclei, [1-5]. The ^{11}Li nucleus has a halo structure consisting of two external weakly bound neutrons. The neutron separation energy in the ^{11}Li nucleus is 0.4 MeV [6], which is significantly less than in stable lithium isotopes: ^{6}Li (5.7 MeV), ^{7}Li (7.3 MeV), ^{9}Li (4 MeV). For this reason, reactions of ^{11}Li nucleus with light nuclei are highly likely to involve neutron transfer and nucleus breakup.

Currently, one of the most widespread theoretical models for describing neutron transfer processes is Distorted Wave Born Approximation method [7]. In this method, particles rearrangement process in the reaction $A(a,b)B$ is divided into three stages: motion of projectile a in optical potential of target A, transfer of nucleon from projectile a to target and motion of ejectile b in field of nucleus *B*. However, this division of interaction process into three stages does not allow analysis of the dynamics of taking place processes.

The use of the time-dependent Schrödinger equation to study rearrangement of neutrons during fusion reactions was proposed by [8,9]. Such approach makes it possible to obtain a microscopic description of dynamics of nuclear fusion [8,9], neutrons stripping and pick-up [10–14], and breakup of nuclei. This paper described neutrons transfer and breakup processes in a collision of ^{11}Li+^{9}Be. As in [8,9], the spin-orbit interaction was not taken into account.

2. Theoretical Calculations

Shell model without spin-orbit interaction was used for calculation of the radial parts of the neutron wave functions of the ^{11}Li nucleus. The radial Schrödinger equation for the central mean field $V(r)$ was solved by the difference method with matrix eigenvalues calculations [15]. Results for normalized radial part $R(r)$ of two wave functions for 1*s* and 1*p* levels neutrons were shown in Figure 1. Wave functions $\Psi(\vec{r})$ of the 1*p* state with projections of angular momentum $m_l = 0, \pm 1$ were used for initial conditions of the time-dependent Schrödinger equation. For used parameters of mean fields of nuclei $V_1(r)$, $V_2(r)$ separation energies of neutron from target and projectile are equal to experimental values [6].

The time-dependent Schrödinger equation describing evolution of wave function of external neutron $\Psi(\vec{r},t)$ in field of colliding nuclei:

$$i\hbar\frac{\partial\Psi}{\partial t} = -\frac{\hbar^2}{2m}\Delta\Psi + \left[V_1\left(\left|\vec{r}-\vec{r}_1(t)\right|\right) + V_2\left(\left|\vec{r}-\vec{r}_2(t)\right|\right)\right]\Psi. \quad (1)$$

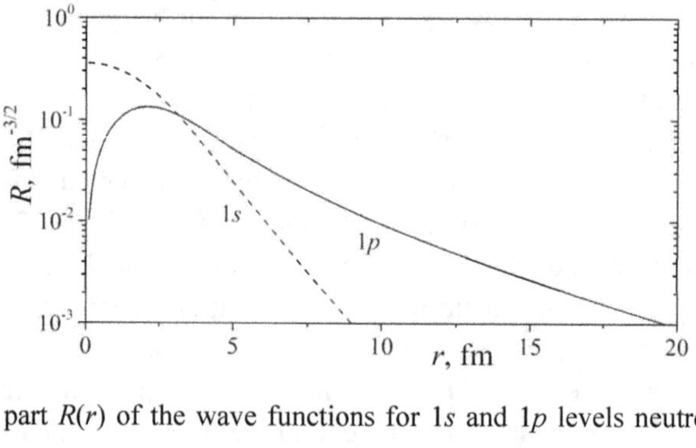

Fig. 1. Radial part $R(r)$ of the wave functions for 1*s* and 1*p* levels neutrons in the ^{11}Li nucleus.

The Eq. (1) was solved by the scheme used in [16]. Here $\vec{r}_1(t)$, $\vec{r}_2(t)$ are the radii of the centers of the colliding nuclei which moved along line trajectories, since at high energies the motion of the centers of nuclei occurs along almost line trajectories.

Figures 2 and 3 present results of a numerical solution of the time-dependent Schrödinger equation. As we see, the time-dependent Schrödinger equation allowed us to assess visually the dynamics of taking place processes. The external neutrons of ^{11}Li nucleus may be removal during two processes: transfer to target nucleus and break up to unbound states of continues energy spectrum. At a distance of about 15–16 fm between the centers of the nuclei, the probability density began to transfer to target (see Figures 2a,b and 3a,b). Transfer process was accompanied by breakup of ^{11}Li nucleus (see Figures 2c,d and 3c,d), due to low binding energy of external neutron of 0.4 MeV [6].

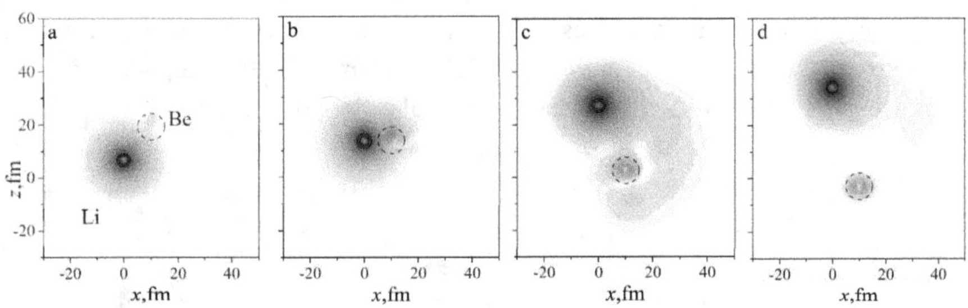

Fig. 2. Evolution of probability densities of external neutron of ^{11}Li in the collision with ^{9}Be at 8 MeV/nucleon. Impact parameter $b = 10$ fm. Time scale from left to right. Dash circle is ^{9}Be nucleus.

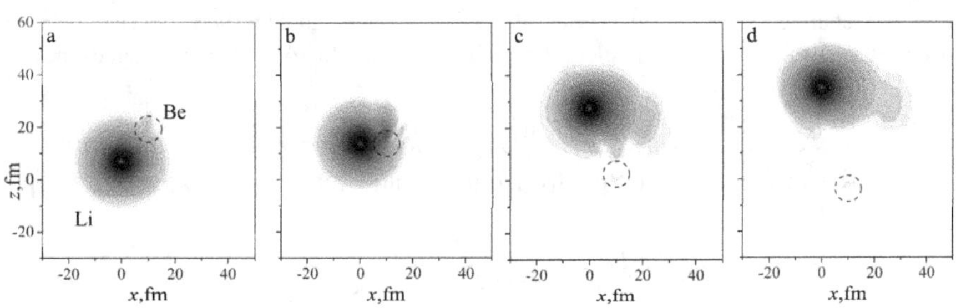

Fig. 3. Evolution of probability densities of external neutron of ^{11}Li in the collision with ^{9}Be at 32.4 MeV/nucleon. Impact parameter $b = 10$ fm. Time scale from left to right. Dash circle is ^{9}Be nucleus.

Probabilities $p(b,E)$ of neutron transfer and breakup of nucleus, what depends on impact parameter b and energy E in the center of mass system, was calculated by integrating probability density over volume:

$$p_{\text{transfer}}(b,E) = \int_T |\Psi|^2 \, dV, \tag{2}$$

$$p_{\text{breakup}}(b,E) = C\int_B |\Psi|^2 \, dV, \tag{3}$$

here C is a parameter that takes into account the uncertainty of the calculation of the breakup probability. It is due to slow separation of the wave function during breakup, see Figures 2d and 3d. For Eq. (2), T is region of integration within the beryllium nucleus volume that corresponds to transfer of neutron. In addition, for Eq. (3), B is region of integration over the volume outside lithium and beryllium that corresponds to the breakup of the ^{11}Li nucleus.

Figure 4 shows linear regression of calculated neutron transfer and nucleus breakup probabilities at 8 and 32.4 MeV/nucleon. As the impact parameter increases, the probabilities decrease by several orders of magnitude.

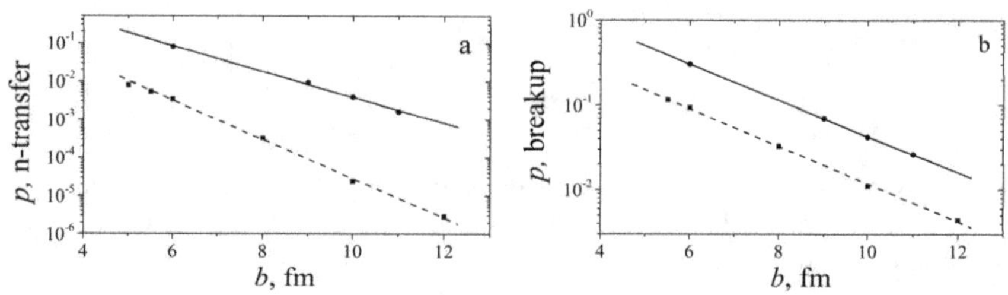

Fig. 4. Dependencies of probabilities of transfer (a) and nucleus breakup (b) on impact parameter b. ● – E = 8 MeV/nucleon (solid line); ■ – 32.4MeV/nucleon (dashed line).

The reaction cross sections (^{11}Li,^{9}Li) were determined by integrating the sum of the probabilities of neutron transfer and nucleus breakup over impact parameter b:

$$\sigma_{-2n} = 2\pi\int\left(p_{\text{breakup}}(b,E) + p_{\text{transfer}}(b,E)\right)b\,db. \tag{4}$$

Figure 5 shows calculated cross sections for neutron transfer and nuclear breakup. For the reaction ^{11}Li+^{9}Be at 29.0 MeV/nucleon, experimental value of breakup two neutrons σ_{-2n} is about 470±100 mb [17].

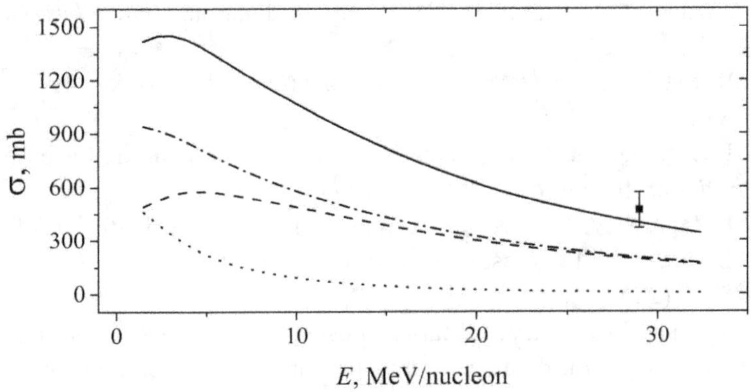

Fig. 5. ■ – σ_{-2n} [17]. Dashed line – calculation for breakup of nucleus; Dotted line – calculation for neutron transfer; Dashed-dotted line – (^{11}Li,^{9}Li) cross section ($C = 1$); Solid line – (^{11}Li,^{9}Li) cross section ($C = 2$).

3. Conclusion

The cross sections of neutron transfer and nucleus breakup in ^{11}Li+^{9}Be was calculated based on the numerical solution of the time-dependent Schrödinger equation. Further theoretical and experimental improvements are needed by the more exactly description the processes of interaction of external neutrons in reactions with weakly bound nuclei.

References

1. A.M. Kabyshev et al., Some peculiarities of interactions of weakly bound lithium nuclei at near-barrier energies, *Journal of Physics G: Nuclear and Particle Physics* **45**, 025103 (2018).
2. R.E. Warner et al., Total reaction and 2n-removal cross section of 20-60A MeV ^{4,6,8}He, $^{6\text{-}9,11}$Li and ^{10}Be on Si, *Phys. Rev. C***54**, 1700 (1996).
3. L. Chen et al., Measurement of total reaction cross section for neutron-rich nucleus ^{11}Li on ^{28}Si target at medium energy, *High Energy Phys. Nucl. Phys.* **31**, 1102 (2007).
4. A.C.C. Villar, Measurements of reaction cross sections for neutron-rich exotic nuclei by a new direct method, *Physics Letters B***268**, 345 (1991).
5. K.A. Kuterbekov, A.M. Kabyshev, A.K. Azhibekov, Peculiarities of interaction of weakly bound lithium nuclei (A=6–11) at low energies: Elastic scattering and total reaction cross sections, *Chinese Journal of Physics* **55**, 2523 (2017).

6. M. Wang et al., The AME2012 atomic evaluation, *Chin. Phys. C***36**, 1603 (2012).
7. G.R. Satchler, *Direct Nuclear Reactions* (Oxford University Press, Oxford, 1983).
8. V.I. Zagrebaev and V.V. Samarin, Role of neutrons in the fusion of nuclei, *Physics of Atomic Nuclei* **70**, 1003 (2007).
9. V.I. Zagrebaev, V.V. Samarin, W. Greiner, Sub-barrier fusion of neutron-rich nuclei and its astrophysical consequences, *Physical Review C***75**, 035809 (2007).
10. V.V. Samarin and K.V. Samarin, Consideration of the spin–orbit interaction in the description of nucleon transfers upon heavy ionic nuclei collisions, *Bull. Russ. Acad. Sci. Phys.***74**, 567 (2010).
11. V.V. Samarin and K.V. Samarin, Dynamic tunnel effect at low-energy nuclear reactions with neutron rich nuclei, *Bull. Russ. Acad. Sci. Phys.* **76**, 450 (2012).
12. V.V. Samarin, Microscopic time-dependent analysis of neutrons transfers at low-energy nuclear reactions with spherical and deformed nuclei, *EPJ Web of Conference* **66**, 03075 (2014).
13. V.V. Samarin, Time-dependent quantum models of the dynamics of neutron transfer reactions near the barrier, *EPJ Web of Conference* **86**, 00040 (2015).
14. V.V. Samarin, Fusion and neutron transfer reactions with weakly bound nuclei within time-dependent and coupled channel approaches, *EPJ Web of Conference* **117**, 08018 (2016).
15. J.H. Wilkinson and C. Reinsch, *Handbook for Automatic Computation Linear Algebra* (Sprunger-Verlag, Berlin, 1971).
16. M.E. Riley and B. Ritchie, Numerical time-dependent Schroedinger description of charge-exchange collisions, *Phys. Rev. A***59**, 3544 (1999).
17. R. Anne et al., Observation of forward neutrons from the break-up of the ^{11}Li neutron halo, *Phys. Lett. B***250**, 19 (1990).

Clusters in Light Stable and Exotic Nuclei

C. Beck
Département de Recherches Subatomiques,
Institut Pluridisciplinaire Hubert Curien,
IN2P3-CNRS and Université de Strasbourg
23, rue du Loess BP 28, F-67037 Strasbourg Cedex 2, France
E-mail: christian.beck@iphc.cnrs.fr

Since the discovery of molecular resonances in $^{12}C+^{12}C$ in the early sixties a great deal of research work has been undertaken to study α-clustering. Our knowledge on physics of nuclear molecules has increased considerably and nuclear clustering remains one of the most fruitful domains of nuclear physics, facing some of the greatest challenges and opportunities in the years ahead. Occurrence of "exotic" shapes and Bose-Einstein Condensates in light α-cluster nuclei are investigated. Various approaches of superdeformed / hyperdeformed shapes associated with quasimolecular resonant structures are discussed. The astrophysical reaction rate of $^{12}C+^{12}C$ is extracted from recent fusion measurements at deep subbarrier energies near the Gamov window. Evolution of clustering from stability to the drip-lines is examined.

Keywords: Nuclear molecular resonances; nuclear clusters.

1. Introduction

In the last decades, one of the greatest challenges in nuclear science is the understanding of the clustered structure of nuclei from both experimental and theoretical perspectives [1–8]. Our knowledge on physics of nuclear molecules has increased considerably and nuclear clustering remains one of the most fruitful domains of nuclear physics. Fig. 1 summarizes the different types of clustering [4]: most of these structures were investigated in an experimental context by using either some new approaches [5] or developments of older methods. The search for resonant structures in the excitation functions for various combinations of light α-cluster (N=Z) nuclei in the energy regime from the Coulomb barrier up to regions with excitation energies of $E_x = 20\div50$ MeV remains a subject of contemporary debate. The question of how nuclear molecules may reflect continuous transitions from scattering states in the ion-ion potential to true cluster states in the compound systems is still unresolved. Clustering in light α-like nuclei is observed as a general phenomenon at high excitation energy close to the α-decay thresholds [4].

This exotic behavior has been perfectly illustrated 50 years ago by the famous "Ikeda-Diagram" for N=Z nuclei [9], which has been modified and

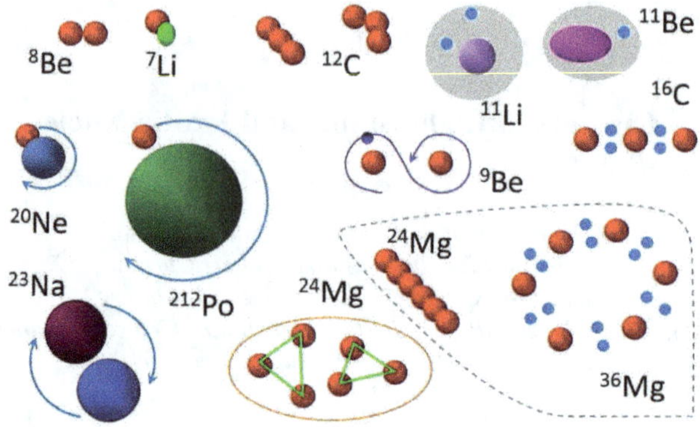

Fig. 1. Different types of clustering behavior identified in nuclei [4], from small clusters outside a closed shell, to complete condensation into α-particles, to halo nucleons outside.

recently extended by von Oertzen [6] for neutron-rich nuclei, as shown in the left panel of Fig. 2. Despite the early inception of cluster studies, it is only recently that radioactive ion beams experiments, with great helps from advanced theoretical works [4, 8], enabled new generation of studies, in which data with variable excess neutron numbers or decay thresholds are compared to predictions with least or no assumptions of cluster cores. Some of the predicted but elusive phenomena, such as molecular orbitals or linear chain structures, are now gradually coming to light.

2. ^{12}C Nucleus "Hoyle" State and BEC in Light Nuclei

The ground state of ^{8}Be is the most simple and convincing example of α-clustering in light nuclei as suggested by several theoretical models and appears naturally in *ab initio* calculations [4, 8]. The picture of the ^{8}Be nucleus predicted as being a dumbbell-shaped configuration of two α particles closely resembles the superdeformed (SD) shapes known to arise in heavier nuclei in the actinide mass region. This dumbbell-like structure gives rise to a rotational band, from which the moment of inertia is found to be commensurate with an axial deformation of 2:1. The large deformations of light α-conjugate nuclei with SD, hyper-deformed (HD) and linear-chain configurations are under discussion.

The renewed interest in ^{12}C was mainly focused to a better understanding of the nature of the so called "Hoyle" state [10, 11], the excited 0^+ state at 7.654 MeV that can be described in terms of a bosonic condensate, a cluster state and/or a α-particle gas [12].

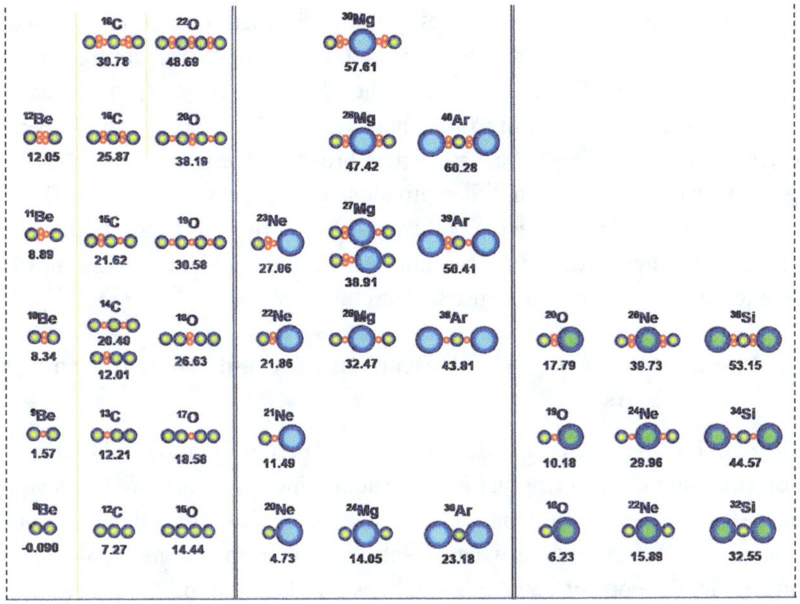

Fig. 2. Schematic illustration of the structures of molecular shape isomers in light neutron-rich isotopes of nuclei consisting of α-particles, ^{16}O- and ^{14}C-clusters plus some covalently bound neutrons (Xn means X neutrons) [6]. The so called "Extended Ikeda-Diagram" with α-particles (left panel) and ^{16}O-cores (middle panel) can be generalized to ^{14}C-cluster cores (right panel). The lowest line of each configuration corresponds to parts of the original "Ikeda-Diagram" [9]. Threshold energies are given in MeV.

The resonant "Hoyle" state [10] is regarded as the prototypical α-cluster state whose existence is of great importance for the nucleosynthesis of ^{12}C within stars. The structure of this state has been thoroughly modelled with both *ab initio* and cluster models [4, 8]. Much experimental progress has been achieved recently as far as the spectroscopy of ^{12}C near and above the α-decay threshold is concerned. More particularly, the 2_2^+ "Hoyle" rotational excitation in ^{12}C has been observed [13]. Another experiment [14] populates a new state compatible with an equilateral triangle configuration of three particles. Still, the structure of the "Hoyle" state remained controversial as experimental results of its direct decay into three α particles are found to be in disagreement until two experiments provided the most precise picture of how a ^{12}C excited state decays into three He nuclei [15, 16].

In the study of Bose-Einstein Condensation (BEC), the α-particle states in light N=Z nuclei [12], are of great interest. The search for an experimental signature of BEC in ^{16}O is of highest priority. Furthermore, *ab initio* calculations [4, 8] predict that nucleons are arranged in a tetrahedral configuration of α clusters. A state with the structure of the "Hoyle" state [10] in ^{12}C coupled to one α-particle is predicted in ^{16}O at about 15.1 MeV (the 0_6^+ state), the energy of which is ≈700 keV above the 4 α-particle breakup threshold.

However, any state in ^{16}O equivalent to the "Hoyle" state [10] in ^{12}C is most certainly going to decay exclusively by particle emission with very small γ-decay branches, thus, very efficient particle-γ coincidence techniques [5] will have to be used in the near future to search for them.

3. Nuclear Molecules, ^{12}C+^{12}C Reaction Rate and Carbon Burning in Massive Stars

The real link between superdeformation / hyperdeformation (SD/HD), nuclear molecules and α-clustering [4] is of particular interest, since nuclear shapes with major-to-minor axis ratios of 2:1–3:1 have the typical ellipsoidal elongation for light nuclei. A further area where electromagnetic transitions would be of great interest in support of cluster models is in the case of the quasi-molecular resonances observed in the ^{12}C+^{12}C reaction.

The widths of these resonances were ≈ 100 keV, indicating the formation of a ^{24}Mg intermediate system with a lifetime significantly longer than the nuclear crossing time. These resonances were subsequently interpreted as ^{12}C+^{12}C cluster states. There has been only one valient attempt to directly observe transitions in this reaction by Haas et al. [4] focusing on transitions between 10^+ and 8^+ resonant states at a bombarding energy $E(^{12}C) = 32$ MeV chosen to populate a known and isolated 10^+ resonance. However, the measurement reported only an upper limit (for the radiative partial width of 1.2 ± 10^{-5}) given the extreme challenges of eliminating all background.

The role of cluster configurations in stellar He burning is well established and, discussion about the nature and the role of resonance structures that characterize the low-energy cross section of the ^{12}C+^{12}C fusion process is underway in recent experimental investigations [17–24]. The resonant structures at very low energies have still been identified as molecular ^{12}C+^{12}C configurations in the ^{24}Mg compound nucleus [4, 17]. However, the reaction rate is calculated on the basis of an average cross section integrating over the molecular resonance components. Indications of possible existence of a pronounced low-energy resonance at $E_{cm} = 2.14$ MeV [17, 23] that can only be explained by strong ^{12}C cluster configurations of the corresponding state in ^{24}Mg.

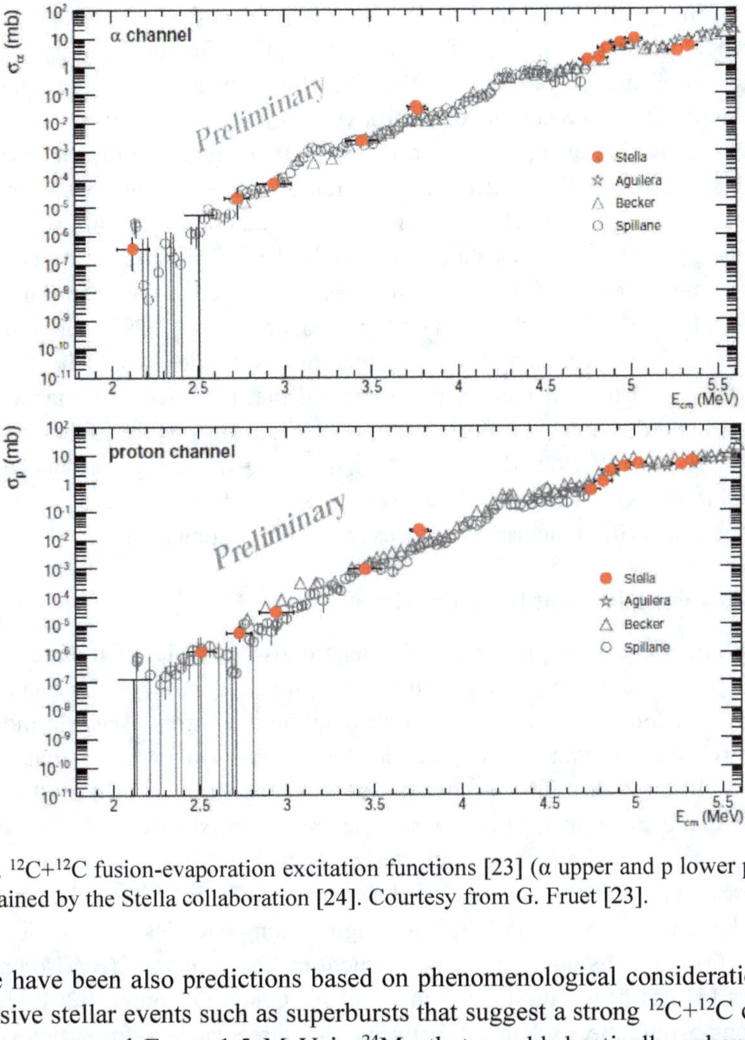

Fig. 3. $^{12}C+^{12}C$ fusion-evaporation excitation functions [23] (α upper and p lower panels) as obtained by the Stella collaboration [24]. Courtesy from G. Fruet [23].

There have been also predictions based on phenomenological considerations of explosive stellar events such as superbursts that suggest a strong $^{12}C+^{12}C$ cluster resonance around E_{cm} = 1.5 MeV in ^{24}Mg that would drastically enhance the energy production and may provide a direct nuclear driver for the superburst phenomenon. However, no indication for such a state was reported. Much of the data collected to date are shown in Fig. 3 taken from [23].

First direct $^{12}C+^{12}C$ measurement [17] seemed to indicate such a resonance. Our preliminary results [23] obtained with Stella [24] confirm the possible occurrence of such a resonant structure in the α channel but not in the proton channel. At higher energies the $^{12}C+^{12}C$ cross sections expressed for Stella in terms of the modified astrophysical S-factor are typically in fair agreement with those measured at Argonne with similar coincidence techniques [20]. Our first conclusions might be summarized such as we confirm the possible fusion hindrance plus persisting resonances near the Gamow energy window. Furthermore, the preliminary Stella S-factors appear to be in qualitative agreement with two recent theoretical investigations [25, 26]. It is not clear that a recent study [21] using the Trojan Horse Method technique (THM) confirmed the cluster level at E_{cm} = 2.1 MeV but rather suggested the existence of the predicted state at E_{cm} = 1.5 MeV for the fusion reaction. Such a discovery of a low energy $^{12}C+^{12}C$ cluster state would indeed have significant impact on the reaction rate; but some doubts [27, 28] have been raised to our attention as far as the validity of the indirect THM is concerned [21, 29]. Obviously an experimental confirmation through direct fusion studies would be of utmost importance. On the other hand, it is expected that if a strong resonance indeed exists around the Gamow energy, then the theoretical structure studies should be able to predict a 0^+ excited state of ^{24}Mg or an L = 0 $^{12}C+^{12}C$ resonance at this particular energy, which plays about the same igniting role as that of the "Hoyle" state [10] in the triple-α process of ^{12}C formation [11].

4. Clustering in Light Neutron-Rich Nuclei

Clustering is a general phenomenon observed also in nuclei with extra neutrons as it is presented in the "Extended Ikeda-Diagram" [9] proposed by von Oertzen [6] (see the left panel of Fig. 2). With additional neutrons, specific molecular structures appear with binding effects based on covalent molecular neutron orbitals. In these diagrams α-clusters and ^{16}O-clusters (as shown by the middle panel of the diagram of Fig. 2) are the main ingredients. Actually, the ^{14}C nucleus may play similar role in clusterization as the ^{16}O one since it has similar properties as a cluster: i) it has closed neutron p-shells, ii) first excited states are well above E* = 6 MeV, and iii) it has high binding energies for α particles. The possibility of extending molecular structures from dimers (Be isotopes) to trimers [6] has been investigated in detail for C and O isotopes [30]. Here the neutrons would be exchanged between the three centers (α-particles). It is possible that the three α-particle configuration can align themselves in a linear fashion, or alternative collapse into a triangle arrangement — in either case the neutrons being localized across the three centers. Possibly the best case for the linear arrangement is ^{16}C. A general picture of clustering and molecular configurations in light nuclei can also be drawn from the detailed investigation of the light O isotopes [30]. The bands of ^{20}O [30] compared with the ones of ^{18}O clearly establishes parity inversion doublets predicted by Antisymmetrized Molecular Dynamics (AMD) [7] calculations for the ^{14}C–^{6}He cluster and ^{14}C–2n–α molecular structures. The corresponding moments of inertia are strongly suggesting large deformations for the cluster structures. We may conclude that the reduction of the moments of inertia of the lowest bands of ^{20}O is consistent with the assumption that the strongly bound ^{14}C nucleus having equivalent properties to ^{16}O, has a similar role as ^{16}O in relevant, less neutron rich nuclei. Therefore, the "Ikeda-Diagram" [9] and the "extended Ikeda-Diagram" consisting of ^{16}O cluster cores with covalently bound neutrons must be further extended to include also the ^{14}C cluster cores as illustrated in Fig. 2.

5. Conclusion

The link of α-clustering, quasimolecular resonances and extreme deformations has been discussed. Several examples emphasize the general connection between molecular structure and deformation effects within *ab initio* models and/or cluster models [8]. We have also presented the BEC picture of light (and medium-light). α-like nuclei that appears to be an alternate way of understanding most of properties of nuclear clusters [4]. New results regarding cluster and molecular states in neutron-rich oxygen isotopes in agreement with AMD predictions are summarized [30]. Consequently, the "Extended Ikeda-Diagram" has been further modified for light neutron-rich nuclei by inclusion of the ^{14}C cluster, similarly to the ^{16}O one. Marked progress has been made in many traditional and novels subjects of nuclear cluster physics and astrophysics (stellar He burning [23, 24]). The developments in these subjects show the importance of clustering among the basic modes of motion of nuclear many-body systems. All these open questions will require precise coincidence measurements [4, 5] coupled with state-of-the-art theory [4, 7, 8].

References

1. C. Beck, *Clusters in Nuclei, Vol. 1*, ed. Springer Verlag Berlin Heidelberg (2010); *Lecture Notes in Physics* **818** (2010).
2. C. Beck, *Clusters in Nuclei, Vol. 2*, ed. Springer Verlag Berlin Heidelberg (2012); *Lecture Notes in Physics* **848** (2012).
3. C. Beck, *Clusters in Nuclei, Vol. 3*, ed. Springer Verlag Berlin Heidelberg (2014); *Lecture Notes in Physics* **865** (2014).
4. C. Beck, *Nuclear Particle Correlations and Cluster Physics*, ed. Wolf- Udo Schroeder (World Scientific Pub.) p.202; arXiv:**1608.03190** (2016).
5. P. Papka and C. Beck, in *Clusters in Nuclei, Vol. 2*, ed. Springer Verlag Berlin Heidelberg; *Lecture Notes in Physics* **848**, 299 (2012).
6. W. von Oertzen et al., *Phys. Rep.* **432**, 43 (2007).
7. Y. Kanada-En'yo and M. Kimura, in *Clusters in Nuclei, Vol. 1*; *Lecture Notes in Physics* **818**, 129 (2010).

8. M. Freer *et al., Rev. Mod. Phys.* **90**, 035004 (2018).
9. H. Horiuchi and K. Ikeda, *Prog. Theor. Phys.* **40**, 277 (1968).
10. F. Hoyle, *Astrophys. J. Suppl. Ser.* **1**, 121 (1954).
11. M. Freer and H.O.U. Fynbo, *Prog. Part. Nucl. Phys*. **71**, 1 (2014).
12. A. Tohsaki *et al., Phys. Rev. Lett.* **87**, 192501 (2001).
13. W. R. Zimmerman *et al., Phys. Rev. Lett.* **110**, 152502 (2014).
14. D.J. Marin-Lambarri *et al.*, *Phys. Rev. Lett.* **113**, 012502 (2014).
15. D. Dell'Aquila *et al.*, *Phys. Rev. Lett.* **119**, 132501 (2017).
16. R. Smith *et al.*, *Phys. Rev. Lett.* **119**, 132502 (2017).
17. T. Spillane *et al.*, *Phys. Rev. Lett.*, **98**, 122501 (2007).
18. C.L. Jiang *et al.*, *Phys. Rev. Lett.*, **110**, 072701 (2013).
19. B. Bucher *et al.*, *Phys. Rev. Lett.*, **114**, 251102 (2015).
20. C.L. Jiang *et al.*, *Phys. Rev. C*, **97**, 012801(R) (2018).
21. A. Tumino *et al.*, *Nature*, **557**, 687 (2018).
22. J. Zickefoose *et al.*, *Phys. Rev. C*, **97**, 065806 (2018).
23. G. Fruet, Ph. D Thesis, Université de Strasbourg (unpublished).
24. M. Heine *et al.*, *Nucl. Instr. Meth. A*, **903**, 1 (2018).
25. A. Diaz-Torres and M. Wiescher, *Phys. Rev. C*, **97**, 055902 (2018).
26. Le Hoang Chien *et al.*, *Phys. Rev. C* **97**; *arXiv:***1810.0788**.
27. A.M. Mukhamedzanov *et al.*, *arXiv:***1806.05921** (2018).
28. A.M. Mukhamedzanov *et al.*, *arXiv:***1806.08828** (2018).
29. A. Tumino *et al.*, *arXiv:***1807.06148** (2018).
30. W. von Oertzen and M. Milin, in *Clusters in Nuclei, Vol. 3*, ed. Springer Verlag Berlin Heidelberg; *Lecture Notes in Physics* **865** 146 (2014).

Search for States with Halo in ^{12}B and ^{12}N[*]

A. N. Danilov[*], A. S. Demyanova and A. A. Ogloblin
National Research Center "Kurchatov Institute",
Moscow, 123182, Russia
[*]*E-mail: danilov1987@mail.ru*
www.nrcki.ru

S. A. Goncharov
Lomonosov Moscow State University,
Moscow, 119991, Russia

T. L. Belyaeva
Universidad Autonoma del Estado de Mexico,
Toluca, 50000, Mexico

On base of ANC analysis of the obtained experimental data neutron halo was confirmed for the 2^-, 1.67 MeV and 1^-, 2.62 MeV states in ^{12}B. An unexpected result was received for the 3^-, 3.39 MeV state, which is unbound and is 19 keV above the decay threshold $^{12}\mathrm{B} \rightarrow {}^{11}\mathrm{B} + \mathrm{n}$. Its halo radius was found to be increased and equal to ~ 5.9 fm. This result can be considered as an evidence of the halo-like structure in this ^{12}B state. It should be noted that last neutron in this state has a nonzero orbital momentum (l=2).

Another interesting question is possible existence of states with halo in other members of the isobaric triplet ^{12}B - ^{12}C - ^{12}N. We can expect the formation of a proton halo in the 2^-, 1.19 MeV and 1^-, 1.80 MeV states of ^{12}N. To check this prediction preliminary MDM analysis of existing (^{3}He,t) experimental data was done.

Keywords: Asymptotic normalization coefficients; neutron and proton halo; MDM model; radii of excited states.

1. Introduction

Recently the evidence of the excited states of light nuclei with nonstandard sizes and enlarged radii, located closely and above the particle-emission threshold, was convincingly demonstrated (see, e.g., Ref. [1] and references therein). The existence of neutron halos in the short-lived excited states of some stable and radioactive nuclei was revealed, in particular, by the ANC analysis of the neutron-transfer reactions [2–4]. Thus Liu et al. [2] reported the observation of halos in the first ($1/2^+$, E_x = 3.089 MeV) excited state of ^{13}C, and the second

[*] The work was supported by the Russian grant RSF 18-12-00312.

(2^-, $E_x = 1.674$ MeV) and third (1^-, $E_x = 2.621$ MeV) excited states of ^{12}B using the (d,p) reactions on ^{12}C and ^{11}B, respectively, by analyzing the data with the ANC method.

Results of the ANC analysis of the ^{11}B(d,p)^{12}B reaction data at $E_{lab} = 21.5$ MeV are shown in [5]. Radii of the valence neutron for the first 5 excited states of ^{12}B were determined. One of the main aims was to estimate the radius of the 3.389 MeV 3^- state located in the continuum spectrum and to answer the question whether there exists a state, in which a neutron halo is formed by the valence neutron occupying the $l = 2$ orbital.

Obtained results for excited states of ^{12}B [5] are compared with preliminary results of MDM analysis of existing (^{3}He,t) literature data for isobar-analogues states in ^{12}N.

2. Results and Discussions

2.1. *^{12}B states*

Compound-nucleus (CN) analysis of the differential cross sections of the ^{11}B(d,p)^{12}B reaction data at $E_{lab} = 21.5$ MeV was carried out [5] within the Hauser-Feshbach formalism of the statistical CN model by using the computer code CNCOR [6]. The analysis showed [5] that the CN mechanism provides less than 0.1% of the cross section values at forward angles and about 1-3% at medium angles 60°-80°. So the transfer mechanism predominates for this reaction at given energy.

The Coupled-reaction-channels (CRC) method calculations were carried out [5] with the code FRESCO [7] to fit the data.

Four first excited states studied belong to the discreet spectrum, whereas the last 3.39 MeV 3^- state lying 0.019 MeV above the neutron-emission threshold is located in the continuum. Nevertheless, in view of the small positive energy and the structure of the proton angular distribution similar to those of the preceding bound states, the effective positive small (0.01 MeV) binding energy of valence neutron was used.

Neutron-transfer reactions with light nuclei are mainly peripheral, that is they actually sensitive only to the surface part of the neutron-core potential.

Calculations of the ANCs with different sp wave functions allowed [5] to answer the question whether there exist states in ^{12}B that are formed due to the peripheral neutron transfer reactions.

As in the sp approximation, the exact overlap function is approximated by the two-body sp wave function, the rms neutron radius is approximately determined by the rms radius of the sp wave function of neutron. To estimate the weight of the asymptotic part of the wave function a coefficient was introduced called D_1. It characterizes the spatial extension of the neutron wave functions to the asymptotic potential region. It is accepted in Ref. [2] that if $D_1(R_N) > 50\%$, than the state is assumed to be a "real" halo.

Calculations [5] showed that the rms radii of the last neutron in the second 1.674 MeV 2^- and third 2.621 MeV 1^- excited states of ^{12}B far exceed those for the g.s. and the first 0.953 MeV 2^+ excited state. Exactly, for the 2^- state, the excess is a factor of 1.7, and for the 1^- state, it is a factor of 2.1, with respect to the rms radius of the ground state. Summary on values of R_h and D_1 for ANC analysis of 21.5 MeV data [5] in comparison with results from [2] can be seen in Table 1.

Table 1. Results of ANC calculations [5] in comparison with results of [2].

State	R_h (fm) [2]	D_1 (%) [2]	R_h (fm) [5]	D_1 (%) [5]
g.s.	3.16 ± 0.32	19.9	3.55 ± 0.20	11
0.953			3.8 ± 0.2	14
1.674	4.01 ± 0.61	53.6	5.9 ± 0.3	53
2.621	5.64 ± 0.90	66.8	7.4 ± 0.4	62
2.723			5.5 ± 0.3	37
3.389			5.9 ± 0.3	39

The large values of D_1 coefficient, more than 50%, indicate that the last neutron spends more than 50% of its time outside the range of the core potential. In this case we can conclude that the rms radius of the last neutron wave function can be associated with the halo radius. From results of [5] condition $D_1(R_N) > 50\%$ is strictly fulfilled for the 1.674 MeV 2^- and 2.621 MeV 1^- states. It confirms existence of halo for these states and a conclusion made by Liu and collaborators in Ref. [2]. Nevertheless, analysis [5] indicated that the halo radii in these states are considerably large.

In the 2.72 MeV 0^+ state located 0.65 MeV below the neutron-emission threshold, the neutron rms radius is also found to be enlarged, but is smaller than that in the 1.67 MeV 2^- state, $R_h(0^+) = 5.5 \pm 0.3$ fm. One can see that $D_1 = 37\%$ is less than 50%. We propose the 2.72 MeV 0^+ state as a neutron halo-like state, in which the halo features are beginning to appear and manifest themselves in the enlarged neutron rms radius.

The fifth 3.389 MeV 3^- excited state is localized 0.019 MeV above the neutron-emission threshold. As this state belongs to the continuum spectrum, it made impossible a correct estimation of the ANC, even though the calculation was carried out with a very small positive neutron binding energy ($\varepsilon = 0.1$ MeV) and the asymptotic behavior of the sp and Hankel functions become the same at very long distance, about 50 fm. An estimation of the rms radius of the last neutron wave function was done [5] by determining the rms radius of the sp wave function. It was found to be $R_h = 5.9$ fm, which is a factor of 1.7 larger than that of the g.s. and is comparable with the rms radii of the last neutron wave function in the 1.674 MeV 2^- state and the 2.621 MeV 1^- state of ^{12}B. But D_l is less than 50%. Thus we propose that the 3.39 MeV 3^- excited state in ^{12}B is a neutron halo-like state with enhanced neutron radius $R_h = 5.9 \pm 0.3$ fm and the orbital momentum $l = 2$ of the last neutron.

2.2. ^{12}N states

To study isobar-analogues states of ^{12}B in ^{12}N we propose to use MDM [8,9]. To study ^{12}N states, this method can be applied to the reaction (^{3}He,t). Its first application made it possible to determine the proton halo in an unbound state of ^{13}N [10].

We have studied existing literature data on ^{12}C(^{3}He,t)^{12}N reaction [11-13]. This data is incomplete and there is no possibility to make definite answer on question about halo in 2^- and 1^- states of ^{12}N. Literature data is present only for 3 energies: 36 [11], 49.8 [12] and 81 MeV [13]. 36 MeV data contains only angular distributions for g.s. and 0.96 MeV states. 49.8 MeV data contains angular distributions for g.s., 0.96 and 1.20 MeV states. 81 MeV data contains all needed states interested for us (1.20 MeV and 1.80 MeV) but the problem is that only one oscillation in angular distributions is present. Also oscillations are indinstinct. And angular distribution for 0.96 MeV state at 81 MeV is probably not fully correct because if we would compare angular distribuions for 0.96 MeV state at all three energies and draw all distributions as a function of linear momentum transfer, then for 81 MeV data it will differ from others. This fact raise a question about correctness of all angular distributions at 81 MeV.

We applied MDM [8,9] to existing experimental data on ^{12}C(^{3}He,t)^{12}N at E(^{3}He) = 36, 49.8 and 81 MeV [11-13]. Preliminary results of analysis are present in Table 2 with comparison to data obtained for ^{12}B [5].

Table 2. Preliminary results of MDM analysis in comparison to data obtained for ^{12}B [5].

State	^{12}B	R_{rms} (fm) [5]	^{12}N	R_{rms} (fm)
1^+	g.s.	2.39 ± 0.02	g.s.	2.47 ± 0.07
2^+	0.95	2.44 ± 0.03	0.96	2.4 ± 0.4
2^-	1.67	2.58 ± 0.11	1.19	2.61 ± 0.37
1^-	2.62	2.86 ± 0.11	1.8	3.2 ± 0.2
0^+	2.72	2.51 ± 0.11	2.44	2.51 ± 0.21

As can be seen from Table 2, close radii were obtained for isobar analog states in ^{12}B and ^{12}N by 2 independent methods: MDM and ANC. This preliminary result can be an argument for possible proton halo in 2^- and 1^- states in ^{12}N. But further analysis is needed and also it will be excellent to obtain new experimental data at energies 40-90 MeV.

3. Conclusion

To conclude, neutron halo was confirmed by ANC for 2^- and 1^- states in ^{12}B. Probable halo-like structure is seen for 0^+ and 3^- states of ^{12}B (the last is in continuum, close to the threshold, l=2). MDM was applied to existing literature ^{12}C(^{3}He,t)^{12}N data. Preliminary results of MDM analysis for ^{12}N states showed that rms radii are close to corresponding values of rms radii obtained by ANC for ^{12}B. This fact can be an argument for possible proton halo in 2^- and 1^- states in ^{12}N.

References

1. A. A. Ogloblin et al., *Nuclear size isomers: The excited states of light nuclei with cluster structure and nonstandard sizes*, Nuclear Particle Correlations and Cluster Physics, ed. by W. Schröder, pp. 311–338 (2017)
2. Z. Liu et al., Phys. Rev. C **64**, 034312 (2001)
3. Z. Liu et al., Chin. Phys. Lett. **19**, 1071 (2002)
4. T. L. Belyaeva et al., Phys. Rev. C **90**, 064610 (2014)
5. T. L. Belyaeva et al., Phys. Rev. C **98**, 034602 (2018)
6. T. L. Belyaeva et al., Comput. Phys. Commun. 70 161 (1992)
7. I. Thompson, fresco user's manual and code, available from the author
8. A. N. Danilov et al., Phys. Rev. C **80**, 054603 (2009)
9. A. S. Demyanova et al., Yad. Fiz. **80**, 427 (2017) [Phys. At. Nucl. **80**, 831 (2017)]
10. A. S. Demyanova et al., JETP Letters **104** (8), 526 (2016)
11. K. P. Artemov et al., Yad. Fiz. **11**, 43 (1970)
12. G. Ball and J. Cerny, Phys. Rev. **177**, 1466 (1969)
13. W. Sterrenburg et al., Nucl. Phys. A **405**, 109 (1983)

Spectroscopic Study in Neutron-Rich Mn Isotopes Around the N = 40 "Island of Inversion"

B. Ding[a], Z. Liu[a,*], X. Y. Liu[a,b], P. Doornenbal[c], A. Obertelli[c,d,e], S. M. Lenzi[f], P. M. Walker[g], L. X. Chung[h], B. D. Linh[h], G. Authelet[d], H. Baba[c], D. Calvet[d], F. Château[d], A. Corsi[d], A. Delbart[d], J.-M. Gheller[d], A. Gillibert[d], T. Isobe[c], V. Lapoux[d], M. Matsushita[i], S. Momiyama[c,j], T. Motobayashi[c], M. Niikura[j], F. Nowacki[k], H. Otsu[c], C. Péron[d], A. Peyaud[d], E. C. Pollacco[d], J.-Y. Roussé[d], H. Sakurai[c,j], M. Sasano[c], Y. Shiga[c,l], S. Takeuchi[c], R. Taniuchi[c,j], T. Uesaka[c], H. Wang[c], K. Yoneda[c], X. D. Tang[a], Y. H. Lam[a], T. H. Huang[a], M. D. Sun[a,b], W. Q. Zhang[a,b], H. Y. Lu[a,b], D. S. Hou[a,b], F. Browne[m], Zs. Dombradi[n], S. Franchoo[o], F. Giacoppo[p], A. Gottardo[o], K. Hadynska-Klek[p], Z. Korkulu[n], S. Koyama[c,j], Y. Kubota[c,i], J. Lee[q], M. Lettmann[e], R. Lozeva[k,r], K. Matsui[c,j], T. Miyazaki[c,j], S. Nishimura[c], C. Louchart[e], L. Olivier[o], S. Ota[i], Z. Patel[g], E. Sahin[p], C. Santamaria[d,c], C. Shand[g], P.-A. Söderström[c], G. L. Stefan[o], D. Steppenbeck[i], T. Sumikama[s], D. Suzuki[o], Zs. Vajta[n], V. Werner[e], J. Wu[c,t], Z. Xu[q], X. H. Zhou[a], Y. H. Zhang[a], H. S. Xu[a] and F. S. Zhang[u,v]

[a]*Key Laboratory of High Precision Nuclear Spectroscopy, Institute of Modern Physics, Chinese Academy of Sciences, Lanzhou 730000, People's Republic of China*
[b]*University of Chinese Academy of Sciences, Beijing 100049, China*
[c]*RIKEN Nishina Center, 2-1 Hirosawa, Wako, Saitama 351-0198, Japan*
[d]*IRFU, CEA, Université Paris-Saclay, F-91191 Gif-sur-Yvette, France*
[e]*Institut für Kernphysik, Technische Universität Darmstadt, 64289 Darmstadt, Germany*
[f]*Universitá di Padova and INFN, Sezione di Padova, I-35131 Padova, Italy*
[g]*Department of Physics, University of Surrey, Guildford GU2 7XH, United Kingdom*
[h]*Institute for Nuclear Science & Technology, VINATOM, 179 Hoang Quoc Viet, Cau Giay, Hanoi, Vietnam*
[i]*Center for Nuclear Study, University of Tokyo, RIKEN Campus, Wako, Saitama 351-0198, Japan*
[j]*Department of Physics, University of Tokyo, 7-3-1 Hongo, Bunkyo Tokyo 113-0033, Japan*
[k]*IPHC, CNRS/IN2P3, Université de Strasbourg, F-67037 Strasbourg, France*
[l]*Department of Physics, Rikkyo University, 3-34-1 Nishi-Ikebukuro, Toshima, Tokyo 172-8501, Japan*
[m]*School of Computing Engineering and Mathematics, University of Brighton, Brighton BN2 4GJ, United Kingdom*
[n]*MTA Atomki, P.O. Box 51, Debrecen H-4001, Hungary*
[o]*Institut de Physique Nucléaire Orsay, IN2P3-CNRS, 91406 Orsay Cedex, France*
[p]*Department of Physics, University of Oslo, N-0316 Oslo, Norway*
[q]*Department of Physics, The University of Hong Kong, Pokfulam, Hong Kong*
[r]*CSNSM, CNRS/IN2P3, Université Paris-Saclay, F-91405 Orsay Campus, France*
[s]*Department of Physics, Tohoku University, Sendai 980-8578, Japan*
[t]*State Key Laboratory of Nuclear Physics and Technology, Peking University, Beijing 100871, People's Republic of China*

[u]*Key Laboratory of Beam Technology and Material Modification of Ministry of Education, College of Nuclear Science and Technology, Beijing Normal University, Beijing 100875, China*
[v]*Beijing Radiation Center, Beijing 100875, China*

Excited states in Mn isotopes with mass number A from 63 to 67 were studied via in-beam γ–ray spectroscopy following knockout reactions from ^{67,68}Fe. In odd-A isotopes ^{63,65,67}Mn, similar level schemes, consisting of the $11/2^-$, $9/2^-$, $7/2^-$, and $5/2^-_{g.s.}$ level sequence, connected by $I \rightarrow I-1$ transitions, were established, the first time for ^{65,67}Mn. Experimental results of odd-A Mn isotopes are compared with large-scale shell-model calculations in the model space that includes the $\nu g_{9/2}$ and $\nu d_{5/2}$ orbitals with the state-of-the-art LNPS effective interaction. The constructed level schemes in odd-odd isotopes ^{64,66}Mn show rotational-like properties. Spin and parity of the excited states, as well as the $\pi f_{7/2} \otimes \nu g_{9/2}$ configuration of the bands are tentatively assigned based on the level systematics in the odd-odd isotopes. These new data extends the knowledge of the excited states in Mn isotopes. Discussion in terms of the systematics helps to deepen our understanding of the N = 40 "island of inversion".

Keywords: In-beam γ-ray spectroscopy; island of inversion; structure evolution; Quansi-rotational band.
PACS: 21.10.-k, 21.10.Re, 23.20.Lv, 21.60.Cs

1. Introduction

With the accumulation of the experimental data on exotic neutron-rich nuclei far from stability, the paradigm of universal shell structure has been abandoned; much progress has been made on the investigation of shell evolution both experimentally and theoretically [1, 2]. A famous phenomenon related to this topic is the so-called "island of inversion" around ^{31}Na [3]. In the last two decades, nuclei in the neutron-rich N ~ 40 region have attracted much attention as they show similar loss of magicity seen in the N ~ 20 "island of inversion" [4]. At the N = 40 harmonic oscillator shell closure, ^{68}Ni has the appearance of doubly magic based on the high 2_1^+ energy (2.34 keV) [5] and low transition probability to the ground state [6]. However, the N = 40 shell closure in Ni was proved to be fragile and disappear rapidly below ^{68}Ni [7]. In isotonic nucleus ^{66}Fe, corresponding to removing only two protons from ^{68}Ni, evidence for enhanced collectivity has been observed: very low excitation of the 2_1^+ state (574 keV), high energy ratio of $R_{4/2} = E(4_1^+)/E(2_1^+) = 2.45$, and large transition rate of the 2_1^+ state [8]. The 2_1^+ energy ($R_{4/2}$) in ^{64}Cr, with two protons less than ^{66}Fe, becomes even lower (larger), indicating increasing collectivity along the isotonic chain. Along the Fe and Cr isotopic chains, the increasing collectivity has also been establish from the measured E_{2+}, $R_{4/2}$ [9], and B(E2) values [8, 10], respectively. The development of quadrupole deformation has been interpreted in the shell model framework in a large model space. The collective character was attributed to the occupancy of the intruder $\nu g_{9/2}$ and $\nu d_{5/2}$ orbitals. Similar to the neutron-rich region around ^{31}Na where the dominant intruder configurations are $\nu f_{7/2}$ and $\nu d_{3/2}$ orbitals, this area has been described as a new "island of inversion" [4, 11]. In addition, the difference in structure between Fe and Cr at N = 40 has been revealed based on knockout reaction cross sections measurements [9, 12]. Mn isotopes lie between Fe and Cr isotopes. Thus, it would be intriguing to study their nuclear structure.

Prior to our work, various experiments aiming at lighter Mn isotopes have been performed. High spin states in $^{57\div60}$Mn have been studied via fusion-evaporation reactions [13]. Using multi-nucleon transfer reactions, spectroscopic study has been extended up to ^{63}Mn [14, 15, 16]. Excited states in ^{64,66}Mn have been established through isomeric decay and β decay, respectively [17]. Recently, ground state properties of $^{53\div63}$Mn have been determined by laser spectroscopy at ISOLDE [18, 19]. So far, in-beam γ spectroscopy of Mn isotopes with $N \geq 40$ has been out of experimental reach. Here, we report the spectroscopic study of the neutron-rich Mn isotopes approaching and beyond N = 40.

2. Experiment Details

The experiment was performed at the Radioactive Isotope Beam Factory (RIBF), operated by the RIKEN Nishina Center and the Center for Nuclear Study of the University of Tokyo. The in-flight fission of a ^{238}U primary beam, accelerated to 345 MeV/u with a mean intensity of 12 pnA, was induced by bombarding a 3-mm-thick ^{9}Be primary target at the entrance of the BigRIPS separator [20]. The secondary beam of interest was selected using the $B\rho$-ΔE-$B\rho$ method [21] to impinge on a 102(1)-mm-thick liquid hydrogen (LH_2) secondary target for about 24 hours. The incident energies at the entrance of the of the LH_2 target were about 275 MeV/u. The secondary reaction residues were produced via proton-induced multi-nucleon knockout reactions. The particle identification of secondary beams and reaction products was performed on the event-by-event basis by a ToF- $B\rho$-ΔE method in the BigRIPS and ZeroDegree spectrometer, respectively.

Surrounding the LH_2 target was a 300-mm-long cylindrical time projection chamber (TPC) known as the device of MINOS, which was used to reconstruct the vertex of the knockout reactions by tracking the emitted protons [22]. De-excitation γ rays were detected by the DALI2 array consisting of 186 NaI(Tl) scintillator detectors [23]. The full-energy peak detection efficiency with no addback was simulated to be 30% for a 500 keV transition emitted in flight at 250 MeV/u; typical energy resolution was about 6% at full width of half maximum (FWHM) for the 1332-keV line, in agreement with the analyses in Refs. [11, 24, 25]. The setup is identical to previous experiments, and has been reported in recent publications [11, 24, 26÷30]. Particularly, the MINOS device is described in greater detail in Ref [31].

3. Results

3.1. *Odd-A* ^{63,65,67}*Mn*

In the present work, excited states in the odd-A Mn isotopes were populated by the multi-nucleon knock out reaction via the ^{68}Fe(p, 2pxn)$^{67\text{-}x}$ Mn reaction channels, with $x = 4$, 2, and, 0 for ^{63,65,67}Mn, respectively.

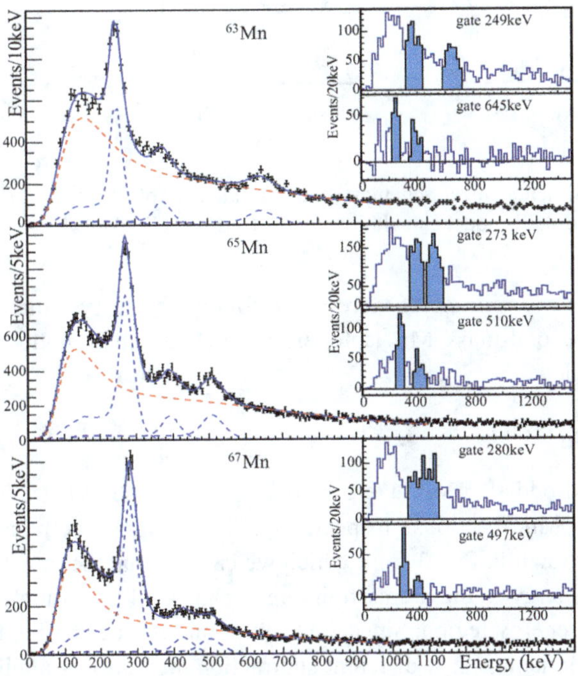

Fig. 1. Doppler-corrected γ-ray spectra of ^{63,65,67}Mn detected by DALI2, and the background subtracted γ-γ coincidence spectra labeled with the gating transitions are displayed in the insets.

The Doppler-corrected γ-ray spectra for the odd-A isotopes are displayed in Fig. 1, where the response functions of DALI2 were simulated with the GEANT4 framework [32] and the bremsstrahlung background shown in the figure was subtracted. Three γ peaks were evident in the spectrum for each isotope. Based on the γ-γ coincidence relationships and the relative intensities, the level schemes have been established and displayed in Fig. 2.

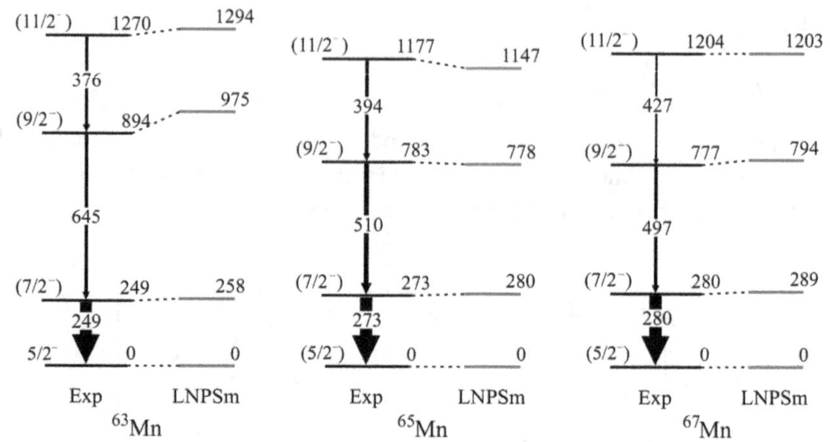

Fig. 2. Experimental level schemes (black solid line) of ^{63,65,67}Mn in comparison with the large-scale shell-model calculation results (red solid line).

The spin-parity assignments were made following the systematics of low-lying levels in the odd-mass Mn isotopes, as well as shell-model calculations as described below.

3.2. *Odd-Od ^{64,66}Mn*

Excited odd-odd ^{64,66}Mn were produced via the ^{67}Fe(p, 2p2n)^{64}Mn and ^{68}Fe(p, 2pn)^{66}Mn reactions, respectively. The Doppler-corrected single γ-ray spectra are shown in Fig. 3 from which we can see that the transitions of interest lie close in energy to the bremsstrahlung background spectrum, especially at the $E < 200$ keV energy region. Since the background is much higher than the that for the odd-A Mn isotopes. A different subtraction method was applied for the odd-odd Mn nuclei; the method was first developed and applied to the data analysis of ^{112}Mo and ^{110}Zr (see Ref. [28] for details). It has been validated that the subtraction method did not create any artificial peaks. With the individual response function, careful fitting has been done to the background subtracted spectra to identify the overlapping γ rays, successfully separating the 220 keV and 281 keV lines in ^{64}Mn, for example. Furthermore, two low-energy γ rays (120 keV line in ^{64}Mn and ~100 keV in ^{66}Mn) and the 595 keV transition in ^{66}Mn have been identified from γ-γ coincidence analysis.

The spin and parity of the observed levels are tentatively assigned following the level systematics in odd-odd Mn isotopes (see Fig. 6).

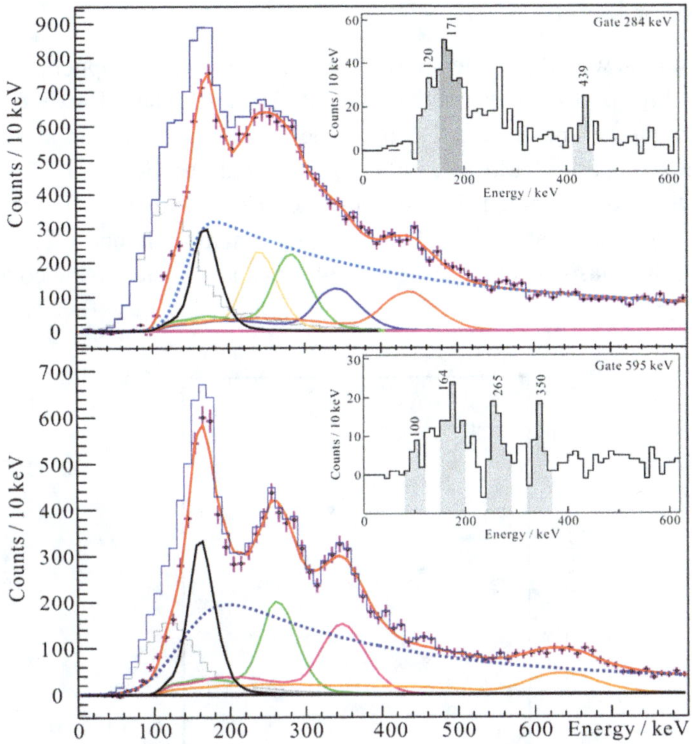

Fig. 3. The same as in Fig. 1, but for the odd-odd ^{64,66}Mn isotopes including the total spectrum (solid blue line), normalized bremsstrahlung component (gray), and the subtracted spectrum (crosses).

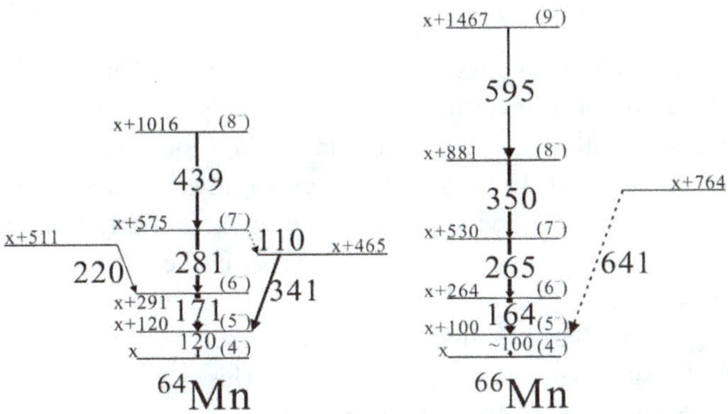

Fig. 4. Level schemes of ^{64,66}Mn deduced from this work.

4. Discussion

For Mn isotopes, with Z=25, the proton fermi surface lies in the upper part of $\pi f_{7/2}$ shell, with three proton-holes relative to the Z=28 closed shell. However, except ^{53}Mn where the ground state is $7/2^-$ of single-particle character, in all the other odd-A Mn isotopes $^{47-65}$Mn, the $J^\pi = 5/2^-$ level drops below the $7/2^-$ state, becoming the ground state. This is the so-called $I = j - 1$ anomaly in the present case of $f^{-3}{}_{7/2}$ configuration [33]. The $I = j - 1$ anomaly is one of the typical elements of the quasi-rotational structure [34]; it can be related to the collectivity of the core [35].

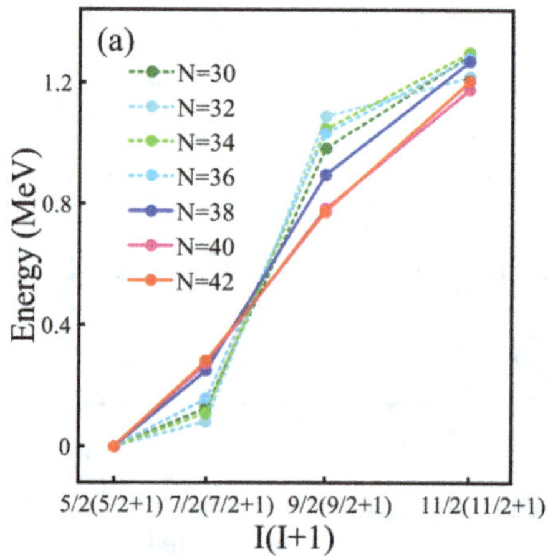

Fig. 5. Experimental excitation energies of low-lying yrast states in even-N Mn isotopes as a function of $I(I+1)$.

Thus, these odd-A Mn isotopes could be of quadrupole character. On the other hand, the neutron Fermi level lie between $f_{5/2}$ and $g_{9/2}$ orbitals. With increasing neutron number the down sloping $g_{9/2}$ orbital could be occupied and will increase the nuclear deformation. In fact, in the neighboring even-even Cr isotopes, the 2_1^+ energies (the ratio $R_{4/2}$) decrease (increase) monotonically from N = 32 (34) to N = 42, indicating the increasing deformation in the neutron-rich Cr nuclei. On theoretical side, shell model calculations show that neutron excitations from pf to $g_{9/2}$ and even $d_{5/2}$ orbitals have to be considered to account for the experimental results when N > 34 [36, 37, 38]. This is also true for Fe isotopes from N = 36(38) to N = 44, and the $g_{9/2}$ orbital has to be considered when approaching N = 40 [8, 7, 11, 17, 39, 40, 41].

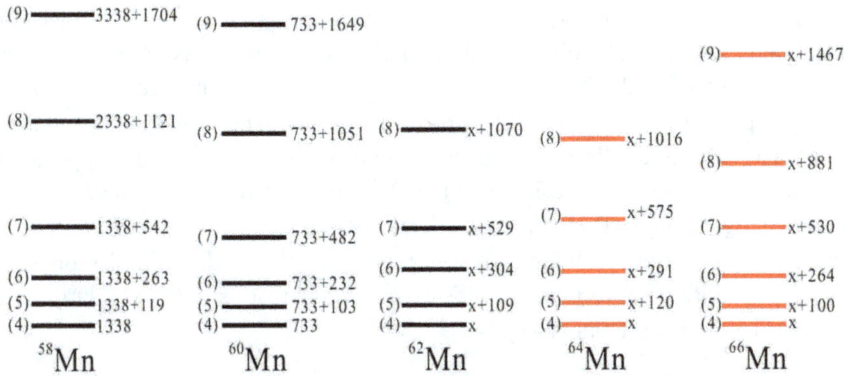

Fig. 6. Level spacing systematics for the $\pi f_{7/2} \otimes \nu g_{9/2}$ bands in doubly odd $^{58\text{-}66}$Mn (The data are from Refs. [13, 15] and the present work (red lines)).

In this work, the level schemes of ^{63,65,67}Mn are very similar, consisting of the $11/2^-$, $9/2^-$, $7/2^-$, and $5/2^-_{\text{g.s.}}$ states, forming a level sequence connected by $\Delta I = 1$ transitions. To probe the properties of the level sequences, excitation energies versus $I(I + 1)$ are presented, including the lighter odd-A isotopes from N = 30 on for a systematic study. With increasing neutron number, the linear relationship between excitation energy and $I(I + 1)$ becomes more clear, indicating more deformed nuclear shape. This evolution is consistent with the behavior in the neighboring even-even nuclei mentioned above. With the nearly linear relationship akin to ideal rotors [42] and the observed dominance of $\Delta I = 1$ transitions, the level sequences in ^{65,67}Mn manifest typical characteristics of strongly coupled bands built on the $K^\pi = 5/2^-$ band head. For ^{63}Mn, a transitional feature between decoupling and strong coupling is seen. Large-scale shell-model calculations for odd-A ^{63,65,67}Mn have been performed using the modified LNPS [4] effective interaction (LNPSm) in the model space of the full *fp* shell for protons and the $p_{3/2}$, $p_{1/2}$, $f_{5/2}$, $g_{9/2}$, and $d_{5/2}$ orbitals for neutrons. Experimental observed levels were reproduced remarkably well. Based on the calculations, proton and neutron occupation numbers have been deduced for all observed states in the present work. The proton occupation numbers are nearly constant: there is about 1 proton excited cross the Z = 28 shell to the $p_{3/2}$ and $f_{5/2}$ orbitals. The neutron configurations are dominated by 4*p*-4*h* neutron excitations from the pf shell into the $g_{9/2}$and $d_{5/2}$ orbitals, and the excited neutron number from the *pf* shell reaches a maximum at N=40. From the above discussion it can be expected that rotational band, based on $\nu g_{9/2}$ coupled to $\pi f_{7/2}$ configuration, should be observed the in odd-odd Mn isotopes, especially for the neutron-rich ones with N > 40. We, therefore, assign a $\pi f_{7/2} \otimes \nu g_{9/2}$ configuration to the quasi-rotational bands observed in

^{64,66}Mn in the present work. Quasi-rotational bands based on this configuration have been systematically observed in the lighter odd-odd Mn isotopes from mass number A = 58 to 62. The relative excitation energies, normalized to the $J = 4$ level, are presented in Fig. 6 for these bands in $^{58\text{-}66}$Mn. The level spacings in these bands decrease gradually with increasing neutron number, and this supports the $\pi f_{7/2} \otimes \nu g_{9/2}$ configuration assignment in ^{64,66}Mn. The decreasing level energy, shown in Fig. 6, suggests increasing collectivity with neutron approaching N ~ 40, which is in agreement with the evolution properties of the odd-A Mn isotopes. To give a quantitative interpretation of the experimental results in ^{64,66}Mn, large shell model calculations are underway [43].

5. Conclusion

In-beam γ spectroscopy studies have been carried out for the neutron-rich Mn isotopes with mass number A=63 to 67. Level schemes for the five nuclei have been established. For odd-A ^{63,65,67}Mn, the level sequences are very similar, and each consists of $11/2^- \rightarrow 9/2^- \rightarrow 7/2^- \rightarrow 5/2^-_{g.s.}$ cascade transitions. With the excitation energies closely proportional with $I(I+1)$ and the decay pattern dominated by $\Delta I = 1$ transitions, the first cases of strongly-coupled rotational bands in the N = 40 "island of inversion" have been identified in ^{65,67}Mn. Large-scale shell-model calculations using the LNPSm effective interaction reproduce the observed excitation energies and decay patterns very well. In addition to the neutron $4p$-$4h$ configuration, the proton $1p$-$1h$ cross-shell excitations are predicted. For odd-odd ^{64,66}Mn, the level schemes form rotational-like bands. Based on the level systematics of the odd-odd isotopes, the spin and parity of the levels, as well as a $\pi f_{7/2} \otimes \nu g_{9/2}$ configuration for the quasi-rotational bands are tentatively assigned.

Acknowledgements

We express our gratitude to the RIKEN Nishina Center accelerator staff for providing the stable and high-intensity uranium beam and to the BigRIPS team for operating the secondary beams. The development of MINOS and the core MINOS team were supported by the European Research Council through the ERC Grant No. MINOS-258567. The work at Institute of Modern Physics was supported by the National Natural Science Foundation of China (Grant Nos.11405224, 11635003, 11675225, U1632144, 11435014, and 1173507), the Key Research Program of the Chinese Academy of Sciences (Grant No. XDPB09), and the National Key Basic Research Development Program of China under Grant No. 2018YFA0404402. All UK authors are supported by STFC. L.X. Chung and B.D. Linh are supported by Ministry of Science and Technology under

Grant Nos. ĐTCB 09/17/VKHKTHN and ĐTĐLCN.25/18. The Hungarian authors are supported by the GINOP-2.3.3-15-2016-00034 project. This project is supported by the national key research and development program (MOST 2016YFA0400501). BMBF Grant Nos. 05P15RDFN1 and 05P12RDFN8 for TU Darmstadt people. The support to X.Y. Liu provided by China Scholarship Council (No. 201700260183) during the visit to Sweden is acknowledged.

Helpful discussions with J. Meng, Y. Sun, B. Cederwall, C. Qi, H. Jin and F.R. Xu are appreciated.

References

1. D. Steppenbeck *et al.*, Evidence for a new nuclear 'magic number' from the level structure of ^{54}Ca, *Nature* **502**, 207-210 (2013), doi:10.1038/nature12522.
2. B. A. Brown, The nuclear shell model towards the drip lines, *Progress in Particle and Nuclear Physics* **47**, 517-599 (2001), doi:10.1016/S0146-6410(01)00159-4.
3. C. Détraz *et al.*, Beta decay of $^{27-32}$Na and their descendants, *Physical Review* C **19**, 164-176 (1979), doi:10.1103/PhysRevC.19.164.
4. S. M. Lenzi *et al.*, Island of inversion around ^{64}Cr, *Physical Review* C **82**, 054301 (2010), doi: 10.1103/PhysRevC.82.054301.
5. R. Broda *et al.*, N = 40 neutron subshell closure in the 68Ni nucleus, *Physical Review Letters* **74**, 868-871 (1995), doi:10.1103/PhysRevLett.74.868.
6. O. Sorlin *et al.*, $^{68}_{28}$Ni$_{40}$: Magicity versus superfluidity, *Physical Review Letters* **88**, 092501 (2002), doi:10.1103/PhysRevLett.88.092501.
7. M. Hannawald *et al.*, Decay of neutron-rich Mn nuclides and deformation of heavy Fe isotopes, *Physical Review Letters* **82**, 1391-1394 (1999), doi:10.1103/PhysRevLett.82.1391.
8. W. Rother *et al.*, Enhanced quadrupole collectivity at N = 40: The case of neutron-rich Fe isotopes, *Physical Review Letters* **106**, 022502 (2011), doi:10.1103/PhysRevLett.106.022502.
9. A. Gade *et al.*, Collectivity at N = 40 in neutron-rich ^{64}Cr, *Physical Review* C **81**, 051304 (2010), doi:10.1103/PhysRevC.81.051304.
10. T. Baugher *et al.*, Intermediate-energy coulomb excitation of ^{58,60,62}Cr: The onset of collectivity toward N = 40, *Physical Review* C **86**, 011305 (2012), doi:10.1103/PhysRevC.86.011305.
11. C. Santamaria *et al.*, Extension of the N = 40 island of inversion towards N = 50: Spectroscopy of ^{66}Cr,^{70,72}Fe, *Physical Review Letters* **115**, 192501 (2015), doi:10.1103/PhysRevLett.115. 192501.

12. P. Adrich *et al.*, In-beam γ-ray spectroscopy and inclusive two-proton knockout cross section measurements at N≈40, *Physical Review* C **77**, 054306 (2008), doi:10.1103/PhysRevC.77. 054306.
13. D. Steppenbeck *et al.*, High-spin structures in the neutron-rich isotopes $^{57\text{-}60}$Mn, *Physical Review* C **81**, 014305 (2010), doi:10.1103/ PhysRevC.81.014305.
14. J. J. Valiente-Dobón *et al.*, Spectroscopy of neutron-rich $^{59\text{-}63}$Mn isotopes, *Physical Review* C **78**, 024302 (2008), doi:10.1103/PhysRevC.78.024302.
15. C. J. Chiara *et al.*, Influence of the $\nu g_{9/2}$ orbital on level structures of neutron-rich $^{61,62}Mn_{36,37}$, *Physical Review* C **82**, 054313 (2010), doi:10.1103/PhysRevC.82.054313.
16. T. Baugher *et al.*, In-beam-ray spectroscopy of ^{63}Mn, *Physical Review* C **93**, 014313 (2016), doi:10.1103/PhysRevC.93.014313.
17. S. N. Liddick *et al.*, Shape coexistence along N = 40, *Physical Review* C **84**, 061305 (2011), doi:10.1103/PhysRevC.84.061305.
18. C. Babcock *et al.*, Evidence for increased neutron and proton excitations between $^{51\text{-}63}$Mn, *Physics Letters* B **750**, 176-180 (2015), doi:10.1016/ j.physletb.2015.09.012.
19. H. Heylen *et al.*, Spins and magnetic moments of ^{58,60,62,64}Mn ground states and isomers, *Physical Review* C **92**, 044311 (2015), doi:10.1103/PhysRevC.92.
20. T. Kubo *et al.*, BigRIPS separator and zero degree spectrometer at RIKEN RI beam factory, *Progress of Theoretical & Experimental Physics*, **2012**, 1 1146-1151 (2012), doi:10.1093/ptep/pts064.
21. N. Fukuda *et al.*, Identification and separation of radioactive isotope beams by the BigRIPS separator at the BigRIPS RIKEN RI beam factory, *Nuclear Instruments and Methods in Physics Research Section* B **317**, 323-332 (2013), doi: 10.1016/j.nimb.2013.08.04.
22. A. Obertelli *et al.*, Minos: A vertex tracker coupled to a thick liquid-hydrogen target for in-beam spectroscopy of exotic nuclei, *The European Physical Journal* A **50**, 8 (2014), doi:10.1140/epja/i2014-14008-y.
23. S. Takeuchi *et al.*, Dali2: A NaI(Tl) detector array for measurements of rays from fast nuclei, *Nuclear Instruments and Methods in Physics Research Section* A **763**, 596-603 (2014), doi:10.1016/j.nima.2014.06.087.
24. S. Chen *et al.*, Low-lying structure and shape evolution in neutron-rich Se isotopes, *Physical Review* C **95**, 041302 (2017), doi:10.1103/Physrevc.95.041302.
25. P. Doornenbal, In-beam gamma-ray spectroscopy at the RIBF, *Progress of Theoretical and Experimental Physics*, **2012**, 1 10 (2012), doi:10.1093/ptep/pts076.

26. X. Y. Liu *et al.*, Spectroscopy of $^{65,67}{}_{25}$Mn Strong coupling in the N = 40 "island of inversion", *Physics Letters* B **784**, 392-396 (2018), doi:10.1016/j.physletb.2018.06.067.
27. C. M. Shand *et al.*, Shell evolution beyond Z = 28 and N = 50: Spectroscopy of ^{81,82,83,84}Zn, *Physics Letters* B **773**, 492-497 (2017), doi:10.1016/j.physletb.2017.09.001.
28. N. Paul *et al.*, Are there signatures of harmonic oscillator shells far from stability? First spectroscopy of ^{110}Zr, *Physical Review Letters* **118**, 032501 (2017), doi:10.1103/PhysRevLett.118. 032501.
29. M. Lettmann *et al.*, Triaxiality of neutron-rich ^{84,86,88}Ge from low-energy nuclear spectra, *Physical Review* C **96**, 011301 (2017), doi:10.1103/PhysRevC.96.011301.
30. F. Flavigny *et al.*, Shape evolution in neutron-rich krypton isotopes beyond N = 60: First spectroscopy of ^{98,100}Kr, *Physical Review Letters* **118**, 242501 (2017), doi:10.1103/PhysRevLett.118.242501.
31. C. Santamaria *et al.*, Tracking with the minos time projection chamber, *Nuclear Instruments and Methods in Physics Research Section* A **905**, 138-148 (2018), doi:10.1016/j.nima.2018.07.053.
32. S. Agostinelli *et al.*, Geant4—A simulation toolkit, *Nuclear Instruments and Methods in Physics Research Section* A **506**, 250-303 (2003), doi:10.1016/S0168-9002(03)01368-8.
33. V. Paar, The solution of the "$i = j$-1" anomaly for ^{107,109}Ag in the alaga model, *Physics Letters* B **39**, 587-590 (1972), doi:https://doi.org/10.1016/0370-2693(72)90003-2.
34. V. Paar, Coupling of a three-particle (hole) valence-shell cluster to quadrupole vibrations (alaga model): The Z = 50 region: odd Ag and I isotopes; and the Z = 28 region: odd Mn and Ga isotopes, *Nuclear Physics* A **211**, 29-76 (1973), doi:10.1016/0375-9474(73)90763-X.
35. T. Rzaca-Urban *et al.*, First observation of excited states in ^{87}Se: Collectivity and j – 1 anomaly at N = 53, *Physical Review* C **88**, 034302 (2013) and references therein, doi:10.1103/PhysRevC. 88.034302.
36. N. Aoi *et al.*, Development of large deformation in ^{62}Cr, *Physical Review Letters* **102**, 012502 (2009), doi: 10.1103/PhysRevLett.102.012502.
37. O. Sorlin *et al.*, New region of deformation in the neutron-rich $^{60}{}_{24}Cr_{36}$ and $^{62}{}_{24}Cr_{38}$, *European Physical Journal* A **16**, 55-61 (2003), doi:10.1140/epja/i2002-10069-9.
38. S. Zhu *et al.*, Level structure of the neutron-rich ^{56,58,60}Cr isotopes: Single-particle and collective aspects, *Physical Review* C **74**, 064315 (2006), doi:10.1103/PhysRevC.74.064315.

39. A. N. Deacon *et al.*, Yrast structures in the neutron-rich isotopes ^{59,60}Fe and the role of the $g_{9/2}$ orbital, *Physical Review* C **76**, 054303 (2007), doi:10.1103/PhysRevC.76.054303.
40. N. Hoteling *et al.*, Structure of ^{60,62}Fe and the onset of $\nu g_{9/2}$ occupancy, *Physical Review* C **82**, 1103 (2010), doi:10.1103/PhysRevC.82.044305.
41. N. Hoteling *et al.*, Yrast structure of ^{64}Fe, *Physical Review* C **74**, 064313 (2006), doi:10.1103/PhysRevC.74.064313.
42. R. Casten, Nuclear structure from a simple perspective (chapter 8). *Oxford University Press*, 2000.
43. S. M. Lenzi, *private communication* 2019.

Emission of High Energy Particles in Heavy Ion Collisions in the Hydrodynamic Approach

A. T. D'yachenko
Emperor Alexander I Petersburg State Transport University, St. Petersburg, Russia
E-mail: dyachenko_a@mail.ru

I. A. Mitropolsky
NRC "Kurchatov Institute", B.P. Konstantinov Petersburg Nuclear Physics Institute Gatchina, Russia
E-mail: mitropolsky_ia@pnpi.nrcki.ru

Yu. G. Sobolev
Joint Institute for Nuclear Research Dubna, Russia
E-mail: sobolev@jinr.ru

A hydrodynamic approach with a non-equilibrium equation of state is used to describe the collisions of heavy ions at medium and intermediate energies. In the development of this approach, with the inclusion of nuclear viscosity effects and the introduction of an amendment to the microcanonical distribution, the double differential cross sections of proton emission in collisions of different nuclei are calculated, which are in agreement with the available experimental data on the emission of high-energy particles, including the cumulative spectral region.

Keywords: Non-equilibrium equation of state; emission of high energy protons; heavy ions.

1. The Model

To describe the collisions of heavy ions we use the non-equilibrium hydrodynamic approach [1,2,3,4], in which the kinetic equation for the nucleon distribution function $f(\vec{r},\vec{p},t)$ is solved jointly with the equations of hydrodynamics, which are essentially local laws of conservation of mass, momentum and energy.

After selecting the region of the local heating, hot spot - the overlap region of the colliding nuclei, we analyze the stages of compression, expansion and freeze-out of matter during the collision of heavy ions. At the compression stage, a collisionless shock wave with a changing front is formed [2]. At the expansion stage [3], when the shock wave reaches the boundaries of hot spot, the initially compressed system is expanded, we describe it in the relaxation approximation taking into account the nuclear viscosity [4]. As the relaxation time we take $\tau = \lambda / \upsilon_T$, where $\lambda = 1/\sigma\rho$ is the mean free path, $\sigma \approx 40$mb is the total nucleon-nucleon cross section, ρ is the nucleon density, and υ_T is the average velocity of the thermal motion of the nucleons. At the freeze-out stage, when the system reaches a critical density ρ^* also called the freeze-out density, the system does not "hold itself" and the secondary particles are formed [1, 2, 3, 4, 5]. As a result, the double differential cross-section of proton emission has the form (where b is the impact parameter, $\hbar$ is a Planck constant, $\vec{r}$ is the radius vector):

$$\frac{d^2\sigma}{dEd\Omega} = \frac{2\pi}{(2\pi\hbar)^3} \int G(b) b db d\vec{r} \gamma(E - \vec{p}\vec{\upsilon}) p f(\vec{r}, \vec{p}, t), \quad (1)$$

where the distribution function of emitted protons

$$f(\vec{r}, \vec{p}, t) = g\left[\exp\left(\frac{\gamma(E - \vec{p}\vec{\upsilon}) - \mu + T\delta}{T}\right) + 1\right]^{-1}. \quad (2)$$

Here the spin factor $g = 2$, $E = \sqrt{p^2 + m^2}$, γ and $\vec{p}$ are respectively the total energy, the Lorentz factor and the proton momentum; $\vec{\upsilon}(\vec{r}, t)$ is the velocity field, $G(b)$ is the factor taking into account that the cross section of the hot spot formation is always greater than the geometric one, μ is the chemical potential, which is found from the conservation of the average number of particles for a grand canonical ensemble, T is the temperature, δ is correction for the microcanonical distribution, which for the kinetic energy $\varepsilon = E - m > E_1$ is equal to

$$\delta = \left[-M \ln\left(1 - \frac{\gamma(E - \vec{p}\vec{\upsilon}) - m}{MT}\right) - \frac{\gamma(E - \vec{p}\vec{\upsilon}) - m}{T}\right], \quad (3)$$

where $M = 3N/2$, N is the number of nucleons in the thermostat, $E_1 (E_1 \gg T)$ is the energy that is close to the energy of the thermostat, i.e. close to the kinematic limit for the energy of the system. We also chose the energy value $E_2 (E_2 < E_1)$, when the distribution function decreases by an order of magnitude compared to its maximum. When $\varepsilon < E_2$ the amendment δ was supposed equal to zero. In the energy interval $E_2 > \varepsilon > E_1$ it was a linear interpolation between zero and expression (3). Here the correction δ is found for the Boltzmann limit of an ideal gas, since deviations from a grand canonical distribution of the Fermi gas are manifested on the "tails" of the energy spectra when the Fermi distribution coincides with the Boltzmann limit.

The probability of a microcanonical distribution in the limit of the Boltzmann limit of an ideal gas is

$$w(\vec{r},\vec{p}) = C_M\left(1-\frac{\varepsilon}{U}\right)^M = C_M \exp\left(M\ln\left(1-\frac{\varepsilon}{MT}\right)\right), \qquad (4)$$

where ε is the kinetic energy of the system, $U = MT$ is the energy of the thermostat, C_M is the normalization factor [6]. As a result, in the limit of a large number of particles N at $M = \frac{3}{2}N \to \infty$, expression (4) becomes a grand canonical distribution $w_0(\vec{r},\vec{p}) = C_M \exp\left(-\frac{\varepsilon}{T}\right)$.

Thus, on the tails of the energy distributions, using formula (4), we find an amendment for the microcanonical distribution (3), which changes the usual Fermi-Dirac distribution, describing the system well away from the tails of the proton spectrum. Moreover, in formulas (2) - (3) it is taken into account that the energy of the system is recalculated in accordance with the Lorentz transformations. The energy in the distribution (2) is reckoned from the value of the self-consistent mean field with allowance for the surface energy, since the nucleons are "locked" by the mean field.

In addition to the contribution of (1) to the cross section for the emission of protons from the hot spot, we also took into account the contribution from the fusion of the non-overlapping parts of the colliding nuclei so called "spectators".

2. A Comparison with Experimental Data

A comparison of the calculated energy spectra of protons emitted in the $^{16}O+^{27}Al \to p+X$ reaction at an angle of 30° at an ^{16}O ion energy of 207 MeV (~ 12 MeV/nucleon) (curves 1) and 310 MeV (~ 19 MeV/nucleon) (curves 2), with experimental data is shown in Fig. 1 (left-hand panel). Our calculation describes the data [7] (which have so far been unexplained) with allowance for the introduced amendment for the microcanonical distribution. Without considering this correction, the spectra turn out to be more harden (corresponding to this variant of the calculation, the dashed curves la and 2a go above the points). The agreement with the experiment is achieved for a non-equilibrium equation of state with a selected compression modulus of $K = 210$ MeV and without breaking the agreement with other data.

In Fig. 1 (right-hand panel) are shown the energy spectra of protons emitted in the $^{14}N+^{124}Sn \to p+X$ reaction at angles of 45° (curve 1), 62° (curve 2) and 90° (curve 3) at an ^{14}N ion energy of 32 MeV/nucleon. The calculated curves

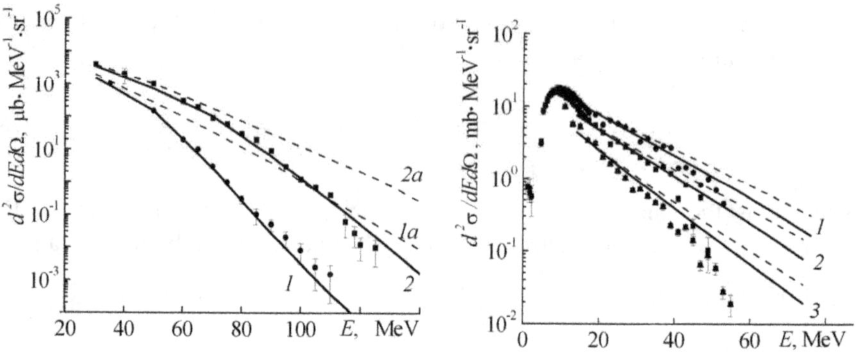

Fig. 1(left-hand panel). The energy spectra of the protons emitted in the $^{16}O+^{27}Al \rightarrow p+X$ reaction at an angle of 30° with an ^{16}O ion energy of 207 MeV (curve 1) and 310 MeV (curve 2). The calculated spectra are shown by solid lines and experimental data [7] are shown by points. The dashed curves la and 2a correspond to calculations without taking into account an amendment for the microcanonical distribution. (right-hand panel). The energy spectra of protons emitted in the reaction $^{14}N+^{124}Sn \rightarrow p+X$ calculated with a non-equilibrium equation of state (solid lines) and experimental data (points from Ref. [8]) at an ^{14}N ion energy of 32 MeV/nucleon at an angle of 45° (curve 1), 62° (curve 2) and 90° (curve 3). The dashed curves correspond to the calculations for the traditional equilibrium equation of state.

corresponding to the non-equilibrium equation of state (solid lines) agree with the experimental data [8], in contrast to the dashed lines corresponding to the choice of the equilibrium equation of state, when they go above the experimental points. At the tails, the experimental spectrum is somewhat "cut off," which is apparently due to systematic experimental errors and is not reproduced in the calculation. An amendment for the microcanonical distribution is in a significantly higher region of proton energies (see the previous figure — left-hand panel)

In Fig. 2 are shown the momentum spectra of protons emitted in the $^{12}C+^{9}Be \rightarrow p+X$ reaction at an angle of 3.5° at an ^{12}C ion energy of 300 MeV/ nucleon. The experimental data [9] are marked by points. The solid curve 1 is our calculation. The dashed curve 1 is our calculation without taking into account an amendment for the microcanonical distribution. Curves 2, 3. 4 are results of calculations for transport codes [9]. Curve 2 corresponds to the cascade model [10], curve 3 corresponds to the transport model of quark-gluon strings [11], curve 4 corresponds to the quantum molecular dynamics (QMD) model built into the GEANT4 package [12]. As can be seen from this figure, in the cumulative region of the spectrum our calculation turned out to be in

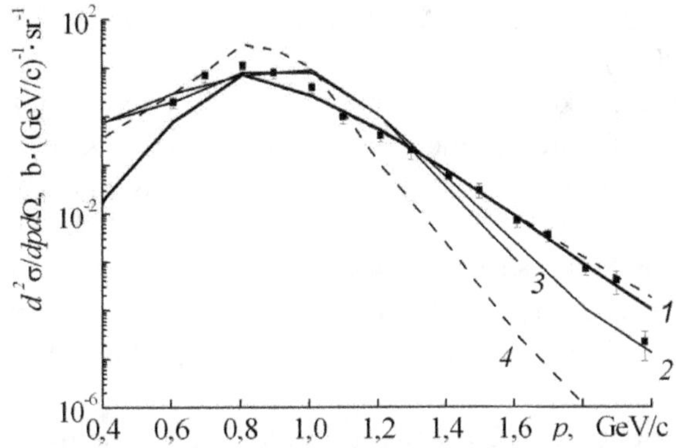

Fig. 2. The momentum distributions of protons emitted in the reaction $^{12}C+^{9}Be \rightarrow p+X$ at an angle of 3.5° at an ^{12}C ion energy of 300 MeV/nucleon. Curve 1 is our calculation, the dashed curve 1 is our calculation without taking into account an amendment for the microcanonical distribution, the points are the experimental data from [9]. Curves 2, 3. 4 are results of calculations for transport codes [9]. Curve 2 corresponds to the cascade model [10], curve 3 corresponds to the transport model of quark-gluon strings [11], curve 4 corresponds to the model of quantum molecular dynamics (QMD) built into the GEANT4 package [12].

agreement with the experimental data [9]. The decline in the cross sections by 5 orders of magnitude is reproduced in our approach no worse than in the Monte Carlo transport codes. Some cascade calculations noticeably underestimate these experimental data in the high-momentum region. However, in the region of small momentums, our calculation underestimates the experimental data, which may be due to the contribution from the protons formed as a result of ^{12}C fragmentation. An amendment for the microcanonical distribution (dashed curve 1) appears in this case only at the very tails of the high-momentum distributions of the protons.

Thus, in this paper, the idea of using of a non-equilibrium equation of state in the hydrodynamic approach to describe the high-momentum proton spectra emitted in heavy-ion collisions over a wide energy range has been further developed.

References

1. A.T. D'yachenko, *Phys. Atom. Nucl.* **57**, 1930 (1994).
2. A.T. D'yachenko, K.A. Gridnev and W. Greiner, *J. Phys. G. Nucl. Part. Phys.* **40**, 085101 (2013).

3. A.T. D'yachenko and I.A. Mitropolsky, *Bull. Russ. Acad. Sci. Phys*. **81**, 1521 (2017).
4. A.T. D'yachenko and I.A. Mitropolsky, Construction of the non-equilibrium equation of state for the description of collisions of heavy ions and its application to finding energy spectra of protons and subthreshold pions. in *the book "In Memory K. A. Gridnev. To the 80th anniversary of his birth."* (Gatchina, NRC "Kurchatov Institute" - PNPI, 2018), p. 63.
5. A.T. D'yachenko, I.A. Mitropolsky and Yu.G. Sobolev, *Proc. VIII Int. Symposium on Exotic Nuclei (Kazan 2016)*, (World Scientific, Singapore, 2017) p. 38.
6. A.I. Anselm, *Fundamentals of statistical physics and thermodynamics* (Moscow: Nauka, 1973).
7. S.I. Gogolev et al, *Proc. Int. School-Seminar on Heavy Ion Physics (Dubna 1993),* (JINR. E7-93-274), **2**, 66 (1993).
8. V.V. Avdeichikov et al. (CHIC-Collaboration), *Proc. Int. School-Seminar on Heavy Ion Physics (Dubna 1993)*. (JINR. E7-93-274), **2**, 238 (1993).
9. B.M. Abramov et al., *Phys. Atom. Nucl*. **78,** 373 (2015).
10. A.V. Dementyev and N.M. Sobolevsky, *Nucl. Tracks Radiat. Meas*. **30**, 553 (1999).
11. S.G. Mashnik et al., *LA-UR-08-2931* (Los Alamos 2008); arXiv.0805.0751 [nucl-th].
12. T. Koi et al., *AIP Conf. Proc*. **896**, 21 (2007).

Total Reaction Cross Sections of Neutron-Rich Light Nuclei Measured by the COMBAS Fragment-Separator

B. Erdemchimeg[1,2,*], A. G. Artukh[1], S. Davaa[2], B. M. Hue[1,3], T. Issatayev[1], S. A. Klygin[1], G. A. Kononenko[1], G. Khuukhenkhuu[2], S. M. Lukyanov[1], T. I. Mikhailova[1], V. A. Maslov[1], K. Mendibayev[1,4], Yu. M. Sereda[1], Yu. E. Penionzhkevich[1] and A. N. Vorontsov[1]

[1]*Joint Institute for Nuclear Research, 141980, Dubna, Russia*
[2]*National University of Mongolia, NRC, Ulaanbaatar, Mongolia*
[3]*Institute of Physics, Hanoi, Vietnam*
[4]*Eurasian Gumilev University, Astana, Kazakhstan*
[*]*E-mail: erdem@jinr.ru*

Preliminary results of measurements of the total reaction cross sections σ_R for nuclei ^{4}He, ^{6}He, ^{8}He, ^{7}Li, ^{8}Li, ^{9}Li, ^{11}Li, ^{7}Be, ^{9}Be, ^{10}Be, ^{11}Be, ^{12}Be, ^{8}B, ^{10}B, ^{11}B and ^{12}B nuclei at energy range (10-50) A MeV with ^{28}Si target is presented. The secondary beams of light nuclei were produced by bombardment of the ^{15}N (50 A MeV) primary beam on Be target and separated by COMBAS fragment-separator. In dispersive focal plane a horizontal slit defined the momentum acceptance as 1% and a wedge degrader of 600 μm Al was installed. The B_ρ of the second section of the fragment-separator was adjusted for measurements in energy range (10-45) A MeV. The strong absorption model reproduces the A-dependence of σ_R, but not the detailed structure. We are comparing our experimental data with predictions of microscopic (dashed line) and strong absorption model, Glauber multiple scattering theory and preliminary results are obtained.

Keywords: Total reaction cross sections; transmission method.

1. Introduction

Nowadays one of the major goals of nuclear science is to learn about the behavior of isotopes with exotic neutron to proton number ratios. The production of radioactive ion beams of light nuclides from the 1980s onwards has contributed towards a substantial increase of information on the properties of nuclei far away from the beta stability line, mainly on new aspects of their nuclear structure such as skins, haloes shells and others, in such a way that heavy-ion reaction cross-section measurements can be considered today as an essential tool to investigate the nuclear structure of exotic nuclei. Measurements of the total reaction cross-section, σ_R, defined as the total, σ_T, minus the elastic, σ_{el}, cross-sections for two colliding nuclei ($\sigma_R = \sigma_T - \sigma_{el}$) are important for the evaluation of the nuclear matter radius, thus providing information about the size of nuclei.

We report σ_R measurements ^{4}He, ^{6}He, ^{8}He, ^{7}Li, ^{8}Li, ^{9}Li, ^{11}Li, ^{7}Be, ^{9}Be, ^{10}Be, ^{11}Be, ^{12}Be, ^{8}B, ^{10}B, ^{11}B and ^{12}B nuclei at energy range (10-50) A MeV with ^{28}Si target is. Most of these nuclei are proton-rich; some are on the proton drip line, and therefore may be proton-halo candidates. The measurements were made by aiming these projectiles at a stack of thin Si elements which served as both targets and detectors.

We interpret our data using phenomenological strong absorption models, as well as more elaborate models such as the optical limit of Glauber multiple-scattering theory.

2. The Experimental Details

In our experiment, a primary 15N beam was accelerated to around 50 A MeV by the U-400M cyclotron of the Flerov Laboratory of Nuclear Reactions, JINR, and sent to a 9Be target of 89 mg/cm^2 thickness. A secondary beam cocktail consisting of particles He, Li, Be and B isotopes was produced and transported by the COMBAS fragment separator system [1] to the detector system (Fig. 1). The secondary beams were partially purified by 600 μm Al wedge. The energy dispersion was limited to less than 1% FWHM by analyzing slits. COMBAS separator a magneto-optic configuration is realized on the rigid focusing principle. Three parameters are important for particle separation and trajectory analysis by COMBAS setup: magnetic rigidity (Bρ), energy loss difference in the degrader (ΔE/Δx) and time of flight (TOF).

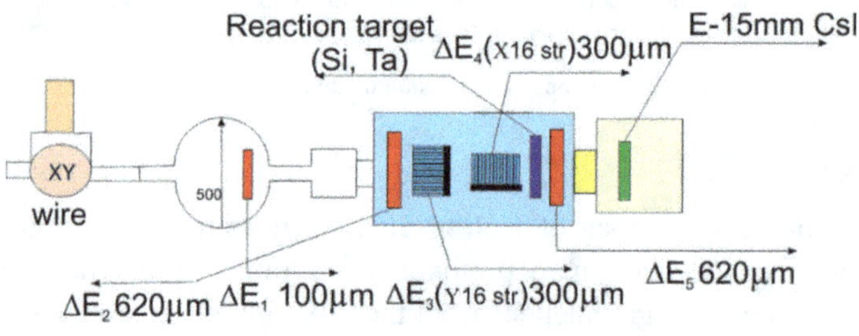

Fig. 1. Experimental schematic view of detection set-up.

The first Si ΔE_1 double-sided strip detector (100 μm thickness, was used to obtain ΔE information and space x − y distribution of incoming secondary beam. Behind this detector, there was a ΔE_2 detector (620 μm thickness, 50 x 50 mm^2) and two other detectors with 16 horizontal X strips and 16 vertical Y strips in rear side of ΔE_3 (300 μm thickness, 55 x 55 mm^2) and ΔE_4 (300 μm thickness, 55 x 55 mm^2), respectively. These detectors were designed for both precise measurement of spectra of the secondary beams and control of divergence by the x-y coordinate distribution. The silicon ΔE_5 (620 μm thickness, 50 x 50 mm^2) detector was acquiring additional ΔE information and performing identification of fragments. In order to measure the remaining energy of particles (Er) a CsI(Tl)) (see Fig. 1) detector with 15-mm thickness was used and mounted at the end of the telescope.

3. Total Reaction Cross Section Measurement

Transmission method [2] is based on one simply counting of the number of beam particles incident on the target N_{inc}, and the corresponding number of outgoing particles, which have not undergone a reaction in the target N_{el}. The difference between these two numbers N_{inc} and N_{el} represents the number of reactions N_{reac}, which occurred in the target for N_{inc} incident particles:

$$\sigma_R = \frac{N_{inc} - N_{el}}{N_{inc}} \cdot \frac{A}{dN_A} \tag{1}$$

where A is mass number, d is target thickness, N_A is Avogadro's number.

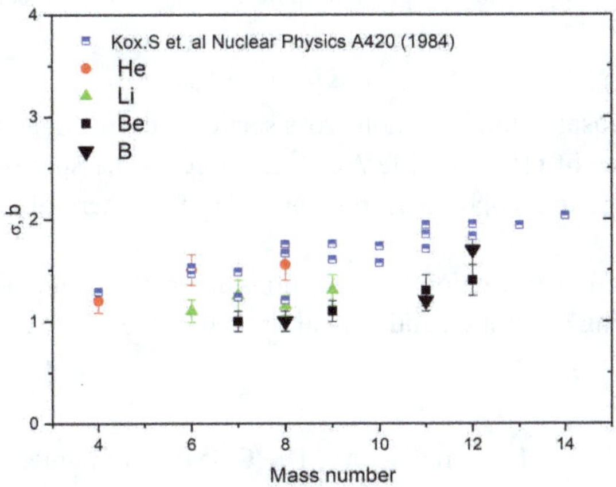

Fig. 2. We compared our measured data's with the A dependence lines of the Kox formula (when the reduced strong absorption radius r_0=1,1 blue line) [3].

The Kox formula gives excellent results for stable nuclei when the reduced strong absorption radius r_0 is fixed at 1.1 fm, and therefore any significant departure from its predictions may disclose a halo structure. The smooth A dependence of the predictions reproduces the data on average, but is unable to explain the scatter of the reaction cross sections observed in the experiment. (see Fig. 2) This observation suggests that a more sophisticated theory is needed to explain these data. We are comparing our experimental data with Glauber multiple scattering calculation [4] and preliminary results are obtained. (see Fig. 3)

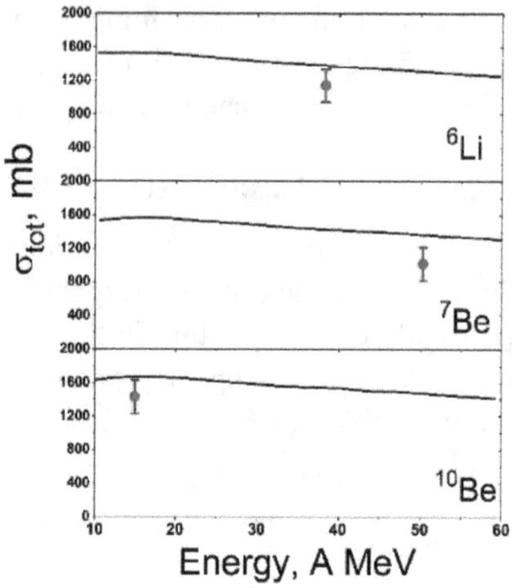

Fig. 3. Experimental reaction cross sections points for ^{6}Li, ^{7}Be, ^{10}Be compared with calculation preliminary results of optical limit of the Glauber model (continues lines).

4. Conclusions

1. We have measured total reaction cross sections (σ_r) for Li and Be, He, B in Si with energies of (10-50)A MeV on ^{28}Si targets. The success of the method shows that Si telescope measurements of σ_r for other solid targets are also feasible.
2. The measured total reaction cross-sections σ_r are compared the calculations by reduced strong absorption and optical limited of the Glauber models.

References

1. A. G. Artukh, Yu. M. Sereda et al., The COMBAS fragment separator, *Nucl. Instrum. Methods. A*. **426**, 605 (1999).
2. R. E. Warner et al., Total reaction and 2*n*-removal cross sections of 20–60*A* MeV ^{4,6,8}He, $^{6-9,11}$Li, and ^{10}Be on Si, *Phys. Rev. C* **54**, 1700.
3. S. Kox, A. Gamp, C. Perrin, J. Arvieux et al., Trends of total reaction cross sections for heavy ion collisions in the intermediate energy range, *Phys. Rev. C*. **35**, 1678 (1987).
4. C. A. Bertulani et al., MOMDIS: A Glauber model computer code for knockout reactions, *Comput. Phys. Commun*. **175** (2006) 372–380.

Break-Up Reactions of ^{6}Li, ^{7}Be and ^{8}B

T. Issatayev*[1,3,8], S. M. Lukyanov[1], B. M. Hue[1,2], K. Mendibayev[1,3,8], A. G. Artukh[1], D. Aznabayev[1,8], C. Borcea[4], S. Calinescu[4], B. Erdemchimeg[1], A. Kabyshev[3], S. A. Klygin[1], G. A. Kononenko[1], K. Kuterbekov[3], V. A. Maslov[1], M. A. Naumenko[1], V. V. Ostashko[5], Yu. E. Penionzhkevich[1,6], F. Rotaru[4], F. Saillant[7], Yu. M. Sereda[1], Yu. G. Sobolev[1], A. N. Vorontsov[1] and T. D. Thiep[2]

[1]*Flerov Laboratory of Nuclear Reactions, Dubna, Russian Federation*
[2]*Institute of Physics, VAST, Hanoi, Vietnam*
[3]*L.N. Gumilyov Eurasian National University, Astana, Kazakhstan*
[4]*National Inst. for Physics and Nuclear Engineering, IFIN-HH, Bucharest Magurele, Romania*
[5]*Nuclear Physics Institute, Kyev, Ukraine*
[6]*National Research Nuclear University "MEPhI", Moscow, Russian Federation*
[7]*GANIL, BP 5027, 14076 Caen Cedex 5, France*
[8]*Institute of Nuclear Physics, Almaty, Kazakhstan*
**E-mail: issatayev@jinr.ru*

Secondary beams (^{6}Li,^{7}Be,^{8}B) obtained by separation of fragmentation products of ^{15}N beam (50 A MeV) impinging on a Be target using COMBAS spectrometer. The secondary products will be detected by a telescope consisting of five Si-detectors (dE) and CsI(Tl) detector (E_r). By the dE-E method, the telescope allowed us to achieve unambiguous particle identification originating from incident secondary beams as well as theirs reaction products resulting from reactions with one of Si detectors chosen as the target. Accordingly, break-up reactions of ^{6}Li, ^{7}Be and ^{8}B were studied in this experiment.

Keywords: Break-up; secondary beam; proton halo.

1. Introduction

In recent years, the study involving loosely bound light nuclei has still attracted intensive experimental and theoretical attention. The experimental investigations for clustering nuclei and halo-like nuclei have been developed quite strongly with the aid of both radioactive ion beams (RIBs) and beams of stable loosely bound nuclei[1]. Utilizing neutron-rich light radioactive projectiles of He, Li and Be isotopes impinging on reaction targets (Be, C and Al) to measure the interaction cross section and the nuclear radii were performed leading to the discovery of the neutron halo[2] being a major breakthrough in nuclear physics and since then, several neutron halos has been confirmed. Soon after, total reaction cross section and neutron-removal cross section (σ_{-xn}) have been measured for $^{4,6,8}He$ and $^{6\text{-}9,11}Li$ nuclei by the silicon detectors used as the active target[3]. On the other hand, the proton halo, is much rarer due to the presence of the Coulomb barrier. The proton drip-line nucleus ^{8}B emerges as one of the best candidates for a proton-halo nucleus since the valence proton is in a p-state with low nuclear charge Z and low separation energy of 137 keV. For many years, proton-halo structure of ^{8}B has been much debated deriving from the first experimental observation of its unusual behavior via the measurement of quadruple moment[4]. Theoretical calculation, however, indicated that the result could also be explained by cluster effects[5]. Furthermore, even if the measurement at low energies (20 and 60 A MeV) showed large total cross sections as an evidence for a proton halo in ^{8}B[6], the results deviated considerably from other experiment at higher energy (790 A MeV)[7]. As another evidence, the measurement of the longitudinal momentum distribution of ^{7}Be via one proton removal channel of ^{8}B[8,9] showed a much narrower distribution than that for stable nuclei whereas no substantial halo contribution was resulted from quasi-elastic scattering of ^{8}B at 40 A MeV[10].

Thus, from the experimental point of view, proton halo structure of ^{8}B is still an open question. Although the theoretical publications can explain the existence of a halo, exclusive experimental data including high-precision proton-removal cross section (σ_{-xp}) and momentum distribution are needed. The proposed study aims at the direct measurement of σ_{-xp} and break-up cross section (σ_{BU}) of ^{8}B and neighboring loosely bound cluster nuclei (^{6}Li and ^{7}Be) respectively as well as momentum distribution of the break-up products.

2. The Experimental Set-Up

In present experiment, the secondary cocktail beam was produced by fragmentation of ^{15}N beam (50 A MeV) irradiated on a Be target before being separated and purified by the achromatic magnetic system of the COMBAS spectrometer[11]. The energy dispersion was limited to less than 1% FWHM by analyzing slits. Two Bρ settings were chosen: one leading to nuclei rather close to stability line and another leading to proton rich nuclei. The secondary products were detected by a telescope consisting of five Si-detectors and a CsI(Tl) detector used to measure the energy loss (dE) and the residual energy (Er), respectively. The telescope was located in the focal plane of the COMBAS spectrometer and its scheme is shown Fig. 1 (not to scale). The telescope served as target, degrader of beam energy and detector system for the reactions.

The first double-sided Si strip detector (DSSD) ΔE_1 was set not only for obtaining the ΔE information and space *x-y* distribution of incoming secondary beam but also for triggering the data acquisition. This one, moreover, was used for the time of flight (TOF) start signal while the stop signal was provided by the cyclotron radio-frequency Behind this detector, there were another single-pad Si (ΔE_2) and two other single-sided Si strip (ΔE_3 and ΔE_4) arranged horizontally and

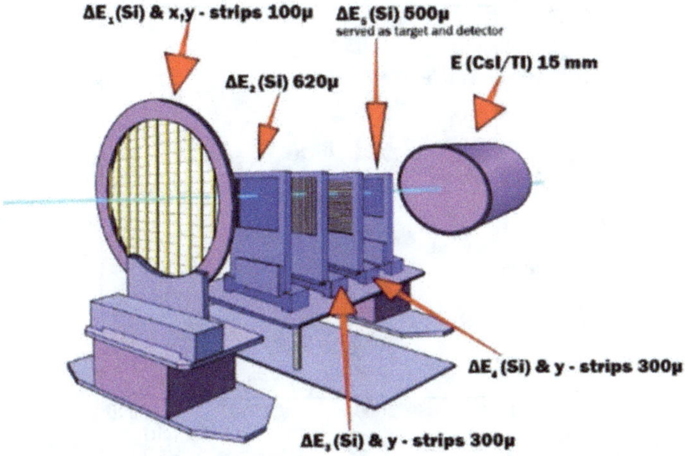

Fig. 1. The telescope (not to scale) used in this experiment.

vertically for both precise measurement of the spectra of the secondary beams and control divergence by measuring x-y coordinate distribution. The last Si detector ΔE_5 (500 μm thickness, 50 x 50 mm^2) was employed for acquiring additional ΔE information and identification for the fragments. It was also used as a secondary target for the goal of the σ_{BU} measurement. In order to measure the remaining energy of particles (E_r), a CsI(Tl) detector (15 mm thickness, 200 mm diameter) was used and mounted at the end of the telescope.

The used telescope allowed us to obtain unambiguous particle identification both for incoming secondary beam and their reaction products by using ΔE-TOF-E method.

3. Data and Data Analysis

The first goal of this section is to identify particles. To investigate the incident beams, a combination of all available information from detectors was used, including the Si detectors in front of ΔE_5 detector. The reaction products and longest range isotopes were identified based on the information of energy loss in the ΔE_5 and E_r in the CsI(Tl) detector. The thickness of CsI(Tl) detector was sufficient to stop all of the reaction products. Two-dimensional ΔE-TOF matrix (Fig. 2) is an example of the applied particle identification method. Multiple ΔE-TOF matrices were built for each of ΔE_1... ΔE_4 detectors and for chosen Bρ setting to identify reaction products. Fig. 2 shows a typical identification spectrum ΔE_2-TOF taken with ΔE_2 for the first Bρ setting. In this Fig. we can see

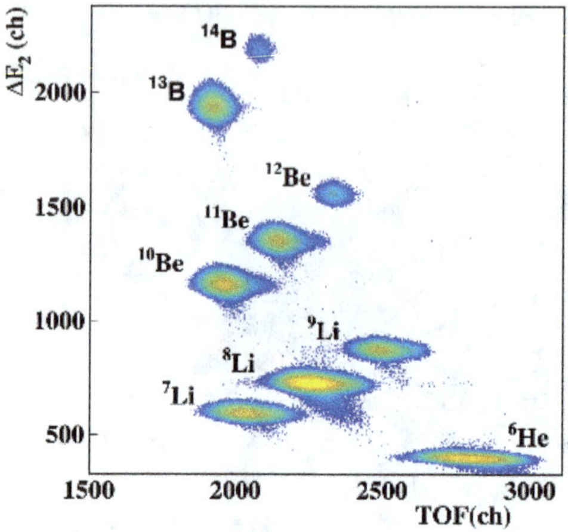

Fig. 2. Typical two-dimensional energy spectrum of ΔE_2 and E_r.

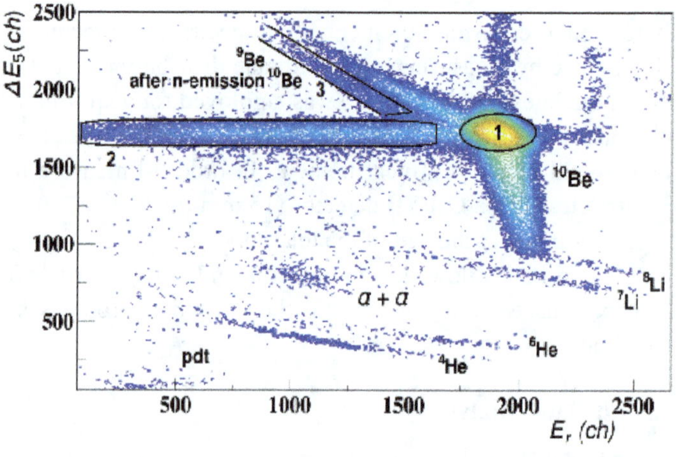

Fig. 3. Two-dimensional energy spectrum of ΔE_5 and E_r selected for 30 A MeV ^{10}Be.

a sharp separation of the following isotopes: ^{6}He, ^{7,8,9}Li, ^{10,11,12}Be and ^{13,14}B. From this cocktail of secondary beams, in the data processing a gate was set around the core of each isotope mentioned above. In the next step, the spectrum ΔE_5-ΔE_r(CsI) was built (Fig. 3) from the cuts of ^{10}Be isotope in the spectra of previous detectors (ΔE_1-ΔE_4). Several regions can be obtained in this figure, which are marked correspondingly:

- (1) unreacted ^{10}Be events
- (2) reacted events in the CsI(Tl) detector

- (3) ^{9}Be after one neutron-removal reaction for ^{10}Be
- Li after stripping one proton
- $\alpha + \alpha$: events resulting from the decay intermediately of ^{8}Be nucleus produced after 2n evaporation from ^{10}Be
- *pdt* and ^{4,6}He obtained from the break-up of ^{10}Be into alphas and hydrogen isotopes

The similar procedure was applied to the second Bρ setting. The condition on the following isotopes: ^{4}He, ^{6}Li, ^{7}Be and ^{8}B was set leading to the spectra in Fig. 4. A various range of reaction products mainly resulted from the break-up reaction of the selected isotopes was observed.

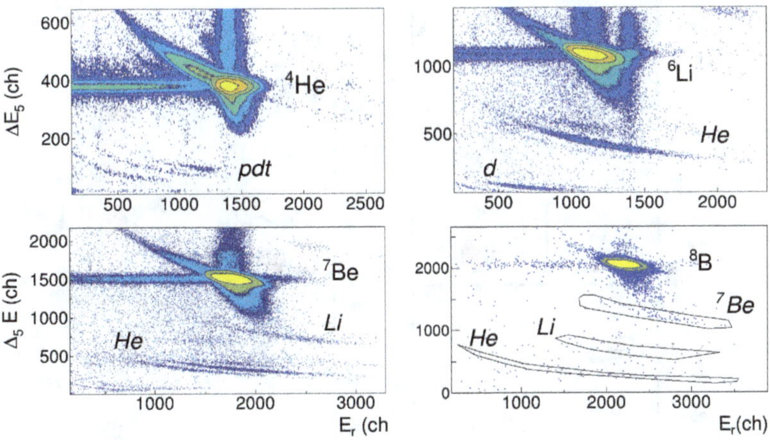

Fig. 4. Two-dimensional energy spectrum of ^{4}He, ^{6}Li, ^{7}Be and ^{8}B after the secondary reaction.

4. Results and Discussion

4.1. *Break-up reaction cross section*

The break-up reaction cross-section for the channels indicated in the Fig. 3 and Fig. 4 was calculated using the following formula:

$$\sigma_{BUchannel} = \frac{N_{select}}{N_{beam} N_{target}}$$

In this formula N_{select} is the number of events corresponding to a particle break-up channel, N_{beam} is the total number of non-reacted (selected isotope) events and N_{target} is the number of target particles per unit volume.

The breakup cross-section values for disintegration into charged fragments have been shown in Fig. 5 for ^{7}Be, ^{8}B, ^{9}Be and ^{10}Be nuclei.

The most interesting result has been observed for ^{8}B as in Fig. 5b. This nucleus has mostly broken into ^{7}Be and proton with a large value of p-removal cross section of 240±20 mb. This fact is a strong indication of ^{7}Be as a core of ^{8}B surrounded by a weakly bound "halo" proton.

For the case of ^{7}Be, the main reaction channels are ^{6}Li + p and ^{3}He + ^{4}He as it is shown in Fig. 5a. The separation energies for these break-up channels are 5.6 MeV and 1.6 MeV respectively. The cross section values, therefore, are correlated with the separation energies.

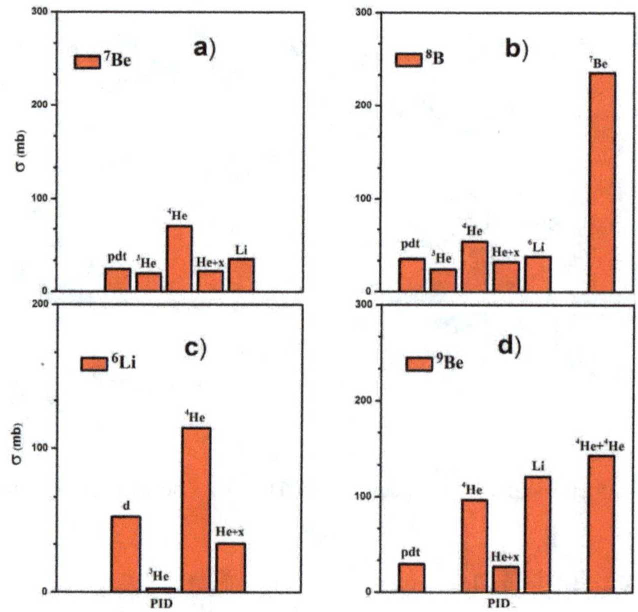

Fig. 5. Measured break-up reaction cross section into various channels for ^{7}Be (a), ^{8}B (b), ^{6}Li (c) and ^{9}Be (d) on Si.

In Fig. 5c, it also illustrated that ^{6}Li has mostly broken into ^{4}He and d with the separation energy of 1.473 MeV, and the breakup probability into ^{3}He is much less, because the separation energy of that reaction is 15.794 MeV. The obtained breakup cross-section of ^{6}Li and ^{7}Be has obviously confirmed the cluster structure of this nuclei, which was studied and established in[12].

The result for ^{9}Be has been shown in Fig. 5c. One can see that ^{9}Be breaks essentially into one neutron ($S_{1n}(^{9}Be)$ = 1.665 MeV) and subsequently two alpha particles.

4.2. *Break-up momentum distribution for ^{6}Li, ^{7}Be and ^{8}B*

The parallel momentum distributions of ^{6}Li, ^{7}Be and ^{8}B can be obtained from the measured energy distribution of the selected breakup products. Fig. 4.2 shows: a) Experimental momentum distribution for d and $^{3,4}He$ from secondary beam of ^{6}Li ions, b) experimental momentum distribution for d and $^{3,4}He$ and ^{6}Li from secondary beam of ^{7}Be ions, and c) experimental momentum distribution for ^{8}B (secondary beam) and ^{7}Be from ^{8}B breakup. Peaks in the right side of the each insertions in Fig. 4.2 corresponding to the unreacted ^{6}Li, ^{7}Be and ^{8}B events.

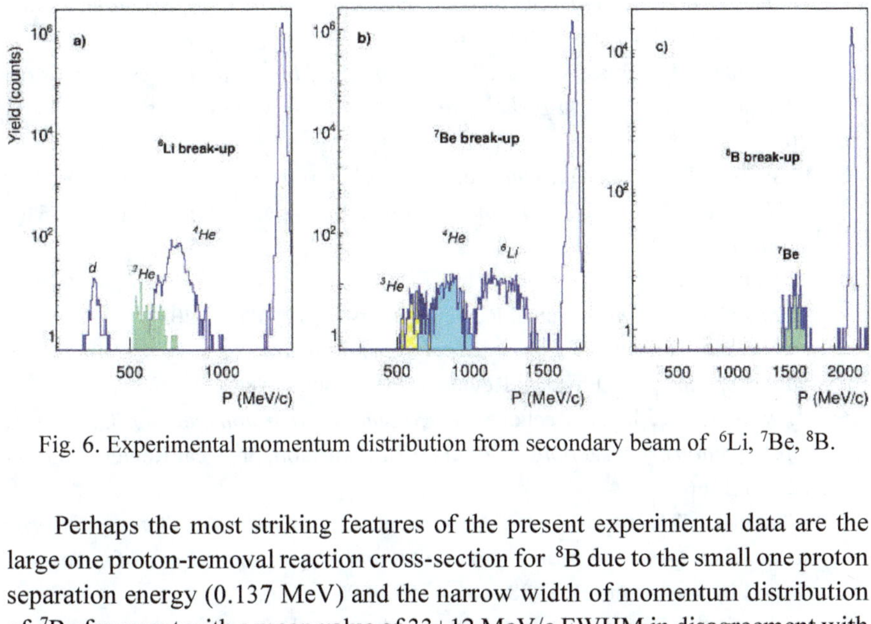

Fig. 6. Experimental momentum distribution from secondary beam of ^{6}Li, ^{7}Be, ^{8}B.

Perhaps the most striking features of the present experimental data are the large one proton-removal reaction cross-section for ^{8}B due to the small one proton separation energy (0.137 MeV) and the narrow width of momentum distribution of ^{7}Be fragment with a mean value of 33±12 MeV/c FWHM in disagreement with result in ref.[8]. These findings support the existence of a one-proton halo in ^{8}B. A new experiment to accumulate larger statistics are planned.

5. Conclusions

The particle identification for secondary beam was performed by employing the ΔE-E method for a various isotope range from He to B isotopes. The ΔE and E were measured by the Si and CsI(Tl) telescope, respectively. The preliminary results on the break-up cross sections of some weakly bound light isotopes ^{6}Li, ^{7}Be and ^{8}B were obtained. The parallel momentum distribution of ^{3,4}He and ^{6}Li produced from the break-up of ^{6}Li, ^{7}Be and ^{8}B was also determined. Large one proton removal cross section for ^{8}B as well as narrow momentum distribution of Be fragment confirm proton-halo structure of ^{8}B. We plan to carry out next experiment on total reaction cross-section, nucleon-removal cross section and elastic scattering measurements of other light neutron-rich nuclei using this technique and COMBAS fragment separator with improved statistics. The authors would like to thank for the support from the Russian Science Foundation (17-12-01170).

References

1. Y.E. Penionzhkevich, *Special features of nuclear reactions induced by loosely bound ^{6}He and ^{6,7}Li nuclei in the vicinity of the Coulomb barrier height*, Physics of Atomic Nuclei (2009) **72**(10) 1617.
2. I. Tanihata, H. Hamagaki et al., *Measurements of interaction cross sections and nuclear radii in the light p-shell region*, Physical Review Letters (1985) **55**(24) 2676.
3. R. Warner, R. Patty et al., *Total reaction and 2n-removal cross sections of 20-60 A MeV He4, 6, 8, Li6-9, 1 1, and Be10 on Si*, Physical Review C (1996) **54**(4) 1700.
4. T. Minamisono, T. Ohtsubo et al., *Proton halo of B8 disclosed by its giant quadrupole moment*, Physical Review Letters (1992) **69**(14) 2058.
5. H. Nakada, T. Otsuka, *E2 properties of nuclei far from stability and the proton-halo problem of B8*, Physical Review C (1994) **49**(2) 886.
6. R. Warner, J. Kelley, P. Zecher et al., *Evidence for a proton halo in B8: Enhanced total reaction cross sections at 20 to 60 MeV/nucleon*, Physical Review C (1995) **52**(3) R1166.
7. I. Tanihata, T. Kobayashi, O. Yamakawa et al., *Measurement of interaction cross sections using isotope beams of Be and B and isospin dependence of the nuclear radii*, Physics Letters B (1988) **206**(4) 592-596.
8. F. Negoita, C. Borcea, F. Carstoiu et al., *B8 proton halo via reaction and breakup cross section measurements*, Physical Review C (1996) **54**(4) 1787.
9. C. Borcea, F. Carstoiu, F. Negoita et al., *B8 studied as a secondary beam at GANIL*, Nuclear Physics A (1997) **616**(1-2) 231-238.
10. I. Pecina, R. Anne, D. Bazin et al., *Quasielastic scattering of B8 and Be7 on C12 at 40 A MeV*, Physical Review C (1995) **52**(1) 191.
11. A. Artukh, G. Gridnev, M. Grushezki et al., *Wide aperture kinematic separator COMBAS realized on the strong focusing principle 1*, Nuclear Instruments and

Methods in Physics Research Section A: Accelerators, Spectrometers, Detectors and Associated Equipment (1999) **426**(2-3) 605-617.

12. Y. Kanada-En'yo, H. Horiuchi, A. Ono, *Structure of Li and Be isotopes studied with antisymmetrized molecular dynamics*, Physical Review C (1995) **52**(2) 628.

Evidence of Enhanced 3α Radius in ^{12}C Probed by Nuclear Reactions

M. Ito*, S. Hirao, M. Nakao

Department of Pure and Applied Physics, Kansai University,
3-3-35 Yamate-cho, Suita, 564-8640, Japan
**E-mail: itomk@kansai-u.ac.jp*

Y. Funaki, T. Yamada

College of Science and Engineering, Kanto Gakuin University,
1-50-1, Yokohama, 236-8501, Japan

Extension of the nuclear radius in 3α state in ^{12}C is probed by the α + ^{12}C inelastic scattering. In the α inelastic scattering, the prominent shrinkage is observed in the differential cross section of the final 2_2^+ channel in comparison to the yrast 2_1^+ channel. This shrinkage is a first evidence of the enhanced nuclear radius, which is originated from the 3α structure in the 2_2^+ state. The nuclear radii of the 3α radius is speculated by applying the Blair's phase rule. The analyses is also extended to the ^{13}C(K^-, π^-)$^{13}{}_{\Lambda}$C reactions.

Keywords: Nuclear radius; α cluster structure; hadronic reaction.

1. Introduction

The cluster structures, in which a nucleus is decomposed into several subunits, are realized in the low excited states of light mass nuclei [1]. In the cluster structure, the sub-units are weakly coupled each other, and their matter radius is prominently extended in comparison to the radius of the ground state, which obeys the law of $\approx A^{1/3}$. In particular, such the enhancement of the matter radius is extensive in the 3α structure of the ^{12}C nucleus [2-4]. The ground state of ^{12}C has a radius of 2.40 fm [2] with the spin-parity of 0^+ (0_1^+), while the excited 0_2^+ state at E_x = 7.65 MeV, which is called the Hoyle state, is considered to have a radius of 3.47 fm with the well-developed 3α cluster structure [2]. Thus, the extension of the radius in 0_2^+ is predicted to be about 1 fm in comparison to the radius of the ground 0_1^+ state.

Unfortunately, a direct measurement of the radius of the Hoyle 0_2^+ states with the developed 3α cluster structure is impossible due to its short life time, but there are several attempts to get an evidence of the enhanced matter radius of the 3α Hoyle states from the inelastic scattering of ^{12}C, which excites the Hoyle 0_2^+ state as a final state. In the inelastic scattering of light ions by ^{12}C, an oscillating pattern in the differential cross section of the scattered ion is discussed in connection to the enhanced radius of the final 0_2^+ state with the 3α structure [5-7]. However, the relation of the extended radius and the oscillating pattern of the cross section of the ^{12}C inelastic scattering still remains unclear [8,9].

In this report, we focus on the 2_2^+ state, which corresponds to the 3α rotational state of the Hoyle 2_2^+ state [2,4] with a life of 10^{-21} sec. (width of $\Gamma \sim 1$ MeV), and demonstrate that a direct evidence of the enhanced matter radius in the 2_2^+ state clearly appears in the α + ^{12}C inelastic scattering [10]. The main results in the present report is compressed from Ref. [10]. The differential cross section of $\alpha + {}^{12}\mathrm{C}(0_1^+) \rightarrow \alpha + {}^{12}\mathrm{C}(2_2^+)$ is compared with the reference reaction of $\alpha + {}^{12}\mathrm{C}(0_1^+) \rightarrow \alpha + {}^{12}\mathrm{C}(2_1^+)$ in which the 2_1^+ state is a rotational state of the ground 0_1^+ state with a spatially compact structure. We will demonstrate that the differential cross section in the 2_2^+ channel is more shrunken than that in the 2_1^+ state, and this shrinkage is a first evidence of the 3α structure in the 2_2^+ state.

We have also done the similar analysis about the $^{13}\mathrm{C}(K^-, \pi^-)^{13}{}_\Lambda\mathrm{C}$. The theoretical calculation predicts that the similar shrinkage occurs in the cross section of the final channel of $1/2_2^+$, which have the developed 3α + Λ structure [16], although its cross section has not been measured yet.

2. Framework

We calculate the differential cross sections of an α particle scattered by ^{12}C in the formulation of the microscopic coupled-channels (MCC) calculation [7,11]. In the MCC calculation, the nuclear potential of and ^{12}C is calculated from the double folding model, which is symbolically written as a function of the α - ^{12}C relative coordinate of **R**,

$$V_{f,i}(\mathbf{R}) = N_R \iint \rho_{f,i}^{(^{12}C)}(\mathbf{r}_1)\rho^{(\alpha)}(\mathbf{r}_2)\upsilon_{NN}(\mathbf{s})\, d\mathbf{r}_1\, d\mathbf{r}_2 \tag{1}$$

with $\mathbf{S} = \mathbf{r}_2 - \mathbf{r}_1 - \mathbf{R}$. Here $\mathbf{r}_1$ ($\mathbf{r}_2$) denotes a coordinate measured from the center of mass in the ^{12}C (α) nucleus. $\rho_{f,i}(\mathbf{r})$ represents the diagonal ($f = i$) or transition ($f \neq i$) densities of ^{12}C, which are calculated by the microscopic 3α cluster model, resonating group method (RGM) [2], while $\rho^{(\alpha)}(\mathbf{r})$ denotes the density of α particle, which reproduce the charge form factor of the electron scattering. In Eq. (1), υ_{NN} represents the effective nucleon-nucleon (NN) interaction which acts between a pair of nucleon contained in the ^{12}C nucleus and the α particle. In the present calculation, we adopt the DDM3Y (Density Dependent Michigan 3-range Yukawa) interaction [12].

The normalization factor N_R is introduced because the folding potential contains a theoretical ambiguity in its strength. Here this factor is set to $N_R = 1.3 \sim 1.4$ over all the incident energy, which is consistent to the MCC calculation in [7]. In addition to the folding potential, we introduce the absorptive potential with the Saxon-Woods form factor in the diagonal ($f = i$) transition in order to simulate other reaction process [7], and the parameter set of Saxon-Woods is tuned so as to reproduce all of the observed differential cross sections as much as possible. As for the internal excitation of ^{12}C, we include the low-lying collective states (2_1^+ and 3_1^-) and the 3α cluster states (0_2^+, 0_3^+, 2_2^+) in addition to the ground 0_1^+ state.

The calculation of the $^{13}C(K^-, \pi^-)^{13}{}_{\Lambda}C$ reaction is also done by applying the distorted wave impulse approximation (DWIA) on the basis of the effective neutron number method [13,14]. The internal wave function of the final $^{13}{}_{\Lambda}$C state is prepared by solving the coupled channel problem of $(^{12}C(0_1^+) + \Lambda) + (^{12}C(0_2^+) + \Lambda)$ [15], while the reduced width amplitude obtained from the sophisticated four-body calculation of $^{13}C = 3\alpha + N$ is used for the initial wave function of $^{13}C(1/2_1^-)$ [17]. The parameters of the calculation of $^{13}{}_{\Lambda}$C is tuned so as to reproduce the recent four-body calculation of Hyper THSR wave function [16]. In this calculation, the excited $1/2_2^+$ state has the enhanced nuclear radius of 0.7 fm in comparison to the ground $1/2_1^+$ state due to the developed $3\alpha + \Lambda$ structure.

3. Results

We have solved the coupled-channels equation for the $\alpha + {}^{12}C$ scattering at $E_\alpha = 386$ MeV [18] with the nuclear interaction in Eq. (1). The MCC calculation reproduce the differential cross sections of the scattering to 0_1^+, 2_1^+, 3_1^- states [10]. Furthermore, the theoretical calculation nicely reproduce the inelastic scattering to the excited state observed at $E_x \sim 10$ MeV, which is considered as the incoherent mixture of 0^+ and 2^+ strength, by taking the summation of the differential cross sections of the 0_3^+ and 2_2^+ states [10].

A comparison of the 2_1^+ cross section (dotted curve) and the 2_2^+ one (solid curve) is shown in the left panel of Fig. 1. The peak position of the 2_2^+ cross section (solid curve) shifts to the forward angular region, and its angular distribution has the shrinkage and rapid fall-down structure in comparison to the 2_1^+ cross section (dotted curve). The shrinkage in the 2_2^+ cross section (solid curve in Fig. 1) is a first evidence of the extended size of the final 2_2^+ states, which can be confirmed in the direct observable, the differential cross section. The inelastic scattering to the 2_1^+ and 2_2^+ states are completely same in the reaction kinematics but only the size of the final state is different: spatially compact structure in 2_1^+ with almost the same radius as the ground 0_1^+ state and developed 3α structure in 2_2^+ having the extended radius. Thus, we can conclude that the shrinkage in the 2_2^+ cross section is originated from the spatially extended structure of the Hoyle rotational 2_2^+ state.

In order to speculate the enhancement in the nuclear radius of the 2_2^+ state, we apply the Fraunhofer diffraction formula, which is called the Blair's phase rule [6,19,20]. We fit the first diffraction peak of the phase rule at $\theta_{c.m.} > 0^0$ to the first peak of $\theta_{c.m.} \sim 5^0$, at which both of the differential cross sections commonly show the prominent peak in the left panel of Fig. 1.

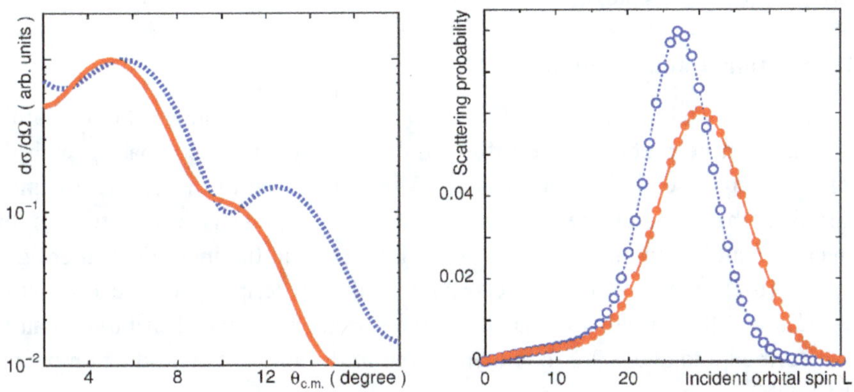

Fig. 1. (Left panel) Theoretical differential cross sections of the final state of 2_1^+ (dotted curve) and 2_2^+ (solid curve) in the α + ^{12}C inelastic scattering. The magnitudes of both the cross sections are normalized by the maximum value around $\theta c.m. \sim 5^0$. (Right panel) Partial wave decomposition of the angle integrated cross sections calculated as the function of the incident orbital spin L (*L*-distribution). The open circles with the dotted curve and the solid circles with the solid curve show the partial cross sections of the 2_1^+ and 2_2^+ final states, respectively. The magnitudes of all the partial cross section is normalized by the total cross section, which is the summations of the partial cross sections. These figures are taken from Ref. [10].

The fit to the peak gives the enhancement of $\Delta\alpha = \alpha(2_2^+) - \alpha(2_1^+) \sim 0.6$ fm, which can be attributed to the extended matter radius of the final 2_2^+ state.

In the right panel of Fig. 1, the partial wave decompositions of the angle integrated cross section are shown. In the comparison of 2_1^+ (open circles) with 2_2^+ (solid circles), we can clearly see the extended *L*-distribution of 2_2^+ to the higher *L* region. This extension of 2_2^+ in the *L*-space (right panel) is just opposite to the shrinkage in the θ-space, the differential cross section (left panel). According to the classical relation of $L = kb$, where k and b denote the incident wave number and the impact parameter, respectively, the extension of the L-distribution clearly means that the production area of the 2_2^+ state is more extended than the area of the 2_1^+ production.

Finally, we briefy report the results of the DWIA calculation for $^{13}C(K^-, \pi^-)^{13}{}_{\Lambda}C$ $(1/2^+{}_{1,2})$ The angular distribution of the π^- emission in the $1/2^+{}_2$ channel with the $3\alpha + \Lambda$ structure reveals the shrunken structure in comparison to the distribution of the ground $1/2^+{}_2$ channel with the $^{12}C(0_1^+) + \Lambda$ structure. This shrinkage is due to the spatial extension of the $1/2^+{}_2$ channel. By applying the Blair's phase rule, we have found that the enhancement in the diffraction radius is about 0.5 fm, which is comparable to the extension of the nuclear radius of the $1/2^+{}_2$ state, 0.7 fm.

4. Summary and Discussion

In summary, we have pointed out that the evidence of the enhanced radius of the 2_2^+ state can be probed in the differential cross section by comparing with the reference cross section of the 2_1^+ state. The distribution of the 2_2^+ state is more shrunken than that of the 2_1^+ state, and this shrinkage is a signature of the enhancement of the matter radius of the final state in the inelastic scattering of $\alpha + {}^{12}C(0_1^+) \rightarrow \alpha + {}^{12}C(2_2^+)$. According to the prescription of the diffraction formula and the partial wave analysis, the shrunken angular distribution can be transformed into about 0.6 fm enhancement in the size of the production area of the final 2_2^+ state.

Although there were the similar analyses in the previous studies in [5-9], the comparison of the final channel with the common spin-parity states, such as the yrast 2_1^+ state and the Hoyle rotational 2_2^+ state, is a new and important insight in the present study about the relation of the inelastic scattering and the nuclear radius. In speculating the enhancement in the nuclear radius correctly, it is quite important to set an appropriate reference channel that should be compared with the cluster exit channel.

The comparison of the common spin-parity channels is the basic technique and possible to apply to the general problem of nuclear reactions probing nuclear radii of excited cluster state. In the present work, we have extended the same analysis to the hyper nucleus production, $^{13}C(K^-, \pi^-)^{13}{}_{\Lambda}C$ $(1/2^+{}_{1,2})$, and predicted that the extended nuclear radius of the 3α state will be probed by the measurement of this differential cross section. The systematic analyses about other cluster states, such as $^{16}O = \alpha + {}^{12}C$, 4α, $^{20}Ne = \alpha + {}^{16}O$ structures, are now underway.

References

1. K. Ikeda *et al., Prog. Theor. Phys.* **68**, 1 (1980), and references therein.
2. Y. Fukushima and M. Kamimura, *Proceedings of the International Conference Nuclear Structure, J. Phys. Soc. Jpn.* **44**, 225 (1977); M. Kamimura, Nucl. Phys. A**351**, 456 (1981).
3. A. Thosaki, H. Horiuchi, P. Schuck, and G. Reöpke, *Phys. Rev. Lett.* **87**, 192501 (2001).
4. Y. Funaki, A. Thosaki, H. Horiuchi, P. Schuck, and G. Reöpke, *Phys. Rev.* C**67**, 051306(R) (2003), and references therein.
5. A. N. Danilov, T. L. Belyaeva, A. S. Demyanova, S. A. Ganchalov, A. A. Ogloblin, *Phys. Rev. C***80**, 054603 (2009).
6. K. Iida, S. Koide, A. Kohama, and K. Oyamatsu, *Mod. Phys. Lett.* A**27**, 1250020 (2012).
7. S. Ohkubo and Y. Hirabayashi, *Phys. Rev.* C**70**, 041602(R) (2007).
8. M. Takashina and Y. Sakuragi, *Phys. Rev.* C**74**, 054606 (2006).
9. M. Takashina, *Phys. Rev*. C**78**, 014602 (2008).
10. M. Ito, *Phys. Rev*. **97**, 044608 (2018).
11. M. Ito, Y. Hirabayashi, and Y. Sakuragi, *Phys. Rev*. C**66**, 034307 (2002), and references therein.
12. A. M. Kobos, B. A. Brown, P. E. Hodgson, G. R. Satchler, A. Budzanowski, *Nucl. Phys*. A**384**, 65 (1982).
13. T. Yamada, K. Ikeda, H. Bandô, T. Motoba, *Phys. Rev*. C**38**, 854 (1988).
14. E. H. Auerbach, A. J. Balitz, C. B. Dover, A. Gal, S. H. Kahana, L. Ludeking and D. J. Millener, *Ann. Phys*. **148**, 381 (1983).
15. T. Yamda, T. Motoba, K. Ikeda and H. Bando, *Prog. Theor. Phys. Supple.* **81**, 104 (1985).
16. Y. Funaki, M. Isaka, E. Hiyama, T. Yamada, K. Ikeda, *Phys. Lett.* B**773**, 336 (2017).
17. T. Yamada and Y. Funaki, *Phys. Rev*. C**92**, 034326 (2015).
18. M. Itoh *et al., Phys. Rev*. C**84**, 054308 (2011).
19. J. S. Blair, *Phys. Rev.* **115**, 928 (1959).
20. A. Kohama, K. Iida, and K. Oyamatsu, *Phys. Rev.* C**69**, 064316 (2004).

Structure of β-Decay Strength Function $S_\beta(E)$ and Wigner Spin-Isospin SU(4) Symmetry

I. N. Izosimov

Joint Institute for Nuclear Research,
Dubna, 141980, Russia
E-mail: izosimov@jinr.ru

It is shown that when the parent nucleus has *nn* Borromean halo structure, then after Gamow-Teller (GT) β^--decay of parent state or after *M*1 γ-decay of Isobar Analogue Resonance (IAR) the states with *np* tango halo structure or mixed *np* tango + *nn* Borromean halo structure can be populated. Resonances in the GT β-decay strength function $S_\beta(E)$ of halo nuclei may have *np* tango halo structure or mixed *np* tango + *nn* Borromean halo structure. Demonstrated that the nuclei with Z/N ≈ 0.6 may correspond to the SU(4) region.

Keywords: Borromean halo; tango halo; mixed halo; beta decay strength function; SU(4) symmetry; SU(4) region.

1. Introduction

The strength function $S_\beta(E)$ governs [1,2] the nuclear energy distribution of elementary charge-exchange excitations and their combinations like proton particle (πp)-neutron hole (νh) coupled into a spin-parity I^π : $[\pi p \otimes \nu h]_I{}^\pi$ and neutron particle (νp)-proton hole (πh) coupled into a spin-parity I^π : $[\nu p \otimes \pi h)]_I{}^\pi$. The strength function of Fermi-type β-transitions takes into account excitations $[\pi p \otimes \nu h]_0{}^+$ or $[\nu p \otimes \pi h]_0{}^+$. Since isospin is a quite good quantum number, the strength of the Fermi-type transitions is concentrated in the region of the isobar-analogue resonance (IAR). The strength function for β-transitions of the Gamow–Teller (GT) type describes excitations $[\pi p \otimes \nu h]_1{}^+$ or $[\nu p \otimes \pi h]_1{}^+$. Residual interaction can cause collectivization of these configurations and occurrence of resonances in $S_\beta(E)$ [1,2].

Generally the term "halo" is used when halo nucleon(s) spend(s) at least 50% of the time outside the range of the core potential, i.e. in the classically forbidden region [3-5]. The necessary conditions for the halo formation are: the small binding energy of the valence particle(s), small relative angular momentum L=0,1 for two-body or hyper momentum K=0,1 for three-body halo systems, and not so high level density (small mixing with non-halo states). Coulomb barrier may suppresses proton-halo formation for Z>10. Neutron and proton halos have been observed in several nuclei [3-5]. In Borromean systems the two-body correlations are too weak to bind any pair of particles while the three-body correlations are responsible for the system binding as a whole. In states with one and only one bound subsystem the bound particles moved in phase and were therefor named “tango states” [4,5].

When the nuclear parent state has the two-neutron (*nn*) Borromean halo structure, then IAR and configuration states (CSs) can simultaneously have *nn*, *np* Borromean halo components in their wave functions [6-9]. After *M*1 γ-decay of IAR with *np* Borromean halo structure or GT β^--decay of parent nuclei with *nn* Borromean halo structure, the states with *np* halo structure of tango type may be populated [7]. It is shown that the nuclei with Z/N ≈ 0.6 may form the region with the Wigner spin-isospin SU(4) symmetry.

2. Beta Decay Strength Function in Halo Nuclei

For the GT β-transitions essential configurations include states made up of the ground state of parent nucleus by the action of the Gamow-Teller operator of the β-transition [1,2] $\boldsymbol{Y}_-$:

$$\boldsymbol{Y}_- = \sum \boldsymbol{\tau}(\boldsymbol{i})_- \boldsymbol{\sigma}(\boldsymbol{i}), \tag{1}$$

where $\boldsymbol{\tau}(\mathbf{i})_-\boldsymbol{\sigma}(\mathbf{i})$ – is a spin-isospin operator. Acting on ground state (g.s.) of parent nuclei by the operator $\boldsymbol{Y}_-$ results in formation of configurations of proton particle (πp)–neutron hole (νh) coupled into a spin-parity I^π=1$^+$. These are [1,2] so called (Fig. 1 – Fig. 3) core polarization (CP), back spin flip (BSF), and spin flip (SF) configurations. Coherent superposition [1,2] of CP, BSF, and SF configurations forms Gamow-Teller (GT) resonance. Non coherent superposition forms resonances in $S_\beta(E)$ at excitation energy *E* lower than

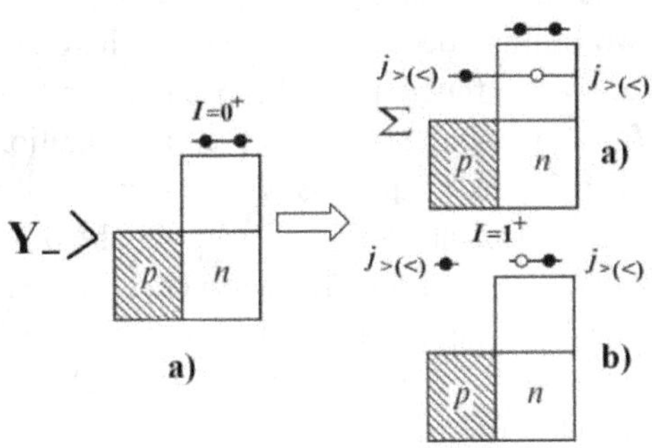

Fig. 1. Proton particle–neutron hole coupled to form the spin-parity I = 1$^+$ and core polarization (CP) states, where $j_> = l + 1/2$, $j_< = l - 1/2$, (a) *nn* Borromean halo component, (b) *np* tango halo component.

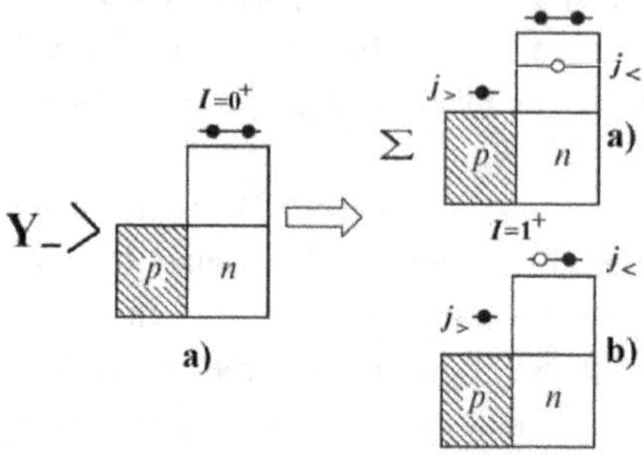

Fig. 2. Proton particle–neutron hole coupled to form the spin-parity $I = 1^+$ and back spin flip (BSF) states: (a) *nn* Borromean halo component, (b) *np* tango halo component.

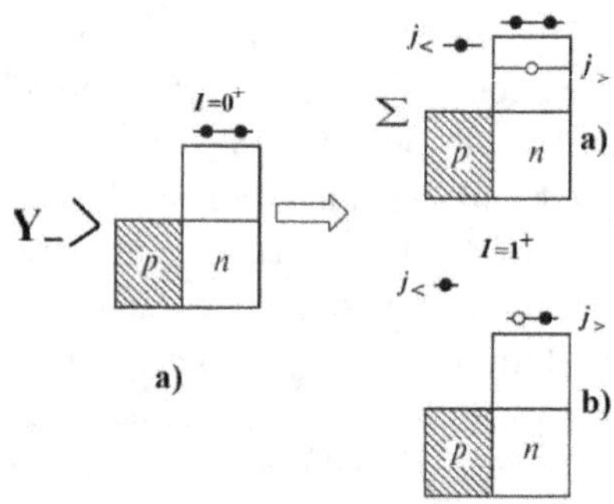

Fig. 3. Proton particle–neutron hole coupled to form the spin-parity $I = 1^+$ and spin flip (SF) states: (a) *nn* Borromean halo component, (b) *np* tango halo component.

energy of GT resonance (so called pigmy resonances). Beacause after action of Y_- operator on *nn* Borromean halo configuration with $I^\pi = 0^+$ the *np* tango halo configurations with $I^\pi = 1^+$ are formed (Fig. 1 – Fig. 3), the GT and pigmy resonances in $S_\beta(E)$ will have components corresponging to *np* tango halo. When neutron excess number is enough high, the SF, CP, and BSF configurations may simultaneously have both *nn* Borromean halo component and *np* tango halo component and form so called mixed halo (Fig. 1 – Fig. 3).

3. Gamow-Teller β^--Decay of ^{6}He, *M*1 γ-Decay of IAR in ^{6}Li and SU(4) Region

Two neurons that form the *nn* halo in ^{6}He ground state (g.s.) occupy the 1*p* orbit ($p_{3/2}$ configuration with a 7% admixture of $p_{1/2}$ configuration). The remaining two neutrons and two protons occupy the 1*s* orbit. Therefore, the action of the operator T_- on the g.s. wave function for the ^{6}He nucleus ($T = 1$, $T_z = 1$) results in the formation of the analogue state with the configuration corresponding to the *pn* halo. This IAR is in the ^{6}Li nucleus ($T = 1$, $T_z = 0$) at the excitation energy 3.56 MeV. The width of this state is $\Gamma = 8.2$ eV, which corresponds to the half-life $T_{1/2} = 6\cdot10^{-17}$ s. The theoretical and experimental data [10-12] indicate that this IAR state has a *np* halo. Since the operators of GT β-decay and *M*1 γ-decay have no spatial components (the radial factor in the *M*λ γ-transition operator is proportional to $r^{\lambda-1}$), GT β-transitions and *M*1 γ-transitions between states with similar spatial shapes are favored. The *M*1 γ-decay of IAS would be hindered [9] if the g.s. of ^{6}Li did not have a halo structure and would be enhanced if the g.s.

of ^{6}Li has a halo structure. A rather large value of the reduced probability of $M1$ γ-transition ($B(M1,\sigma)$ = 8.2 W.u.) for $M1$ γ-decay from IAS and large B(GT) $\approx 5g_A^2/4\pi$ (Σ(Ikeda sum rule) = $6g_A^2/4\pi$) value for β^--transition to the ground state is evidence [9] for the existence of tango halo structure in the ^{6}Li ground state. Because of large B(GT) value (B(GT) $\approx 5g_A^2/4\pi$), the ^{6}Li g.s. has structure corresponding to the low-energy Gamow-Teller phonon [13,14] and the energy of this GT phonon is lower than ($E_{GT} < E_{IAR}$) the energy of IAR (E_{IAR} = 3562.88 keV). In heavy and middle nuclei (Fig. 4), because of repulsive character of the spin-isospin residual interaction [1,15], the energy of Gamow-Teller resonance is larger than the energy of IAR ($E_{GT} > E_{IAR}$).

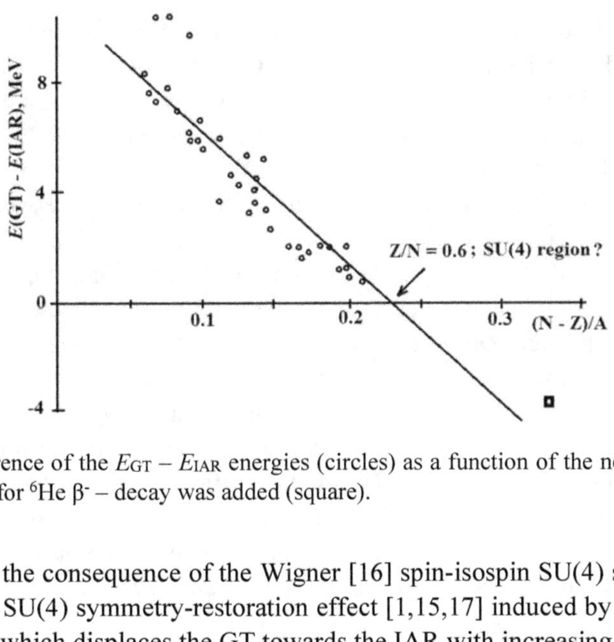

Fig. 4. Difference of the $E_{GT} - E_{IAR}$ energies (circles) as a function of the neutron excess [15]. Datum for ^{6}He β^- – decay was added (square).

One of the consequence of the Wigner [16] spin-isospin SU(4) symmetry is $E_{GT} = E_{IAR}$. SU(4) symmetry-restoration effect [1,15,17] induced by the residual interaction, which displaces the GT towards the IAR with increasing (N-Z)/A. In tango halo nucleus ^{6}Li (g.s.) for low energy GT phonon we have $E_{GT} < E_{IAR}$, $E_{GT} - E_{IAR}$= -3562.88 keV, and (N-Z)/A = 0.33 for ^{6}He (^{6}He g.s. is the parent state). Such situation may be connected with contribution of the attractive [13,14] component of residual interaction in this nucleus. It will be very interesting to find a region of atomic nuclei, where the $E_{GT} \approx E_{IAR}$ and spin-isospin SU(4) symmetry determine the nuclear properties (SU(4) region). As it follows from estimation (Fig. 4), value Z/N $\approx$ 0.6 corresponds to the SU(4) region.

4. Conclusion

Gamow-Teller resonance and pygmy resonances in GT beta decay strength function $S_\beta(E)$ for halo nuclei may have structure corresponging to *np* tango halo. When neutron excess is high enough, resonances in $S_\beta(E)$ may simultaneously have both *nn* Borromean halo component and *np* tango halo component and form so-called mixed halo. Structure of resonances in $S_\beta(E)$ is manifested in charge exchange reactions. Halo structure of some pygmy resonances is important for beta-decay analysis in halo nuclei. Demonstrated that the nuclei with $Z/N \approx 0.6$ may belong to the SU(4) region.

References

1. Yu.V. Naumov, A.A. Bykov and I.N. Izosimov, Structure of β-decay strength functions, *Sov. J. Part. Nucl.* **14**, 175 (1983).
2. I.N. Izosimov, V.G. Kalinnikov and A.A. Solnyshkin, Fine Structure of Strength Functions for Beta Decays of Atomic Nuclei, *Phys. of Part. and Nucl.* **42**, 963 (2011).
3. I. Tanihata, Neutron halo nuclei, *J. Phys. G: Nucl. Part. Phys.* **22**, 157 (1996).
4. A.S. Jensen, K. Riisager, D.V. Fedorov and E. Garrido, Structure and reactions of quantum halos, *Rev. Mod. Phys.* **76**, 215 (2004).
5. B. Jonson, Light Dripline Nuclei, *Phys. Rep.* **389**, 1 (2004).
6. I.N. Izosimov, Structure of the isobar analog states (IAS), double isobar analog states (DIAS), and configuration states (CS) in halo nuclei, *Procceedings of the International Symposium on Exotic Nuclei (EXON2012),* Vladivostok, Russia, 2012, (World Scientific, 2013) pp. 129-133.
7. I.N. Izosimov, Isobar Analog States (IAS), Double Isobar Analog States (DIAS), Configuration States (CS), and Double Configuration States (DCS) in Halo Nuclei. Halo Isomers, *AIP Conference Proceedings* **1681**, 030006 (2015).
8. I.N. Izosimov, Borromean Halo, Tango Halo, and Halo Isomers in Atomic Nuclei, *EPJ Web of Conf.* **107**, 09003 (2016).
9. I.N. Izosimov, Isospin in Halo Nuclei: Borromean Halo, Tango Halo, and Halo Isomers, *Phys. of At. Nucl.* **80**, 867 (2017).
10. Y. Suzuki and K. Yabana, Isobaric analogue halo states, *Phys. Lett. B* **272**, 173 (1991).
11. L. Zhihong, L. Weiping, B. Xixiang, W. Youbao, L. Gang, L. Zhichang and Z. Sheng, First observation of neutron-proton halo structure for the 3.563 MeV 0^+ state in ^6Li via $^1H(^6He,^6Li)n$ reaction, *Phys. Lett. B* **527**, 50 (2002).

12. National Nuclear Data Center, Brookhaven National Laboratory. http://www.nndc.bnl.gov
13. Y. Fujita, et al., Observation of Low- and High-Energy Gamow-Teller Phonon Excitations in Nuclei *Phys. Rev. Lett.* **112**, 112502 (2014).
14. Y. Fujita, et al., High-resolution study of Gamow-Teller excitations in the ^{42}Ca(^{3}He,*t*) ^{42}Sc reaction and the observation of a "low-energy super-Gamow-Teller state", *Phys. Rev. C* **91**, 064316 (2015).
15. I.N. Izosimov, Non-Statistical Effects Manifestation in Atomic Nuclei, *Phys. of Part. and Nucl.* **30**, 131 (1999).
16. E.P. Wigner, Supermultiplets, *Phys. Rev.* **51**, 106 (1937).
17. Yu.S. Lutostansky, V.N. Tikhonov, Charge-exchange resonances and restoration of Wigner's supersymmetry in heavy and super-heavy nuclei, *Phys. of At. Nucl.* **79**, 929 (2016).

Study of Cluster Structure of ^{9}Be in d-^{9}Be Interaction

E. Konobeevski*, A. Afonin, A. Kasparov, V. Mitcuk, M. Mordovskoy and S. Zuyev

Institute for Nuclear Research of the Russian Academy of Sciences,
Moscow, 117312, Russia
**E-mail: konobeev@inr.ru*
www.inr.ru

V. Lebedev and A. Spassky

Skobeltsyn Institute of Nuclear Physics, Moscow State University
Moscow, 119991, Russia

A kinematical simulation of inelastic scattering of deuterons from ^{9}Be nucleus has been performed. The simulation results showed a possibility of determining the structure of ^{9}Be excited states in the exclusive experiment with registration of both the scattered deuteron and the particle (neutron or alpha particle) from the breakup of the excited state. The preliminary results obtained in a test inclusive ^{9}Be(d,d') experiment are presented.

Keywords: ^{9}Be; inelastic scattering; cluster structure; deuterons.

1. Introduction

Recently, Beryllium nuclei attract attention of experimenters and theorists because of their cluster structure [1–3]. It is known that the ground state of ^{8}Be is unbound by 92 keV and decays into two alpha particles with time about 10^{-16} s. The only stable beryllium isotope is ^{9}Be. The additional neutron acts as a covalent particle between α-cluster centers, increasing the stability of the ^{9}Be structure. Thus, such states are called as nuclear molecules. In addition to the α+n+α structure it is possible to assume that two-body configurations ^{8}Be+n and ^{5}He+α also may play an important role for the excited states of ^{9}Be. Although studies of the ^{9}Be cluster structure have been performed for a long time, it is necessary to obtain quantitative data on the probability of the breakup of excited states via different cluster channels. We propose to determine the cluster structure of ^{9}Be excited states by studying the inelastic scattering of deuterons with registration of a scattered deuteron and a decay particle emitted during the breakup of these states. Since the thresholds of breakup via the ^{8}Be+n, α+n+α and α+^{5}He channels are low enough, transitions of many excited states into these channels are energetically possible.

2. Simulation

For the kinematical simulation of the inelastic scattering reaction, we have used the program of nuclear reactions kinematics [4] in which events with arbitrarily chosen (or partially defined parameters, for example, the angles or energies of some secondary particles) are selected according to the criterion of compliance with energy and momentum conservation laws. As a result, an array of output parameters (energies and emission angles) corresponding to a certain reaction is obtained. At the first stage, a two-particle inelastic scattering reaction of the deuteron from ^{9}Be with excitation of a certain low-lying level E_x is simulated. By choosing the detection angle of the deuteron Θ_d, we fix deuteron energy at this angle E_d. This value of secondary deuteron energy was used as an input parameter in the simulation of reactions with three and four particles in the final state

$$d + {}^{9}Be \rightarrow d' + {}^{9}Be^{*} \rightarrow d' + {}^{4}He + {}^{5}He \rightarrow d' + {}^{4}He + {}^{4}He + n$$
$$d + {}^{9}Be \rightarrow d' + {}^{9}Be^{*} \rightarrow d' + n + {}^{8}Be \rightarrow d' + {}^{4}He + {}^{4}He + n$$
$$d + {}^{9}Be \rightarrow d' + {}^{9}Be^{*} \rightarrow d' + n + {}^{8}Be \rightarrow d' + {}^{4}He + {}^{4}He + n.$$

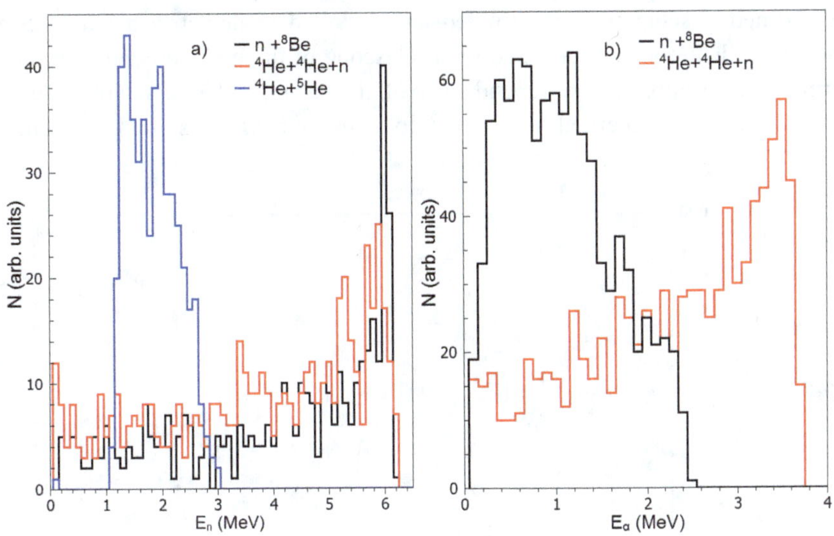

Fig. 1. Simulation of reactions ^{9}Be(d,d'n)X (a) and ^{9}Be(d,d'α)X (b) for various breakup channels at deuteron beam energy of 15 MeV. (a) – Neutron spectrum from the breakup of E^{*}=6.76 level, $\Theta_{d'}$=30°, Θ_n=20°; (b) – Alfa spectrum from the breakup of E^{*}= 3.05 MeV level, $\Theta_{d'}$=50°, Θ_α=55°.

A kinematic simulation of the $^9Be(d,d'n)X$ for various excited levels of ^9Be and breakup channels was carried out. Figure 1a shows the spectra of neutrons emitting in the 10°–30° angular interval, from the breakup of 6.76 MeV level for various breakup channels. The parameters used in the calculation are shown in the figure. It is seen that the energy behavior of the simulated neutron spectrum for the $^4He+^5He$ configuration differs significantly from the same for the $n+^8Be$ and $^4He+^4He+n$ configurations. However, the breakup via the last two configurations leads to almost similar energy spectra.

Simulation of inelastic scattering of deuterons with registration of scattered deuteron and alpha particle was also carried out. Figure 1b shows a simulated spectrum of alpha particles emitted in angular interval of 50°–60°, from the breakup of 3.05 level for various decay channels. The detection angle of the scattered deuteron is 50°. It is seen that registering alpha particles it is possible to distinguish the decay channels $n+^8Be$ and $^4He+^4He+n$.

3. Experiment

In order to study deuteron scattering from ^9Be detecting both the scattered deuteron and the breakup particle (neutron or alpha) in coincidences, an experimental setup was constructed and installed at deuteron beam of SINP. Scattered deuterons are detected by a telescope of three silicon detectors in one arm of the setup. In the second arm of the setup, either a neutron detector or a detector of charged particles (alpha particles) is installed. Preliminary

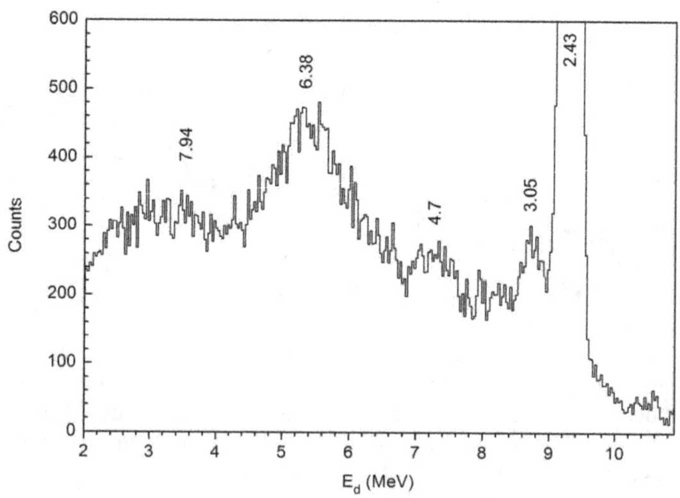

Fig. 2. Energy spectrum of deuterons scattered from ^9Be at 60°.

measurements were performed in the inclusive variant of the experiment (with registration of single deuteron). Two-dimensional diagrams for the three-detector system ΔE-E_1-E_2 are used to separate events of deuteron detection. In Figure 2 the energy spectrum of deuterons scattered from ^{9}Be at 60° is shown. Energies of ^{9}Be excited levels are indicated in the figure.

4. Summary

A kinematic simulation of the ^{9}Be(d,d′n)X and ^{9}Be(d,d′α)X reactions for various excited levels of ^{9}Be and breakup channels was carried out. The results of the simulation show that recording a scattered deuteron with excitation of a specific excited ^{9}Be state in coincidence with a particle from the decay of this state will make it possible to determine the contributions of various cluster configurations to the structure of these states. A study of the ^{9}Be cluster structure in inclusive and exclusive experiment was started on the deuteron beam of the Skobeltsyn Institute of Nuclear Physics. Preliminary data are obtained on the energy spectra and differential cross sections of d-^{9}Be scattering for several excited states.

References

1. S. M. Lukyanov, M. N. Harakeh, M. A. Naumenko *et al.*, *World Journal of Nuclear Science and Technology*. **5**, 265 (2015).
2. T. A. D. Brown, P. Papka, B. R. Fulton *et al.*, *Phys. Rev.* **C76**, 054605 (2007).
3. Y. Hirayama, T. Shimoda, H. Miyatake *et al.*, *Phys. Rev.* **C91**, 024328 (2015).
4. S. V. Zuyev, A. A. Kasparov and E. S. Konobeevski, *Bull. Russ. Acad. Sci. Phys*. **78**, 345 (2014).

New Nuclear Structure Data Beyond ^{132}Sn*

R. Lozeva
CSNSM, CNRS/IN2P3, Université Paris-Saclay, F-91405 Orsay Campus, France
E-mail: radomira.lozeva@csnsm.in2p3.fr

Nuclear structure studies beyond ^{132}Sn on the shell evolution, the competition or coexistence of single-particle and collective structures and the GT strength are performed using prompt and decay spectroscopy. Here, we present an overview of our recent data from such experiments on neutron-induced fission products in the mass chain A=136 using the EXILL and Lohengrin spectrometers at the ILL in the framework of other studies we have recently performed in the region around ^{132}Sn.

Keywords: Prompt spectroscopy; decay spectroscopy; neutron-rich nuclei.

1. Physics

One of the most interesting regions of the nuclear chart as the one around the doubly-magic ^{132}Sn, attracts a lot of experimental and theoretical attention. Close to the r-process path, and to the A=130 peak, where experimental data are insufficient for the theoretical description [1,2], one observes domination of single-particle excitations in excited isomeric states around [3] and close beyond ^{132}Sn [4]. On the other hand, the evolution of shapes below ^{132}Sn [5], is expected to develop collectivity beyond ^{132}Sn [6] only few particles away in the Sn chain with seniority mixing [7], deformation in Te [8] to octupole shapes [2]. However, no experimental data on such collective mode excitations is available below Xe [9,10]. The intriguing orbital evolution beyond ^{132}Sn was recently discussed in the framework of isomer and decay investigations [11,12]. Several works report on the neutron orbital evolution such as between the low-lying $\nu f_{7/2}$ and $p_{3/2}$ and $h_{9/2}$ and on the proton side the low-lying $\pi g_{9/2}$ and $d_{5/2}$ [4,6,11,13-15]. As a consequence of this evolution beyond ^{132}Sn, certain $T_{1/2}$ [16], multiple β-delayed neutron emissions [17] and the opening of GT trength in the β-decay is expected [18], with experimental evidences found in the I isotopes from A=136 [19] to A=140 [12].

2. Experimental Techniques and Results

We performed several experimental studies in the region beyond ^{132}Sn [3,4,9,11,12]. Relevant to the above context are the experiments, performed on

---

* This work was presented with an oral contribution at the EXON18 symposium.

thermal-neutron induced fission products at the PF1B and PN1 lines of the ILL research reactor with collimated neutron flux of about 10^8 n/s/cm^2. Detailed prompt spectroscopy was possible as fission can populate rather high spins of the resulting fragments. In a subsequent experiment, in addition, we measured the β-decay of some of these products, first separated by the Lohengrin spectrometer and then implanted in a tape station, adapted to the lifetime of the nuclei of interest. As beyond ^{132}Sn, the probability for β-delayed neutron emission is rather high [17], we were able to detect also these channels using specific experimental setup including five user neutron detectors. The data analysis from the neutron tagging investigation will be reported in a forthcoming article [20]. Here, we give a brief overview of the results without necessity for neutron tagging such as from the A=136 chain from our decay work, compared to the prompt measurement and extending our experimental spectra beyond our recent work [19]. The prompt measurements of A=136 were done within EXILL [21], using ^{235}U and ^{241}Pu targets and γγγ coincidences recorded by Clover HpGe detectors. These data complement the above findings, especially at high spin and yrast excitations. The observed non-yrast excitations could be compared to our β-decay measurements, where we used only ^{235}U production target. The population of prompt, predominantly yrast with some non-yrast states, was seen in the first case, while mostly non-yrast low-spin excitations were populated and observed in the β-decay case.

Results on the A=136 chain are recently published in [19], with yrast and non-yrast schemes from prompt fission and decay of fission products. It is interesting to underline, that the data from complementary studies, e.g. different fission targets or β-decay data are consistent with each other and are used for coherent extension of the yrast and non-yrast level schemes. Furthermore, both the non-yrast low-spin states from decay of fission products and direct β-decay data were used for energies detected up to 3.5 MeV. In an additional β-decay data set, an energy range between 3.5 and 6 MeV was also covered. Therefore, it was possible to detect γ-rays to up to about 6.5 MeV for the A=136 chain, composed of I and Xe. The known Q_β value of ^{136}Te β-decay is 5.120(14) MeV, while the for ^{136}I decay it is 6.884(14) MeV [22].

It is visible in the spectrum shown in Fig. 1, that several transition candidates exist beyond the earlier data threshold of 3.5 MeV and these can be assigned to decay transitions in ^{136}I and ^{136}Xe from respectively ^{136}Te and ^{136}I β-decay. ^{136}Xe can be considered as stable with negative Q_β value, resulting in no electron or positron emission energetically possible. From the sharp end-point energy detected here, it is reasonable to assume that only these β-decays take place and that no strong neutron channel is present, consistent with the low P_n reported for the ^{136}Te decay [17].

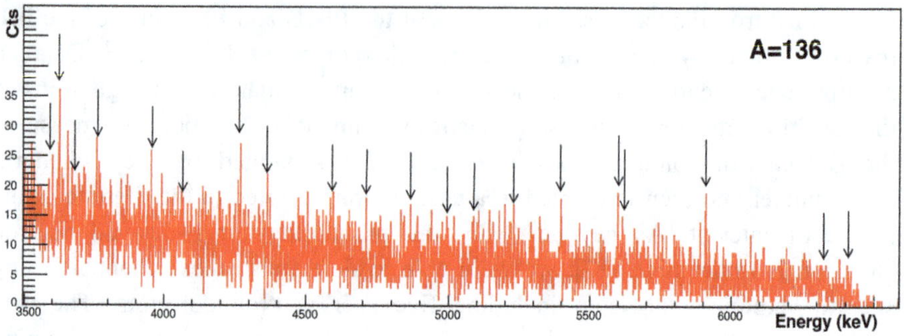

Fig. 1. Decay spectrum for A=136 above 3.5 MeV.

3. Discussion

The new results in the A=136 chain from our recent measurements trigger several discussions. The position of the proton $\pi g_{7/2}$ and $\pi d_{5/2}$ orbitals at the beginning of the $\nu g_{7/2}$ shell (N>82) could be investigated, especially in the context of our recent work at its end, where at N=89 a crossing between the two orbitals was proposed [11]. The new results at N=83 are consistent with large energy spacing at this N and show to be reasonable with the proposed orbital evolution [11]. One has to note that this is very valuable as, a fast drop of the $\pi d_{5/2}$ at N=83 was proposed earlier to explain data in ^{136}I [23], inconsistent with our data [19]. Furthermore, from the high excitation energy above 2.5 MeV and large spacings between states containing neutron $\nu i_{13/2}$ orbital, one can conclude that the placement of this orbital is also rather high in energy, consistent with slowly developing collectivity close to ^{132}Sn e.g. with the increase of both Z and N. In the decay data, we have also identified some GT strength for A=136 and compared it to the GT strength for A=140 [12] iodine nuclei, based on the strong occupation of the $\nu h_{9/2}$ orbital and transitions of type $\nu h_{9/2} \rightarrow \pi h_{11/2}$ for which we used also β-decay data from ^{136}Te→^{136}I→^{136}Xe. These give rise to several 1^+ states in ^{136}I and 2^- states in ^{136}Xe, populated with GT transitions [19]. One cannot exclude the existence of other such states in ^{136}I, extended to higher excitation energy above 3.5 MeV, though they may have more mixed configurations analyzing the shell-model results in [19].

References

1. T. Kratz *et al.,* TAJ 792:6, 7 (2014) Astrophys. J. 662, 39-52 (2007)
2. P. Möller *et al.,* At. Dat. Nucl. Dat. Tabl. 109, 1 (2016)
3. R. Lozeva *et al.,* Phys. Rev. C 77, 064313 (2008)
4. R. Lozeva *et al.,* Phys. Rev. C 92, 024304; 92, 029901 (2015)

5. T. Togashi *et al.,* Phys. Rev. Lett. 121, 062501 (2018)
6. H. Naïdja *et al.,* Phys. Rev. C 96, 034312 (2017)
7. Naïdja *et al.,* JPCS580 (2015) 012030
8. W. Urban *et al.,* Phys. Rev. C 62, 044315 (2000); 93, 034326 (2016)
9. S. Ilieva *et al.,* Phys. Rev. C 94, 034302 (2016)
10. B. Bucher *et al.,* Phys. Rev. Lett. 116, 112503 (2016)
11. R. Lozeva *et al.,* Phys. Rev. C 93, 014316 (2016)
12. B. Moon *et al.,* Phys. Rev. C 95, 044322; 96, 014325 (2017)
13. L. Coraggio *et al.,* Phys. Rev. C 87, 034309 (2013)
14. A. Covello *et al.,* JPCS 267, 012019 (2011)
15. S. Sarkar *et al.,* JPCS 267, 012040 (2011)
16. Y.F. Niu *et al.,* Phys. Lett. B 780, 325 (2018)
17. R. Caballero-Folch *et al.,* Phys. Rev. C 98, 034310 (2018)
18. I. Borzov *et al.,* Phys. Rev. C 71, 965801 (2005)
19. R. Lozeva *et al.,* PRC98, 024323 (2018)
20. R. Lozeva *et al.,* in preparation
21. M. Jentschel *et al.,* J. Instr. 12, P11003 (2017)
22. nndc.bnl.gov
23. W. Urban *et al.,* Eur. Phys. J A 27, 257 (2006)

New Insight into the Cluster Structure of ^{9}Be by Reactions with Deuteron Beam

S. M. Lukyanov[1], B. Urazbekov[1], D. M. Janseitov[2], V. Burjan[3], A. S. Denikin[1], W. H. Trzaska[4], M. Harakeh[5], D. Etasse[6], T. Issatayev[1,7], V. Kroha[3], V. A. Maslov[1], J. Mrazek[3], K. Mendibayev[1,7], M. A. Naumenko[1], I. Siváček[1,3], V. Glagolev[3], Š. Piskor[3], Yu. E. Penionzhkevich[1,8], N. K. Skobelev[1], I. Stefan[9], D. Verney[9], K. Kuterbekov[7] and T. Zholdybayev[10]

[1]*Flerov Laboratory of Nuclear Reactions, JINR, Dubna, Russian Federation*
[2]*Bogolyubov Laboratory of Theoretical Physics, JINR, Dubna, Russian Federation*
[3]*Nuclear Physics Institute, Rez, Czech Republic*
[4]*Department of Physics, University of Jyvaskyla, Jyvaskyla, Finland*
[5]*KVI-CART, University of Groningen, Groningen, the Netherlands*
[6]*LPC-Caen, ENSICAEN, Universite de Caen, CNRS/IN2P3-ENSI, Caen, France*
[7]*Eurasian Gumilev University, Astana, Kazakhstan*
[8]*International University "Dubna", Dubna, 141980, Russian Federation*
[9]*Institut de Physique Nucleaire, Univ. Paris-Sud, Universite Paris-Saclay, Orsay, France*
[10]*Nuclear Physicanamcs Institute, Almaty, Kazakhstan*

Angular distributions of protons, deuterons, tritons, and alpha particles emitted in the reaction ^{2}H+^{9}Be at E_{lab}=19.5, 25, and 35 MeV were measured to study the structure of ^{9}Be, especially to shed light on the internal clusters and possible cluster transfer of ^{5}He. The experiments were performed at sufficiently high energies to ensure suppression of compound nucleus contribution. Thus, the direct reaction mechanism should be mainly responsible for the measured five-nucleon transfer cross section. The analysis suggests a significant contribution of simultaneous five-nucleon transfer in the reaction channel ^{9}Be (d,^{4}He) ^{7}Li.

Keywords: Elastic and inelastic scattering; transfer reactions; clusters in light nuclei; coupled channels; DWBA; optical model.

1. Introduction

Due to the Borromean structure, special attention has been focused on the ^{9}Be nucleus, the breakup of which can proceed directly into two α-particles and a neutron or via one of two unstable intermediate nuclei such as ^{8}Be or ^{5}He [1, 2].

Scattering of a projectile, such as ^{1,2}H or ^{3,4}He, on a target is a standard tool to study the structure of nuclei. This method involves the angular distribution measurement of elastic and inelastic scattering. Energy and angular distributions of the products provide information on the internal structure of the interacting nuclei. The angular distributions for the ^{9}Be(^{3}He,^{3}He)^{9}Be, ^{9}Be(^{3}He,^{5}He)^{7}Be, ^{9}Be(^{3}He,^{5}Li)^{7}Li, ^{9}Be(^{3}He,^{6}Be)^{6}He, and ^{9}Be(^{3}He,^{6}Li)^{6}Li reaction channels were measured and reported in [3, 4]. The experimental data were described within the framework of the optical model, the coupled-channel approach and the distorted-wave Born approximation. The performed analysis of the experimental data shows that the potential parameters are quite sensitive to the exit channel and, hence, to the cluster structure of the populated states. The experiment [3, 4] was an attempt to determine the contribution of the ^{8}Be + n and ^{5}He + α channels by inclusive measurements.

We found that the ratio about 2.7 : 1 may be assigned to the contributions of these two channels, respectively. That justifies that the ^{5}He + α breakup channel plays an important role.

Another motivation was an attempt to find not only the cluster structure (in particular, ^{5}He), but also to clarify how the cluster structure is involved into the nuclear reaction mechanism. Indeed, starting from Detraz [5, 6], multiparticle - multihole structures were expected to occur at rather low excitation energies in nuclei. Four-nucleon transfer reactions are being extensively studied. Despite the complexity of such transfer, one may assume that the nucleons are transferred as a whole, strongly correlated in a cluster with internal quantum numbers of a free particle.

This article is aimed at obtaining new experimental data and their preliminary analysis for the reaction ^{2}H+^{9}Be at various projectile energies: E_{lab}=19.5, 25, and 35 MeV. Compared with the experiment [7], the present experiments were performed at sufficiently high energies to ensure suppression of compound nucleus contribution. Thus, the direct reaction mechanism should provide the main contribution to the reaction channels.

2. Experimental Method

The experiments were performed using the beam of ^{2}H ions on the cyclotron of Nuclear Physics Institute of the Czech Academy of Sciences (Rez, Czech Republic), Institut de Physique Nucleaire, CNRS-IN2P3, Universite Paris-Sud, Universite Paris-Saclay, and the Cyclotron facility of the Accelerator Laboratory of the Physics Department of Jyvaskyla University at energies 19.5, 25, and 35 MeV, respectively. The average beam current during the experiment was maintained at 20 nA. The self-supporting Be target was prepared from a thin beryllium foil. To measure (in)elastically scattered ions, a set of four telescopes, each consisting of ΔE_0, ΔE, E_r detectors with thicknesses respectively 12 μm, 100 μm, and 3 mm were used. The telescopes were mounted at a distance ~20 cm from the target in the reaction chamber. Particle identification was performed based on the measurements of energy loss ΔE and residual energy E_r, i.e., by the so-called $\Delta E - E$ method. An example of the two-dimensional plot (yield vs. energy loss ΔE and residual energy E_r) is shown in Fig. 1. This

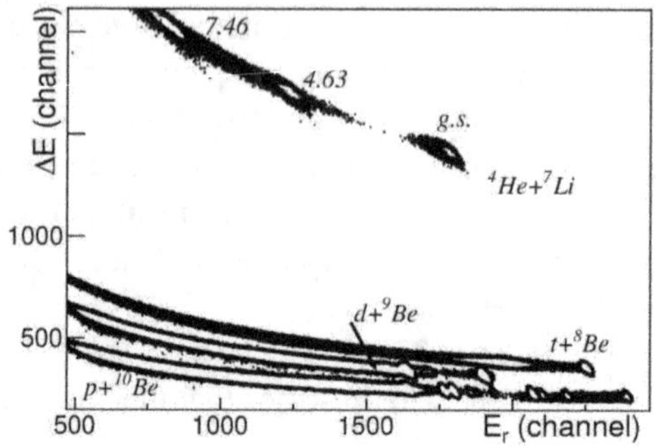

Fig. 1. Typical particle identification plot for the products of the $^2H+^9Be$ reaction: *p, d, t,* and ^4He. ΔE is the energy loss and E_r is the residual energy. Excited states for the ^7Li reaction channel $^7Li+\alpha$ are indicated.

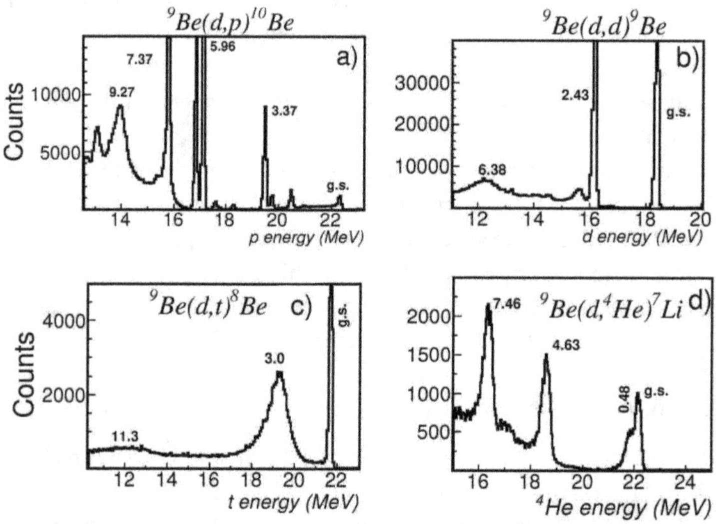

Fig. 2. Total deposited energy spectra measured at $\Theta_{lab}=32^0$ for the detected *p* (panel a), *d* (panel b), *t* (panel c), and ^4He (panel d). The ground and the excited states of ^7Li in the case of the detected complementary product ^4He as well as the ground states and the excited states of ^8Be, ^9Be, and ^{10}Be in the case of the detected complementary products respectively *t, d,* and *p* were unambiguously identified.

experimental technique with good resolution ~150 keV allowed us to identify the particles *p, d, t,* and ^4He and determine their total deposited energy. The spectra of total deposited energy are shown in Fig. 2. All peaks observed in the histograms in Fig. 2 were identified and found to belong to the ground and the excited states of ^{10}Be, ^9Be, ^8Be, and ^7Li as the complementary products to the detected particles *p, d, t,* and ^4He, respectively.

That confirms our previous finding about one-step multinucleon transfer and the two body reaction mechanism in the exit reaction channels.

3. Results and Data Analysis

3.1. *Elastic and inelastic scattering*

The differential cross sections for elastic scattering of deuterons on the beryllium target measured at various projectile energies are presented in Fig. 3. The obtained elastic scattering cross sections were analyzed within the optical model available in the NRV web knowledge base on low-energy nuclear physics [9, 10]. Theoretical results are shown as the solid curves.

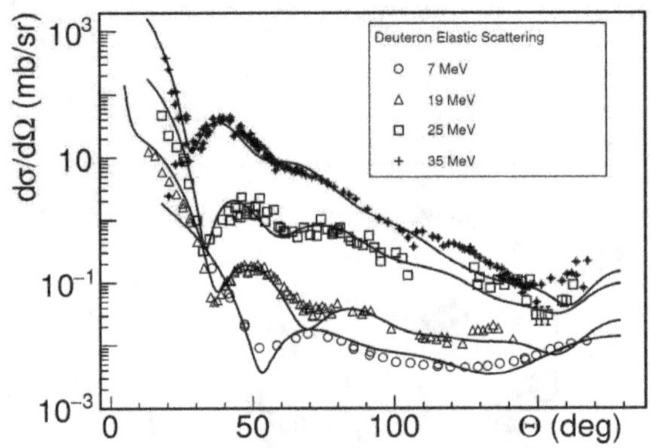

Fig. 3. Differential angular distributions for elastic scattering of d+^{9}Be at various projectile energies. Data obtained in the present work at 35, 25, and 19 MeV are shown respectively by (+), (□), and (△) symbols; data from [7] are denoted by (○) symbols.

3.2. *Transfer reaction channels*

Fig. 4a shows experimental angular distributions for $t + {}^8$Be, ^{4}He + ^{7}Li, and ^{4}He + ^{7}Li reaction channels as well as the channel leading to the production of boron isotopes. For boron isotopes, only the atomic number Z was identified; we were not able to determine exactly the atomic mass A due to the limitations connected with kinematic broadening and the detector thickness.

Experimental angular distributions for the detected ^{7}Li and ^{6}Li are shown in Fig. 4b. Differential angular distribution in the case of the detected ^{7}Li as a supplementary product to ^{4}He was compared with 90^0- reflected angular distribution for the reaction channel ^{4}He + ^{7}Li. One may see good overlapping of the differential cross sections for ^{7}Li+x and ^{4}He+x reaction channels. That confirms the two-body reaction mechanism in exit channel.

Special attention was paid to the ^{4}He + ^{7}Li reaction channel which can proceed via two different mechanisms, either by transfer of d (dashed line in Fig. 5) or by transfer of ^{5}He (solid line in Fig. 5) from the target to the projectile. In the exit channel both these mechanisms are indistinguishable. However, in the former case, the ^{4}He nuclei are expected to fly forward in the center of mass system, whereas in the latter case, they must preferably fly into the backward angles.

The calculations carried out within the DWBA method for these two mechanisms at 19 MeV are shown in Fig. 5.

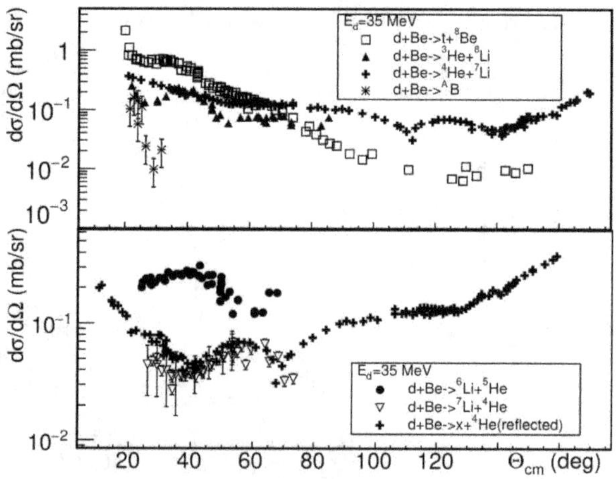

Fig. 4. Experimental angular distributions for $t + {}^8$Be (□), ^{4}He + ^{7}Li (+), ^{4}He + ^{7}Li (Δ) reaction channels. The channel leading to the production of boron isotopes is denoted by (∗).

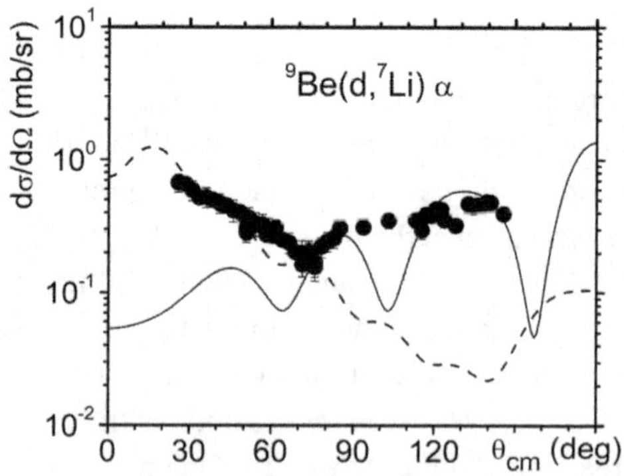

Fig. 5. Experimental angular distributions for reaction channels ^{9}Be(^{2}H, ^{4}He)^{7}Li (•) and ^{9}Be(^{2}H,^{7}Li)^{4}He (○) at 19.5 MeV (for details, see text).

Their coherent sum shown as the black curve is in good agreement with the experimental points. It is interesting to note that for the correct description of the magnitude of the data, it is necessary to use sufficiently large values of the spectroscopic factors for the reaction, $S_f S_i$ = 1.2. In addition, to describe the structure of the angular distributions, it is also necessary to assume a 30% admixture of the *d*-state in the structure of ^{7}Li. Data analysis for the projectile energy 7 and 35 MeV is in progress.

Convincing evidence for direct five-nucleon transfer processes has been presented for the first time in the papers [7, 11]. Data analysis for these reactions has been usually performed under the assumption of the single-step transfer of the ^{5}He cluster. Such an analysis was treated as a proof of the presence of ^{5}He clusterization in light nuclei. However, it is possible that five nucleons are transferred not as ^{5}He, but most likely via the transfer of α-particle and a neutron. Such a mechanism can be quite important because transfer of nucleons as well as α-particles is known to proceed with large values of the cross sections. The open question is how these particles can be transferred – simultaneously in a one-step process or sequentially in a two-step process. In [11], it was mentioned that simultaneous transfer is a more general mechanism than transfer of a single five-nucleon cluster because the α-particle and the nucleon can change the state of their relative motion during the reaction. Only a specific simultaneous transfer of a nucleon and α-particle, i.e., a correlated transfer, in which the transferred nucleons form the cluster with the quantum numbers of ^{5}He, is equivalent to a single five-nucleon cluster transfer.

Our results confirm finding of [7, 11] that the largest contribution corresponds to the transfer of α-particle and a nucleon correlated to ^{5}He as a whole cluster.

In summary, angular distributions for the ^{9}Be (*d, d*) ^{9}Be*, ^{9}Be (*d, p*) ^{10}Be, ^{9}Be (*d, t*) ^{8}Be, and ^{9}Be (*d*, ^{4}He) ^{7}Li channels were measured. Experimental angular distributions were described within the optical model, the coupled channel approach, and the distorted wave Born approximation. The optical model provides good agreement with the elastic scattering data. The DWBA calculations agree well with the transfer reaction data. The spectroscopic factors for the systems $^9\text{Be} = \alpha + {}^5\text{He}$ and $^7\text{Li} = d + {}^5\text{He}$ are close to unity, which confirms the contribution of the considered cluster configurations to the structure of ground states.

The analysis shows that the contribution of the compound nucleus mechanism is negligible. In the (*d*, ^{4}He) channel, the deuteron transfer provides only a small contribution, whereas a relatively large contribution of ^{5}He transfer is found, in agreement with the result [7]. This demonstrates that the specific structure of the ^{9}Be nucleus as a weakly bound system of two α-particles and a neutron strongly favors five-nucleon transfer compared to deuteron transfer.

Acknowledgments

Authors thank the CANAM project (http://users.canam.ujf.cas.cz) and the mobility grant from the Academy of Finland for providing beam time for the experiments and support. This work was also supported by the Russian Science Foundation, Grant No. 17-12-01170.

References

1. T. A. D. Brown, P. Papka, B. R. Fulton *et al.*, Decay studies for states in ^{9}Be up to 11 MeV: Insights into the n+^{8}Be and α+^{5}He cluster structure, *Phys. Rev. C* **76**, 054605 (2007).
2. P. Papka, T. A. D. Brown, B. R. Fulton *et al.*, Decay path measurements for the 2.429 MeV state in ^{9}B, *Phys. Rev. C* **75**, 045803 (2007).
3. S. M. Lukyanov, A. S. Denikin, E. I. Voskoboynik *et al.*, Study of internal structures of ^{9,10}Be and ^{10}B in scattering of ^{4}He from ^{9}Be, *J. Phys. G* **41**, 035103 (2014).
4. S. M. Lukyanov, M. Harakeh, M. A. Naumenko *et al.*, Some Insights into Cluster Structure of ^{9}Be from ^{3}He+ ^{9}Be Reaction, *World Journal of Nuclear Science and Technology* **5**, 265 (2015). doi: 10.4236/wjnst.2015.54026.
5. C. Detraz, H. H. Duhm and H. Hafner, The (^{3}He, ^{7}Be) Reaction on Light Nuclei, *Nucl. Phys. A* **147**, 488 (1970).
6. C. Detraz, F. Pougheon, M. Bernas *et al.*, Selectivity in the ^{11}B(^{3}He,^{6}Li)^{8}Be reaction, *Nucl. Phys. A* **228**, 39 (1974).
7. A. Szczurek, K. Bodek, J. Krug *et al.*, Mechanism of Reactions Induced by 7 MeV Deuterons on ^{9}Be[(*d, p*), (*d, d*), (*d, t*), (*d*, ^{4}He)], *Zeitschrift fur Physik A Atomic Nuclei*, **333**, 271 (1989).
8. http://www.ianthompson.org/surrey and http://nrv.jinr.ru
9. V. I. Zagrebaev, A. S. Denikin and A. P. Alekseev, http://nrv.jinr.ru/nrv/ *"Optical model of elastic scattering"*.
10. A. Denikin, S. Lukyanov, S. Khlebnikov *et al.*, Inelastic scattering and clusters transfer in ^{3,4}He + ^{9}Be reactions, *Physics of Particles and Nuclei Letters*, Vol. **12**, No. 5, 703712 (2015).
11. L. Jarczyk, B. Kamys, M. Kistryn *et al.*, Five-Nucleon Simultaneous and Sequential Transfer in the ^{12}C(^{11}B, ^{6}Li)^{17}O and ^{12}C(*d*, ^{7}Li)^{7}Be Reactions, *Phys. Rev. C* **54**,v.3, 1302 (1996).
12. B. Kamys, Z. Rudy, J. Kisiel, E. Kwasniewicz *et al.*, Simultaneous and Sequential Transfer of Proton and Alpha-Particle in Elastic ^{11}B + ^{16}O Scattering, *Zeitschrift für Physik A Atomic Nuclei*, **342**, 149 (1992).

Coulomb Shift in Two-Center Cluster Systems

M. Nakao and M. Ito*

Department of Pure and Applied Physics, Kansai University, Yamatecho, 3-3-35, Suita 564-8680, Japan

**Research Center for Nuclear Physics (RCNP), Osaka University, Mihogaoka 10-1, Suita 567-0047, Japan*

**E-mail: itomk@kansai-u.ac.jp*

H. Umehara

Department of Physics, Osaka University, Machikaneyama-cho, 1-1, Toyonaka 560-0043, Japan

S. Ebata

Laboratory for Advanced Nuclear Energy, Institute of Innovative Research, Tokyo Institute of Technology, 2-12-1-N1-16 Ookayama, Meguro-ku 152-8550, Japan

Coulomb shift is analyzed for the mirror cluster systems of ^{18}O = α + ^{14}C and ^{18}Ne = α + ^{14}O by applying the orthogonality condition model (OCM). The OCM calculation clearly predicts the suppressed excitation energy of the higher 0^+ states in the proton-rich system of ^{18}Ne. This results can be interpreted in terms of the extension of the Thomas-Ehrman shift (TES), which is discussed in ^{17}O = ^{16}O + N and ^{17}F = ^{16}O + P, to the cluster degrees of freedom.

Keywords: α-cluster structure; mirror systems; Coulomb shift; monopole transition.

1. Introduction

α-cluster structures are well known to appear in the excited states of light mass systems. α particle is a building block in constructing the intrinsic structures of nuclei because of its stable and inert property. The α-cluster structures have been extensively discussed in the 4N systems, which have the multiple mass number of the α particle [1], such as ^{8}Be = 2 α, ^{12}C = 3 α, ^{16}O = α + ^{12}C and ^{20}Ne = α + ^{16}O. These α-cluster structures mainly appear in the excited 0^+ states below 15 MeV in the excitation energy. In the α-cluster structure, the α particle and the residual nuclei are weakly coupled. This weak coupling feature leads to a large extension of the wave function of the cluster relative motion with a large mixture of the S-wave component. In this situation, the Coulomb interaction is not so effective, and the suppressed Coulomb shift, which is similar to the so-called Thomas-Ehrman shift (TES) discussed in ^{17}O - ^{17}F [2], is expected to be observed. Therefore, it is interesting to extend the viewpoint of TES, which is discussed in the single particle picture, to α clustering phenomena. If TES occurs in the α cluster states, we will observe the compressed excitation energy of the excited 0^+ states in the proton-rich system. In fact, such the compression of the excitation energy for the excited 0^+ states has been briefly discussed in ^{10}Be - ^{10}C from the viewpoint of the covalent molecular picture [3].

In the present report, we extend the analysis of the Coulomb shift studied in ^{10}Be - ^{10}C to the heavier systems, such as ^{18}O - ^{18}Ne, which are handled by the cluster models of α + ^{14}C and α + ^{14}O, respectively. Since the full microscopic model is not necessarily successful in reproducing the energy levels in ^{18}O [4, 5], we introduce the semi-microscopic model, called the orthogonality condition model (OCM) [6, 7]. The OCM application to ^{18}O was done by Furutani *et al.* but the scattering boundary condition for the unbound continuum is not considered [7]. Here we employ the OCM plus absorbing boundary condition (ABC) [8], which is powerful tool to handle the unbound continuum, and analyze the continuum states in the mirror cluster-systems of ^{18}O - ^{18}Ne. The results in the present report are compressed from Refs. [9, 10].

2. Framework

We apply the orthogonality condition model (OCM) to the α + ^{14}C and α + ^{14}O systems. In the following, we explain the OCM formulation for the α + ^{14}C system because the extension to the α + ^{14}O can be achieved in a straight forward manner. We construct the Pauli allowed states in a simple manner based on the Elliott's SU(3) algebra [11]. The internal configuration of ^{14}C is set to the lowest shell model configuration, such as $(\lambda, \mu) = (0, 2)$ in the SU(3) irreducible representation of the harmonic oscillator (HO) wave function [11]. The Pauli allowed states of the total system, which are composed of α and ^{14}C, must have the total HO oscillator quanta of $N \geq 6$ in the α - ^{14}C relative motion. The $N = 6$ state corresponds to the partially allowed state, while the quantum states with $N \geq 8$ are the completely allowed states for the positive parity. The SU(3) representation for the partially allowed states can be generated according to the following coupling scheme of the SU(3) irreducible representation:

$$(0, 2) \otimes (6, 0) = (4, 0) \qquad (1)$$

The totally coupled state of (4, 0) corresponds to the lowest shell model state of $^{18}\mathrm{O} = {}^{16}\mathrm{O} \otimes \nu(1s0d)^2$.

After constructing the partially allowed state, the total wave function is expanded by the channel wave function of α + ^{14}C$(0^+, 2^+)$, and the coupled-channel (CC) equations are solved. In the CC calculations, the nuclear potential of α and ^{14}C is calculated from the double folding (DF) model [12]. We use the matter density of the α particle, which reproduces the charge form factor of the electron scattering, while the density of ^{14}C is calculated from the HO-type distribution. As for the effective nucleon-nucleon interaction, we employ the DDM3Y (density dependent Michigan 3-range Yukawa) interaction [13] with the target density approximation (TDA) [14]. The DF+TDA potential nicely reproduces the angular distribution of the $\alpha + {}^{14}C(0^+)$ elastic scattering [9].

We assume the same DF potential in the excited channel of $\alpha + {}^{14}C(2^+)$ for simplicity. The coupling potential of $^{14}C(0^+) \rightarrow {}^{14}C(2^+)$ is constructed from the differential function of the DF potential with the strength of -0.2 fm. Above the α decay threshold, the absorbing boundary condition (ABC) [8] is applied to identify the resonance parameters, such as the resonance energy E_R and the decay width Γ_R.

3. Results

The calculated energy spectra for the $J^\pi = 0^+$ state is shown in Fig. 1. The OCM calculation reproduces the binding energy of the ground 0^+_1 state (−6.2 MeV) with respect to the α decay threshold (dashed line). The first excited 0^+_2 state appears at the binding energy of E = −1.95 MeV, which corresponds to the observed 0^+_2 level (−2.60 MeV). The bound 0^+_1 and 0^+_2 states are reproduced but the 0^+_3 state is missing in the present calculation because this state has the $\nu(1s_{1/2})^2$ configuration with the ^{16}O core.

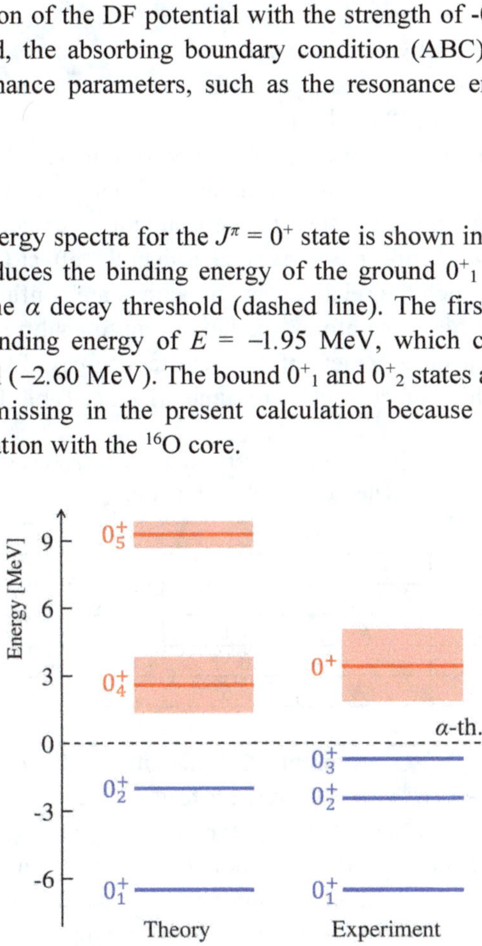

Fig. 1. Energy spectra in $^{18}O = \alpha + {}^{14}C$. The left and right levels are the results of the theoretical calculation and the experimental observations, respectively. The dashed line shows the α threshold, while the shades attached to the resonant levels represent the decay width. This figure is taken from Ref. [10]

On the contrary, in the unbound region, two resonant 0^+ states are obtained in the present calculation: the resonances at $E_R \sim 2.5$ MeV and at $E_R = 9.3$ MeV. We label the former and latter resonances the 0^+_4 and 0^+_5 states, respectively, because of missing 0^+_3 in our calculation. The 0^+_4 level, which was not obtained in the previous calculation [7], exists around the Coulomb barrier of the $\alpha + {}^{14}C_{g.s.}$ ($L = 0$) channel with a broad decay width of $\Gamma_R \sim 2.5$ MeV. A broad 0^+ resonance is observed in the same energy region of the $\alpha + {}^{14}C$ elastic scattering, which is plotted by 0^+ with a shade above the α threshold (Experiment) [15]; the observed energy and width are $E_R = 3.7$ MeV ($E_{ex} = 9.9$ MeV) and $\Gamma_R = 3.2$ MeV, respectively. Moreover, a new resonant state has just been measured at the excitation energy close to the energy position of the 0^+_5 state in the reaction of ${}^9Be({}^{13}C, {}^{18}O)\,\alpha$ [16].

In the resonant $0^+_{4,5}$ states, the relative wave functions have the one higher node in comparison to the bound states, and their tails are extended to the outer region. In this situation, the effect of the Coulomb interaction for the resonant states (0^+_4 and 0^+_5) is expected to be weaker than the effect for the bound states (0^+_1 and 0^+_2), in which the relative wave functions are confined in the region of the nuclear interaction. Therefore, it is interesting to analyze the Coulomb shift in connection to the development of the α-cluster degrees of freedom.

The energy shifts of ${}^{18}O$ - ${}^{18}Ne$ are summarized in Table 1.

Table 1. Energy shift in ${}^{18}O$ - ${}^{18}Ne$ systems. See text for details. This table is taken from Ref. [10]

	ΔE_{Coul}		ΔE_{ex}	
	Expt.	Th.	Expt.	Th.
0^+_1 (B.S.)	1.12	1.12	0	0
0^+_2 (B.S.)	1.06	1.14	0.06	0.03
0^+_4 (Res.)	-	0.53	-	0.58
0^+_5 (Res.)	-	0.65	-	0.46

In this table, ΔE_{Coul} represents the magnitude of the energy difference, $\Delta E_{Coul} = [E_R({}^{18}O) - E_R({}^{18}Ne)]$ with respect to the α decay threshold. As for the two bound states, 0^+_1 and 0^+_2, the experimental energy shifts (Expt.) are reproduced by the theoretical calculation (Th.), which amounts to about 1.1 MeV. The ΔE_{Coul} for the unbound resonances, 0^+_4 and 0^+_5, are reduced to about half of the bound state, just about 0.6 MeV. This reduction is originated from the extended and dilute structure of the α - ${}^{14}C$ relative wave function. This reduced energy shift with respect to the α threshold just corresponds to the reduction of the excitation energy, which is measured from the ground 0^+_1 state. In Table 1, the shift of the excitation energies are shown by ΔE_{ex}, which is defined by

$\Delta E_{ex} = [E_{ex}(^{18}O) - E_{ex}(^{18}Ne)]$ with the excitation energy of $E_{ex} = E(0^+_{ex}) - E(0^+_1)$. ΔE_{ex} is almost negligible for the bound 0^+_2 state but it is enhanced to be about 0.5 MeV for the unbound resonant 0^+_4 and 0^+_5 states. The shift of the excitation energy corresponds to the so-called Thomas-Ehrman shift (TES), which is originally discussed in the single particle picture of $^{17}O = {}^{16}O + n$ and $^{17}F = {}^{16}O + p$ [2].

4. Summary

In sum, we have investigated the $J^\pi = 0^+$ energy spectra in $^{18}O = \alpha + {}^{14}C$ by applying the orthogonality condition model (OCM) under the absorbing boundary condition (ABC). The OCM + ABC calculation for ^{18}O nicely reproduces the observed $J^\pi = 0^+$ levels from the bound to unbound region except for the 0^+_3 state, which is out of the α-cluster model space.

The OCM + ABC calculation is also extended to $^{18}Ne = \alpha + {}^{14}O$, and the Coulomb shift for the mirror systems of ^{18}O - ^{18}Ne is investigated. The Coulomb shift, which is measured from the α decay threshold, is large for the bound 0^+_1 and 0^+_2 states but it is reduced to be half for the 0^+_4 and 0^+_5 states. This reduction of the Coulomb shift is originated from the extended and dilute structure in the α - ^{14}C (^{14}O) relative wave function, which involves a large mixture of S-wave component. The reduced Coulomb shift in the cluster states corresponds to the compression of the excitation energy in the proton-rich ^{18}Ne. The origin of this compressed energy is just similar to the Thomas-Ehrman shift (TES), which is discussed in $^{17}O = {}^{16}O + n$ and $^{17}F = {}^{16}O + p$ [2]. Thus, the energy compression predicted in the cluster system should be called "Cluster TES". We propose that Cluster TES is a new probe to identify the extended and dilute structure of the cluster configuration.

The present result about Cluster TES is also possible to occur in more general $N \neq Z$ systems. For example, it is interesting to consider the 4N nuclear system with two extra nucleons, such as ^{10}Be - ^{10}C, ^{14}C - ^{14}O, and ^{22}Ne - ^{22}Mg. In advancing the study of Cluster TES, it is essential to investigate the level structure, especially the unbound resonances, experimentally. The experimental analysis of the resonant structures in the proton-rich side is strongly desired.

References

1. H. Horiuchi *et al.*, Suppl. Prog. Theor. Phys. **192**, 1 (2012), and references therein.
2. A. Bohr and B. R. Mottelson, *Nuclear Structure* Vol. I (New York: Benjamin, 1969) p. 320.
3. M. Ito, EPJ Web of Conf. **122** (2016).

4. P. Descouvemont and D. Baye, Phys. Rev. C **31**, 2274 (1985).
5. N. Furutachi *et al.*, Prog. Theor. Phys. **119**, 403 (2008).
6. S. Saito, Suppl. Prog. Theor. Phys. **62**, 11 (1977).
7. H. Furutani *et al.*, Suppl. Prog. Theor. Phys. **68**, 279 (1980).
8. M. Iwasaki, R. Otani, Y. Takenaka and M. Ito, PTEP **2015**, 023D01.
9. M. Nakao, H. Umehara, S. Sonoda, S. Ebata and M. Ito, EPJ Web of Conf. **163**, 00040 (2017).
10. M. Nakao, H. Umehara, S. Sonoda, S. Ebata and M. Ito, Phys. Rev. C **98**, 054318 (2018).
11. J. P. Elliott, Proc. Roy. Soc. A **245** (1958), 128, J. P. Elliott, Proc. Roy. Soc. A **245** (1958), 562, M. Harvey, Adv. in Nucl Phys. Vol. 1, eds. M. Baranger and E. W. Vogt (Plenum Press, New York, 1968), p. 67.
12. M. Ito, Y. Hirabayashi and Y. Sakuragi, Phys. Rev. C **66**, 034307 (2002), and references therein.
13. M. E. Brandan and G. R. Satchler, Nucl. Phys. A **487**, 477 (1988), and references therein.
14. K. Egashira *et al.*, Phys. Rev. C **89**, 064611 (2014).
15. M. L. Avila *et al.*, Phys. Rev. C **90**, 024327 (2014).
16. Y. Ye, Private communication (2018).

Shape Coexistence ^{31}Mg Revealed by β-γ and β-n-γ Spectroscopy with Spin-Polarized ^{31}Na

H. Nishibata
RIKEN
Saitama, 351-0198, Japan
E-mail: hiroki.nishibata@riken.jp

A. Odahara, S. Kanaya and T. Shimoda
Department of Physics, Osaka University
Osaka, 560-0043, Japan

M. R. Pearson and C. D. P. Levy
TRIUMF
Vancouver, BC V6T 2A3, Canada

The β-decay spectroscopy with a spin-polarized nucleus is a powerful tool, which unambiguously assigns spins and parities of the daughter levels by measuring the asymmetric β-decay of the polarized parent nucleus. This method has been applied to structure investigations of neutron-rich Mg isotopes. In our recent study of ^{31}Mg, which is located at a border of the region of the $N = 20$ "island of inversion", we firmly assigned spins of all positive-parity excited states below the neutron separation energy at 2.3 MeV. This results clearly showed the shape coexistence in ^{31}Mg. Following the results on ^{31}Mg, we plan to apply this method to neutron-rich ^{31}Al and ^{33}Al.

Keywords: Spin-polarized beam; β-decay spectroscopy; island of inversion; ^{31}Mg; shape coexistence.

1. Introduction

One of the most important subjects in nuclear physics is the structure of nuclei far from the β-stability line. In particular, in neutron-rich nuclei with the neutron number close to the magic number $N = 20$, namely in nuclei located in or near the "island of inversion", shape change of the ground states from spherical to deformed shape has been investigated as a function of the neutron number for many years by a variety of experimental and theoretical studies. However, up to now, the detailed experimental information, such as spins and parities of the excited states, transition probabilities, is not available in most of the island-of-inversion nuclei. For example, in ^{31}Mg, there is no firm spin-parity assignment except for the $1/2^+$ ground state [1].

In the neutron-rich Mg isotopes, the theoretical calculation of the AMD+GCM framework has predicted not only the ground-state deformation but also shape coexistence in a low excitation energy region [2]. Later the shell model calculation pointed out the important role of competition between the mean field, which favors spherical shape, and the nuclear correlations, which cause deformation [3]. Very recently, the large-scale shell model calculation showed configuration mixings of spherical 0p0h and deformed *npnh* configurations to a large extent [4]. Therefore, it is expected that various types of structure coexist in a narrow excitation energy region in such nuclei. However, because of the above-mentioned situation in the experimental data, it has been difficult to discuss the nuclear structure in detail.

In the present proceedings, we introduce our unique and powerful method of β-delayed spectroscopy using spin-polarized nucleus to assign the spin-parity of levels of the daughter nucleus [5], and as one of the examples, we report the successful results on the structures of ^{31}Mg [6, 7].

2. Experiment on ^{31}Mg

In the present work, we take advantage of the anisotropic β decay of spin-polarized ^{31}Na to determine spins and parities of the excited states in the daughter nucleus ^{31}Mg. For an allowed transition, the emission of the β rays shows anisotropic angular distribution as shown by

$$W(\theta) \sim 1 + AP\cos\theta,$$

where θ, A and P are the emission angle of the β rays with respect to the polarization direction, the asymmetry parameter for each β transition, and the polarization of the parent nucleus. The important point is that this asymmetry parameter shows discrete values depending on the spins of the initial parent state (I_i) and the final daughter state (I_f) according to

$$A \begin{cases} = 1 & (I_f = I_i - 1), \\ \sim -\dfrac{1}{I_i + 1} & (I_f = I_i), \\ = -\dfrac{I_i}{I_i + 1} & (I_f = I_i + 1), \end{cases}$$

assuming pure Gamow-Teller transitions. In the case of the β decay of ^{31}Na (I_i = 3/2), the asymmetry parameters for possible ^{31}Mg states with 1/2, 3/2, and 5/2 are listed in Table 1. The parities of the final ^{31}Mg states are positive because the parity of the initial ^{31}Na is positive. Therefore, we can assign the spins and parities of the ^{31}Mg levels from the experimental A values.

Table 1. Asymmetry parameters for allowed β decay of ^{31}Na.

I_i (^{31}Na)	I_f (^{31}Mg)	A
3/2	1/2	−1
	3/2	−0.4
	5/2	+0.6

The AP value for each β transition can be obtained from the β–γ or β–n coincidence counted by the β-ray detectors placed at 0 and 180 degrees with respect to the polarization direction. In order to cancel out the instrumental asymmetry due to β-ray detection efficiencies, the direction of polarization is periodically flipped by 180 degrees. Note that the polarization P is common for all β-transitions. Therefore, once P is obtained by comparing two AP values, the asymmetry parameters for other levels are obtained. Details of this method are described in Ref. [7].

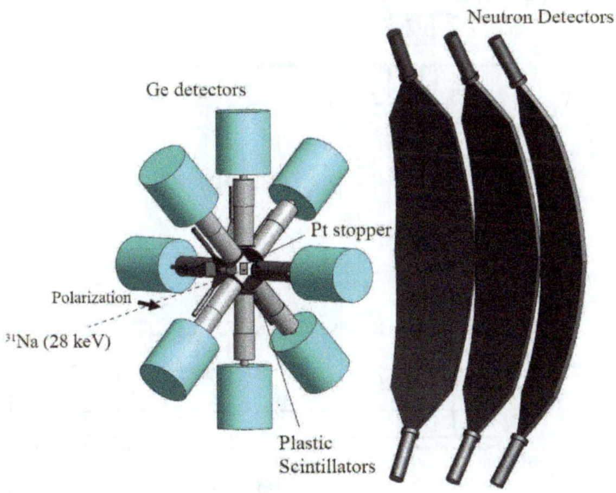

Fig. 1. Experimental setup.

The spin-polarized ^{31}Na beam was produced at the polarizer beamline at TRIUMF. The polarization of ~30% was obtained by the collinear optical pumping method using two laser frequencies to pump both ^{31}Na atomic ground states. The ^{31}Na beam was stopped at the end of the Osaka beamline, where our experimental setup was placed.

The experimental setup was shown in Fig. 1. In the figure, the ^{31}Na beam at 28 keV comes from left to right and stops in a 20-μm thick Pt foil. A static magnetic field of 0.5 T is applied to the Pt foil. The β and γ rays associated with the ^{31}Na β-decay are distinguishably measured by 8 telescopes, each consisting of a high purity germanium (Ge) detector and one (or two) 1.5 mm-thick plastic scintillator(s). The β-decay asymmetry was measured by the two telescopes placed along the polarization axis. The β-delayed neutrons were measured by three 20 mm-thick plastic scintillator placed 1.5 m downstream from the Pt foil. The total γ-ray detection efficiency was 2.9% at 1332 keV. The β-ray efficiency was ~16% for the telescopes placed at 0 and 180 degree, and ~32% for all telescopes. The beam intensity of ^{31}Na at the Pt foil was ~200 pps, and 1.4×10^7 β-γ coincidence events were in 108 hours.

3. Experimental Results

Figure 2 shows the revised decay scheme of ^{31}Na in the present work.

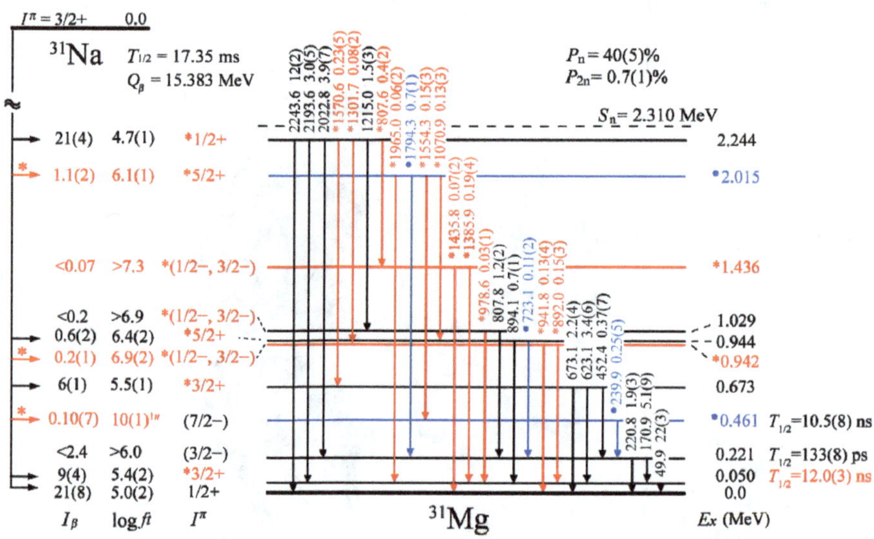

Fig. 2. Revised decay scheme of ^{31}Na in the present work. The β and γ transition, levels, and spins and parities labeled with * (displayed in red) are newly found in the present work. The levels and transitions labeled with • (displayed in blue) are those observed in the β decay for the first time.

By detailed γ-γ and β-γ analyses, 26 γ-ray peaks, including 11 new γ rays and 2 γ rays observed for the first time in β-decay of ^{31}Na, were identified as γ rays in ^{31}Mg, and 10 excited levels, including 2 new levels, were proposed. The spin-parity assignments were performed by comparing the experimental asymmetry parameters with the expected one for each spin-parity, as shown in Fig. 3. In the figure, the experimental A values are consistent with either expected value −1 for $1/2^+$, −0.4 for $3/2^+$, or +0.6 for $5/2^+$, hence the spin-parity assignments are firmly performed as $1/2^+$ for the 2.244-MeV level, $3/2^+$ for the 0.050- and 0.673-MeV levels, and $5/2^+$ for the 0.944- and 2.015-MeV levels.

In the preset work, we can also investigate neutron-unbound levels in ^{31}Mg by analyzing the β-delayed neutron data. These analyses are still on-going.

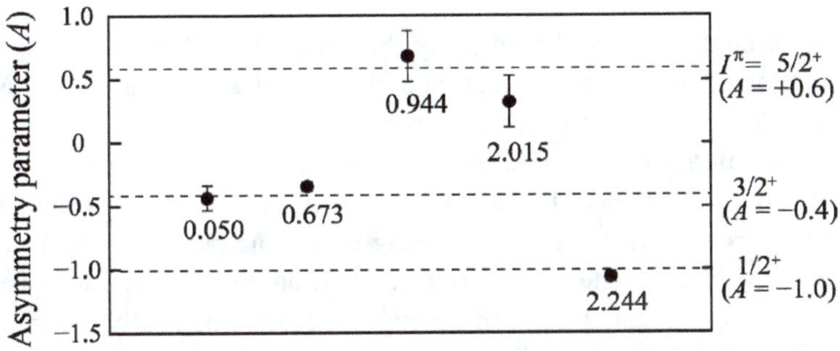

Fig. 3. Experimental asymmetry parameters A for the levels at 0.050, 0.673, 0.944, 2.015, and 2.224 MeV. The horizontal lines show the expected A values for $1/2^+$, $3/2^+$, and $5/2^+$.

4. Discussion and Perspective

Figure 4 shows the experimental levels displayed according to their structures.

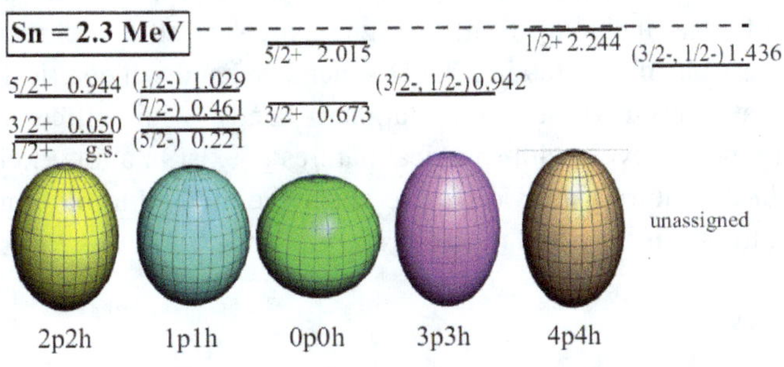

Fig. 4. Summary of structure in ^{31}Mg.

By firmly assigning spins and parities of the levels of ^{31}Mg, we can compare the experimental levels with theoretical predictions, such as by the AMD+GCM [2] and the shell model [4] on a level-by-level basis.

As the result, we found that various nuclear structures such as four rotational bands with different degrees of deformation and levels with spherical natures coexist in a low excitation energy region. These facts show clear evidence for the shape coexistence in ^{31}Mg. We note that the 2.244-MeV 1/2$^+$ level was puzzling level in Ref. [6], however, by comparing with the shell model with the EEdf1 interaction, we can assign this level as a level with 4p4h configuration [7]. It suggests that further investigation on the neutron-unbound states are important because the other member of the rotational band built on the 2.244-MeV 1/2$^+$ level may be unbound. The present work on ^{31}Mg clearly demonstrates powerfulness of the β-decay spectroscopy of spin-polarized nucleus.

As the next step, we plan to apply this method to further neutron-rich Mg isotope ^{32}Mg. The hyperfine structure and the nuclear spin are not known for ^{32}Na, therefore, first we prepare to measure them.

In the polarized beam line at TRIUMF, spin-polarized non-alkali beams have also been developed. For example, the spin-polarized Mg beam, such as ^{31}Mg [8], was already available. By using these spin-polarized Mg beams, we plan to study excited states of neutron-rich ^{31}Al and ^{33}Al, which are located the north-side border of the "island of inversion". Study on such border nucleus is important to understand exotic features, such as shape evolution and shape coexistence which appear in the "island of inversion".

5. Summary

Our unique method of β–delayed spectroscopy with a spin-polarized nucleus has been applied to neutron-rich Mg isotopes. In the present proceedings, we showed the recent results on ^{31}Mg. In ^{31}Mg, we proposed the detailed decay scheme of ^{31}Na, including firmly assigned spins of all 5 positive-parity levels. By comparison of the detailed level scheme with the theoretical ones, it was found that various types of rotational bands with different degrees of deformation and levels with spherical natures coexist in a low excitation energy region. In the future plan, we plan to apply the method to other nuclei such as more neutron-rich ^{32}Mg and neutron-rich ^{31}Al and ^{33}Al.

References

1. C. Quellet *et al.*, Nuclear Data Sheets for $A = 31$, *Nucl. Data Sheets* **114**, 209 (2013).
2. M. Kimura, Intruder features of ^{31}Mg and the coexistence of many-particle and many-hole states, *Phys. Rev. C* **75**, 014302(R) (2007).
3. E. Caurier *et al.*, Merging of the islands of inversion at $N = 20$ and $N = 28$, *Phys. Rev. C* **90**, 014302 (2014).
4. N. Tsunoda *et al.*, Exotic neutron-rich medium-mass nuclei with realistic nuclear forces, *Phys. Rev. C* **95**, 021304(R) (2017).

5. Y. Hirayama *et al.*, Study of ^{11}Be structure through β-delayed decays from polarized ^{11}Li, *Phys. Lett. B* **611**, 239 (2005).
6. H. Nishibata *et al.*, Shape coexistence in the N = 19 neutron-rich nucleus ^{31}Mg explored by β–γ spectroscopy of spin-polarized ^{31}Na, *Phys. Lett. B* **767**, 81 (2017).
7. H. Nishibata *et al.*, Structure of ^{31}Mg: Shape coexistence revealed by β–γ spectroscopy with spin-polarized ^{31}Na, *Phys. Rev. C* (in press).
8. C.D.P. Levy *et al.*, Development of a polarized ^{31}Mg^{+} beam as a spin-1/2 probe for BNMR, *Hyperfine Interact.* **237**, 162 (2016).

Energy Dependence of the Total Cross Section for the Reaction $^{8}He + ^{28}Si$

Yu. E. Penionzhkevich[1,2,*], Yu. G. Sobolev[1], V. V. Samarin[1,3], M. A. Naumenko[1], S. S. Stukalov[1], A. A. Bezbakh[1], S. A. Krupko[1], A. Kugler, V. A. Maslov[1] and I. Siváček[1,4]

[1]*Flerov Laboratory of Nuclear Reactions, Joint Institute for Nuclear Research, Dubna, 141980, Russia*

[2]*Department of Experimental Methods in Nuclear Physics, National Research Nuclear University, Moscow 115409, Russia*

[3]*Department of Nuclear Physics, Dubna State University, Dubna 141982, Russia*

[4]*Department of Nuclear Reactions, Nuclear Physics Institute of the Czech Academy of Sciences, Řež, 25068, Czech Republic*

[*]*E-mail: pyuer@jinr.ru*

In this paper the experimental results on the total cross sections for the $^{8}He + ^{28}Si$ reaction in the beam energy range 6–25 A MeV are reported. The experimental cross sections were obtained by registration of the prompt gamma and neutron radiation accompanying the interaction of ^{8}He neutron-rich nuclei with ^{28}Si nuclei. The analysis of reaction channels was made using fusion-evaporation model and numerical solution of the time-dependent Schrödinger equation for the external weakly bound neutrons of the projectile nucleus ^{8}He.

Keywords: Neutron-rich nuclei; total reaction cross section; fusion-evaporation processes; time-dependent Schrödinger equation.

1. Introduction

The study of nuclear reactions involving neutron-rich weakly bound nuclei makes it possible to obtain information on the structure of the investigated nuclei (clusters, neutron halo, neutron skin, etc.) and its manifestation in reactions [1–3]. The main reaction channels for the $^{8}He + ^{28}Si$ collision at above barrier energies in the studied energy range 6–25 A MeV are complete and incomplete fusion (capture of an α-cluster of ^{8}He by the target nucleus), stripping of outer weakly bound neutrons of ^{8}He with transfer to the target nucleus and with their removal from both nuclei (breakup of the ^{8}He nucleus). Gamma rays and neutrons produced in various reaction channels can be registered, e.g., by a scintillation spectrometer. In [4], differential cross sections for neutron emission in reactions of ^{11}Li with Au, Ni, Be targets at the beam energy 29 A MeV were measured. It was shown that the emission of neutrons is strongly anisotropic with a maximum yield in a narrow range of forward angles. These neutrons are not registered if they are emitted outside of the solid angle of the spectrometer.

For qualitative analysis of evolution and rearrangement of four outer neutrons of ^{8}He during collision, we use a time-dependent approach with a quantum description of neutrons in combination with motion of the centers of colliding nuclei along classical trajectories [2, 5, 6]. The two-component spinor wave function of independent neutrons in the initial state $1p_{3/2}$ was calculated by numerical solution of the time-dependent Schrödinger equation taking into account spin-orbit interaction. An example of the evolution of the probability density of external neutrons of the ^{8}He nucleus in the collision with the ^{28}Si nucleus is shown in Fig. 1.

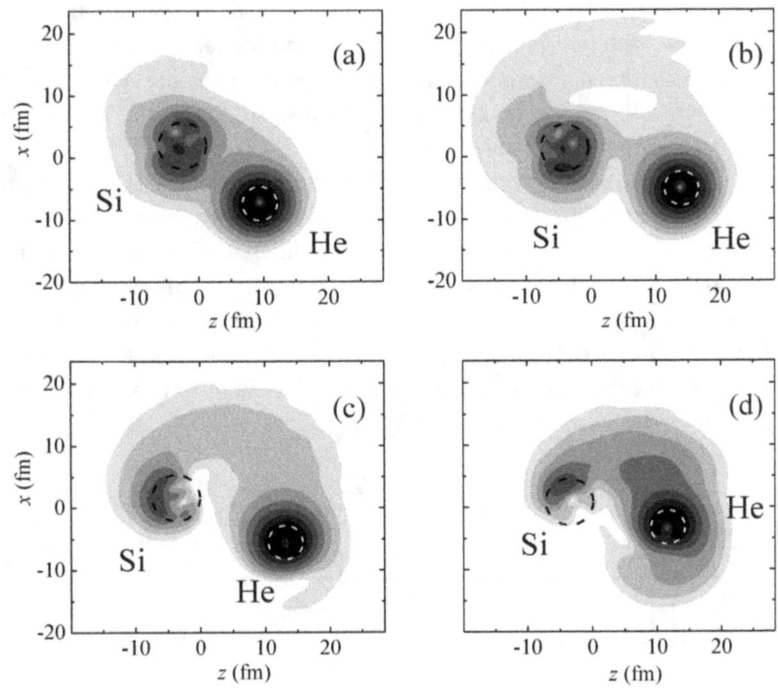

Fig. 1. An example of the probability density for the outer neutrons of the ^{8}He nucleus in the final stage of collision with the ^{28}Si nucleus at energy E_{lab} = 5 A MeV (a), 10 A MeV (b), 20 A MeV (c), and 35 A MeV (d).

It can be seen that, at energies ≈5 A MeV, the outer neutrons lost by the ^{8}He nucleus are mainly transferred to excited states of the target nucleus ^{28}Si. At energies ≈10–20 A MeV, the outer neutrons are transferred to excited states and quasistationary states of the target nucleus in the continuous spectrum with comparable probabilities. From Fig. 1(a), (b), (c), it can be seen that only a small fraction of neutron probability density moves into forward angles at energies less than 20 A MeV. This fraction grows with increasing energy [Fig. 1(d)].

2. Experiment

Experiments on measuring the energy dependence of the total cross section for the reaction $^{8}He + {}^{28}Si$ were carried out by the transmission method using the CsI(Tl) γ-spectrometer. The experimental setup and the method for measuring reaction cross sections by the detection of prompt *n*, γ radiation were described in Refs. [3, 7, 8]. The gamma spectrometer used in the first experiments consisted of six CsI(Tl) scintillators [3, 8]. In this work, we present the results of measurements with an improved gamma spectrometer consisting of twelve CsI(Tl) scintillators. The experiment was carried out at the accelerator U-400M of the Flerov Laboratory of Nuclear Reactions, JINR, on the channel of the achromatic fragment separator ACCULINNA [9]. The primary ^{15}N beam with energy 49.7 A MeV was focused on the producing ^{9}Be target. The secondary beam of fragmentation reaction products was formed and purified by the magnetic system of the fragment separator. At the exit of the last dipole magnet of the separator, the beam entered the straight section of the ion conductor, where a system for measuring the time of flight T_{TOF} consisting of two scintillation detectors ΔE_{TOF1}, ΔE_{TOF2} was located. Polyethylene plates were installed behind the detector ΔE_{TOF2} to reduce the energy of the beam particles. The scheme of the setup is shown in Fig. 2.

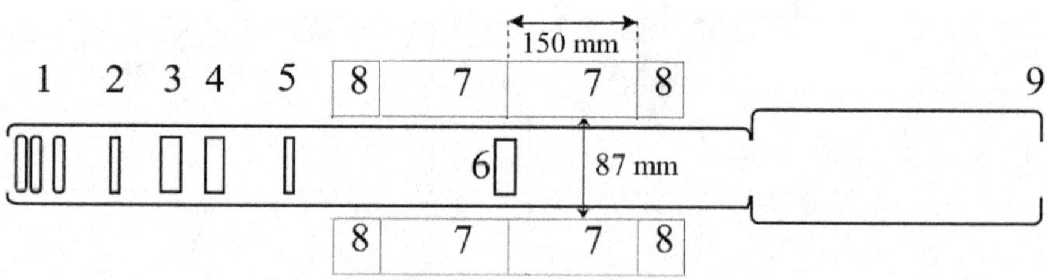

Fig. 2. The experimental setup with dimensions: (1) polyethylene plates for reducing the energy of the beam particles; (2) scintillation detector AC1; (3) position-sensitive Si detector ΔE (16X:16Y); (4) start ΔE_0 Si detector; (5) scintillation detector AC2; (6) target ΔE_T Si PIN detector; CsI(Tl) scintillation detectors (7) and photomultipliers (8) of γ spectrometer; (9) window for the exit of the beam (stainless steel foil).

The beam was focused on a position-sensitive two-layer strip ΔE Si detector with a thickness of 300 μm (detector 16X:16Y, see Fig. 2) located in the focal plane of the fragment separator. In the experiment, this detector was used for preliminary tuning of the beam parameters (intensity, profile, as well as isotopic composition).

Detector ΔE_0 for identification of incident particles was located upstream of the target. The two-dimensional spectrum $\Delta E_0 \times T_{TOF}$ for identification of secondary beam particles (Fig. 3) showed that the isotopes ^{8}He and ^{11}Li of the beam particles formed well separated regions, which made it possible to reliably select a certain group of particles for subsequent offline analysis.

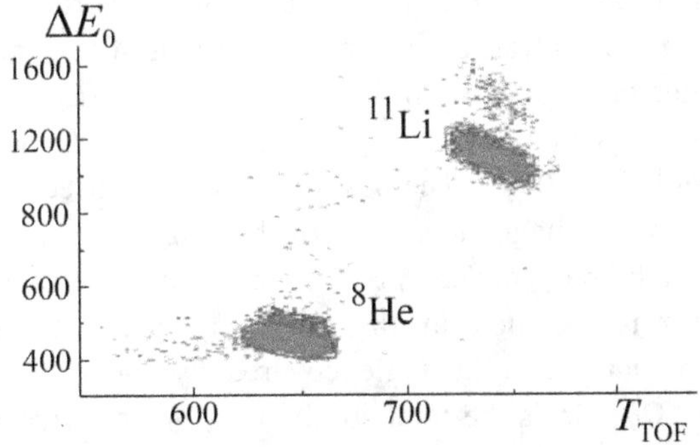

Fig. 3. Two-dimensional spectrum of the secondary beam consisting of ^{8}He and ^{11}Li nuclei. ΔE_0 is the ionization loss of particles in the detector. T_{TOF} is the time of flight of beam particles at a distance 8.5 m of the time of flight base.

The scintillation detectors AC1, AC2 were used as active collimators and served for selecting the events of flight of beam particles in a given solid angle with the axis at the center of the target. The detectors AC1 and AC2 were installed in such a way that the trajectories of particles passed through the sensitive region of the target detector ΔE_T (natural Si, 243-μm thick) and exit window 9 without touching their holders. The ΔE_0 Si detector (380-μm thick) was used to generate the start signal for the information acquisition system for each event of entering of the beam particles into the ΔE_0 detector as well as for subsequent offline identification of the beam particles by their ionization losses ΔE_0 and time of flight T_{TOF}. The resulting time resolution was ~140 ps.

The target was located in the reaction chamber which was a thin-walled vacuum cylinder of stainless steel. On the outside, the cylinder was surrounded by the gamma spectrometer of twelve CsI(Tl) scintillators (see Fig. 2) which were prisms with the base in the form of a regular hexagon. From the end, each scintillator was optically connected to the photomultiplier (PMT) and surrounded on the outside by a reflective coating consisting of three layers of Cd, Pb, and Cu plates (each 1-mm thick) for protection from the background gamma quanta. Other detectors were located outside the sensitive zone of the γ-spectrometer, which provided a minimum level of background triggering for the detectors. The spectrometer with the reaction chamber was located inside a lead cube with a wall thickness of 5 cm, the outer sides of which were covered by plates of 10-cm-thick boron-containing polyethylene.

Each measurement session for a certain beam energy was conducted both with the target and without the target. The irradiation time was chosen in such a way that the number of events in both cases was approximately the same. The experimental information from all the detectors was recorded on a storage disk for subsequent offline analysis of each event of flight of a beam particle through the start ΔE_0 detector, regardless of the reaction in the ΔE_T detector. To reduce the effect of possible superposition of pulses in ΔE_0 and ΔE_T detectors, the beam intensity was limited to 10^3 s^{-1}. The beam energy was varied without significant loss of intensity by the magnetic system of the fragment separator in the range 20–30 A MeV and by polyethylene plates in the range 6–20 A MeV.

The target was installed in the center of the sensitive zone of the spectrometer, so that the solid angle covered by the CsI(Tl) detectors was $\Omega_i = \Omega_0\eta_i$, where $\Omega_0 = 4\pi$ is the total solid angle, $\eta_i \approx \cos\theta_{\min,i}$, and $\theta_{\min,i}$ is the minimum angle between the axis of the setup and the direction of emission of a γ-quantum or a neutron from the center of the target with the entry into the volume of CsI(Tl) scintillators. For the setup with six CsI(Tl) scintillators [3, 8], $\theta_{\min,1} \approx 30^0$, and for the setup shown in Fig. 2 with twelve CsI(Tl) scintillators, $\theta_{\min,2} \approx 16^0$. The gaps between the detectors accounted totally for solid angles $0.06\Omega_0$ and $0.09\Omega_0$ for 6 and 12 detectors, respectively. Taking into account the gaps between the detectors, the solid angles covered by the γ-spectrometer were $\Omega_1 = \Omega_0\eta_1$, $\eta_1 = 0.79$ for 6 detectors and $\Omega_2 = \Omega_0\eta_2$, $\eta_2 = 0.87$ for 12 detectors.

In our experiments, we measured the integral cross section of $x\gamma$ and yn channels in two solid angles Ω_1 and Ω_2

$$\sigma_{\text{int}}^{(i)} = \sum_{x,y} \sigma_{x\gamma,yn}^{(\Omega_i)} = \int_{\varepsilon>\varepsilon_{\min}} d\varepsilon_{\gamma,n} \int_{\Omega_i} d\Omega \frac{d^2\sigma_{x\gamma,yn}}{d\Omega d\varepsilon_{\gamma,n}}, \tag{1}$$

where the energy threshold is $\varepsilon_{\min} = 150$ keV. Taking into account the fact that the fraction of neutrons emitted into forward angles is small [Fig. 1(a), (b), (c)], the total reaction cross section σ_R may be reconstructed as

$$\sigma_R = \sigma_{\text{int}}^{(i)} \frac{4\pi}{\Omega_i} = \sigma_{\text{int}}^{(i)} \frac{1}{\eta_i}. \tag{2}$$

The number ΔI of reaction events from the number I_0 of preselected events was determined using the logical conditions for registering a γ-quantum or a neutron at least in one of the twelve detectors of the spectrometer.

The condition for registering a quantum or a neutron in each CsI(Tl) detector was described by a contour on the two-dimensional amplitude-time $E_{\text{Cs}} \times T_{\text{Cs}}$ spectrum.

The uncertainties δ_E associated with the spread of the beam energy before the target were determined from the analysis of one-dimensional spectra ΔE_0 and T_{TOF}. The energy losses E_1 of the projectile nuclei at the exit from the target were calculated using the LISE++ program [10].

The spread of energy losses in the target was determined by the expression $\Delta E = (E_0 - E_1) / 2 >> \delta_E$. The value of the cross section was attributed to the value of energy $E_0 - \Delta E$.

The beam transmission method used in the experiment consisted in measuring the flux I_0 of particles incident on the target with respect to its part ΔI corresponding to inelastic reaction channels. The quantity ΔI is equal to the difference in the flux I_0 of particles incident on the target and the flux I of particles passing through the target without interaction. In experiments with thin targets, when the condition $\Delta I = I_0 - I << I_0$ applies, the experimentally measured integral cross section of the $x\gamma$ and yn channels in two solid angles Ω_1 and Ω_2 is determined by formula

$$\sigma_{\text{int}} = \Delta I / (I_0 n), \tag{3}$$

where n is the number of target nuclei per unit area. The value ΔI was determined taking into account the detection efficiency of γ-quanta and neutrons by detectors.

3. Response Function of Gamma Spectrometer

In order to determine the detection efficiency P of gamma quanta with different multiplicities M by the CsI(Tl) spectrometer (spectrometer response), measurements were performed with a ^{60}Co calibration source installed instead of the target and an additional scintillation detector $CeBr_3$ of dimensions $51 \times 51 \times 51$ mm^3 positioned at the beam axis at a distance of 9 cm from the ^{60}Co source. The solid angle covered by the detector was $\Omega_3 = 0.021 \cdot 4\pi = (1 - \rho)4\pi$, where $\rho = 0.979$.

After beta decay of ^{60}Co, the excited state 4^+ of ^{60}Ni is formed. With probability 0.988, it emits the first gamma quantum with energy $E_{\gamma,1} = 1173$ keV in the transition from the 4^+ state to the 2^+ state and the second gamma quantum with energy $E_{\gamma,2} = 1332$ keV in the transition to the ground state 0^+. The lifetime of 2^+ state is about 0.7 ps.

A set G_1 of events of registration of particles with energy release 1173 ± 10 keV in the $CeBr_3$ scintillation detector was recorded. These events, in the overwhelming number of cases, correspond to the emission of gamma quantum with energy $E_{\gamma,1}$ from ^{60}Ni. The exceptions are background events that form a pedestal in the energy spectrum under the total absorption peak. The total number of events in the set G_1 is denoted by n_1. According to the ^{60}Co decay scheme, these events should correspond to the emission from the source of the second gamma quantum with energy $E_{\gamma,2}$. Events in the set G_1 included the cases when the second gamma quantum was emitted in the solid angle $\Omega_4 = 4\pi - \Omega_3 = \rho 4\pi$ (excluding situations when the second gamma quantum was registered by the $CeBr_3$ detector). Taking into account averaging over the angles of emission of the first gamma quantum entering the $CeBr_3$ detector, in the first approximation, the emission of the second gamma quantum can be considered isotropic. Events from the set G_1 correspond to the processes of isotropic emission of a single gamma quantum into solid angle Ω_4 with energy $E_{\gamma,2}$ from the position of the target. The experimental efficiency $p\,(1)$ of detection of gamma radiation into solid angle Ω_4 with energy $E_{\gamma,2}$ and multiplicity $M = 1$ by the CsI(Tl) spectrometer was determined as

$$p(1) = \frac{1}{n_1} \sum_{k=1}^{12} N_k^{(1)}, \tag{4}$$

where $N_k^{(1)}$ is the number of events from G_1 in which k detectors of the spectrometer were triggered with energy release above the threshold (150 keV).

By successive combination of events from the set G_1 in pairs 1-2, 3-4, 5-6, etc., a set G_2 of events was obtained corresponding to the emission of two gamma quanta with energy $E_{\gamma,2}$ into Ω_4. In the general case, a set G_M of events was formed in which isotropic emission of M gamma quanta with energy $E_{\gamma,2}$ occurred; the number n_M of such events and the number $N_k\,(M)$ of events from the set G_M in which triggering of at least one of the CsI(Tl) detectors with energy release above the threshold was recorded. Combining events in groups of M, we modeled the process of emission of M gamma quanta into Ω_4. The experimental detection efficiency of gamma radiation with energy $E_{\gamma,2}$ and multiplicity M by the CsI(Tl) spectrometer was determined as

$$p(M) = \frac{1}{n_M} \sum_{k=1}^{12} N_k^{(M)} , \tag{5}$$

where we restricted ourselves to $M \leq 8$.

Results of the measurements for the solid angle Ω_4 may be reduced to the total solid angle 4π. For instance, the total number of events n_1 of emission of one gamma quantum, reduced to the total solid angle 4π will be equal to $n'_1 = n_1 (4\pi/\Omega_4) = n_1/0.979$ and formula (4) is converted to

$$P(1) = \frac{1}{n'_1} \sum_{k=1}^{12} N_k^{(1)} = p(1)u_1(1) = \rho p(1) , \tag{6}$$

where $u_1(1)$ is the probability that the direction of flight of the γ-quantum isotropically emitted from the target be within the solid angle Ω_4. When the multiplicity is $M = 2$, the detection efficiency is equal to

$$P(2) = p(1)u_2(1) + p(2)u_2(2) , \tag{7}$$

where $u_2(1)$ is the probability of emission of only one of the two gamma quanta into Ω_4, $u_2(2)$ is the probability of emission of both gamma quanta into Ω_4,

$$u_2(1) = 2\rho(1-\rho),\ u_2(2) = \rho^2 . \tag{8}$$

For the multiplicity M, the reduced detection efficiency is equal to

$$P(M) = \sum_{j=1}^{M} u_M(j) p(j) , \tag{9}$$

where $u_M(j)$ is the probability that only j of the M isotropically emitted from the target γ-quanta be within the solid angle Ω_4,

$$u_M(j) = C_M^j \rho^j (1-\rho)^{M-j} , \tag{10}$$

where C_M^j are binomial coefficients.

The obtained dependence of the experimental detection efficiency of the spectrometer on the multiplicity $M \leq 8$ of gamma radiation with an energy of $E_{\gamma,2} = 1332$ keV is shown in Fig. 4(a). It is seen that with increasing multiplicity, $P(M) \to 1$.

The relative frequencies $w_M(k)$ of the number k of triggered spectrometer detectors in the registration of M photons in Ω_ρ were determined as

$$w_M(k) = \frac{N_k^{(M)}}{n_M\, p(M)} . \tag{11}$$

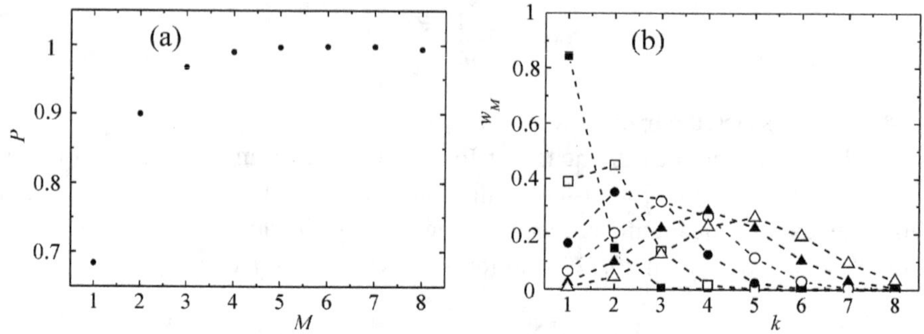

Fig. 4. (a) Detection efficiency of a group of M photons by the spectrometer (spectrometer response): experimental results for gamma energy 1332 keV. (b) The relative frequencies $w_M(k)$ of the number of triggered spectrometer detectors in the registration of M photons for $k = 1$ (filled squares), $k = 2$ (empty squares), $k = 3$ (filled circles), $k = 4$ (empty circles), $k = 5$ (filled triangles), $k = 6$ (empty triangles).

The relative frequencies $w_M(k)$ of the number k of triggered spectrometer detectors in the registration of M photons reduced to the total solid angle 4π are

$$W_M(k) = \sum_{j=1}^{M} u_M(j) w_j(k) \tag{12}$$

are shown in Fig. 4(b).

The probability of the number k of triggered γ-spectrometer detectors in the registration of M photons is equal to $P(M)W_M(k)$.

4. Analysis of Results of Measurements

Sessions without the target were used to measure the background. Let N'_k be the number of triggering of k detectors for measurements without the target. The relationship between the number N'_k and the flux I_0 can be approximated by a linear dependence

$$N'_k = N'_{0k} + \beta_k I_0 \,. \tag{13}$$

The values of the coefficients β_k and their uncertainties $\delta\beta_k$ can be determined using linear regression. In practice, the values of the parameters N'_{0k} are small in magnitude, so we will use a simpler expression

$$N'_k = \beta_k I_0 \tag{14}$$

with the coefficients β_k found by the least squares method from the results of m measurements

$$\beta_k = \left(\sum_{j=1}^{m} I_{0j}^2 \right)^{-1} \sum_{j=1}^{m} I_{0j} N'_{kj} \,. \tag{15}$$

The numbers N'_{0k} as function of flux I_0 for several energies E with exposures without the target and plots of expressions (13) are given in Fig. 5(a). The relative frequencies $w'(k)$

$$w'(k) = N'_k \left(\sum_{k=1}^{12} N'_k \right)^{-1} \tag{16}$$

of the number of triggered spectrometer detectors for exposures without the target are given in Fig. 5(b). Comparison with Fig. 4(b) shows that in most cases a single γ-quantum or neutron is detected. Denote N_k the number of registered events with the triggering of k detectors. The relative frequencies $w(k)$

$$w(k) = N_k \left(\sum_{k=1}^{12} N_k \right)^{-1} \tag{17}$$

of the number of triggered spectrometer detectors for exposures with the target [see Fig. 5(b)] show that multiplicities $1 \le M \le 6$ of emission of γ-quanta and/or neutrons are more probable compared to exposures without the target. The dependence $w(k)$ for six values of E_{lab} in the range 6–21 A MeV is practically identical.

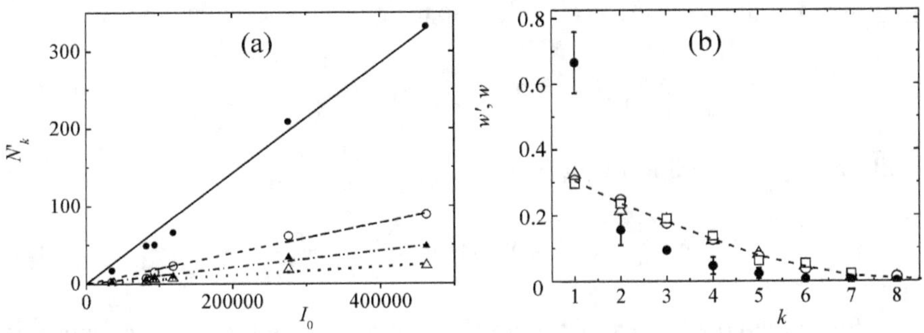

Fig. 5. The numbers N'_k as function of flux I_0 for several energies E with exposures without the target and plots of expression (13): (a) for $k = 1$ (dots and solid line), $k = 2$ (circles and dashed line), $k = 3$ (triangles and dash-dotted line), $k = 4$ (empty triangles and dotted line). (b) The average relative frequencies $w'(k)$ of the number of triggered spectrometer detectors for exposures without the target (filled circles) and relative frequencies $w(k)$ for $E_{lab} = 6$ A MeV (empty squares), $E_{lab} = 13$ A MeV (empty circles), $E_{lab} = 21$ A MeV (empty triangles); dashed curve is average dependence for six values of E_{lab} in the range 6–21 A MeV.

The result of measuring the reaction cross section was determined taking into account the numbers of triggering of detectors in the following order. Let M photons and/or neutrons be emitted into the total solid angle 4π in the reaction with a probability $\Gamma(M)$. Then the probability of registration, after the reaction, of k photons and/or neutrons (with triggering of $1 \le k \le k_{\max}$ detectors) is equal to

$$\sum_{M=1}^{M_{\max}} \Gamma(M)P(M)W_M(k), \tag{18}$$

where $M_{\max}$ is the maximum number of photons and/or neutrons taken into account. We used the value $M_{\max} = k_{\max} = 5$ for gamma spectrometer with six CsI(Tl) scintillators and $M_{\max} = k_{\max} = 8$ for twelve CsI(Tl) scintillators. For the total number of interactions $\Delta I = I_0 \sigma_R n$, the estimated number of their registrations with triggering of k detectors in the total solid angle 4π will be

$$\Delta I \sum_{M=1}^{M_{\max}} \Gamma(M)P(M)W_M(k) = I_0 \sigma_R n \sum_{M=1}^{M_{\max}} \Gamma(M)P(M)W_M(k). \tag{19}$$

The estimated number of their registrations with triggering of k detectors in the solid angles $\Omega_i = \eta_i 4\pi$, $i = 1,2$ will be

$$\eta_i I_0 n \sum_{M=1}^{M_{\max}} \sigma_M P(M)W_M(k), \tag{20}$$

where $\sigma_M = \sigma_R \Gamma(M)$. From the condition of equality of the number of registered events $N_k - N'_k = N_k - \beta_k I_0$ to their calculated value

$$N_k - \beta_k I_0 = \eta_i I_0 n \sum_{M=1}^{M_{\max}} \sigma_M P(M)W_M(k) \tag{21}$$

follows a system of linear equations

$$\sum_{M=1}^{M_{\max}} \sigma_M P(M)W_M(k) - \frac{N_k - \beta_k I_0}{I_0 n \eta_i} = 0. \tag{22}$$

Since the coefficients of system (22) are determined with uncertainties, its exact solution may yield non-physical values $\sigma_M < 0$. Therefore, it is more correct to find these values from the condition that the sum of the squares of the left-hand sides is minimum

$$F\left(\sigma_1, \ldots, \sigma_{M_{\max}}\right) = \sum_{k=1}^{k_{\max}} \left[\sum_{M=1}^{M_{\max}} \sigma_M P(M)W_M(k) - \frac{N_k - \beta_k I_0}{I_0 n \eta_i} \right]^2 \tag{23}$$

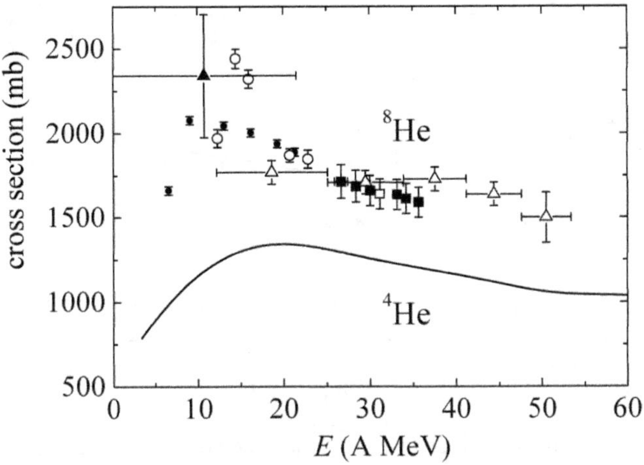

Fig. 6. Experimental (symbols) total cross sections σ_R for the reaction $^8He + {}^{28}Si$: results for the gamma spectrometer with six (○) and twelve (●) CsI(Tl) scintillators (this work), results from [11] ▲, [12] △, [13] ■, [14] □; curve is the result of spline smoothing of the total cross section for the $^4He + {}^{28}Si$ reaction, see review of data in [2].

under the constraint $\sigma_M \geq 0$. The total reaction cross section can be found as

$$\sigma_R = \sum_{M=1}^{M_{max}} \sigma_M \ . \tag{24}$$

The uncertainty in $\Delta\sigma_R$ due to the uncertainty $\delta\beta_k$ of the coefficients β_k was calculated by formula

$$\Delta\sigma_R = \frac{1}{2}\left|\sigma_R^{(+)} - \sigma_R^{(-)}\right|, \tag{25}$$

where $\sigma_R^{(+)}$ and $\sigma_R^{(-)}$ are the values for the set of parameters $\beta_k + \delta\beta_k$ and $\beta_k - \delta\beta_k$, respectively.

The total cross sections σ_R determined for gamma spectrometer with six and twelve CsI(Tl) scintillators are compared with other data for total cross section in Fig. 6. The results of new analysis of experimental data are consistent with the results for the total cross section from works [11–14]. It may be interpreted as the fact that the fraction of neutrons flying into forward angles is small. The estimated mean value $\langle M\rangle$ of photons and/or neutrons emitted in the total solid angle 4π in the reaction was calculated by formula

$$\langle M\rangle = \frac{1}{\sigma_R}\sum_{M=1}^{M_{max}} M\sigma_M \tag{26}$$

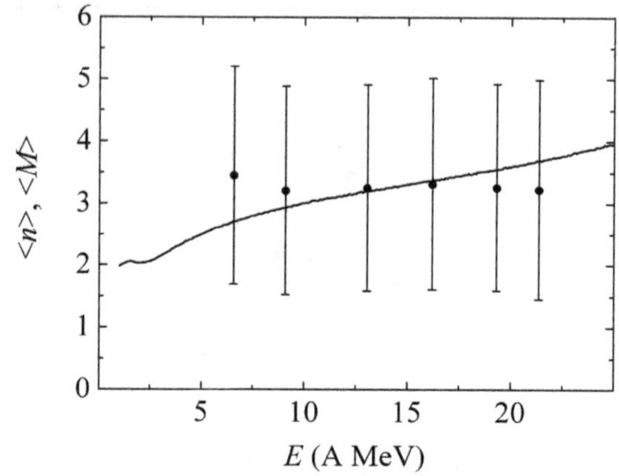

Fig. 7. The estimated mean value $\langle M \rangle$ of photons and/or neutrons emitted in the reaction (dots); the mean number $\langle n \rangle$ of emitted neutrons (curve) calculated using the code of the NRV web knowledge base [15–17].

and is shown in Fig. 7. Estimation of the number $\langle n \rangle$ of emitted neutrons from the compound nucleus in the fusion channel was made using the code of the NRV web knowledge base [15–17]. The values $\langle M \rangle$ and $\langle n \rangle$ are close in the range of E_{lab} from 6 to 21 A MeV.

In Refs. [12, 18], it was shown experimentally that the total cross sections for reactions involving weakly bound ^{6}He and ^{11}Li nuclei can be represented in the form $\sigma_R(^6\text{He}) \approx \sigma_R(^4\text{He}) + \sigma_{-2n}(^6\text{He})$ and $\sigma_R(^{11}\text{Li}) \approx \sigma_R(^9\text{Li}) + \sigma_{-2n}(^{11}\text{Li})$, respectively. The difference between the total cross sections σ_R for ^{8}He and ^{4}He can be explained by the channels of multi-neutron transfer and ^{8}He breakup as well as by the effect of additional nucleus-nucleus attraction because of the outer neutrons rearrangement between surfaces of approached nuclei [2].

5. Conclusions

The energy dependence of the total cross section for the $^8\text{He} + ^{28}\text{Si}$ reaction has been measured in the beam energy range 6–25 A MeV by the direct and model-independent transmission method. An unusual significant wide enhancement in the total cross sections for the $^8\text{He} + ^{28}\text{Si}$ reaction in comparison with the $^4\text{He} + ^{28}\text{Si}$ reaction was observed in the entire investigated energy range. The maximum of the measured cross section was observed at $\approx$10 A MeV. The time-dependent qualitative analysis performed in the work shows that the maximum in the total cross section is due to the processes of neutron transfer from the external skin shell to the target nucleus and the redistribution of the appreciable part of the inner skin shell into the region between the surfaces of the approached nuclei. Such an increase in the total cross section of the reaction is most noticeable in the energy range at which the relative velocity of the nuclei is close in magnitude to the average velocity of outer neutrons in the investigated weakly bound nuclei.

Acknowledgments

The authors express their gratitude to the scientific group of the ACCULINNA setup for all possible assistance in carrying out experiments on the beams of the fragment separator. The work was supported by the Russian Science Foundation, Grant No. 17-12-01170.

References

1. Yu. E. Penionzhkevich, Reactions involving loosely bound cluster nuclei: heavy ions and new technologies, *Phys. Atom. Nucl.* **74**, 1615 (2011).
2. Yu. E. Penionzhkevich, Yu. G. Sobolev, V. V. Samarin, and M. A. Naumenko, Peculiarities in total cross sections of reactions with weakly bound nuclei ^{6}He, ^{9}Li, *Phys. Atom. Nucl.* **80**, 928 (2017).
3. Yu. E. Penionzhkevich, Yu. G. Sobolev, V. V. Samarin, M. A. Naumenko, N. A. Lashmanov, V. A. Maslov, I. Siváček, and S. S. Stukalov, Energy dependence of the total cross section for the ^{11}Li + ^{28}Si reaction, Phys. Rev. C **99**, 014609 (2019).
4. R. Anne, S. Arnell, R. Bimbot, H. Emling, D. Guillemaud-Mueller, P. Hansen, L. Johannsen, B. Jonson, M. Lewitowicz, S. Mattsson, A. Mueller, R. Neugart, G. Nyman, F. Pougheon, A. Richter, K. Riisager, M. Saint-Laurent, G. Schrieder, O. Sorlin, and K. Wilhelmsen, *Phys. Lett. B* **250**, 19 (1990).
5. V. V. Samarin, Description of nucleon-transfer and fusion reactions within time-dependent approaches and coupled-channel method, *Phys. Atom. Nucl.* **78**, 128 (2015).
6. V. Samarin, Time-dependent quantum models of the dynamics of neutron transfer reactions near the barrier, *EPJ Web Conf.* **86**, 00040 (2015).
7. Yu. G. Sobolev, Yu. E. Penionzhkevich, D. Aznabaev, E. V. Zemlyanaya, M. P. Ivanov, G. D. Kabdrakhimova, A. M. Kabyshev, A. G. Knyazev, A. Kugler, N. A. Lashmanov, K. V. Lukyanov, A. Maj, V. A. Maslov, K. Mendibayev, N. K. Skobelev, R. S. Slepnev, V. V. Smirnov, and D. Testov, Experimental study of the energy dependence of the total cross section for the ^{6}He + ^{nat}Si and ^{9}Li + ^{nat}Si reactions, *Phys. Part. Nucl.* **48**, 922 (2017).

8. Yu. G. Sobolev, Yu. E. Penionzhkevich, V. A. Maslov, M. A. Naumenko, V. V. Samarin, I. Siváček, and S. S. Stukalov, Measuring the total cross sections for reactions in collisions of ^{6,8}He + ^{28}Si and ^{9}Li + ^{28}Si, *Bull. Russ. Acad. Sci.: Phys.* **83**, 402 (2019).
9. A. Rodin, S. Stepantsov, D. Bogdanov, M. Golovkov, A. Fomichev, S. Sidorchuk, R. Slepnev, R. Wolski, G. Ter-Akopian, Yu. Oganessian, A. Yukhimchuk, V. Perevozchikov, Yu. Vinogradov, S. Grishechkin, A. Demin, S. Zlatoustovskiy, A. Kuryakin, S. Fil'chagin, and R. Ilkaev, Status of ACCULINNA beam line, *Nucl. Instrum. Methods Phys. Res. Sect. B* **204**, 114 (2003).
10. LISE++ code, http://lise.nscl.msu.edu/.
11. A. C. C. Villari, W. Mittig, E. Plagnol, Y. Schutz, M. Lewitowicz, L. Bianchi, B. Fernandez, J. Gastebois, A. Gillibert, C. Stephan, L. Tassan-Got, G. Audi, W. Zhan, A. Cunsolo, A. Foti, A. Belezyorov, S. Lukyanov, and Yu. Penionzhkevich, Measurements of reaction cross sections for neutron-rich exotic nuclei by a new direct method, *Phys. Lett. B* **268**, 345 (1991).
12. R. E. Warner, R. A. Patty, P. M. Voyles, A. Nadasen, F. D. Becchetti, J. A. Brown, H. Esbensen, A. Galonsky, J. J. Kolata, J. Kruse, M. Y. Lee, R. M. Ronningen, P. Schwandt, J. von Schwarzenberg, B. M. Sherrill, K. Subotic, J. Wang, and P. Zecher, Total reaction and 2n removal cross sections of 20–60 A MeV ^{4,6,8}He, $^{6-9,11}$Li, and ^{10}Be on Si, *Phys. Rev. C* **54**, 1700 (1996).
13. Li Chen, Zhan Wen-Long, Xiao Guo-Qing, Guo Zhong-Yan, Sun Zhi-Yu, Li Jia-Xing, Chen Zhi-Qiang, Wang Meng, Wang Wu-Sheng, Mao Rui-Shi, Chen Li-Xin, Bai Jie, Hu Zheng-Guo, Measurement of total reaction cross sections for neutron-rich nucleus ^{8}He on ^{28}Si, *High Energy Physics and Nuclear Physics* **29**, 944 (2005).
14. Li Chen, Ye Yan-Lin, Zhan Wen-Long, Xiao Guo-Qing, Xu Hu-Shan, Guo Zhong-Yan, Wang Jian-Song, Sun Zhi-Yu, Li Jia-Xing, Wang Meng, Chen Zhi-Qiang, Jiang Dong-Xing, Zheng Tao, Hua Hui, Li Zhi-Huan, Li Xiang-Qing, Ge Yu-Cheng, Wu Cui-E, Pang Dan-Yang, Lou Jian-Ling, Lu Fei, and Fan Feng-Ying, Measurement of neutron-removal cross section of neutron-rich nucleus ^{8}He by using the transmission method, *High Energy Physics and Nuclear Physics* **31**, 52 (2007).
15. V. I. Zagrebaev, A. S. Denikin, A. V. Karpov, A. P. Alekseev, M. A. Naumenko, V. A. Rachkov, V. V. Samarin and V. V. Saiko, NRV web knowledge base on low-energy nuclear physics, http://nrv.jinr.ru/nrv/.

16. A. V. Karpov, A. S. Denikin, A. P. Alekseev, V. I. Zagrebaev, V. A. Rachkov, M. A. Naumenko, and V. V. Saiko, *Phys. Atom. Nucl.* **79**, 749 (2016).
17. A. V. Karpov, A. S. Denikin, M. A. Naumenko, A. P. Alekseev, V. A. Rachkov, V. V. Samarin, V. V. Saiko and V. I. Zagrebaev, NRV web knowledge base on low-energy nuclear physics, *Nucl. Instrum. Methods Phys. Res. Sect. A* **859**, 112 (2017).
18. I. Tanihata, D. Hirata, T. Kobayashi, S. Shimoura, K. Sugimoto, and H. Toki, Revelation of thick neutron skins in nuclei, *Phys. Lett. B* **289**, 261 (1992).

Study of Ground States of ^{7,9,11}Li Nuclei and Dynamics of External Neutrons in Reactions ^{7,9,11}Li + ^{28}Si

V. V. Samarin[1,2,*] and M. A. Naumenko[1]

[1]*Flerov Laboratory of Nuclear Reactions, Joint Institute for Nuclear Research, Dubna, 141980, Russia*
[2]*Department of Nuclear Physics, Dubna State University, Dubna, 141982, Russia*
**E-mail: samarin@jinr.ru*

This work is devoted to studying light nuclei ^{7,9,11}Li. The probability density for the ground states of ^{7,11}Li nuclei is calculated by Feynman's continual integrals method. The dynamical approach based on the numeric solution of the time-dependent Schrödinger equation is applied to the description of dynamics of outer neutrons in the reaction ^{11}Li + ^{28}Si and to the calculation of total reaction cross sections.

Keywords: Feynman's continual integrals method; time-dependent Schrödinger equation; total reaction cross sections.

1. Introduction

The study of nuclear reactions involving neutron-rich weakly bound nuclei makes it possible to obtain information on the structure of the investigated nuclei (clusters, neutron halo, etc.) and its manifestation in reactions [1]. It also provides an opportunity for testing various microscopic models. One of the criteria for the limits of applicability and the degree of accuracy of theoretical models is the quantitative agreement between the values of the calculated and the experimentally measured total cross sections of nuclear reactions.

The results of experiments on measuring total cross sections for the ^{9}Li + ^{28}Si reaction as a function of energy showed that in the energy range 10 ÷ 20 A MeV the values of the total cross section are much larger than those for the ^{7}Li + ^{28}Si reaction [2, 3]. In [2], it was assumed that the reason for the observed behavior was related to the properties of the shell of weakly bound external neutrons and their adiabatic and diabatic rearrangement in the process of collision with the target nucleus. In the ^{11}Li nucleus, the external neutrons are even more weakly bound, which determines interest in studying reactions involving this nucleus. In Refs. [4, 5], it was shown experimentally that the total cross sections for reactions involving weakly bound ^{6}He and ^{11}Li nuclei can be represented in the form $\sigma_R(^6\mathrm{He})\approx\sigma_R(^4\mathrm{He})+\sigma_{-2n}(^6\mathrm{He})$ and $\sigma_R(^{11}\mathrm{Li})\approx\sigma_R(^9\mathrm{Li})+\sigma_{-2n}(^{11}\mathrm{Li})$, respectively. In this work, calculations of the total cross sections for the ^{11}Li + ^{28}Si reaction are based on the model of the ^{11}Li nucleus as a system of a ^{9}Li core and two neutrons. The evolution of the wave functions of external neutrons is calculated using a numerical solution of the time-dependent Schrödinger equation. The initial conditions for the wave functions are obtained based on the shell model calculations with the parameters providing neutron separation energies close to the experimental values as well as results of calculations of probability densities using Feynman's continual integrals method [6].

2. Structure of ^{7,9,11}Li Nuclei

2.1. *Structure of ^{7}Li nucleus in four-body model*

For studying the structure of ^{7,11}Li nuclei in the few-body model, we used Feynman's continual integrals method [6]. It provides a mathematically more simple possibility for calculating the energy and the probability density for the ground states of few-body systems compared to other approaches, *e.g.*, hyperspherical harmonics method, because it does not require expansion of the wave function into a system of functions. The squared modulus of the ground-state wave function $|\psi_0(q)|^2$ in the points q of the finite region corresponding to finite motion can be determined using Monte Carlo method and NVIDIA CUDA technology as described in work [6]. Parallel calculations were performed on the Heterogeneous Cluster of LIT, JINR.

In the our model, neutrons (n) and protons (p) in the nuclei interact with each other by nucleon-nucleon potentials with repulsive cores

$$V_{N-N}(r) = \sum_{k=1}^{3} u_k \exp(-r^2/b_k^2), \tag{1}$$

where N = $\{n, p\}$.

The nuclear part of the α-nucleon potentials used for ^{6}He, ^{6}Li, ^{9}Be nuclei (see [6]) has a repulsive core for excluding the forbidden (internal) 1s state in these nuclei.

$$V_{\alpha-N}(r) = \sum_{i=1}^{3} U_i[1+\exp((r-R_i)/a_i)]^{-1}. \tag{2}$$

The values of parameters of potentials (1)-(2) are given in Ref. [6]. An example of the probability density for the ground state of ^{7}Li in the 4-body model is shown in Figure 1. The most probable configuration is α + triton, and the neutron separation energies from ^{7}Li nucleus and triton equal to close values:

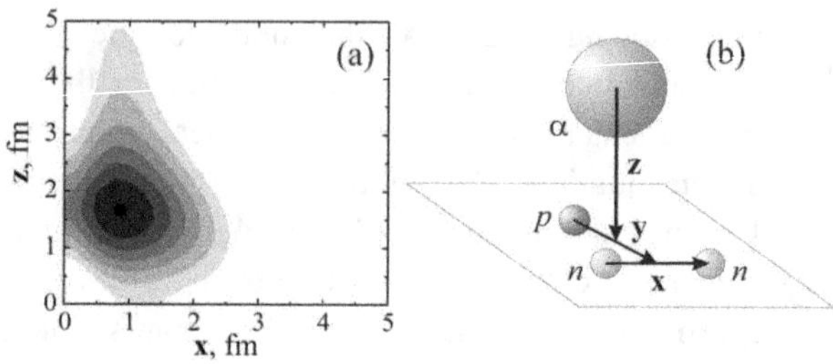

Fig. 1. The probability density for the ^{7}Li nucleus in the model $\alpha + p + n + n$ with regular triangle configuration of nucleons and the vectors **x**, **y**, **z** in the Jacobi coordinates; nucleons and α-core are denoted as small and large spheres, respectively. The most probable configuration is α + triton.

7.25 MeV and 6.3 MeV, respectively. These values for outer neutrons may be obtained in the shell model of deformed nucleus ^{7}Li.

2.2. *Structure of Li nuclides in shell model of deformed nuclei*

It is well known that the ^{7}Li, ^{9}Li, ^{11}Li nuclei are deformed; the experimental values of the quadrupole deformation parameter β_2 are from -0.9 to -1.5 for ^{7}Li, from -0.6 to -0.8 for ^{9}Li, -0.6 for ^{11}Li [7, 8].

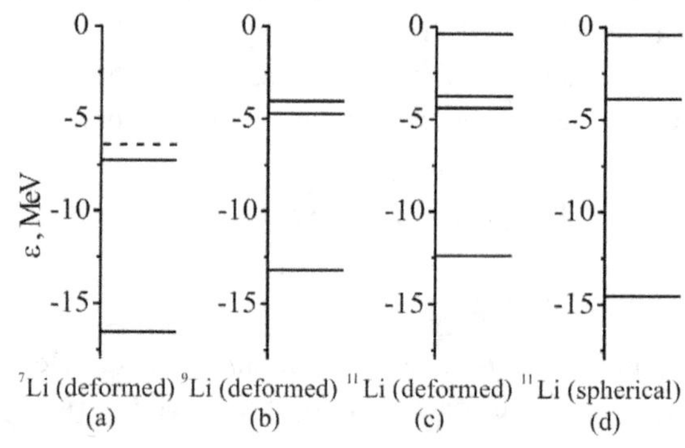

Fig. 2. Schemes of neutron levels for the nuclei ^{7}Li (a), ^{9}Li (b), ^{11}Li (c) in the shell model of the deformed nucleus and in the shell model of the spherical nucleus ^{11}Li (d).

Calculations in the shell model of the deformed nucleus by the method of Ref. [9] with nonspherical Woods-Saxon potential from Ref. [10] provided the energies of the upper occupied levels of the nucleus ^{7}Li approximately equal to the experimental values of the neutron separation energy taken with the opposite sign. The obtained neutron-level diagram is shown in Figure 2(a). Two neutrons and two protons at deep lower levels corresponding to the level $1s_{1/2}$ of the spherical nucleus with the projection of the total angular momentum on the axis of symmetry of the nucleus $|m_j| = 1/2$ belong to a nuclear core similar to an α-cluster. In the few-body model, the two external neutrons of the ^{7}Li nucleus on the sublevel with the projection of the total angular momentum on the axis of symmetry of the nucleus $|m_j| = 3/2$ corresponding to the level $1p_{3/2}$ of the spherical nucleus can be considered quite strongly bound to the triton cluster. The obtained neutron-level diagram for the ^{9}Li nucleus is shown in Figure 2(b). In the ^{9}Li nucleus, the neutron separation energy for sublevel $|m_j| = 3/2$ is noticeably lower than in the ^{7}Li nucleus, and for the higher-lying sublevel with $|m_j| = 1/2$ the separation energy is 4.06 MeV. Thus, for the four external neutrons of the 9Li nucleus, the bond with the α-cluster core is weakened. In collisions with heavy nuclei, the probability density distribution for these neutrons can change more significantly than for the two external neutrons of the ^{7}Li nucleus [2]. The level scheme for the neutrons of the deformed ^{11}Li nucleus is shown in Figure 2(c). The energies of the sublevels with $|m_j| = 3/2$ and $|m_j| = 1/2$, corresponding to the level $1p_{3/2}$ of the spherical nucleus, turn out to be close. This makes it possible, with sufficient accuracy, to use the spherical shell model for the ^{11}Li nucleus with three filled neutron shells: $1s_{1/2}$ (the α-cluster core shell), $1p_{3/2}$ (the inner skin shell), and $1p_{1/2}$ (the outer halo shell).

2.3. *Structure of ^{11}Li nucleus in three-body model*

The separation energies of one and two outer neutrons from the ^{11}Li nucleus are very low, 0.40 MeV and 0.37 MeV, respectively, therefore, the ^{11}Linucleus is considered as a system of a ^{9}Li-core and two neutrons.

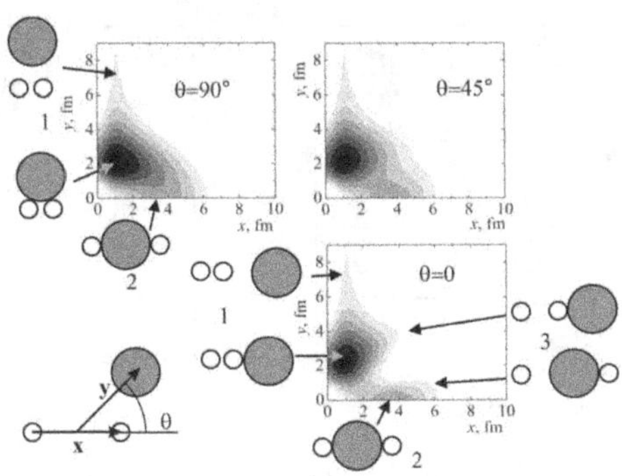

Fig. 3. The probability density for the ^{11}Li nucleus (configuration ^{9}Li+n+n) and the vectors in the Jacobi coordinates; neutrons and ^{9}Li-core are denoted as small empty circles and large filled circles, respectively. The most probable configurations are ^{9}Li + dineutron (1) and the cigar configuration (2).

An example of the probability density for the ground state of ^{11}Li in this 3-body model is shown in Figure 3. The outer neutrons forming spread weakly-bound dineutron may be considered as independent particles in the shell model.

3. Calculation of Total Reaction Cross Sections

As in Refs. [4, 5], we consider two main groups of reaction channels, those that are the consequence of the interaction of the ^{9}Li-like core of the ^{11}Li nucleus with the ^{28}Si nucleus and the consequence of neutron loss from the outer shell $1p^2_{1/2}$ of the ^{11}Li nucleus. The loss of one neutron (with some probability) leads to a subsequent loss of the second neutron by the unbound ^{10}Li nucleus.

The independent probabilities P_{core} of the reaction because of the interaction with the ^{9}Li-like core of the ^{11}Li nucleus and P_{loss} of the neutron loss from the outer shell can be determined as functions of energy E and the impact parameter b in the semiclassical model: $P_{\text{core}}(b,E)$, $P_{\text{loss}}(b,E)$. The semiclassical expression for the total cross section of the reaction ^{9}Li + ^{28}Si

$$\sigma_{\text{R}} = 2\pi \int_0^\infty P_{\text{core}}(b, E) b db = \frac{\pi}{k^2} \sum_{l=0}^{\infty} (2l+1) \dot{P}_{\text{core}}(l, E) \tag{3}$$

corresponds to the sum over the orbital angular momenta in the quantum approach taking into account the relation l ~ kb, where k is the modulus of the wave vector.

The calculation of the total cross section σ_{R} for the reaction ^{9}Li + ^{28}Si and the probability $\dot{P}_{\text{core}}(l,E)$ in the optical model with the energy dependent optical potential was performed in Ref. [2] based on the solution of the time-dependent

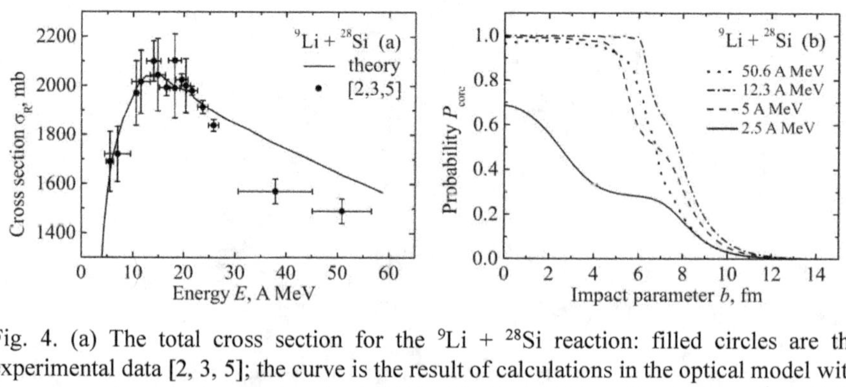

Fig. 4. (a) The total cross section for the ^{9}Li + ^{28}Si reaction: filled circles are the experimental data [2, 3, 5]; the curve is the result of calculations in the optical model with the energy-dependent optical potential [2]. (b) Probabilities $P_{\text{core}}(b,E) = \dot{P}_{\text{core}}(kb,E)$ depending on the impact parameter b for energies 2.5 A MeV (solid line), 5 A MeV (dashed line), 12.3 A MeV (dash-dotted line), 50.6 A MeV (dotted line).

Schrödinger equation for the external neutrons of the ^{9}Li nucleus. The comparison of the results of calculations with the experimental data is shown in Figure 4(a). The dependence of the probabilities $P_{core}(b,E) = \dot{P}_{core}(kb,E)$ on b is shown in Figure 4(b).

An increase in the cross section of the reaction is most noticeable in the energy range at which the relative velocity of the nuclei is close in magnitude to the average velocity of external neutrons in the investigated weakly bound nuclei [2].

The total reaction cross section σ_R can be expressed in terms of probabilities P_{core} and P_{loss}. The probability of absence of the reaction involving the core is equal to 1- $P_{core}(b,E)$, the probability of absence of loss of at least one neutron is $[1 - P_{loss}(b,E)]^2$. The probability of the reaction involving the core or because of the loss of a neutron from the outer shell of the ^{11}Li nucleus is equal to

$$P_R(b, E) = 1 - [1 - P_{core}(b, E)][1 - P_{loss}(b, E)]^2. \tag{4}$$

In the semiclassical approach, the total cross section for the ^{11}Li + ^{28}Si reaction is represented by an integral over impact parameters

$$\sigma_R = 2\pi \int_0^\infty P_R(b, E) b db. \tag{5}$$

For calculation of the probability P_{loss} of neutron loss from the outer shell, we use a time-dependent approach with a quantum description of neutrons in combination with motion of the centers of colliding nuclei along classical trajectories [9, 11]. The two-component spinor wave function $\psi(\boldsymbol{r}, t)$ of each of the two independent neutrons with the radius vector $\mathbf{r}$ and the initial state $1p_{1/2}$ was calculated by numerical solution of the time-dependent Schrödinger equation (TDSE) taking into account spin-orbit interaction [2, 12, 13].

The lattice spacing in the TDSE method is 0.15 fm; the colliding nuclei are enclosed in a box of typical dimensions 90×75×40 fm^3.

An example of the evolution of the probability density of external neutrons of the ^{11}Li nucleus in the collision with the ^{28}Si nucleus is shown in Figure 5.

It can be seen that at energies ≈ 50 A MeV the external neutrons lost by the ^{11}Li nucleus leave both nuclei with energy in the continuous spectrum (mainly) or they are transferred to the target nucleus ^{28}Si with probabilities P_c and P_d, respectively.

The probability P_{loss} of neutron loss from the outer shell was determined by the expression

$$P_{loss}(b, E) = \min\{P_d(b, E) + P_c(b, E), 1\}. \tag{6}$$

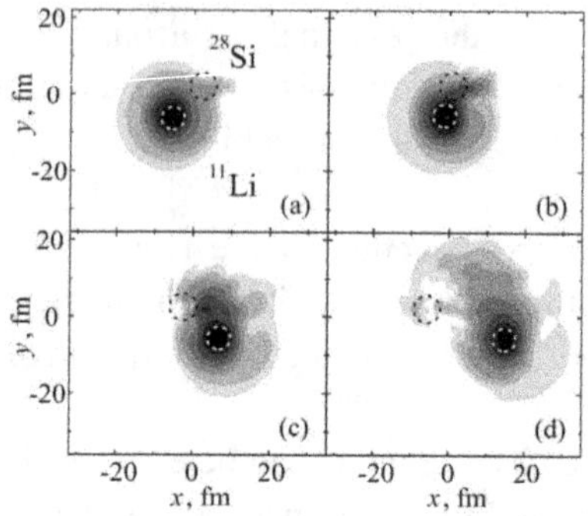

Fig. 5. An example of the evolution of the probability density for external neutrons of the ^{11}Li nucleus in the collision with the ^{28}Si nucleus at energy $E = 50.6$ A MeV and impact parameter $b = 8$ fm. The location of the panels (a) - (d) corresponds to time.

The probabilities $P_d(b,E)$ of neutron transfer to unoccupied bound states of the discrete spectrum in the ^{28}Si nucleus determined in the same way as in Refs. [12, 13] are shown in Figure 6(a) as functions of the distance of the closest approach $R_{min}(b,E)$ between the centers of the nuclei.

The probabilities P_c of transfer to the states of the continuous spectrum can be determined by integrating the probability density outside the vicinity of the nuclei. As an estimate for P_c, we can use the expression $P_c = CP_{max}$, where P_{max} is the maximum value of the probability of neutron presence in the spherical layer D around the ^{28}Si nucleus with boundary radii $r_1 = R_{Si} + \Delta R_1$ and $r_2 = R_{Si} + \Delta R_2$; R_{Si} is the radius of the target nucleus ^{28}Si; C is a variable (adjustable)

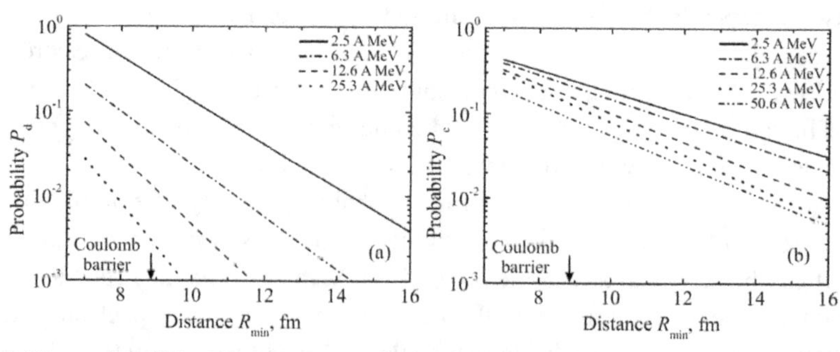

Fig. 6. The probabilities of neutron transfer to the unoccupied bound states of the discrete spectrum in the ^{28}Si nucleus (a) and the probabilities of transfer to the states of the continuous spectrum with $C = 2$, $\Delta R_1 = 3$ fm, and $\Delta R_2 = 10$ fm (b) for energies 2.5 A MeV (solid line) 6.3 A MeV (dash-dotted line), 12.6 A MeV (dashed line), 25.3 A MeV (dotted line), 50.6 A MeV (dash-dot-dotted line). Arrows indicate the position of the Coulomb barrier.

parameter. It is assumed that the released neutrons initially appear in the layer D in the form of a compact three-dimensional wave packet (see Figure 5) and then gradually leave it when the packet spreads. The dependence of the probabilities P_c on $R_{min}(b,E)$ for $C = 2$, $\Delta R_1 = 3$ fm, and $\Delta R_1 = 10$ fm is shown in Figure 6(b).

The results of calculations of total reaction cross section for the value of the adjustable parameter $C = 2$ for the transition probability to the states of the continuous spectrum are shown in Figure 7. Good agreement with the experimental data is obtained.

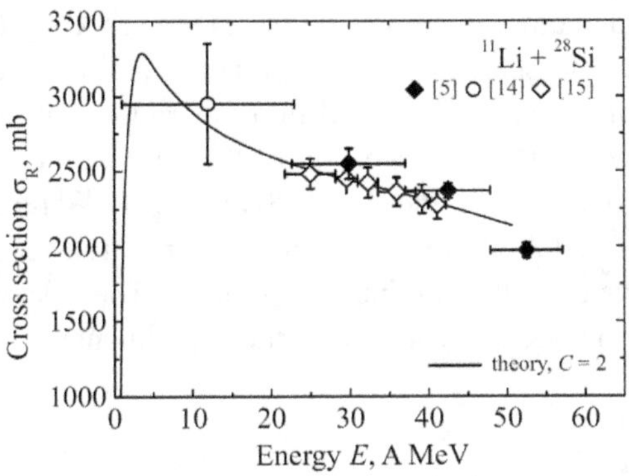

Fig. 7. The total cross section for the $^{11}Li + ^{28}Si$ reaction: symbols are experimental data [5, 14, 15], curves are the results of calculations for the value of the adjustable parameter $C = 2$ (solid curve) with the probability P_c of transition to the states of the continuous spectrum.

In the energy dependence of the total cross section, a sharp maximum is observed at energies near 5 A MeV. On the lower energy side, it is largely because of a sharp increase in the reaction probability P_{core} of the interaction of the target nucleus with the ^9Li-like core of the ^{11}Li nucleus and an increase in the probability of transfer to the target nucleus of the neutron from the extended halo shell $1p^2_{1/2}$ of the ^{11}Li nucleus. A fairly sharp decrease in the total cross section from the high-energy side at $E \approx 10$ A MeV is because of the rapid decrease of the probability of neutron transfer to the target nucleus and the decrease of the reaction probability P_{core} of the interaction of the target nucleus with the ^9Li-like core of the ^{11}Li nucleus in the region of sharp enhancement of the total cross section for the $^9Li + ^{28}Si$ reaction. The probability P_c of transfer to the states of the continuous spectrum from the extended halo shell $1p^2_{1/2}$ changes (decreases) with increasing energy fairly smoothly. This is the reason for the enhancement of the total cross section for the reaction ^{11}Li + ^{28}Si in comparison with the ^{9}Li + ^{28}Si and ^{7}Li + ^{28}Si reactions.

4. Conclusions

The experimental data on the total cross sections for the ^{7,9,11}Li + ^{28}Si reactions have been compared with theoretical calculations. The total cross sections for the ^{11}Li + ^{28}Si reaction have been calculated based on the numerical solution of the time-dependent Schrödinger equation for the external weakly bound neutrons of the projectile nucleus ^{11}Li. The model shows that the sharp maximum in the total cross section results from the processes of neutron transfer from the external halo shell to the target nucleus and the redistribution of the appreciable part of the inner skin shell into the region between the surfaces of the approached nuclei. The enhancement of the cross section for the ^{11}Li + ^{28}Si reaction compared to those for the ^{9}Li + ^{28}Si and ^{7}Li + ^{28}Si reactions in the entire energy range (up to 50 A MeV) results from neutron transfer from the extended halo shell to the states of the continuous spectrum. The calculated total reaction cross sections are in good agreement with the experimental data.

Acknowledgment

The work was supported by the Russian Science Foundation, Grant No. 17-12-01170.

References

1. Yu. E. Penionzhkevich, Reactions involving loosely bound cluster nuclei: heavy ions and new technologies, *Phys. Atom. Nucl.* **74**, 1615 (2011).
2. Yu. E. Penionzhkevich, Yu. G. Sobolev, V. V. Samarin and M. A. Naumenko, Peculiarities in total cross sections of reactions with weakly bound nuclei ^{6}He, ^{9}Li, *Phys. Atom. Nucl.* **80**, 928 (2017).
3. Yu. G. Sobolev, Yu. E. Penionzhkevich, D. Aznabaev, E. V. Zemlyanaya, M. P. Ivanov, G. D. Kabdrakhimova, A. M. Kabyshev, A. G. Knyazev, A. Kugler, N. A. Lashmanov, K. V. Lukyanov, A. Maj, V. A. Maslov, K. Mendibayev, N. K. Skobelev, R. S. Slepnev, V. V. Smirnov and D. Testov, Experimental study of the energy dependence of the total cross section for the ^{6}He + ^{nat}Si and ^{9}Li + ^{nat}Si reactions, *Phys. Part. Nucl.* **48**, 922 (2017).
4. I. Tanihata, D. Hirata, T. Kobayashi, S. Shimoura, K. Sugimoto and H. Toki, Revelation of thick neutron skins in nuclei, *Phys. Lett. B* **289**, 261 (1992).

5. R. E. Warner, R. A. Patty, P. M. Voyles, A. Nadasen, F. D. Becchetti, J. A. Brown, H. Esbensen, A. Galonsky, J. J. Kolata, J. Kruse, M. Y. Lee, R. M. Ronningen, P. Schwandt, J. von Schwarzenberg, B. M. Sherrill, K. Subotic, J. Wang and P. Zecher, Total reaction and 2n- removal cross sections of 20-60 A MeV ^{4,6,8}He, $^{6-9,11}$Li, and ^{10}Be on Si, *Phys. Rev. C* **54**, 1700 (1996).
6. V. V. Samarin, M. A. Naumenko, Study of ground states of ^{3}H, ^{3,4,6}He, ^{6}Li, and ^{9}Be nuclei by Feynman's continual integrals method, *Phys. Atom. Nucl.* **80**, 877 (2017).
7. V. I. Zagrebaev, A. S. Denikin, A. V. Karpov, A. P. Alekseev, M. A. Naumenko, V. A. Rachkov, V. V. Samarin and V. V. Saiko, NRV Web Knowledge Base on Low-Energy Nuclear Physics, http://nrv.jinr.ru/.
8. Centre for Photonuclear Experiments Data, http://cdfe.sinp.msu.ru/.
9. V. V. Samarin, Description of nucleon-transfer and fusion reactions within time-dependent approaches and coupled-channel method, *Phys. Atom. Nucl.* **78**, 128 (2015).
10. V. I. Zagrebaev and V. V. Samarin, Near-barrier fusion of heavy nuclei: coupling of channels, *Phys. Atom. Nucl.* **67**, 1462 (2004).
11. V. Samarin, Time-dependent quantum models of the dynamics of neutron transfer reactions near the barrier, *EPJ Web Conf.* **86**, 00040 (2015).
12. M. A. Naumenko, V. V. Samarin, Yu. E. Penionzhkevich and N. K. Skobelev, Near-barrier neutron transfer in reactions with ^{3}He nucleus, *Bull. Russ. Acad. Sci.: Phys.* **80**, 264 (2016).
13. M. A. Naumenko, V. V. Samarin, Yu. E. Penionzhkevich and N. K. Skobelev, Near-barrier neutron transfer in reactions ^{6}He + ^{45}Sc, ^{64}Zn, and ^{197}Au, *Bull. Russ. Acad. Sci.: Phys.* **81**, 710 (2017).
14. A. C. C. Villari, W. Mittig, E. Plagnol, Y. Schutz, M. Lewitowicz, L. Bianchi, B. Fernandez, J. Gastebois, A. Gillibert, C. Stephan, L. Tassan-Got, G. Audi, W. Zhan, A. Cunsolo, A. Foti, A. Belezyorov, S. Lukyanov and Yu. Penionzhkevich, Measurements of reaction cross sections for neutron-rich exotic nuclei by a new direct method, *Phys. Lett. B* **268**, 345 (1991).
15. Li Chen, Ye Yan-Lin, Zhan Wen-Long, Xiao Guo-Qing, Xu Hu-Shan, Guo Zhong-Yang, Wang Jian-Song, Sun Zhi-Yu, Li Jia-Xing, Wang Meng, Chen Zhi-Qiang, Jiang Dong-Xing, Zheng Tao, Hua Hui, Li Zhi-Huan, Li Xiang-Qing, Ge Yu-Cheng, Pang Dan-Yang, Lou Jian-Ling, Lu Fei, Fan Feng-Ying and Lu Lin-Hui, Measurement of total reaction cross section for neutron-rich nucleus ^{11}Li on ^{28}Si target at medium energy, *High Energy Physics and Nuclear Physics* **31**, 1102 (2007).

Study of Nucleon Transfer in Reactions $^{3}He + ^{194}Pt$, ^{45}Sc Within Time-Dependent Approach

V. V. Samarin[1,2,*], Yu. E. Penionzhkevich[1,3], M. A. Naumenko[1] and N. K. Skobelev[1]

[1]*Flerov Laboratory of Nuclear Reactions, Joint Institute for Nuclear Research, Dubna, 141980, Russia*

[2]*Department of Nuclear Physics, Dubna State University, Dubna, 141982, Russia*

[3]*Department of Experimental Methods in Nuclear Physics, National Research Nuclear University MEPhI, Moscow, 115409, Russia*

[*]*E-mail: samarin@jinr.ru*

Theoretical description of the experimental data on the formation of various isotopes in reactions $^{3}He + ^{194}Pt$ and $^{3}He + ^{45}Sc$ requires taking into account neutron and proton transfer channels, as well as fusion-evaporation channels. To calculate the probabilities of nucleon transfer as well as transfer cross sections the time-dependent Schrödinger equation (TDSE) has been solved numerically. Fusion-evaporation was taken into account using the statistical model code of the NRV web knowledge base. Results of calculations are in agreement with experimental data.

Keywords: Neutron and proton transfer; fusion-evaporation; time-dependent Schrödinger equation.

1. Introduction

The processes of neutron and proton transfer have been extensively studied both experimentally and theoretically. The theoretical description of experimental data on formation of gold isotopes in the $^{3}He + ^{194}Pt$ reaction [1] and the ^{45}Ti isotope in the $^{3}He + ^{45}Sc$ reaction [2] requires taking into account proton transfer channels. Considering these channels in the DWBA approximation provides (for the appropriate choice of parameters) satisfactory description of the experimental angular distributions for energies much higher than the Coulomb barrier [3]. For near-barrier energies, however, oscillations in angular distributions are weakly pronounced, complicating determination of the optical potential parameters. It is easier in this case to compare theoretical calculations and experimental data on the formation cross sections of the reaction products.

The cross section for the ^{194}Au isotope formation in the $^{3}He + ^{194}Pt$ reaction was measured by means of activation in [1]. Possible mechanisms of the ^{194}Au isotope formation include proton capture by the target nucleus with the loss of a neutron due to its evaporation (after the excitation associated with proton capture), or to its capture by the projectile nucleus (with the transformation of ^{3}He into ^{3}H). Another possible mechanism of ^{194}Au isotope formation, fusion of nuclei with evaporation of two neutrons and a proton, is of low probability because of the high Coulomb barrier for protons. For the ^{3}He + ^{45}Sc reaction [2], the probability of ^{45}Ti isotope formation in the fusion-evaporation channel is much higher because of the lower Coulomb barrier.

In this work, fusion-evaporation was considered using the numerical code of the NRV web knowledge base [4, 5]. The numerical solution of the time-dependent Schrödinger equation taking into account spin-orbit interaction [6] was successfully used to calculate the cross sections of neutron transfer [7] and proton transfer [8] in reactions with the participation of ^{3}He nuclei. In this work, the time-dependent quantum approach is used to describe proton stripping from the ^{3}He nuclei in collisions with the ^{194}Pt and ^{45}Sc nuclei combined with neutron pickup from the target nuclei.

2. Theory

The time-dependent Schrödinger equation for protons or neutrons [6–8]

$$i\hbar\frac{\partial}{\partial t}\Psi(\vec{r},t)=\left\{-\frac{\hbar^2}{2m}\Delta+W(\vec{r},t)+\hat{V}_{LS}(\vec{r},t)\right\}\Psi(\vec{r},t) \tag{1}$$

in combination with the classical equations of motion for nuclear centers of mass

$$m_1\ddot{\vec{r}}_1=-\nabla_{\vec{r}_1}V_{12}\left(\left|\vec{r}_1-\vec{r}_2\right|\right),\ m_2\ddot{\vec{r}}_2=-\nabla_{\vec{r}_2}V_{12}\left(\left|\vec{r}_2-\vec{r}_1\right|\right) \tag{2}$$

was used to describe transfer channels in collisions of atomic nuclei. Here, $\vec{r}_1(t)$, $\vec{r}_2(t)$ are the centers of nuclei with masses m_1, m_2, and $V_{12}(r)$ is the potential energy of interaction of these nuclei, m is a proton (neutron) mass, Ψ is a spinor wave function. Before touching of the surfaces of nuclei, the potential energy $W(\vec{r},t)$ of the proton (neutron) and the operator $\hat{V}_{LS}(\vec{r},t)$ of spin-orbit interaction took the form of the sum of the central potentials $V_i(r)$ and the operators $\hat{V}_{LS,i}(\vec{r},t)$ for a ^{3}He projectile nucleus ($i=1$) and for a heavy target nucleus ($i=2$). Residual interaction and pairing forces were not taken into account our calculations.

The initial condition for the proton wave function of the ^{3}He projectile nucleus was defined taking into account the long-range character of the Coulomb interaction with the target nucleus. The proton wave function in an isolated projectile nucleus at a finite distance from the target nucleus was preliminarily subjected to slow (adiabatic) switching of the Coulomb interaction with the target nucleus. Thus, the effects of the proton cloud polarization were taken into account in the initial condition.

3. Results for ^{3}He + ^{194}Pt Reaction

The initial condition for the proton wave function of the ^{3}He projectile nucleus was based on the shell model with the mean field potential $V_1(r)$ of the ^{3}He nucleus given in [7]. It differs from the Woods–Saxon potential by the maximum at zero providing better agreement with the experimental charge distribution [4]. The proton and neutron energy levels ε of the ^{3}He nucleus are shown in Figs. 1(a) and 1(d), respectively. The Woods–Saxon form $V_2(r)$ was used as the nuclear part of the potential of proton (neutron) interaction with the heavy target nucleus. The values of the parameters were chosen from the condition of equality of the theoretical and experimental values of the energy E_s of proton or neutron separation from the target [8].

After proton transfer to the ^{194}Pt nucleus, the $^{195}_{79}\mathrm{Au}_{116}$ nucleus is formed. The upper single-particle proton levels of the ^{195}Au nucleus are shown in Fig. 1(b). The neutron levels of the ^{194}Pt nucleus near the Fermi level are shown in Fig. 1(e).

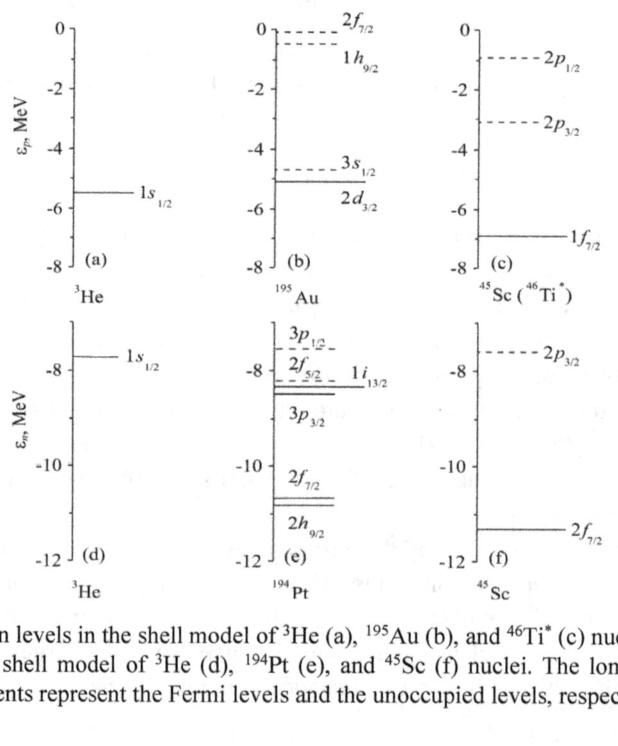

Fig. 1. Proton levels in the shell model of ^{3}He (a), ^{195}Au (b), and ^{46}Ti* (c) nuclei. Neutron levels in the shell model of ^{3}He (d), ^{194}Pt (e), and ^{45}Sc (f) nuclei. The longest and the dashed segments represent the Fermi levels and the unoccupied levels, respectively.

The typical pattern of the change in the proton probability density upon its stripping in the ^{3}He + ^{194}Pt collision is shown in Fig. 2. Here and below, the effective radius of the ^{3}He nucleus is $\bar{R}_1 = 2.2$ fm, which is somewhat larger than its root-mean-square charge radius $R_{1,\text{ch}} \approx 2$ fm [4], and the effective radius of the ^{194}Pt nucleus was determined according to the usual formula $\bar{R}_2 = 1.25A_2^{1/3} \approx 7.2$ fm.

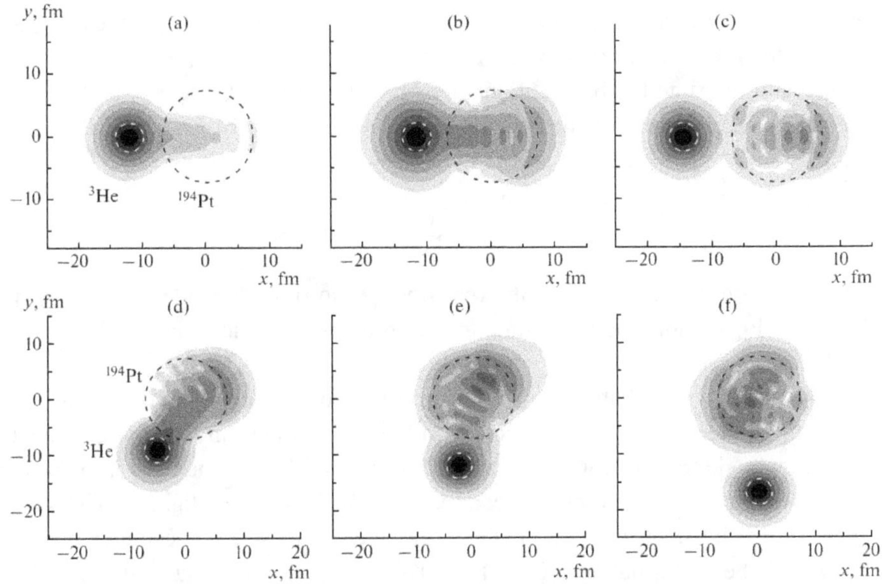

Fig. 2. Evolution of the probability density for protons of the ^{3}He projectile nucleus in the collisions with the ^{194}Pt target nucleus for different values of center-of-mass energy E and impact parameter b: (a)–(c) $E = 19$ MeV, $b = 0$; (d)–(f) $E = 25$ MeV, $b = 4.6$ fm. The radii of the circles are equal to the effective radii of the nuclei. The time goes from left to right: (a)–(b)–(c), (d)–(e)–(f).

For near-barrier energies [Figs. 2(a)–2(c)], the structure of the probability density in the target nucleus corresponds to the dominant stripping of the neutron to the state with close energy $3s_{1/2}$ [see Fig. 1(b)] of the produced ^{195}Au nucleus. Under the conditions of the slow relative motion of the colliding nuclei and the fast motion of the transferred proton, we observe an adiabatic regime [9] of probability density redistribution with the population of two-center states. In this case, only the high-lying states $2f_{7/2}$ and $1h_{9/2}$ can be populated, along with the state $3s_{1/2}$. The population of such states is higher in the grazing collisions at above-barrier energies [Figs. 2(d)–2(f)], as can be seen from the increased probability density extending beyond the boundaries of the nucleus. As nuclei fly away and the distance between them grows, excitation can be reduced by the emission of both protons and neutrons. This gives the first mechanism of ^{194}Au isotope formation. The probability can be estimated using the formula

$$p_1 = C\left(p_0 - p^{(-)}\right), \tag{3}$$

where $C \leq 1$ is a variable (adjustable) parameter, p_0 is the total probability for the proton from the ^{3}He nucleus to enter the region of the ^{194}Pt nucleus (not farther than the approximate radius of nuclear forces $R_{NN} \approx 3$ fm from its surface) in the time the projectile nucleus passes the target nucleus; and $p^{(-)}$ is the sum of populations of the initially free levels with energy $\varepsilon < 0$ in the ^{195}Au nucleus:

$$p(b,E) = \lim_{t\to\infty} \sum_k |a_k(t)|^2, \tag{4}$$

where a_k are the coefficients of expansion of the time-dependent proton wave function (Fig. 2) into the wave functions of the single-particle states of the target nucleus.

The second mechanism of ^{194}Au nucleus formation is a combination of neutron transfer from the ^{194}Pt nucleus to the ^{3}He nucleus and proton transfer in the opposite direction. During the transformation of the ^{3}He nucleus into ^{3}H, the mean field of the shell model can be assumed to vary slightly. The close energies of the neutron level in the projectile nucleus and the upper neutron levels of the ^{194}Pt nucleus [see Figs. 1(d), 1(e)] in this case increases the probability of transfer between them. The most probable is neutron pickup from level $3p_{3/2}$ due to lower centrifugal barrier. Although the probability of neutron pickup from the broad potential well of the target nucleus to the narrow well of the projectile nucleus is low, the large number of neutrons (16) on the levels near the Fermi level increases the probability of pickup of one of them, and thus makes the contribution of the second mechanism of ^{194}Au formation noticeable.

The probabilities of the above processes are shown in Fig. 3(a) as functions of minimum distance $R_{\min}(b,E)$ between the centers of the colliding nuclei, where b is the collision impact parameter, and E is the center-of-mass energy. The probability logarithms can be smoothed by linear dependence

$$\lg p_i(b,E) \approx A_i(E) - B_i(E) R_{\min}(b,E) \tag{5}$$

with parameters A and B depending weakly on energy. The cross section of proton transfer (or a combination of proton and neutron transfer) was calculated by integrating the corresponding probability over the impact parameters of grazing collisions $b > b_{min}$:

$$\sigma(E) = \int_{b_{min}}^{\infty} p(b,E)bdb. \tag{6}$$

The results of calculating the cross section of ^{194}Au isotope formation via proton stripping with neutron evaporation from the target nucleus and via proton stripping with neutron pickup are shown in Fig. 3(b) along with the negligible contribution of the fusion-evaporation mechanism.

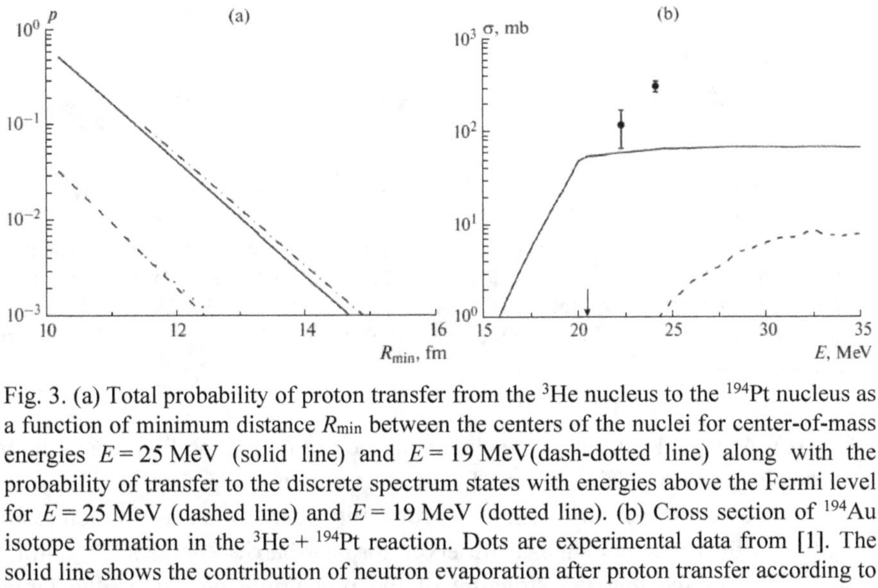

Fig. 3. (a) Total probability of proton transfer from the ^{3}He nucleus to the ^{194}Pt nucleus as a function of minimum distance R_{min} between the centers of the nuclei for center-of-mass energies $E = 25$ MeV (solid line) and $E = 19$ MeV(dash-dotted line) along with the probability of transfer to the discrete spectrum states with energies above the Fermi level for $E = 25$ MeV (dashed line) and $E = 19$ MeV (dotted line). (b) Cross section of ^{194}Au isotope formation in the ^{3}He + ^{194}Pt reaction. Dots are experimental data from [1]. The solid line shows the contribution of neutron evaporation after proton transfer according to formula (3) with $C = 1$; the dashed line shows the contribution of the fusion-evaporation channel calculated using the code of the NRV web knowledge base [4]. The arrow shows the position of the Coulomb barrier.

4. Results for ^{3}He + ^{45}Sc Reaction

The nucleus $^{45}_{21}\mathrm{Sc}_{24}$ has an unpaired proton on shell $1f_{7/2}$. The experimental angular distributions for inelastic scattering given in [3] show that after proton transfer, the ground state of the $^{46}_{22}\mathrm{Ti}_{24}$ nucleus with zero spin and two paired outer protons (on shell $1f_{7/2}$) is produced with rather low probability. Therefore, the Ti nucleus is formed in exited states, $^{46}_{22}\mathrm{Ti}^*_{24}$, with unpaired outer protons. In calculations, we used upper single-particle proton levels of the $^{46}_{22}\mathrm{Ti}^*_{24}$ nucleus which were similar to the levels of the $^{45}_{21}\mathrm{Sc}_{24}$ nucleus; they are shown in

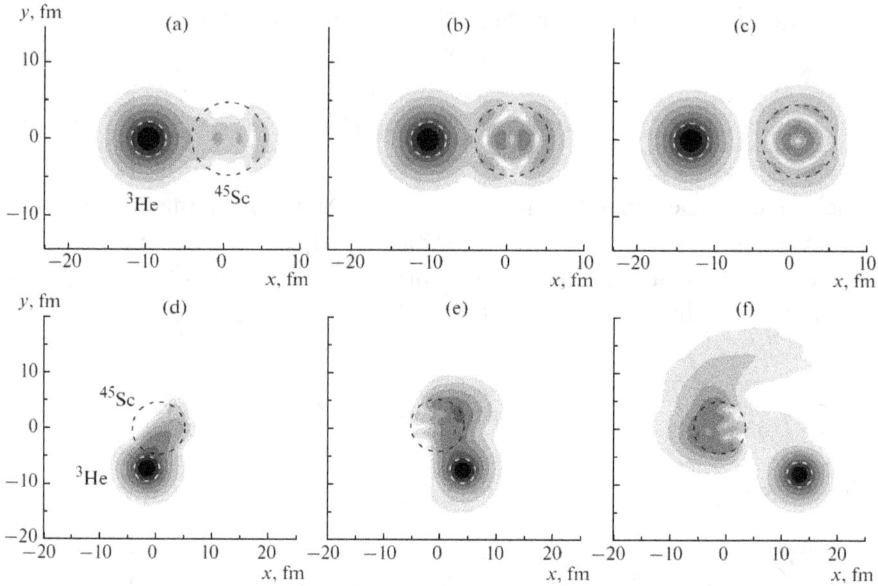

Fig. 4. Evolution of the probability density for protons of the ^{3}He projectile nucleus in the collisions with the ^{45}Sc target nucleus for different values of center-of-mass energy E and impact parameter b: (a)–(c) $E = 6$ MeV, $b = 0$; (d)–(f) $E = 35$ MeV, $b = 7$ fm. The radii of the circles are equal to the effective radii of the nuclei. The time goes from left to right: (a)–(b)–(c), (d)–(e)–(f).

Fig. 1(c). A typical pattern of the change in the proton probability density upon its stripping in the collision ^{3}He + ^{45}Sc nuclei is shown in Fig. 4.

An adiabatic scenario [9] of probability density redistribution with population of two-center states is observed for near-barrier energies [Figs. 4(a)–4(c)]. We can see that the structure of the probability density in the target nucleus corresponds to the dominant proton stripping to states $2p_{3/2}$, $2p_{1/2}$ of the formed ^{46}Ti nucleus. Similar behavior was observed in [7] for neutron stripping from the state $1s_{1/2}$ of the ^{3}He nucleus to the state $2p_{3/2}$ of the Sc nucleus. Since the energies of these levels [see Figs. 1(d), 1(f)] are close, the probability of neutron transfer is high. The probability of proton transfer is much lower because of the larger difference between the energies of the initial and final levels [see Figs. 1(a), 1(c)]. The probability of neutron pickup by the ^{3}He nucleus from the lower levels of the ^{45}Sc nucleus in combination with proton stripping is extremely low for the same reason. In the case of grazing collisions with above-barrier energies [Figs. 4(d)–4(f)], the rate of proton probability density redistribution is comparable to the relative velocity of the nuclei. This results in a nonadiabatic proton transfer regime [9] with the formation of highly excited states. As nuclei go away from each other, excitation can be reduced by the emission of both protons and neutrons. The probabilities of these processes are shown in Fig. 5(a) as functions of the minimum distance between the centers of the colliding nuclei. The results of calculating the cross section of ^{45}Ti isotope formation via proton stripping with neutron evaporation from the target nucleus and via proton stripping with neutron pickup are shown in Fig. 5(b) along with the contribution from the fusion-evaporation mechanism which is the main mechanism in this case.

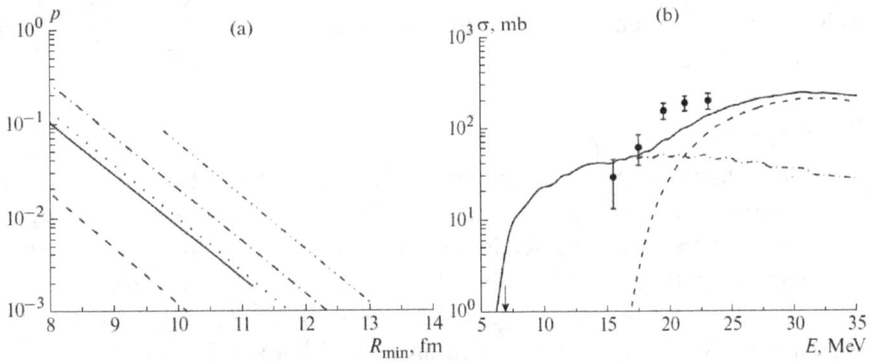

Fig. 5. (a) Total probability of proton transfer from the ^{3}He nucleus to the ^{45}Sc nucleus as a function of the minimum distance R_{min} between the centers of nuclei for center-of-mass energies $E = 35$ MeV (solid line), $E = 16$ MeV (dash-dotted line), and $E = 6$ MeV (dashed-dot-dotted line), along with the probability of transfer to the discrete spectrum states with energies above the Fermi level: $E = 35$ MeV (dashed line), $E = 16$ MeV (dotted line), and $E = 6$ MeV (coincides with the dashed-dot-dotted line). (b) Cross section of ^{45}Ti isotope formation in the ^{3}He + ^{45}Sc reaction. Dots are the experimental data from [1]. The dash-dotted line shows the estimated contribution of the neutron evaporation channel after proton transfer according to formula (3) with $C = 0.5$; the dashed line shows the contribution of the fusion-evaporation channel calculated using the code of the NRV web knowledge base [4]; the solid line shows the sum of contributions of both channels. The arrow shows the position of the Coulomb barrier.

5. Conclusions

The experimental difference in the near-barrier energy dependence of formation cross sections for the ^{194}Au isotope in the ^{3}He + ^{194}Pt reaction and the ^{45}Ti isotope in the ^{3}He + ^{45}Sc reaction was explained by the difference in the proton and neutron shells of the target nuclei determining different character of evolution of the probability density for the proton of the projectile nucleus and the neutron of the target nucleus during the collision. Study of proton and neutron transfer can provide information on the properties of nuclear states of predominantly single-particle nature. Experimental data on the formation cross sections for a number of isotopes can be used for validation and refinement of theoretical models of nucleon transfer, nuclear fusion, and pre-equilibrium processes. The time-dependent approach used in this work can also be applied to calculation of the proton transfer cross sections in reactions with nuclei having proton halos (e.g., ^{8}B) and the cross sections of charged cluster (deuterons, α-clusters) transfer.

Acknowledgment

This work was supported by the Russian Science Foundation, Grant No. 17-12-01170.

References

1. N. K. Skobelev, Yu. E. Penionzhkevich, E. I. Voskoboinik, V. Kroha, V. Burjan, Z. Hons, J. Mrázek, Š. Piskoř, E. Šimečkova and A. Kugler, Fusion and transfer cross sections of ^{3}He induced reaction on Pt and Au in energy range 10–24.5 MeV, *Phys. Part. Nucl. Lett.* **11**, 114 (2014).
2. N. K. Skobelev, A. A. Kulko, Yu. E. Penionzhkevich, E. I. Voskoboynik, V. Kroha, V. Burjan, Z. Hons, J. Mrázek, Š. Piskoř and E. Šimečkova, Cross sections of ^{43}Sc, ^{44}Sc, ^{46}Sc isotopes formed in the $^{45}Sc + {}^{3}He$ reaction, *Bull. Russ. Acad. Sci.: Phys.* **77**, 795 (2013).
3. R. W. Barnard and G. D. Jones, A study of ^{46}Ti by the $^{45}Sc(^{3}He, d)^{46}Ti$ and $^{46}Ti(p, p')^{46}Ti$ reactions, *Nucl. Phys. A* **111**, 17 (1968).
4. V. I. Zagrebaev, A. S. Denikin, A. V. Karpov, A. P. Alekseev, M. A. Naumenko, V. A. Rachkov, V. V. Samarin and V. V. Saiko, NRV web knowledge base on low-energy nuclear physics, http://nrv.jinr.ru/nrv/.
5. A. V. Karpov, A. S. Denikin, M. A. Naumenko, A. P. Alekseev, V. A. Rachkov, V. V. Samarin, V. V. Saiko and V. I. Zagrebaev, NRV web knowledge base on low-energy nuclear physics, *Nucl. Instrum. Methods Phys. Res. Sect. A* **859**, 112 (2017).
6. V. V. Samarin, Description of nucleon-transfer and fusion reactions within time-dependent approaches and coupled-channel method, *Phys. Atom. Nucl.* **78**, 128 (2015).
7. M. A. Naumenko, V. V. Samarin, Yu. E. Penionzhkevich and N. K. Skobelev, Near-barrier neutron transfer in reactions with ^{3}He nucleus, *Bull. Russ. Acad. Sci.: Phys.* **80**, 264 (2016).

8. V. V. Samarin, Yu. E. Penionzhkevich, M. A. Naumenko and N. K. Skobelev, Near-Barrier Proton Transfer in Reactions with ^{3}He Nucleus, *Bull. Russ. Acad. Sci.: Phys.* **82**, 637 (2018).
9. Yu. E. Penionzhkevich, Yu. G. Sobolev, V. V. Samarin and M. A. Naumenko, Peculiarities in total cross sections of reactions with weakly bound nuclei ^{6}He, ^{9}Li, *Phys. Atom. Nucl.* **80**, 928 (2017).

The Low-Lying States of ^{9}Be from ^{9}Be+α Scattering at 30 and 90 MeV

V. I. Starastsin*, A. S. Demyanova, A. A. Ogloblin, A. N. Danilov and S. V. Dmitriev

Russian Research Center Kurchatov Institute, Moscow, 123182, Russia
**E-mail: starastsinvi@yandex.ru*
www.nrcki.ru

The differential cross-sections of the ^{9}Be(α,α)^{9}Be* scattering (E_{lab}(α) = 30 and 90 MeV) with excitation of ^{9}Be states 2.43 MeV, 2.78 MeV, 5.59 MeV and 7.94 MeV were obtained. The question of the formation of the third rotational band based on the 2.78 MeV state is considered. The new excited state 3.82 MeV in the ^{9}Be with an assumed spin-parity Jπ = 3/2- were observed.

Keywords: Differential cross-sections; excited states of ^{9}Be; rotational band.

1. Introduction

In this paper the differential cross sections for elastic and inelastic scattering 30 [1] and 90 MeV α-particles on ^{9}Be for excited states at 2.43 MeV, 2.78 MeV, 5.59 MeV and 7.94 MeV are analyzed. The purpose of these studies is to determine spin-parities of states which form the assumed third rotational band of the ^{9}Be. The third rotational band based on 2.78 MeV state and also should include 5.59 MeV and 7.94 MeV states [1].

In the levels scheme [2] for excited states 2.78 MeV, 4.7 MeV, 5.59 MeV and 7.94 MeV in ^{9}Be the following spin-parity values $1/2^-$, $3/2^+$, $3/2^-$, and $5/2^-$ have been given correspondingly.

However, in article [3] new spin-parity values for the above-mentioned excited states were determined: $J^\pi = 3/2^-$, $1/2^-$, $5/2^-$ and $3/2^-$ respectively. In our article, we confirmed the preservation of the spin-parity value of the state 4.7 MeV [1, 4].

However, the question of the existence of the third rotational band remains open. Nevertheless, we are analyzing experimental data, obtained in experiments on α+^{9}Be scattering induced by 30- and 90-MeV α-particles in order to determine the spin-parity values of existed states that form the assumed third rotational band in the 9Be. The analysis is based on MDM method [5] and by comparing the behavior of the obtained differential cross sections with differential cross sections with differential cross sections at well studied states.

2. Experiment and Data Analysis

2.1. *Experiment*

The measurements of α+^{9}Be scattering induced by 90-MeV α-particles have been done on the K 130 cyclotron of Jyvaskyla University, Finland using the Large Scattering Chamber with several sets of Δ*E*-*E* telescopes.

In Fig. 1 we can see typical spectrum from the α+^{9}Be scattering with $E_{lab}(\alpha)$ = 90 MeV.

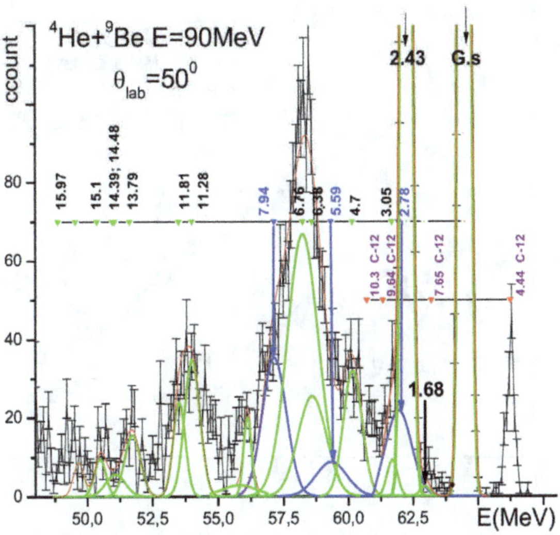

Fig. 1. Typical spectrum from the ^{9}Be+α scattering induced by 90-MeV α-particles. Blue peaks denote states, that form the assumed third rotational band of the ^{9}Be. Green peaks are the rest peaks in ^{9}Be. Red triangles denote impurity states of ^{12}C.

2.2. *Data analysis*

2.2.1. 2.78 MeV and 7.94 MeV states in ^{9}Be

Analysis, performed for 2.78 state in [4], showed that the spin-parity value for this existence state remains the same ($1/2^{-}$), as in [2].

As we can see on the Fig. 2, the behaviour of the differential cross section at 7.94 MeV coincide with the curve, which corresponding to momentum transfer L = 2. That means that the value of spin-parity should be $5/2^{-}$, i.e. remains generally accepted as shown in [2].

The additional factor in favor of keeping the generally accepted spin-parity value at 7.94 MeV is to compare the behaviour of differential cross section at 7.94 MeV with behaviour at well studied 2.43 MeV state (momentum transfer $L = 2$). This comparison shows that the position of extrema in diffraction part coincides.

Moreover, in article 6 the question of spin-parity values at 7.94 MeV has been raised. The authors have not confirmed the change of spin-parity value of this state.

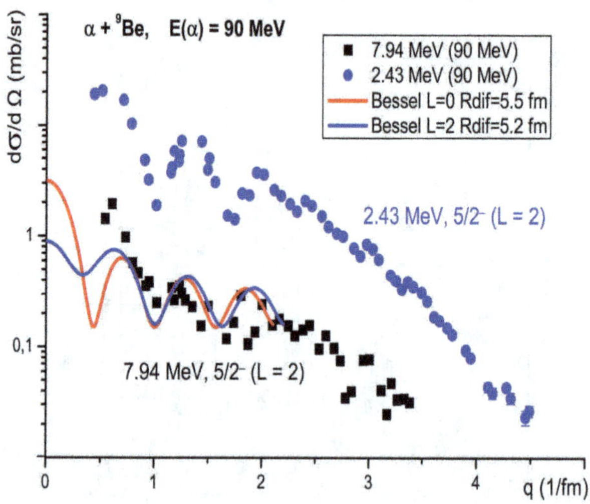

Fig. 2. Differential cross sections for inelastic scattering induced by 90-MeV α-particles with excitation of 7.94 MeV (black dots) and 2.43 MeV (blue dots) states in ^{9}Be are shown as a function of momentum transfer q. The curves correspond to the diffraction mechanism, described by the squared Bessel function with $L = 0$ (red curve) and $L = 2$ (blue curve).

2.2.2. 5.59 MeV state in ^{9}Be

As for 5.59 MeV state, we have not got definite answer. The differential cross section for the 5.59 MeV state is compared with squared Bessel functions with $L = 0$ and $L = 2$ on Fig. 3.

From this comparison we could not determine which linear momentum transfer corresponds to this state. In case when L equals 0, the spin-parity value remains generally accepted as $3/2^-$ [2]. However, if L equals 2, the spin-parity value will equal $5/2^-$, as predicted in [3].

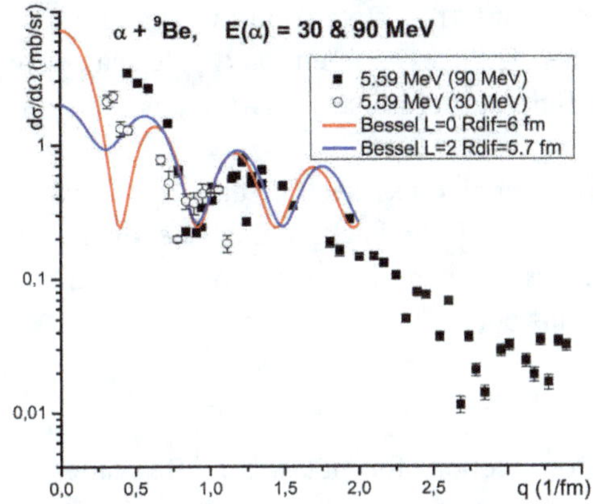

Fig. 3. Differential cross sections for inelastic scattering induced by 30-MeV (open dots) and 90-MeV (black dots) α-particles at 5.59 MeV excited state in ^{9}Be. The curves correspond to the diffraction mechanism at the $L = 0$ (red solid curve) and $L = 2$ (blue solid curve).

2.2.3. 3.82 MeV state in ^{9}Be

In article of R. Smith [7] the question of existent of a new state in ^{9}Be nuclei at 3.82 MeV ($3/2^-$) exited state has been raised. The value of spin-parity allows us to suppose, that 3.82 MeV exited state may be the second member of assumed third rotational band.

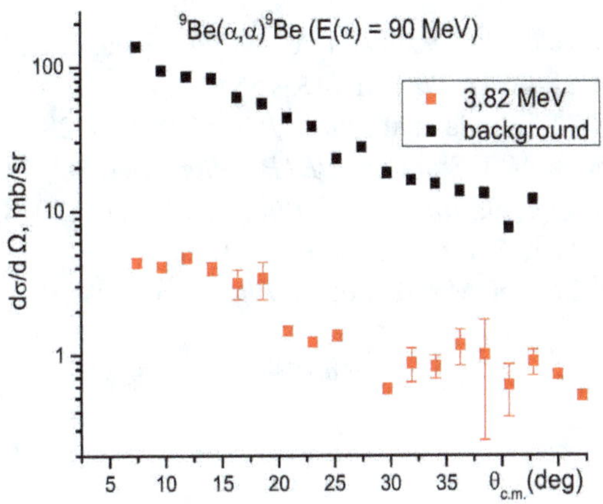

Fig. 4. Differential cross sections for inelastic α+9Be scattering induced by 30-MeV α-particles at 3.82 MeV state in ^{9}Be (red dots). The black dots correspond to the angular distribution of the background.

We could not identify this state from the experimental data at 90 MeV. As for the data, obtained in the experiment on α+^{9}Be scattering induced by 30-MeV α-particles, the differential cross sections for this state have been obtained and shown on Fig. 4.

However, differential cross section at 3.82 MeV state was close to the differential cross section of the background, as shown on Fig. 4. In this case we cannot claim that this state can be seen. However, the question of existence of this state still remains open.

3. Conclusion

The differential cross sections for inelastic α+^{9}Be scattering induced by 90-MeV α-particles were obtained for the following states in ^{9}Be: 2.78 MeV, 5.59 MeV and 7.94 MeV. The question of existence of the third rotational band, based on the 2.78 MeV, $1/2^-$ state, remains open. It has been shown that for the 5.59 MeV state the spin-parity value could be changed from $3/2^-$ to $5/2^-$. The assumption was made that in case of changing spin-parity value at 5.59 MeV, the second member of the third rotational band may be the 3.82 MeV state with assumed spin-parity value $3/2^-$.

Acknowledgements

This work was partly supported by grant of RFBR 18-52-53025.

References

1. A. S. Demyanova et al., *EPJ Web of Conferences* **66**, 02026 (2014).
2. D. R. Tilley et. al., *Nucl. Phys. A* **745**, 155 (2004).
3. Y. Hirayama, T. Shimoda et al., *Phys. Rev. C* **91**, 024328 (2015).
4. A. S. Demyanova, V. I. Starastsin *JETP Letters* **104**, 289 (2016).
5. A. S. Demyanova et al., *Int. J. Mod. Phys. E* **17**, 2118 (2008).
6. J. Casal et al., *Phys. Rev. C* **92**, 054611 (2015).
7. R. Smith, C. Wheldon, M. Freer et al., *Phys. Rev. C* **94**, 014320 (2016).

Studies of ^{22}C and ^{50,52}Ca at SAMURAI

Y. Togano[1,2,3,*] for SAMURAI DayOne and SAMURAI09 Collaborations
[1]*Department of Physics, Rikkyo University, Tokyo 171-8501, Japan*
[2]*Department of Physics, Tokyo Institute of Technology, Tokyo 152-8551, Japan*
[3]*RIKEN Nishina Center, Saitama 351-0198, Japan*
[*]*E-mail: togano@rikkyo.ac.jp*

The matter radius of the two-neutron halo nucleus ^{22}C and the electric dipole response of neutron-rich calcium isotopes ^{50,52}Ca have been studied with the SAMURAI spectrometer at RIKEN RIBF. For ^{22}C, the present experimental matter radius is compared with the recent theoretical estimates. For ^{50,52}Ca, the brief summary of the experiment and the preliminary result is discussed.

Keywords: Two neutron halo nuclei; electric dipole response; unstable nuclei.

1. Matter Radius of Two-Neutron Halo Nucleus ^{22}C

Recently, the most neutron-rich carbon isotope ^{22}C (two-neutron separation energy $S_{2n} = -0.14 \pm 0.46$ MeV [1] has paid much attention, due to a possible largely extended two-neutron halo structure suggested by a reaction cross section ($\sigma_R = 1.338 \pm 0.274$ b) measured on a proton target at 40 MeV/nucleon [2]. A root-mean-squared (rms) radius r_m of 5.4 ± 0.9 fm was deduced with relatively large uncertainty.

The large uncertainties on the earlier measurement [2] of σ_R and the deduced r_m do not significantly constrain the theoretical models. A mean-field model using an adjusted Skyrme interaction yielded r_m of 3.89 fm [3]. Three-body models yielded r_m in the range of 3.5 ~ 3.7 fm [4, 5]. Given the large uncertainty in the experimental values in Ref. [2] and that the theoretical values are within $\sim 2\sigma$ of the experimental value, more definitive conclusions require data with higher precision.

With this motivation, the interaction cross section σ_I of ^{22}C has been measured with a carbon target at 235 MeV/nucleon at RIKEN RIBF. σ_I is the cross section for a change of nuclide in the collision, while $\sigma_R = \sigma_I + \sigma_{inel}$ includes any inelastic cross sections to the bound excited states. Since σ_{inel} is small at the present experimental condition [6], σ_I can be approximated to σ_R.

A cocktail beam of ^{19,20,22}C was produced via the projectile fragmentation of a 345 MeV/nucleon ^{48}Ca beam. The beams ware separated and transported by BigRIPS [7] and impinged on a carbon target with a thickness of 1.789 g/cm^2. The incident particles were identified event-by-event by using the detectors located at the upstream of the carbon target. The reaction products from the $^{22}C + C$ reaction were identified using the SAMURAI spectrometer [8]. The multi-wire drift chambers at the entrance (FDC1) and exit (FDC2) of the SAMURAI magnet enabled one to reconstruct the momentum of the charged reaction products. The plastic scintillator wall located at the downstream of FDC2 measured the time of flight and energy loss of the fragments. The time-over-threshold from FDC1 were also used to measure the energy loss. The particle identification of the reaction products were performed by combining the measured momentum, time of flight, and energy loss. The details of the experimental setup can be found in Ref. [6].

The interaction cross section σ_I can be written as,

$$\sigma_I = -\frac{1}{N_t}\log\left(\frac{\Gamma}{\Gamma_0}\right), \tag{1}$$

where N_t is the number of target nuclei per unit area and Γ is the ratio of the number of non-interacting outgoing particles to the number of incoming particles. Γ_0 corresponds to Γ for an empty target, to take into account the loss of particles in the detectors.

The obtained σ_I of the $^{22}C + C$ reaction is $1.280 \pm 0.022 \pm 0.007$ b, where the first and second errors correspond the statistical and systematic ones, respectively. The present σ_I value is analyzed using a four-body (three-body projectile plus target) Glauber reaction mode [14]. In the following discussion σ_I and σ_R are compared directly by assuming σ_{inel} is negligible. The details of the interaction between each constitutions are described in Ref. [4]. The matter density of the ^{20}C core is obtained from the Hartree-Fock calculation with the SkX interaction [9], by assuming to have a filled $1d_{5/2}$ sub-shell bound by S_n = 2.93 MeV. From the comparison with the model calculations and the present σ_I value, r_m of ^{22}C is obtained to be 3.44 ± 0.08 fm. The present r_m is smaller than the previously reported value (5.4 ± 0.9 fm [2]), while they are consistent by 2σ level.

Figure 1 shows the mass number dependence of r_m for the carbon isotopes from the present study (Filled circles). The open circles corresponds to the data obtained with higher beam energy [10], and the dashed line connects the theoretical prediction by Abu-Ibrahim *et al.* [11]. The range of theoretical predictions by Horiuchi [5], Shulgina [12], Suzuki [13], and Sun [14] are shown respectively blue, orange, red and green bands. The present r_m of $^{19,20}C$ are consistent with the previous measurement. The theoretical predictions are in agreement with the present r_m of ^{22}C, while each theories gives different S_{2n} of ^{22}C. The range of S_{2n} predicted by Horiuchi *et al.* [5] is $0.39 \div 0.57$ MeV with a dominant *s*-wave halo neutrons. Shulgina's calculation [12] can reproduce the present r_m with $S_{2n} = 0.07$ and 0.4 MeV by changing the contribution of *s*- and *d*-wave configuration of halo neutrons. The three-body model by Suzuki *et al.* [13] takes the polarization of ^{20}C core into account, and can reproduce the present result at $S_{2n} < 0.3$ MeV. The Hartree-Fock Bogoliubov calculation by Sun *et al.* [14] predicts ^{22}C as a deformed halo nucleus, and r_m to be 3.25 fm with $S_{2n} = 0.43$ MeV.

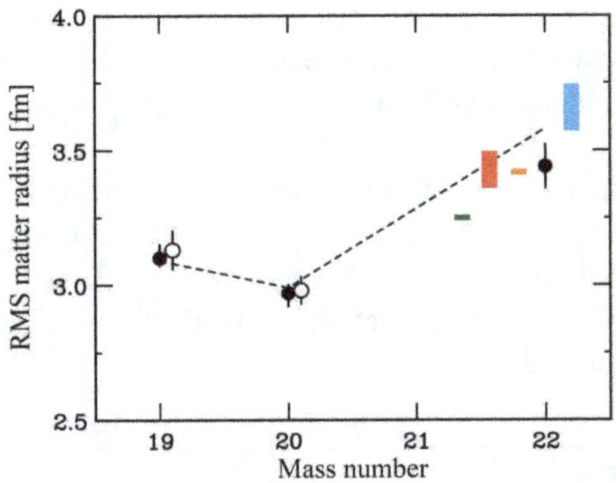

Fig. 1. Mass number dependence of the root-mean-square radii of carbon isotopes obtained in the present study together with the previous studies. Colored bands correspond to the predictions for ^{22}C by several theoretical calculations.

While the calculated r_m values are similar, the structure of ^{22}C and the separation energy S_{2n} are not settled. To solve this situation, the more exclusive reaction experiments, such as Coulomb breakup, and more precise mass measurements are required.

2. Electric Dipole Responses of $^{50,52}Ca$

The equation of state (EOS) of neutron-rich matter is important understand the properties of neutron-rich nuclei and astrophysical objects such as neutron star mergers. The neutron skin thickness of neutron-rich nuclei is related to the symmetry energy, the isospin-asymmetric part of EOS, close to the saturation density [15], while the precise measurement of skin thickness of neutron-rich nuclei is difficult.

Recent theoretical work showed that the low-lying dipole strength, sometimes called as the pygmy dipole resonance, and the dipole polarizability α_D is correlated to the symmetry energy and its density dependence. The low-lying dipole strength locates at the excitation energies of 6 to 10 MeV, and the strength of that in neutron-rich Calcium isotopes ^{52,54}Ca are indicated to be well correlated to the neutron skin thickness and the symmetry energy [16]. In addition, the sudden increase of the low-energy dipole strengths at $N = 28\sim 34$ in Ca isotopes is predicted [17].

The polarizability α_D is the inversely energy weighted sum of $E1$ strength distribution, and known to be correlated to the skin thickness and the symmetry energy with less model dependence [18]. Given that the low-energy dipole strength in ^{52}Ca and α_D are correlated to the symmetry energy and the sudden increase of the low-energy dipole strength at $N = 28 \sim 34$ regions, we have performed the Coulomb excitation of ^{50}Ca and ^{52}Ca to measure their electric dipole responses.

The experiment was performed using the SAMURAI spectrometer at RIKEN RIBF. The secondary beams of ^{50}Ca and ^{52}Ca were produced by the projectile-fragmentation reaction of a 345 MeV/nucleon ^{70}Zn beam on a Be target. The beams were separated by using the BigRIPS [7] and impinged on Pb and C secondary targets.

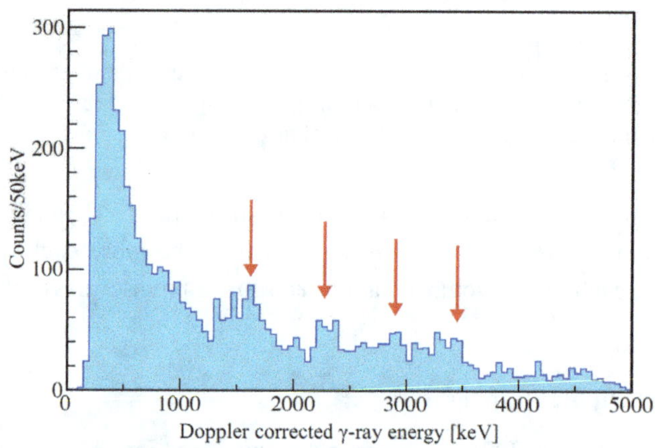

Fig. 2. Doppler corrected γ-ray energy spectrum of ^{51}Ca produced in the ^{52}Ca + Pb → ^{51}Ca + n reaction. Red arrows show the peaks corresponding to the known excited states.

The particle identification of secondary beams were performed by using the information taken by plastic scintillators and ionization chambers located at the focal planes of BigRIPS. The hit position and incident angles on the secondary target were measured by two drift chambers. The newly constructed γ-ray detector CATANA [19] and 8 large-volume $LaBr_3$ detectors were placed around the target to measure the de-excitation γ-rays from the reaction residues.

The momentum of outgoing charged particles were reconstructed using the hit positions and angles at drift chambers located at the entrance and exit of the SAMURAI magnet. The time of flight and the energy loss of the outgoing particles were measured by the plastic scintillator wall at the downstream of the drift chamber located at the exit of the magnet. The outgoing neutrons were detected by the combination of the NeuLAND demonstrator [20] and NEBULA [21].

Figure 2 shows the preliminary Doppler corrected γ-ray energy spectrum from ^{51}Ca produced in the ^{52}Ca + Pb → ^{51}Ca + n reaction. The several known excited states were observed. The analysis on the neutron decay channel is ongoing to obtain the electric dipole strength distributions.

References

1. L. Graudefroy *et al.*, Direct mass measurements of ^{19}B, ^{22}C, ^{29}F, ^{31}Ne, ^{34}Na and other light exotic nuclei, *Phys. Rev. Lett.* **109**, p. 202503 (2012).
2. K. Tanaka *et al.*, Observation of large reaction cross section in the drip-line nucleus ^{22}C, *Phys. Rev. Lett.* **104**, p. 062701 (2010).
3. T. Inakura, W. Horiuchi, Y. Suzuki and T. Nakatsukasa, Mean-field analysis of ground-state and low-lying electric dipole strength in ^{22}C, *Phys. Rev. C* **89**, 064316 (2014).
4. Y. Kucuk and J. A. Tostevin, Intermediate-energy four-body breakup calculations for ^{22}C, *Phys. Rev. C* **89**, 034607 (2014).
5. W. Horiuchi and Y. Suzuki, an *s*-wave two-neutron halo nucleus, *Phys. Rev. C* **74**, 034311 (2006).
6. Y. Togano *et al.*, Interaction cross section study of the two-neutron halo nucleus ^{22}C, *Phys. Lett. B* **761**, 412 (2016).
7. T. Kubo, In-ight RI beam separator BigRIPS at RIKEN and elsewhere in Japan, *Nucl. Instrum. Methods Phys. Res. Sect. B* **204**, 97 (2003).
8. T. Kobayashi *et al.*, SAMURAI spectrometer for RI beam experiments, *Nucl. Instrum. Methods Phys. Res. Sect. B* **317**, 294 (2013).
9. B. A. Brown, New skyrme interaction for normal and exotic nuclei, *Phys. Rev. C* **58**, 220 (1998).
10. A. Ozawa, T. Kobayashi, T. Suzuki, K. Yoshida and I. Tanihata, New magic number, $N = 16$, near the neutron drip line, *Phys. Rev. Lett.* **84**, 5493 (2000).
11. B. Abu-Ibrahim *et al.*, Reaction cross sections of carbon isotopes incident on a proton, *Phys. Rev. C* **77**, 034607 (2008).
12. N. B. Shulgina, S. N. Ershov, J. S. Vaagen and M. V. Zhukov, Superhalo of ^{22}C reexamined, *Phys. Rev. C* **97**, 064307 (2018).

13. T. Suzuki, T. Otsuka, C. Yuan and N. Alahari, Two-neutron halo from the low-energy limit of neutron-neutron interaction: Application to drip-line nuclei ^{22}C and ^{24}O, *Phys. Lett. B* **753**, 199 (2016).
14. X.-X. Sun, J. Zhao and S.-G. Zhou, Shrunk halo and quenched shell gap at N = 16 in ^{22}C: inversion of s_d states and deformation effects, *Phys. Lett. B* **785**, 530 (2018).
15. F. J. Fattoyev and J. Piekarewicz, Neutron skins and neutron stars, *Phys. Rev. C* **86**, 015802 (2012).
16. T. Inakura, T. Nakatsukasa and K. Yabana, Low-energy $E1$ strength in selected nuclei: possible constraints on neutron skin and symmetry energy, *Phys. Rev. C* **88**, 051305(R) (2013).
17. T. Inakura, T. Nakatsukasa and K. Yabana, Emergence of pygmy dipole resonances: Magic numbers and neutron skins, *Phys. Rev. C* **84**, 021302 (2011).
18. P.-G. Reinhard and W. Nazarewicz, Information content of a new observable: The case of the nuclear neutron skin, *Phys. Rev. C* **81**, 051303(R) (2010).
19. Y. Togano *et al.*, Test of prototype crystals of the γ-ray detector CATANA, *RIKEN Accel. Prog. Rep.* **48**, 207 (2015).
20. NeuLAND Technical Design Report, *https://edms.cern.ch/document/1865739*.
21. T. Nakamura and Y. Kondo, Large acceptance spectrometers for invariant mass spectroscopy of exotic nuclei and future developments, *Nucl. Instrum. Methods Phys. Res. Sect B* **376**, 156 (2016).

A New Explanation of Heavy Ions Sub-Barrier Fusion Enhancement Phenomenon

R. Wolski

Flerov Laboratory of Nuclear Reactions, JINR, Dubna, Russia
Henryk Niewodniczański Institute of Nuclear Physics PAS, Cracow, Poland

A global systematics for experimental sub-barrier fusion data is discussed. The approach is based on the observation of a strong correlation between fusion cross sections at sub-barrier energies and the Q values for fusing nuclei. Selected samples of the fusion data were analysed in terms of the proposed rule, and an anomaly was found in the fusion of alpha particles and protons.

Keywords: Sub-barrier fusion, systematics, fusion Q-value.

1. Introduction

Fusion of two atomic nuclei occurs only when the interacting bodies can somehow overcome the barrier formed by the sum of the attractive nuclear potential and the repulsive Coulomb and centrifugal potentials. If the kinetic energy of the centre of mass (c.m.) is below the height of this barrier, fusion is classically forbidden. It is believed, however, that a merging of two heavy ions can happen as a single, instantaneous act of barrier quantum tunnelling, a concept that is borrowed from α decay theory. In α decay, a small variation in the barrier height can lead to a large change in the lifetime. The tunnelling probability therefore strongly depends on the barrier parameters. It has been often observed that the fusion cross sections (c.s.) differ dramatically for the nuclei of isotopes of the same element and with almost the same barrier. This effect is known as sub-barrier fusion enhancement, and has hitherto been explained as being caused by the intrinsic properties of individual nuclei, their susceptibility to collective excitations and nucleon or cluster transfers (see e.g. [1]). The excitation of these degrees of freedom by direct reactions can modify the barrier through a coupling mechanism of the reaction channels. Due to this coupling, the fusion can be either enhanced or suppressed, depending on the specific case. Obviously, enhancement or suppression of fusion assumes a similitude to reference values. These values have to be provided either by predictions obtained from a simple model with no nuclear structure effects included or from fusion c.s. measurements for a reference system.

2. *Q*-Value Rule for Sub-Barrier Fusion

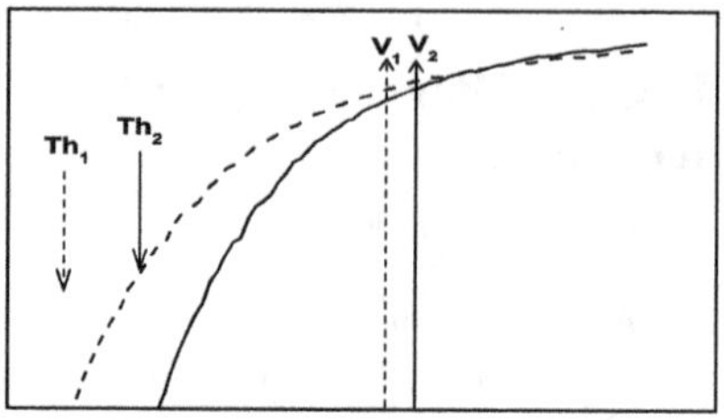

Fig. 1. Fusion c.s. vs. kinetic c.m. energy for two fictitious fusing systems of similar barrier heights, V_1 and V_2, but different reaction thresholds, Th_1 and Th_2. Unlike the above barrier fusion, at low enough energies the fusion c.s. is always larger for the system with the lower threshold (see text).

In the physical approach presented above, the Q value is ignored, which is the equivalent of a tacit assumption of $Q = 0$. However, if we inspect Fig. 1, it becomes clear that the energy balance of sub-barrier fusion is of fundamental importance. This fact was first pointed out in [2], and was discussed further in [3]. Two fictitious fusing systems are shown in Fig.1, with almost the same barrier heights, V_1 and V_2, but different reaction thresholds, Th_1 and Th_2, ($Th = -Q$). At low enough kinetic energy, the fusion c.s. will be always higher for the system with the lower threshold. Moreover, neglecting the non-zero threshold leads to an unphysical low energy asymptotic for the fusion excitation function. It is obvious that at positive energies that are below the respective thresholds, the fusion c.s. must be equal to zero, otherwise the energy is not conserved. This aspect is not reproducible in any fusion model based exclusively on barrier penetration effects. Most experimental fusion data confirm the correlation between the Q value for fusion and the fusion c.s. for a corresponding system [2,3]. A higher fusion Q value (lower threshold) for a given fusion system offers greater opportunity for the system to fuse at energies far below the barrier. In order to provide a qualitative criterion for comparing various data, we introduce a definition and make an assumption. Firstly, the sub-barrier fusion energy region is defined as the difference in energy between the threshold and the barrier height. Secondly, it is assumed that the shapes of the different fusion excitation functions can be reduced to a single quasi-universal curve if presented vs. normalised sub-barrier energy. The simplest way to examine this assumption may be a substitution of the c.m. energy (E) by the reduced energy parameter (scaling factor) E_r, which could be defined for example as:

$$E_r = (E - \mathrm{Th})/(\mathrm{V} - \mathrm{Th}) \tag{1}$$

If the energy fusion threshold is ignored, the variable E_r is reduced to the ratio of the energy to the barrier height, a quantity that is frequently used in discussions of fusion. Of the many model prescriptions for the potential barrier height V, the simplest and most widely known formula is used here:

$$V = Z_1Z_2/r_0(A_1^{(1/3)} + A_2^{(1/3)}) \tag{2}$$

where the reduced radius r_0 is in units of 1.44fm.

For given sets of experimental fusion data, the value of r_0 is adjusted to give minimal divergence between these data points sets. This treatment of the barrier height V is compared with those provided by nuclear models in [4]. To our knowledge, no one-step, two-body barrier penetration model has used the concept of the fusion energy threshold as a limiting factor for sub-barrier fusion to occur and as a possible origin of sub-barrier fusion enhancement.

A classical example [5] of sub-barrier enhancement is shown in Fig. 2. The left-hand side of Fig. 2 shows fusion c.s. for $^{16}O + ^{144,148,154}Sm$ systems, complemented by a $^{17}O + ^{144}Sm$ system. These data and others presented later are taken from the compilation in [6], unless specified otherwise.

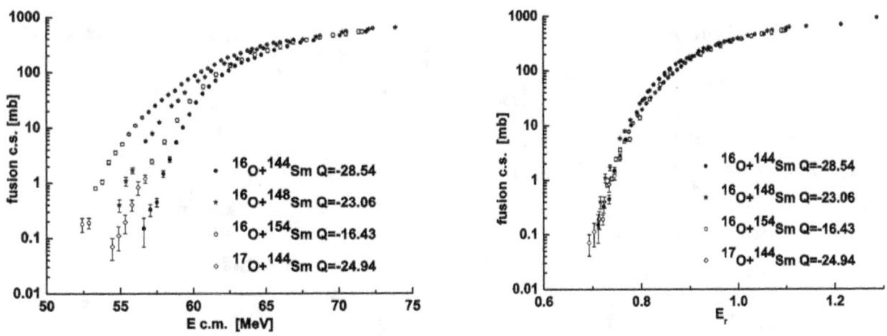

Fig. 2. Left: fusion excitation functions for $^{16}O + ^{144,148,154}Sm$ and $^{17}O + ^{144}Sm$ systems [6]; right: the same data vs. the reduced energy parameter E_r for $r_0 = 0.94$ (see text).

The $^{16}O + {}^{A}Sm$ data are treated as an evidence of structural effects in fusion below the barrier. The fusion of a ^{16}O projectile with a deformed ^{154}Sm target is enhanced by more than two orders of magnitude with respect to that for the spherical ^{144}Sm target, whereas for ^{148}Sm nuclei the situation is an intermediate one. However, it can be seen from Fig. 2, which shows the fusion Q values for these processes, that there is a strong correlation between the fusion c.s. and the fusion Q value. The lowest c.s. corresponds to the lowest Q value, and the highest fusion probably corresponds to the system with the highest Q value. Therefore, a reasonable question arises with regard to the physical origin of the observed enhancement. The right-hand side of Fig. 2 shows these four fusion data sets vs. the reduced energy variable E_r, where the value of the barrier parameter r_0 is 0.94 for all systems. It can be seen that the enhancement is greatly reduced or even removed entirely, since the experimental systematic inaccuracies are not taken into account.

A straightforward implication of the proposed scaling is its predictive power. The sub-barrier fusion c.s. for a pair of nuclei can easily be predicted if the c.s. data measured for a neighbouring system are known. A single adjustable parameter in the data treatment discussed here is the barrier height V or the radius parameter r_0 (see Eq. (2)). To allow sets of data to be compared, a value of r_0 should be constant or weakly varying in a way that can be predicted. It would be unreasonable to expect that a constant value for this parameter could properly describe a large body of fusion data. The constant value of 1.0 for r_0 was applied to the energy-reduced fusion data shown in Fig. 3. The colliding pairs selected for this graph had various total masses and Z_1Z_2 products but rather low mass asymmetry. The most asymmetrical case in terms of mass presented in Fig. 3 is the fusion between $^{40}Ca + {}^{96}Zr$.

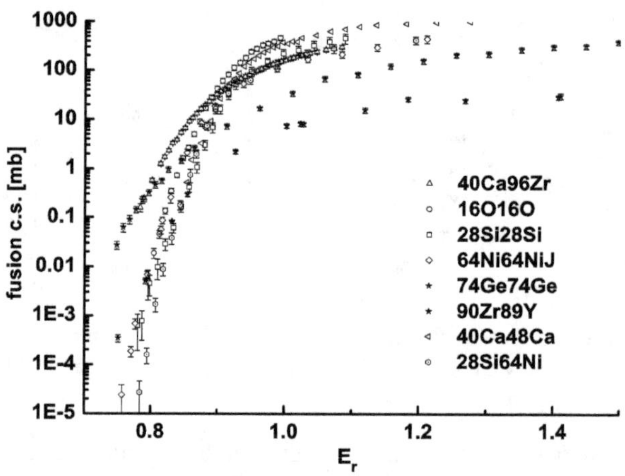

Fig. 3. Energy-reduced fusion excitation curves for systems with different Z_1Z_2. Outlined symbols are used for $^{40}Ca+{}^{96}Zr$, $^{16}O+{}^{16}O$, $^{28}Si+{}^{28}Si$, $^{64}Ni+{}^{64}Ni$, $^{40}Ca+{}^{48}Ca$, $^{28}Si+{}^{64}Si$, and solid symbols for $^{74}Ge+{}^{74}Ge$, $^{90}Zr+{}^{89}Y$. The radius parameter r_0=1.0 for all systems.

As it can see from Fig. 3, the systems with low Z_1Z_2 products, from $^{16}O + {}^{16}O$ to $^{64}Ni + {}^{64}Ni$, seem to obey our fusion Q value rule with a common value of r_0. In the heavier systems, the fusion excitation functions have different shapes. It should be noted that only the evaporation residue yields were measured for these systems, meaning that it is possible that fusion-fission contributions are absent from the relevant data for the heavier systems shown in Fig. 3.

A commonly accepted opinion is that the coupling of relative motion to collective degrees of freedom generates an enhancement of sub-barrier fusion. It has also been asserted that experimental data have revealed an additional enhancement of the sub-barrier fusion c.s. caused by neutron transfer with positive Q values (see [7] and the discussion therein). The authors of [7] analysed $^{32}S + ^{90,94,96}Zr$ and $^{40}Ca + ^{90,94,96}Zr$ fusion data using the framework of their theoretical approach, which included neutron transfer effects, and stated that the fusion of each of these projectiles with ^{96}Zr nuclei was enhanced with respect to the fusion with other Zr isotope nuclei due to neutron rearrangement. According to these authors, an even larger impact of neutron transfer on fusion is expected for ^{6}He and ^{18}O projectiles.

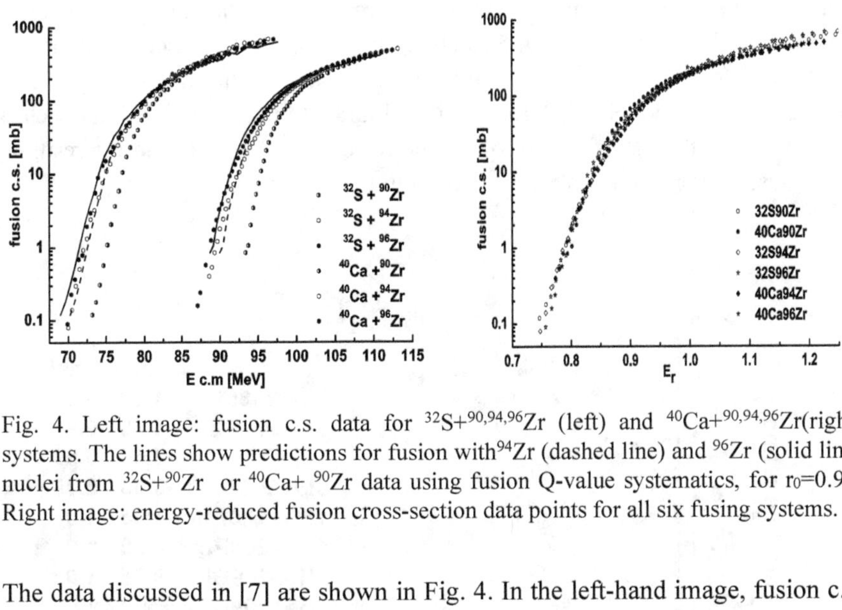

Fig. 4. Left image: fusion c.s. data for $^{32}S+^{90,94,96}Zr$ (left) and $^{40}Ca+^{90,94,96}Zr$(right) systems. The lines show predictions for fusion with ^{94}Zr (dashed line) and ^{96}Zr (solid line) nuclei from $^{32}S+^{90}Zr$ or $^{40}Ca+ ^{90}Zr$ data using fusion Q-value systematics, for $r_0=0.99$. Right image: energy-reduced fusion cross-section data points for all six fusing systems.

The data discussed in [7] are shown in Fig. 4. In the left-hand image, fusion c.s. data points for $^{32}S + ^{90,94,96}Zr$ and $^{40}Ca + ^{90,94,96}Zr$ systems are presented. The lines correspond to predictions obtained using our systematic for the fusion of ^{32}S or ^{40}Ca with ^{94}Zr (dashed line) and ^{96}Zr (solid line) nuclei. With this aim, the $^{32}S + ^{90}Zr$ or $^{40}Ca + ^{90}Zr$ data were taken as reference values, with a common parameter r_0 of 0.99. The right-hand image in Fig. 4 shows fusion c.s. vs. reduced energy E_r for all six systems. The r_0 values have been slightly adjusted (within 1%) for the right-hand image. We can conclude that on the ground of our fusion Q value rule, the concept of neutron transfer as a factor facilitating fusion is redundant.

According to this rule, we can expect that for ^{3}He or ^{6}He ions, the sub-barrier fusion c.s. should be much larger than for alpha particles fusing with the same nuclei due to the very large Q differences for the reaction, particularly for heavy targets such as Pb and Bi nuclei. Fig. 5 shows the available low-energy fusion c.s. data for highly mass asymmetrical systems of light projectiles on heavy targets, which are compiled and presented vs. the E_r variable. The solid symbols in Fig. 5 represent the fusion c.s. for ^{4}He + ^{208}Pb,^{209}Bi, p + ^{209}Bi and ^{181}Ta [8,9]. The outlined symbols indicate the ^{9,6}Li + ^{208}Pb, ^{6,7}Li + ^{209}Bi, ^{3}He, d + ^{209}Bi [10,11] and ^{16}O + ^{208}Pb fusion systems. The values of fusion energy gain (Q) and the barrier parameter r_0 are given in Fig. 5 for each dataset presented. The fusion data for ^{3}He, p, d were obtained off-line using the stack activation method, and are inferior in quality with respect to those for the alpha particles. Nevertheless, the general overall trend for all the data points is clear. The range of data points in Fig. 5 can be grouped into two distinct areas: the first contains ^{4}He and proton data, while the second contains all the other data, including those for ^{3}He, d, ^{6,7}Li, exotic ^{9}Li and ^{16}O projectiles. Within each of these regions, the data seem to follow the fusion Q value rule, and the resulting two quasi-universal curves can be seen in Fig. 5.

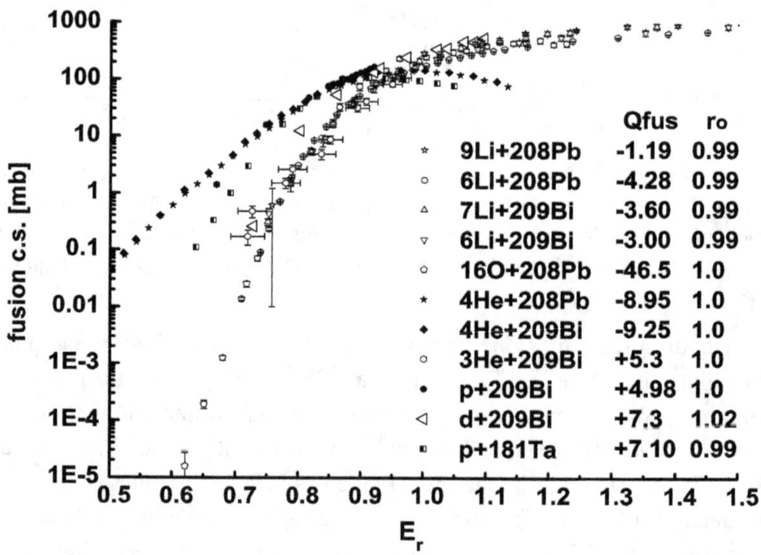

Fig. 5. Energy-reduced fusion c.s. for light particles on heavy targets. The solid symbols represent the fusion of ^{4}He + ^{208}Pb,^{209}Bi, p + ^{209}Bi and ^{181}Ta, while the outlined symbols represent ^{9,6}Li + ^{208}Pb, ^{6,7}Li + ^{209}Bi, ^{16}O + ^{208}Pb and ^{3}He, d + ^{209}Bi. The values of fusion Q and r_0 are given for each system.

The only reasonable explanation for this division of the data is the fact that ^{4}He and protons are structureless projectiles in sub-barrier fusion, whereas all the others are complex objects in this process.

To our knowledge, the fusion Q value rule, in which the concept of structurally dependent energy scaling is used, (the fusion Q value is a structural parameter), is the first generalisation of a vast body of sub-barrier fusion data. This approach allows us to estimate the consistency of the existing data and to roughly predict the probability of unmeasured fusion, if data are available for a closed system. The success of this approach demonstrates that sub-barrier fusion is a multi-step, long-lasting process.

References

1. A.B. Balantekin, N. Takigawa, *Rev. Mod. Phys.* **70,** 77 (1998),
2. R. Wolski, *Phys. Rev. C* **88**, 041603 (2013).
3. G.L. Zhang et al., *Eur. Phys. J. A* **52**: 39 (2016).
4. W.W. Qu et al., *Phys. Rev. C* **90**, 064603 (2014).
5. R.G. Stokstad, Y. Eisen, S. Kaplanis et al., *Phys. Rev. Lett.* **41**, 465 (1978).
6. http//nrv.jinr.ru.
7. A.V. Karpov, V.A. Rachkov, V.V. Samarin, *Phys. Rev. C* **92,** 064603 (2015).
8. C.G. Andre, J.R. Huizenga et al., *Phys. Rev.* **101,** 645 (1956).
9. R.G. Thomas, W. Bartolini et al., *Nucl. Phys.* **30**, 389 (1962).
10. Z. Szucs et al., *Radio-Chem. Acta* **65**, 87, (1994).
11. W.J. Ramler et al., *Phys. Rev.* **114**, 154 (1959).

Topological Typology of Nuclei*

A. V. Yushkov, V. V. Dyachkov, Y. A. Zaripova† and Ya. V. Sidorov
Scientific Research Institute of Experimental and Theoretical Physics, Almaty, Republic of Kazakhstan
†*E-mail: zjkaznu2016@gmail.com*

The classification of nuclei according to their topological type is performed within the framework of the hypothesis of non-Euclidean nuclear curvilinear spaces. One of the parameters that determine the topology of isotopes is the quadrupole nuclear nonsphericity parameter and its sign. That is, this classification takes into account the effects of prolateness- oblateness of atomic nuclei. An attempt was made to distinguish between topological types of nuclei by Euler's characteristic.

Keywords: Shapes of nuclei; sizes of nuclei; topology of non-euclidean geometries.

Introduction

With the experimental discovery of spatially isolated multiclusters in nuclei [1-3], nuclear physics and its main section - the physics of the structure of the nucleus - acquires a new impulse and a new content that differs sharply from the historical neutron-proton model of Ivanenko-Heisenberg. The generally accepted hypothetical homogeneity of the radial distribution of nuclear matter in our days is fragmented into a whole series of new topological typifications (Table 1). Remaining within the framework of the Schrödinger equation of motion, in search of optimal parameters of the geometric structure of nuclei in the paradigm of the S-matrix Heisenberg representation, in this paper a new classification of nuclear geometry for seven types of the topological structure of the atomic nucleus, which are experimentally justified with varying degrees of uniqueness and reliability.

1. Mathematical Formalization of Topological Analysis

The spatial structure of nuclei with the necessary accuracy is experimentally probed on the beams of accelerated light and heavy ions. The most sensitive result of studying the 3D distribution of nuclear matter using nuclear reactions

* Work partially supported by grant IRN AP05131884 of the Ministry of Education and Science of the Republic of Kazakhstan.

are the angular distributions of differential cross sections (DCS) – $d\sigma/d\Omega\,(\theta) \equiv \sigma(\theta)$. And the most accurate and dominant nuclear process, indicating a complex spatial morphology of the nucleus, is the reaction of elastic scattering under conditions of nuclear diffraction. In this case, the shape of the angular distributions of DCS is quite complex and one-to-one reflects the whole gamut of the mechanisms of elastic scattering of incident ions with the nucleus - from a variety of direct processes to reactions with the formation of pre-equilibrium and equilibrium composite nuclei. In angular distributions of DCS, these mechanisms, in accordance with the necessary impact parameters, are distributed over the scattering angles θ and demonstrate a non-trivial structure and complex behavior.

From a mathematical point of view, the angular distributions of DCS, measured point by point with a certain angular interval Δθ, are the set (vector) of functions $\sigma = \{\sigma(\theta)$, where $\theta = (\theta_1, \theta_2, \dots, \theta_d) \in Z^d$, measured in some the final domain $D \in R^3$, accumulated in a certain time interval. Such a set is called a space-time scene [4]. The grid of the arguments of the scene of the angular distributions of DCS is formed by localizing the measurements along the angle θ with a certain, sometimes chaotic, step Δθ. The smoothness of the function $\sigma(\theta)$ in the angular distributions of DCS is determined by the asymptotic character of the macro measurements, although in the microcosm the whole process is quantized. Therefore, $\sigma(\theta)$ can be considered as samples of a regular physical field and, thus, the scene allows for a geometric representation in the form of a hypersurface $f(x)$: $D \to R, x \in R^3, f \in R^p$, where p and d – pseudometrics In the framework of nuclear measurements on beams of accelerated ions $d = 2$, $p = 1$, then the scene is described by a 3D graph f^1:

$$Gr(f) = \{(x, y, z) \in R^3 \mid (x, y) \in U, z = f\}, \qquad (1)$$

where U - open coordinate neighborhood D. The orthogonal projection of $Gr(f)$ onto the (x, y) plane is a map of the section field on which the level sets $f(x, y) = \sigma$. These levels are depicted by a system of isolines in correlation measurements and straight lines in measurements on one side of the beam axis in experiments on the study of differential cross sections.

In this paper, a topological approach is used to construct a morphological measure for the scene $\sigma(\theta)$ in the angular distributions of DCS as distributions over the angle θ of intensity emitted from a nuclear target on which a beam of ions of secondary products of nuclear reactions falls. As far as the authors of this article are aware, in the world literature such a topological method of theoretical analysis of angular distributions of DCS and internal morphology of the nucleus was used for the first time. Its purpose is to determine the topological type of the target nucleus by the nature of the $\sigma(\theta)$ scenes.

For the specified morphological analysis, the Euler characteristic is used. In it, the set of sections $\sigma(\theta)$ determines the distances between them ними $\sigma_1(\theta), \sigma_2(\theta) \to d(\sigma_i(\theta))$ and turns $\sigma(\theta)$ into a pseudometric space. The number of topological components $n(\sigma_i, d)$ formed by the set of levels $\{x \in D \mid \sigma(\theta_i) > \Sigma\}$ 2D field graph at height Σ is used as an indicator function of the data of physical fields $\sigma(\theta, d)$. As a result of such sections, elementary lines *M* are generated, which are formed by their final topological intersection and union. On this set *M*, the Euler characteristic χ is given, for which each segment $A \in M$ is assigned a number $\chi(A)$, satisfying all the necessary topological axioms [4] in 2*D*.

If *A* is formed by the set $\{A_1, \ldots, A_n \in M\}$, then [4]

$$\chi(A) = \Sigma^{(1)}\chi(A_i) - \Sigma^2\chi(A_{i1} \cap A_{i2}) + \cdots + (-1)^{n-1}\Sigma^{(n)}(A_1 \cap \ldots \cap A_n).$$

In this expression, $\Sigma^{(i)}$ denotes sums over all *i*-elementary parts [4].

To calculate these sums, taking into account the specifics of the angular distributions of DCS for various mechanisms of nuclear reactions, reflecting indirectly different morphological structures of specific complex nuclei, we used the following computer procedure. Consider the set of elementary segments $\{C_1, \ldots, C_n\}$ и $A \cup C_j, j = 1,2, \ldots, n$. Let the segments be the vertices of *A* lying at different heights and numbered from below (in $\sigma(\theta)$) upwards. Let us draw the base horizontal line L_0 and let h_i $(i = 1, \ldots, m)$ distance from the vertex вершин V_i to this line. Consider the second horizontal line *L*, moving up from L_0; denoting the intersection of the vertex by V_i and L_i, we put between them the values V_{i-1} and V_i. The intersection $C_j \cap L$ is a segment, and the intersection $A \cap L$ gives a finite number of elementary segments in *2D*. Denote $\varphi(h_i) = \chi(A \cap L_i)$ and $\varphi(h_{i-0}) = \chi(A \cap L_{i-0})$ as the values of the Euler characteristic in $L \cap V_i$. Then, finally, the Euler characteristic for the angular distributions of DCS will be the expression

$$\chi(A) = \sum_{i=1}^{m}[\varphi(h_i) - \varphi(h_{i-1})].$$

2. Classification of Topological Types of Atomic Nuclei

For the first time, we propose the following topological classification of atomic nuclei, whose angular distributions of DCS are specially measured under conditions of nuclear diffraction of Fresnel, Fraunhofer and Wulf-Bragg types with an increased angular resolution of the main spectrometer. The statistical error of each point in the measured angular distributions of DCS is also given special attention. Our experimental data on the angular distributions of DCS were obtained on alpha-particle beams with energies of approximately 10 MeV/nucleon at the U-150m Kazakhstan accelerator in 1965-1985, and their numerical values were included in world databases such as the National Nuclear Database Center (NNDC). In addition, the indicated numerical values of angular distributions of DCS can be found in the preprints of the Kazakhstan Institute of Nuclear Physics [9-11].

Figure 1 show all seven types with their modifications in the S-matrix representation and the parameters of quadrupole non-sphericity β, obtained in the experiment [5, 6]. Numerical data on the parameter β can also be found in the annexes to the monograph [7]. The quadrupole nonsphericity parameter for the ^{11}Li nucleus, in which the planetoid topology was found experimentally for the first time [6], has not yet been measured.

The topological type I – Spherides: $^{4}_{2}He_{2}$, $^{14}_{6}O_{8}$, $^{16}_{8}O_{8}$, $^{18}_{10}O_{8}$, $^{20}_{12}O_{8}$.
Topological type II – *Prolate spheroids:* ^{18}Ne, ^{22}Ne, ^{22}Mg, ^{24}Ne, ^{24}Mg.
Topological type III – *Oblate spheroids:* ^{12}C, ^{26}Si, ^{28}Si, ^{30}Si, ^{30}S, ^{32}Si, ^{32}S.
The topological type IY – *Toroid:* ^{32}S, ^{34}S.
The topological type Y – Saturnoids* [8]***:*** ^{70}Ge, ^{72}Ge, ^{78}Se, ^{80}Se.
Topological type YI – Simplexoids (polyhedra): ^{9}Be, ^{11}B, ^{12}C, ^{13}C, ^{14}C.
Topological type YII – Planetoids (halo nuclei): ^{6}He, ^{6}Be, ^{10}Be, ^{10}C, ^{11}Li.

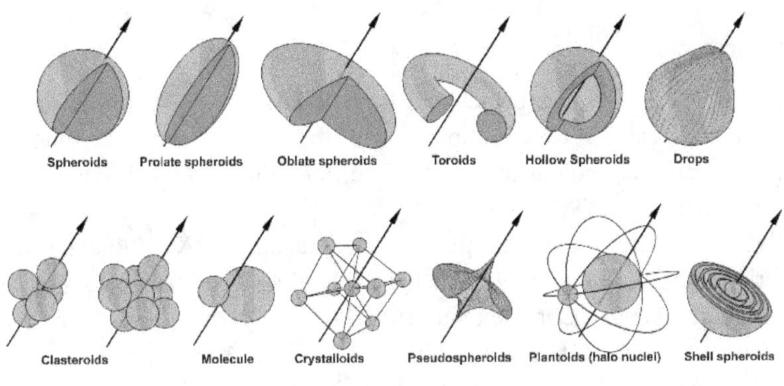

Fig. 1. Topological types of specific nuclei.

Figure 2 shows the Euler characteristics of spherical nuclei $^{58}_{30}Ni_{28}$, $^{92}_{50}Mo_{42}$ and clustered nuclei $^{12}_{6}C_{6}$, $^{24}_{12}Mg_{12}$.

Thus, the most recent discoveries of new forms in the radial distributions of nuclear density impose significantly new requirements for conducting beam studies of the structure of the nucleus. This is an increased angular accuracy of the spectrometers, a reduced angle step for increasing the accuracy of the

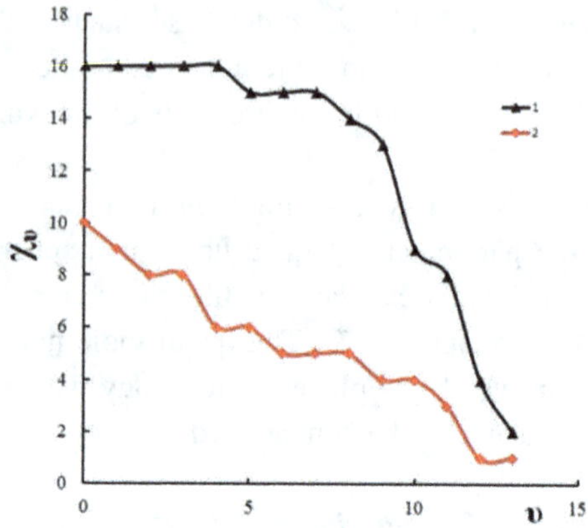

1 – cluster nuclei; 2 – spherical nuclei

Fig. 2. Euler characteristics for clustered deformed nuclei and spherical nuclei without an internal structure.

angular distributions of differential cross sections, the minimum statistical error to detection thin angular fluctuations of cross sections that carry information about exotic nuclei and their clusters.

References

1. K.A. Gridnev, V.V. Dyachkov, A.V. Yushkov, Phenomenon of the diffraction rise of cross sections in the forward angular hemisphere as an effect of nuclear and cluster interference, *Bulletin of the Russian Academy of Sciences: Physics* **79(7)**, 856 (2015).
2. V.V. Dyachkov, Yu.A. Zaripova, A.V. Yushkov, T.K. Zholdybaev, Zh.K. Kerimkulov, Kinematic Method for Separating Dominant Types of Cluster Configurations in a Complex Nucleus, *Bulletin of the Russian Academy of Sciences: Physics* **81(10)**, 1174 (2017).
3. V.V. Dyachkov, Y.A. Zaripova, A.V. Yushkov, T.K. Zholdybayev, D.K. Gridnev, Direct experimental detection of spatially localized clusters in nuclei on alpha-particle beams, *International Journal of Modern Physics E-Nuclear Physics* **27(2)**, 1850017 (2018).
4. N.G. Makarenko, L.M. Karimova, A.V. Kardashev, T.V. Nagay, A.V. Yushkov, D.N. Belyashov, I.I. Komarov, I.L. Aristova, Topological method for the analysis of areal contamination of areas of former nuclear test sites, *Radiation biology. Radioecology* **39(5)**, 521 (1999).

5. A.V. Yushkov, The nuclelar-deformation surface β(Z,N) for nuclei with Z=2-102, *Physics of Elementary Particles and Atomic Nuclei* **24(2)**, 348 (1993).
6. Yu.E. Penionzhkevich, R.G. Kalpakchieva, *Light nuclei at the neutron stability boundary* (JINR, Dubna, 2016).
7. M.A. Zhusupov, A.V. Yushkov, *Physics of Atomic Nuclei* (Parus, Almaty, 2006).
8. A.B. Migdal, *Theory of finite Fermi systems and properties of atomic nuclei* (Science, Moscow, 1965).

Study of ^{6}He-d Reactions as a First Experiment at ACCULINNA-2 Separator

B. Zalewski[1,2], S. G. Belogurov[1,3], A. A. Bezbakh[1,4], V. Chudoba[1,5], A. S. Fomichev[1,6], E. M. Gazeeva[1,6], M. S. Golovkov[1,6], A. V. Gorshkov[1,4], G. Kamiński[1,2], S. A. Krupko[1,4], B. Mauyey[1,7,8], I. A. Muzalevsky[1,5,6], P. Napiorkowski[2], E. Yu. Nikolskii[1,9], W. Piątek[2], P. Pluciński[2], K. Rusek[2], A. Serikov[1,6], S. I. Sidorchuk[1], R. S. Slepnev[1], P. G. Sharov[1,4], N. Sokołowska[10], G. M. Ter-Akopian[1,6], A. Trzcińska[2] and R. Wolski[1,11]

[1]Flerov Laboratory of Nuclear Reactions, JINR, 141980 Dubna, Russia
[2]Heavy Ion Laboratory, University of Warsaw, 02-093 Warsaw, Poland;3 National
[3]Research Nuclear University, MEPhI, Moscow, 115409 Russia
[4]SSC RF ITEP of NRC "Kurchatov Institute", 117218 Moscow, Russia
[5]Institute of Physics, Silesian University in Opava, Opava 74601, Czech Republic
[6]Dubna State University, 141982 Dubna, Russia
[7]Institute of Nuclear Physics, 050032 Almaty, Kazakhstan
[8]L.N. Gumilyov Eurasian National University, 010000 Astana, Kazakhstan
[9]National Research Center "Kurchatov Institute", 123182 Moscow, Russia
[10]Faculty of Physics, University of Warsaw, 02-093 Warszawa, Poland
[11]Institute of Nuclear Physics PAN, 31-342 Krakow, Poland

ACCULINNA-2 is a newly assembled fragment separator at Flerov Laboratory of Nuclear Reactions, Joint Institute for Nuclear Research, Russia. As a first experiment, ^{6}He-d elastic and inelastic scattering as well as other reactions have been measured in inverse kinematics at 26 AMeV with beam intensity of $\approx 10^5$ pps. Angular distributions for elastic and inelastic scattering were obtained for a range of 25 - 130 CM degrees. Preliminary data analysis for elastic scattering as well as isotope identification based on angular correlations between reaction products is presented.

Keywords: Fragment separator; elastic scattering; optical potential model; halo nuclei.

Introduction

Elastic scattering differential cross-section is an essential attribute of all processes occurring during interaction of atomic nuclei. In addition to information on the properties of nuclei and nuclear interaction directly extracted from the elastic scattering data, the optical potentials are widely used in almost all models related to the description of the dynamics of nuclear interactions (DWBA, Coupled Channels method, etc.). Knowledge of the optical potentials significantly increases the reliability of the information on the structure of the nucleus obtained from experiments studying various nuclear reactions. It is well known that the global parameterization of optical potential successfully applied in the field of heavy and medium nuclei, is facing serious problems when examining light nuclei. Obviously, this is related to relatively small number of nucleons and the manifestation of the cluster structure of light nuclei, which leads to specific features of the nuclear interaction in the transition from one nucleus to another.

Up to date, experimental data on the elastic scattering of deuterons from the ^{6}He nucleus in the literature is non-existent, while the study of the interaction of this pair of nuclei represents undoubted interest. Both nuclei are weakly bound and are characterized by an extended spatial structure that can manifest itself in the specifics of the differential cross-section of elastic scattering, for example, in comparison with elastic scattering of protons from ^{6}He.

^{6}He nuclei is loosely bound (thresholds of one- and two-neutron decay are 1.87 and 0.97 MeV, respectively) and does not have nucleon-stable excited states.

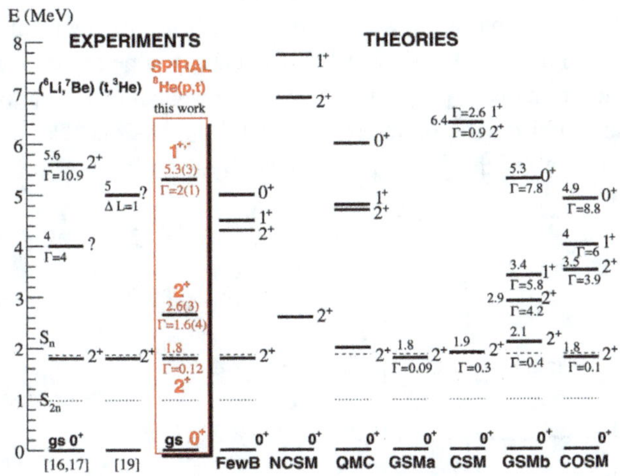

Fig. 1. Comparison of ^{6}He excited state levels by Mougeot [1] with results from other experiments and from several theories: few-body model (FewB) [2], QMC [3], NCSM [4], CSM [5], GSMa [6], GSMb [7] and the COSM [8].

The first excited state (2+, 1.8 MeV) is well known. However, the question of the existence of low-lying (<10 MeV) excited states is not answered yet. There are theoretical predictions indicating the manifestation of a soft dipole excitation mode in the region of ~ 4-6 MeV, but this phenomenon is not confirmed experimentally. A number of experimental works report the observation of resonances in ^{6}He in various reactions (transfer, charge-exchange, absorption of slow pions) but, consensus both in experimental data and in theoretical calculations has not been reached. Figure 1 taken from the work of Mougeot [1] illustrates a compilation of existing experimental and theoretical data [2-8] and confirms this situation.

In this experiment, the measurement of inelastic scattering is based on the registration of the deuteron in the coincidence with alpha particles from the decay of ^{6}He excited states. It may allow to observe not only the spectrum of excited states, but also get an information about spin and parity of those states from the analysis of angular correlations.

Experimental Details

ACCULINNA-2 is the new fragment separator installed at the primary beam line of U-400M cyclotron at Flerov Laboratory of Nuclear Reactions at Joint Institute for Nuclear Research in Dubna, Russia. ACCULINNA-2 offers secondary beams of particles with atomic number ranging from 1 to 36 and energy range from 5 to 50 AMeV [9]. Two-stage isotope separation with dipole magnets is additionally supported with beryllium wedge. This allows to obtain cleaner beams of isotopes of interest. ACCULINNA-2 is ready to work with solid targets as well as the gas or liquid ones.

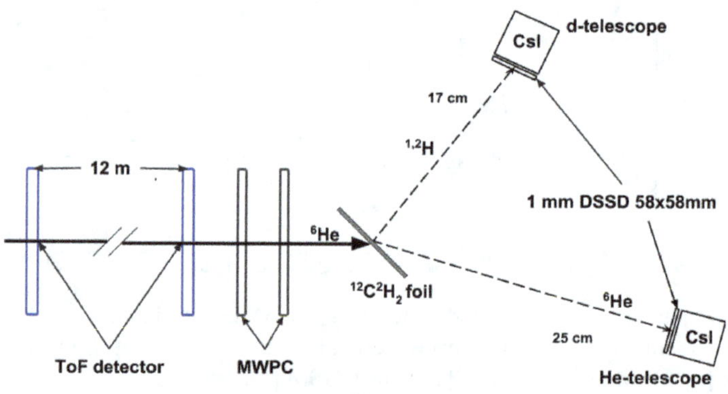

Fig. 2. Schematic view of the experimental setup.

In the experiment, primary beam of ^{15}N from U-400M cyclotron with energy of 49.7 MeV/n was bombarding a beryllium production target. The separator was used to form secondary ^{6}He beam with the intensity of ~10^5 pps. The beam energy was determined by time-of-flight technique at 12 m base, while ion identification was performed with the use of dE-ToF method. Beam tracking on target was performed with Multi-Wire Proportional Chambers (MWPC). Beam spot on the target had 17 mm diameter (FWHM). As the target, 100 μm foil made of deuterized polyethylene has been used. As later analysis showed, the target was contaminated with ^{1}H atoms. The schematic drawing of the experimental setup is presented in Fig. 2. The reaction products have been measured with two telescopes: one registering hydrogene (d-telescope) and one registering He ions (He-telescope). Both telescopes consisted of 1 mm thickness Double Side Strip Detectors as a dE detector and an array of 16 CsI(Tl) crystals each coupled with PMT R9880U Hamamatsu and used as a E detector. The measurement has been divided into three runs, each of them with d-telescope covering following angular ranges in the laboratory frame: 25° - 45°, 40° - 60° and 55° - 75°. He-telescope was stationary and covered the range from 5° to 25°. Data acquisition system has been described in detail in contribution of A.V. Gorshkov [10] "Current status of new fragment separator ACCULINNA-2 and the first day experiments" in this proceedings. Coincidence of deuterons in the d-telescope with ^{6}He in the He-telescope was determined as a best way to identify elastic scattering process. Since even first excited state of ^{6}He immediately decays into ^{4}He and 2 neutrons, one can distinguish between elastic and inelastic scattering. Isotope identification is obtained through dE - E spectra analysis. In the range of lower deuterium energies dE - E particle identification method is not possible and only correlation between angles of scattered particles can be used as

Results

The coincidences of ^{6}He in He-telescope with particles in d-telescope are presented in Fig. 3.

The data from each run is presented with a different color. The locus corresponding to elastic scattering of ^{6}He on deuterium is on the right side of graph and locus corresponding for elastic scattering on protons is on the left side. The black lines are indicating calculated relation between angles for corresponding process. Events of elastic scattering of ^{6}He on deuterium were identified with the use of plot presented in Fig. 3. The statistics obtained after normalization to particles impinged on target is presented on Fig 4 as a function of CM angle. Geometrical distribution of particles on the target is causing decrease in detection efficiency, which in turn leads to decrease in event count on the edges of each of the ranges. Monte Carlo simulation of the experiment with use of Geant4 package will be used to calculate detection efficiency, in order to obtain differential cross-section for elastic scattering of ^{6}He on deuterium.

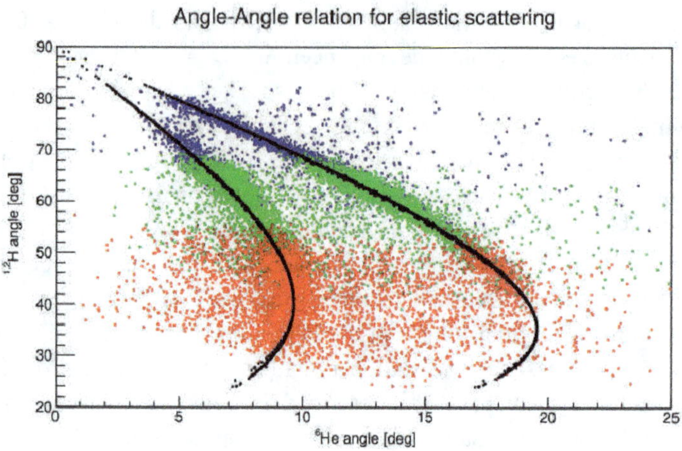

Fig. 3. The angle – angle correlation of ^{6}He with particles in the d-telescope. Black lines indicate calculated relation between ^{6}He and ^{1,2}H for elastic scattering process. Each run is marked with different color.

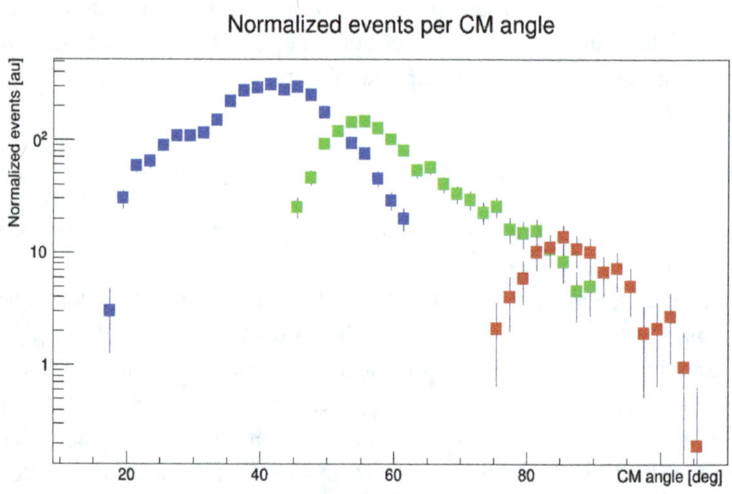

Fig. 4. Normalized events per CM angle for elastic scattering.

Conclusions

The data collected during experimental studying of ^{6}He-d reactions investigations shows that the ACCULINNA-2 separator is ready for full scale scientific programme. The experimental setup used for the study allows for in-depth investigation of optical potential through the elastic scattering measurements and subsequently extracted differential cross-sections. As preliminary analysis showed, we will be able to obtain also information about higher lying excited states of ^{6}He nuclei.

This work was partly supported by the Russian Science Foundation, Grant № 17-12-01367.

References

1. X. Mougeot et al., *Phys. Lett. B* **718**, 441 (2012).
2. B. V. Danilin et al., *Phys. Rev. C* **55**, 577 (1997).
3. S. C. Pieper, R. B. Wiringa, J. Carlson, *Phys. Rev. C* **70**, 054325 (2004).
4. P. Navrátil, W. E. Ormand, *Phys. Rev. C* **68**, 034305 (2003).
5. A. Volya, V. Zelevinsky, *Phys. Rev. Lett.* **94**, 052501 (2005).
6. N. Michel, W. Nazarewicz, M. Ploszajczak, *Phys. Rev. C* **82**, 044315 (2010).
7. G. Hagen, M. Hjorth-Jensen, J. S. Vaagen, *Phys. Rev. C* **71**, 044314 (2005).
8. T. Myo, K. Kato, K. Ikeda, *Phys. Rev. C* **76**, 054309 (2007).
9. A. S. Fomichev et al., *Eur. Phys. J. A,* **54**, 97 (2018).
10. A. V. Gorshkov et al., This proceedings.

Microscopic Analysis of the ^{12,14}Be Scattering on ^{12}C and Protons

E. V. Zemlyanaya*, V. K. Lukyanov and K. V. Lukyanov

Joint Institute for Nuclear Research,
141980 Dubna, Russia
**E-mail: elena@jinr.ru*

D. Kadrev, M. K. Gaidarov, A. N. Antonov and K. Spasova

Institute for Nuclear Research and Nuclear Energy,
Bulgarian Academy of Sciences,
Sofia 1784, Bulgaria

Differential cross sections of elastic scattering of ^{12,14}Be on ^{12}C and protons have analyzed within the microscopic model of the optical potential (OP). The microscopic OP consists of the double folding real part and the imaginary part which is constructed using the high-energy approximation theory. The OP depends on the nuclear density distributions of ^{12,14}Be and thus, their microscopic models have tested in our study.

Keywords: Exotic nuclei; microscopic optical potential; differential cross sections.

1. Introduction

The present study aims to analyze the elastic scattering of the neutron-rich isotopes ^{12,14}Be on ^{12}C and proton targets by using of microscopically calculated optical potential (OP), where the nucleon density distributions of the exotic nuclei ^{12,14}Be are being of the main physical interest. The experimental data of ^{12,14}Be + ^{12}C scattering have been published in1 and interpreted with the help of phenomenological OP [1, 2]. However, a reasonable agreement with the data was obtained by fitting of more than 10 phenomenological parameters. Moreover, the values of these parameters in [1, 2] occurred to be very different. Thus, the problem to explain the experimental data on the basis of a realistic theoretical approach is still open.

The elastic ^{12,14}Be + p scattering cross sections at 700 MeV/nucleon have been measured and explained in [11] within the Glauber theory by using the phenomenological ^{12,14}Be density distributions in the form of the symmetrized Fermi function (SF). In our study, the hybrid microscopic OP is used [3, 4] where the real part of OP is constructed within the double folding model (DF) [5] accounting for the antisymmetrization of the whole wave function. As to the imaginary part of OP, it is calculated on the basis of the high-energy approximation (HEA) [6].

2. Theoretical Framework

The DF OP consists of the direct and exchange terms, V^D and V^{EX} [3 - 5]:

$$V^{DF}(r) = V^D(r) + V^{EX}(r). \tag{1}$$

Both potentials are composed from the isoscalar and isovector terms and the former one is determined as follows:

$$V^D(r) = \int d^3r_p d^3r_t \rho_p(\mathbf{r}_p)\rho_t(\mathbf{r}_t)\upsilon_{NN}^D(s), \tag{2}$$

$$V^{EX}(r) = \int d^3r_p d^3r_t \rho_p(\mathbf{r}_p, \mathbf{r}_p + \mathbf{s})\rho_t(\mathbf{r}_t, \mathbf{r}_t - \mathbf{s})\upsilon_{NN}^{EX}(s)\exp\left[\frac{i\mathbf{K}(r)s}{M}\right]. \tag{3}$$

Here $\mathbf{s} = \mathbf{r}+\mathbf{r}_t-\mathbf{r}_p$ is the vector between two nucleons, one of which belongs to the projectile and another one to the target nucleus. $\rho_{p,t}$ — projectile and target densities, $\mathbf{K}(r)$ — local momentum of the nucleus-nucleus relative motion, $v^{D,EX}{}_{NN}$ — effective Paris NN CDM3Y potentials parameterized in [7]. The isovector potential is determined by the same formulas (2,3) but ρ_i ($i = t, p$) should be exchanged by $\delta\rho_i$, the difference between proton and neutron densities for every i-nucleus.

At comparably high energies, the *NN* potential is expressed through its explicit form [6]. In this framework, the HEA OP is determined as follows [3]:

$$U_{opt}^H(r) = -\frac{E}{k}\underline{\sigma}_N(i+\underline{\alpha}_N)\frac{1}{(2\pi)^3}\int e^{-i\mathbf{qr}}\rho_p(q)\rho_t(q)f_N(q)d^3q. \tag{4}$$

Here $\underline{\sigma}_N$ is the isospin averaged *NN* total cross section, $\underline{\alpha}_N$ is the ratio of real to imaginary part of the forward nucleon-nucleon amplitude, and $f_N(q) = \exp(-\beta_N q^2/2)$, where β_N is the slope parameter.

The hybrid form of the microscopic nucleus-nucleus OP is:

$$U(r) = N_R V^{DF}(r) + iN_I W^H(r), \tag{5}$$

where N_R and N_I are the renormalization factors of the real and imaginary OPs which are adjusted to the experimental data. In the case of the known densities of interacting nuclei, there are no more parameters to be fitted.

The standard DWUCK4 [8] code is used for calculation of the cross sections. The Coulomb potential is taken in the standard form of the uniformly charged sphere of the radius R_C.

We use the following densities of ^{12,14}Be:

- Microscopic density calculated within the generator coordinate method (GCM) [9]. In this framework, the ^{14}Be nucleus is considered as a three-cluster nucleus, involving several ^{12}Be+n+n configurations. The ^{12}Be core nucleus is described in the harmonic oscillator model with all possible configurations in the p shell.
- In the variational Monte Carlo model (VMC) [10], the proton and neutron densities are computed with the so called AV18+UX Hamiltonian, in which the Argonne v18 two-nucleon and Urbana X three-nucleon potentials are used.
- Phenomenological density in the form of the symmetrized Fermi function (SF):

$$\rho_{SF}(r) = \frac{A}{\frac{4}{3}\pi R^3}\left[1+\left(\pi\frac{a}{R}\right)^2\right]^{-1} \times \frac{\sinh\left(\frac{R}{a}\right)}{\cosh\left(\frac{r}{a}\right)+\cosh\left(\frac{R}{a}\right)}. \tag{6}$$

Here the parameters, radius R and diffuseness a, were established in [11] by fitting (within the Glauber approach) to the data of $^{12,14}\mathrm{Be}+\mathrm{p}$ elastic scattering at 700 MeV/nucleon: $R = 1{:}37$ fm, $a = 0{:}67$ fm for ^{12}Be and $R = 0{:}99$ fm, $a = 0{:}84$ fm for ^{14}Be.

The ^{12}C density is taken in the modified SF form:

$$\rho(r) = \rho_{SF}(r) + \rho_{SF}^{(1)}(r). \tag{7}$$

where the ρ_{SF} is determined by Eq. (6) with radius $R = 2.275$ fm and diffuseness $a = 0.393$ fm and the surface term $\rho^{(1)}$ is calculated via the 1st derivative of ρ_{SF}. The parameters of this density were obtained in [12] by fitting to the eA scattering data.

3. Results

It was suggested in [1] and [2] that the experimental data should be considered as quasielastic scattering, i.e., the contribution of inelastic channels related to excitations of the low-laying collective states of a nucleus, should be accounted for. Within our microscopic approach, the inelastic OP was calculated via the derivative of the microscopic OPs: $U_{inel} = -\underline{R}dU/dr$, where U is microscopic OP in the form (5), $\underline{R}$ is the potential radius (we put $\underline{R} = 4.25$ fm as in [1]). At this stage, we only accounted for excitation of the 2^+ state ($E2^+ = 4.436$ MeV). In this case, there is one more fitting parameter apart from the factors N_R and N_I,

Table 1. Values of parameters of microscopic OP

Nucleus	Density	Figure 1			Figure 2		
		N_R	N_I	β_2	N_R	N_I	χ^2/N
^{12}Be	GCM	0.42	1.09	0.66	1.51	0.93	12.63
	SF	0.57	0.91	0.78	0.93	1.17	0.99
	VMC	0.42	1.10	0.59	1.41	1.01	2.39
^{14}Be	GCM	0.36	1.32	0.42	1.08	0.92	12.43
	SF	0.43	0.62	0.48	0.77	1.10	1.10

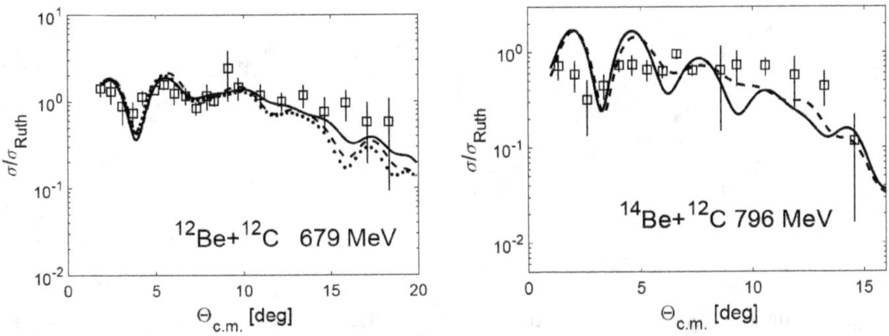

Fig. 1. The differential cross sections of the ^{12,14}Be + ^{12}C quasielastic scattering with accounting for the 2^+ inelastic channel. Solid, dashed, and dotted lines correspond, respectively, to the SF, GCM, and VMC densities of ^{12,14}Be used in calculations of OP.

the deformation parameter β_2. The results are shown in Fig. 1. The calculations have been performed with different densities of ^{12,14}Be. The values of the parameters N_R, N_I, and β_2 are given in Table 1. It is seen that the SF density provides better agreement of theoretical curve (solid line) with the experimental data on ^{12}Be + ^{12}C scattering in comparison with the GCM and VMC model densities of ^{12}Be. In the case of ^{14}Be + ^{12}C scattering, a noticeable discrepancies with experimental data are observed in both cases of SF and GCM densities. One expects that accounting for excitation of the 3^- state in the inelastic channel can provide an agreement with the experimental data to be comparable with results in [1, 2] on the basis of phenomenological approach.

The results of calculation of differential cross sections of the ^{12,14}Be + p elastic scattering are presented in Fig. 2. Here, the HEA OP [Eq. (4)] was used with parameters α_n, β, σ_n as they done in [11]. The cross sections are calculated by means of numerical solution of the respective relativistic equation, see [13] for details. The calculations can reasonably reproduce the data of ^{12,14}Be + p elastic scattering even in case of N_R=N_I=1. Having in mind that the cross sections in our study and in [11] are calculated on different backgrounds, we have fitted parameters N_R and N_I of the microscopic OP and improved the agreement of calculations with experimental data, see Fig. 2. The values of the best fit parameters N_R, N_I, and χ^2/N, where N stands for the number of the experimental points, are listed in the Table 1. One sees that the calculations with SF and VMC densities well reproduce the experimental data, while the GCM model does not provide so good agreement at $\Theta > 5^0$.

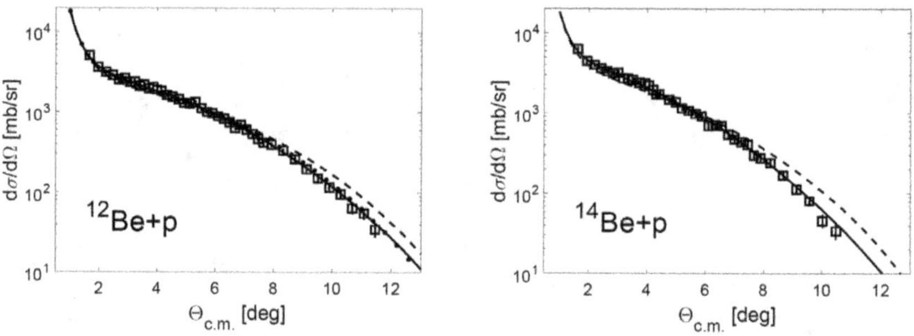

Fig. 2. The differential cross sections of the $^{12,14}Be + p$ elastic scattering at 700 MeV/nucleon. Calculations are performed with the modified SF-density of ^{12}C (7) and the different densities of $^{12,14}Be$. Solid, dashed and dotted lines correspond, respectively, the SF, GCM, and VMC densities of $^{12,14}Be$.

4. Summary

The differential cross sections of the $^{12,14}Be + {}^{12}C$ elastic scattering at energy 56 MeV/nucleon and of the $^{12,14}Be + p$ at energy 700 MeV/nucleon have been analyzed within the hybrid model of the microscopic OP. Three models of the $^{12,14}Be$ density distributions are tested. It is shown that the inclusion of both elastic and inelastic channels in the calculations allows to explain the experimental data on the $^{12,14}Be + {}^{12}C$ elastic scattering at energy of 56 MeV/nucleon with the given resolution. Also, the microscopic Ops for the $^{12,14}Be + p$ scattering calculated with the VMC and SF densities of $^{12,14}Be$ provide an agreement with the experimental data at energy about 700 MeV/nucleon.

Acknowledgements

This work is partly supported by the grant of the "JINR-Bulgaria" Program, by the DFNI-T02/19 and DNTS/Russia 01/3 grants of the Bulgarian Science Foundation and by the Russian Foundation for Basic Research (Grants No. 17-52-18057-bolg-A, 17-01-00661A). This financial support is gratefully acknowledged.

References

1. M. Zahar *et al.*, Quasielastic scattering of ^{12,14}Be on ^{12}C, *Phys. Rev. C* **49**, 1540 (1994).
2. M. C. Mermaz, Optical model analysis of ^{12}Be and ^{14}Be quasielastic scattering on a ^{12}C target at 56 Mev per nucleon, *Phys. Rev. C* **50**, 2620 (1994).
3. V. K. Lukyanov, E. V. Zemlyanaya, K. V. Lukyanov, Nucleus-nucleus scattering in the high-energy approximation and optical folding potential, *Phys. At. Nucl.* **69**, 240 (2006).
4. K. V. Lukyanov. *JINR Communication R11-2007-38* (Dubna, 2007).
5. D. T. Khoa and G. R. Satchler. *Nucl. Phys. A* **668**, 3 (2000).
6. R. J. Glauber, *Lectures in Theoretical Physics* (N.Y.: Interscience, 1959).
7. N. Anataraman, H. Toki, G. F. Bertsch, An effective interaction for inelastic scattering derived from the Paris potential, *Nucl. Rev. A* **398**, 269 (1983).
8. P. D. Kunz and E. Rost, *Computational Nuclear Physics* **2**, 88 (Eds. Langanke K. *et al.*, Springer-Verlag, 1993).
9. P. Descouvemont, Halo structure of ^{14}Be in a microscopic ^{12}Be$n + n$ cluster model, *Phys. Rev. C* **52** 704 (1995).
10. S. C. Pieper, private communication.
11. S. Ilieva, F. Aksouh, G. D. Alkhazov *et al.*, Nuclear-matter density distribution in the neutron-rich nuclei ^{12,14}Be from proton elastic scattering in inverse kinematics *Nucl. Phys. A* **875**, 8 (2012).
12. V. V. Burov, V. K. Lukyanov, D. N. Kadrev, Yu. S. Pol', Analysis of charge-density distributions in nuclei, *Phys. At. Nucl.* **61**, No.4, 525 (1998).
13. V. K. Lukyanov, E. V. Zemlyanaya, K. V. Lukyanov. Analysis of elastic scattering and breakup of ^{11}Li nuclei colliding with protons on the basis of a dynamical two-cluster model, *Phys. At. Nucl.* **78**, No.1, 142 (2015).

2

Superheavy elements. Synthesis and properties

Dynamical Approach for Synthesis of Superheavy Elements —Fusion Mechanism and Nuclear Structure—

Y. Aritomo*, N. Liyana and B. Yanagi

Department of Electric and Electronic Engineering, Faculty of Science and Engineering, Kindai University, Higashi-Osaka, 577-8502, Japan
**E-mail: aritomo@ele.kindai.ac.jp*

H. Hachikubo

Graduate School of Engineering, Nagoya University Furo-cho, Chikusa-ku, Nagoya 464-8603, Japan

To synthesis of superheavy elements, the shell structure is very important not only the stability of nuclei, but also the fusion process, especially the cold fusion reaction. We employ the Langevin equation with the microscopic transport coefficients and calculate the fusion cross section for the reaction $^{70}Zn+^{208}Pb\rightarrow^{278}Cn$. In the dynamical process, the effect of nuclear structure is discussed.

Keywords: Superheavy elements; Langevin equation; nuclear structure.

1. Introduction

More than 50 years ago, the existence of the Island of Stability in the nuclear chart surrounding the doubly magic superheavy nucleus containing 114 protons and 184 neutrons was predicted [1]. It is well known that the stability and the decay properties of superheavy nuclei strongly depends on the shell structure. However, it was pointed out that this shell structure also dramatically affects the fusion process, especially the cold fusion reaction [2-6]. To synthesis of new superheavy elements $Z\geq119$, we try to clarify the mechanism of the dynamical process and take advantage of the shell structure in fusion process.

According to the macroscopic-microscopic model, the single-particle level diagram for the reaction $^{70}Zn+^{208}Pb\rightarrow^{278}Cn$ was calculated by P. Moller *et al.* [5, 6], and it showed that the magic-fragment gap combination 28+82=110 remains far inside the touching point. The original fragment cluster or shell structure is presented during most of the fusion process. In this case, the kinetic energy in the entrance channel does not dissipate into internal energy, and finally the fusion is enhanced. Moreover, in the cold fusion reaction, due to the shell effect, there are so-called cold fusion valley, which leads to fusion. It is expected that the nuclear friction would be small along the valley, too [5, 6]. To describe such aspects, the dynamical studies with a microscopic dissipation model should be required. Here, we employ the Langevin equation with the microscopic transport coefficients, friction and mass tensors, which are calculated by the linear response theory [7-9]. We estimate the fusion probability in cold fusion reaction and investigate the effect of it. Moreover, we try to clarify the dynamical mechanism of cold fusion reaction. The dissipative mechanism during the fusion process is discussed.

The paper is organized as follows. In Section 2, we detail the framework of the model. In Section 3, we show the results for fusion cross section by dynamical model. In Section 4, we present a summary of this study and further discussion.

2. Model

We use the fluctuation-dissipation model and employ Langevin equations [10] to investigate the dynamics of the fission process. The nuclear shape is defined by the two-center parametrization [11, 12], which has three deformation parameters, z_0; δ, and α to serve as collective coordinates: z_0 is the distance between two potential centers, while $\alpha = (A_1 - A_2)/(A_1 + A_2)$ is the mass asymmetry of the two fragments, where A_1 and A_2 denote the mass numbers of heavy and light fragments. The symbol δ denotes the deformation of the fragments. The detail of the definition is explained in [10]. We assume in this work that each fragment has the same deformation. In order to reduce the computational time, we employ the coordinate z defined as $z = z_0/(R_{CN}B)$, where R_{CN} denotes the radius of a spherical compound nucleus and B is defined as $B = (3 + \delta)/(3 - 2\delta)$. We use the neck parameter $\epsilon = 1.0$ in fusion process. The three collective coordinates may be abbreviated as q, $q = \{z, \delta, \alpha\}$. The potential energy is defined as a sum of the liquid-drop part $V_{\rm LD}$ and a microscopic part $E^0_{\rm shell}$ [10].

The multidimensional Langevin equations [10] are given as.

$$\frac{dq_i}{dt} = (m^{-1})_{ij}\, p_j,$$

$$\frac{dp_i}{dt} = -\frac{\partial V}{\partial q_i} - \frac{1}{2}\frac{\partial}{\partial q_i}(m^{-1})_{jk}\, p_j p_k - \gamma_{ij}(m^{-1})_{jk}\, p_k + g_{ij} R_j(t), \tag{1}$$

where $i = \{z, \delta, \alpha\}$. and $p_i = m_{ij}\, dq_j/dt$ is a momentum conjugate to coordinate q_i. The summation is performed over repeated indices. In the Langevin equation, m_{ij} and γ_{ij} are the shape-dependent collective inertia and the friction tensors, respectively, which are called as the transport coefficient. Generally, the wall-and-window one-body dissipation is adopted for the friction tensor [13]. A hydrodynamical inertia tensor is adopted with the Werner-Wheeler approximation for the velocity field [14]. So, we use the macroscopic transport coefficients.

The normalized random force $R_i(t)$ is assumed to be that of white noise, *i.e.*, $\langle R_i(t)\rangle = 0$ and $\langle R_i(t_1)R_j(t_2)\rangle = 2\,\delta_{ij}\,\delta(t_1 - t_2)$. The strength of the random force g_{ij} is given by Einstein relation $\gamma_{ij}T = \Sigma_k\, g_{ij}g_{jk}$. T is the nuclear temperature. The fusion events are determined in our model calculation by identifying the different trajectories in the deformation space.

3. Fusion Probability Using Microscopic Transport Coefficients

To calculate the fusion probability, we have used the macroscopic transport coefficients, which do not take into account the nuclear structures. As mentioned in section 1, to treat cold fusion reaction, we need to introduce the microscopic transport coefficients. It is well-known that the microscopic transport coefficients are calculated by the linear response theory [7-9]. The transport coefficients depend on the nuclear temperature and the friction is smaller than the one-body friction (macroscopic model) at the low temperature.

Figure 1 shows the friction tensors calculated by the liner response theory for ^{278}Cn, depend on the nuclear temperature T.

The black-solid line denotes the friction tensor calculated by macroscopic model. In the cold fusion reaction, the excitation energy of compound nuclei is about 10~15 MeV, which corresponds to about T = 0.5~1.0 MeV. We can

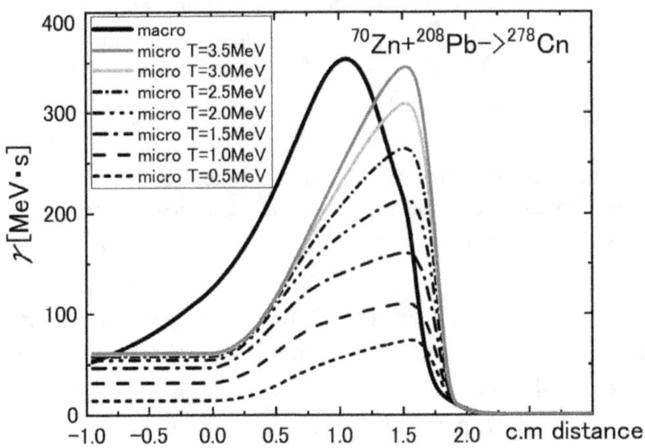

Fig. 1. Friction tensors calculated by the liner response theory for ^{278}Cn, depend on the nuclear temperature T. The black-solid line denotes the friction tensor calculated by macroscopic model.

see that the magnitude of the microscopic friction tensor is smaller than the macroscopic one at the low temperature [7-9]. To investigate the effects of the microscopic transport coefficients along the cold fusion valley, we calculate the fusion cross section in the reaction $^{70}Zn+^{208}Pb \rightarrow ^{278}Cn$ at the low excitation energy. The fusion cross sections with the macroscopic and microscopic transport coefficients are denoted by the solid and the dashed lines in Fig. 2, respectively. Here, for simplicity, we use the microscopic transport coefficients in $T = 1.0$ MeV case, though the temperature should be changed at each point. We obtain that the fusion cross section is enhanced one order of magnitude, comparison with the case using the macroscopic transport coefficients.

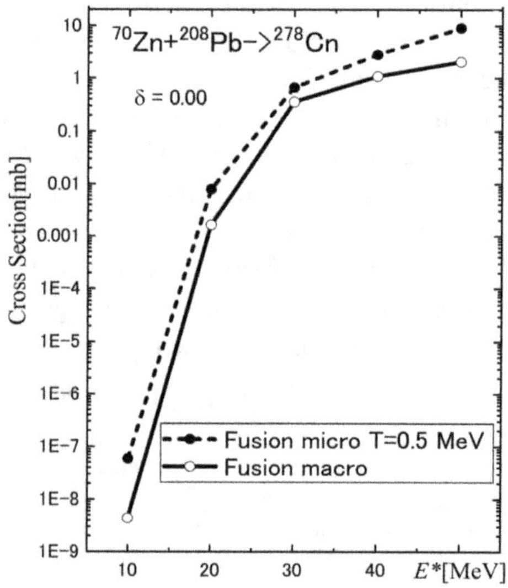

Fig. 2. Calculated fusion cross section in the reaction $^{70}Zn+^{208}Pb \rightarrow ^{278}Cn$, with the macroscopic and microscopic transport coefficients, which are denoted by the solid and the dashed lines, respectively.

It is clearly shown thc shell effects in the fusion process. Figure 3 shows the time evolution of the kinetic energy in the reaction $^{70}Zn+^{208}Pb \rightarrow ^{278}Cn$. The black and the gray lines denote the results with the microscopic and macroscopic transport coefficients, respectively. Because of the smaller friction tensor in the microscopic one, the dissipative speed is slower than that in the macroscopic case. We can see that the dynamics of fusion process in $t \leq 1.5 \times 10^{-21}$s is very important. The fusion probability is affected significantly till this time.

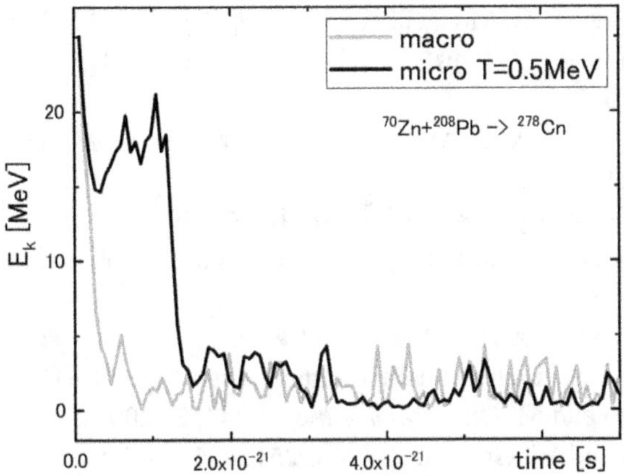

Fig. 3. Time evolution of the kinetic energy in the reaction $^{70}Zn+^{208}Pb \to ^{278}Cn$. The black and the gray lines denote the results with the microscopic and macroscopic transport coefficients, respectively.

4. Summary

In this study, we investigate the effect of nuclear structure in the cold fusion reaction using the dynamical model. In Langevin equations, instead of the macroscopic transport coefficients, the microscopic one is introduced.

The fusion cross section $^{70}Zn+^{208}Pb \to ^{278}Cn$ is enhanced one order of magnitude at the low excitation energy, comparison with the case using the macroscopic transport coefficients. In the future, we plan to calculate the evaporation residue cross section in cold fusion reaction and compare with the experimental data.

Acknowledgments

The authors are grateful to Prof. T. Wada (original Langevin code) and Dr. F. Ivanyuk (microscopic transport coefficients) for their helpful suggestions and valuable discussion throughout the present work. The authors would like to thank Dr. S. Yamaji and his collaborators who developed the calculation code for potential energy with a two-center parameterization.

References

1. W.D. Myers and W.J. Swiatecki, *Nucl. Phys.* **81**, 1 (1966); A. Sobiczewski, F.A. Gareev, and B.N. Kalinkin, *Phys. Lett.* **22**, 500 (1966).
2. Yu.Ts. Oganessian, *Lecture Notes in Physics* (Springer Heidelberg) **33**, 221 (1975).

3. S. Hofmann and G. Munzenberg, *Rev. Mod. Phys.* **72**, 733 (2000).
4. K. Morita *et al.*, *J. Phys. Soc. Jpn.* **73**, 2593 (2004); *J. Phys. Soc. Jpn.* **81**, 103201 (2012).
5. P. Moller, J.R. Nix, P. Armbruster, S. Hofmann, G. Munzenberg, *Z. Phys. A* **359**, 251 (1997).
6. P. Armbruster, *Rep. Prog. Phys.* **62**, 465 (1999).
7. F.A. Ivanyuk, H. Hofmann, V.V. Pashkevich, S. Yamaji, *Phys. Rev. C* **55**, 1730 (1997).
8. H. Hofmann, *Phys. Rep.* **284**, 137 (1997).
9. S. Yamaji, F.A. Ivanyuk, H. Hofmann, *Nucl. Phys. A* **612**, 1 (1997).
10. Y. Aritomo and M. Ohta, *Nucl. Phys. A* **744**, 3 (2004).
11. J. Maruhn and W. Greiner, *Z. Phys.* **251**, 431 (1972).
12. K. Sato, A. Iwamoto, K. Harada, S. Yamaji, and S. Yoshida, *Z. Phys. A* **288**, 383 (1978). *C* **20**, 992 (1979).
13. J. Blocki, Y. Boneh, J.R. Nix, J. Randrup, M. Robel, A.J. Sierk, and W.J. Swiatecki, *Ann. Phys.* **113**, 330 (1978).
14. K.T.R. Davies, A.J. Sierk, and J.R. Nix, *Phys. Rev. C* **13**, 2385 (1976).

About a Possible New Paradigm in the Search for SHE

C. Borcea
IFIN-HH "Horia Hulubei" National Institute for Research and Development in Physics and Nuclear Engineering
Bucharest - Magurele, ROMANIA
E-mail: borcea@nipne.ro

Few recent developments in the field of search for super-heavy elements (SHE) are reviewed from the perspective of finding new ways to access this part of the Chart of Nuclides.

Keywords: Nuclear reactions; rare and new isotopes.

1. Current Situation

The search for superheavy elements (SHE) plays an important role in our understanding of the limits of stability for nuclei. For this reason, many efforts have been invested in synthesizing these elements. The term SHE refers usually to nuclei with $Z \geq 106 \div 111$. Nowadays, this race arrived at Z=118, Og, a noble gas. It was soon observed that with increasing Z, the cross section for their synthesis, obtained as a result of a fusion-evaporation reaction, becomes lower. The practice indicated also that using the rare isotope ^{48}Ca in the so-called hot fission reactions offers a consistent advantage compared to other combinations, even though the cross sections are of the order of pb or lower. Once the advantage of ^{48}Ca was demonstrated, in order to access even higher Z values, it was necessary to use radioactive targets ($^{240, 242, 244}$Pu, ^{243}Am, ^{245,248}Cm, ^{249}Bk, ^{249}Cf). The latter ones are not only very active and therefore difficult and risky to handle, but also diminish during long irradiation periods. The situation would be even more complicated with ^{254}Es or ^{257}Fm. All these facts are described in the vast overview [1].

Another aspect of the situation in the SHE region of the chart of nuclides, is that due to the limited number of available target-projectile combinations most of the already synthesized SHE are neutron-deficient nuclei with the consequence that the predicted island of stability at Z=114, N=184 is difficult to reach. One can observe also that along the isotopic chains many of the even-even nuclei are not yet known. The reason is that the only convincing argument for assigning a SHE is to observe its α-decay chain and to arrive along this chain to an already observed isotope. This constraint seriously limits the capacity to access new isotopes, especially those that only decay by SF (spontaneous fission).

It is customary to discuss the phenomenology of SHE synthesis in terms of formation probability depicted as $P = P_b P_{fus} P_{surv}$, where P_b is the Coulomb barrier penetrability in the entrance channel, P_{fus} is the probability of fission and P_{surv} is the survival probability of the formed compound. Obviously, the three factors are correlated. Of them, P_{fus} is probably the most complex and also critical. It seems, and some studies confirmed [2] that it takes a relatively long time ($>10^{-20}$s) to fuse the two colliding nuclei. One example in the above reference is the reaction ^{64}Ni + ^{207}Pb at 281 MeV (CM), for which the lifetime of a nuclear molecule formed prior to complete fusion is of the order of few units of 10^{-20} s and only a very small fraction of these molecules lead to CN fusion. This first stage is dominated by the kinetic energy dissipation prior to the onset of nucleon exchange. A common theoretical treatment of all the three factors is given in [3].

As a summary of the present challenges of the SHE synthesis one can mention:

- Use of n-rich incident beams above ^{48}Ca.
- The quasi-fission and fast fission processes dominate the decay of the di-nuclear system formed after the collision, leaving only a very small fraction for the formation of the CN.
- Get closer to the island of stability, precisely establishing its existence and position.
- Present identification of the SHE is essentially based on decay chains landing in a known isotope. Elements decaying by SF are not yet identified with sufficient precision.
- Many cases on the chart of nuclides in the Z>104 region and on the n-rich side are still empty.

Faced with these challenges, the studies developed in other directions, aiming to bring clarifications to the process of synthesis. Many of them are oriented towards the study of the different aspects of the fast fission and quasi-fission as competing processes [4-6].

There are also interesting new approaches. One of them refers to the observation of α-decaying heavy elements created in ^{238}U+^{232}Th reactions at 6.1-7.5 MeV/A [7]. The study mentions α-activities specific for very heavy elements, without indicating explicitly which ones.

Another interesting result is the observation of 200 keV X-rays resulting from the collision of ^{238}U (6.6 MeV/A) with ^{64}Ni [8]. These high-energy X-rays coming from the Z=120 compound nucleus would need at least 2.5×10^{-18}s for an internal electron vacancy to decay emitting a K X-ray, which means that the compound nucleus lives, or survives fission, for a longer period than the mentioned one.

2. A New Possible Solution

In the following we will present a new possible solution for the synthesis of SHE. The idea stems from previous experiments in which the emission of α-particles was observed in reactions induced by heavy ions at energies below 10 MeV/A [9-11]. The idea of obtaining a cold residue after the α emission with an energy close to the kinematic limit was already expressed at that time. More recently, measurements of ^{48}Ca (261 MeV) +Ta [12] and ^{48}Ca + ^{238}U [13] were performed.

The key elements in these measurements are: spectra of α-particles were measured as a function of α-energy at 0^0 (lab) and were pursued up to α-energies close to the two-body ***kinematic limit***. This last concept needs to be explained in more detail. Fig. 1 (from Ref. 10) shows such spectra for the reaction ^{40}Ar (220 MeV) + ^{232}Th and ^{22}Ne (178MeV) + ^{232}Th. The heavy residues are

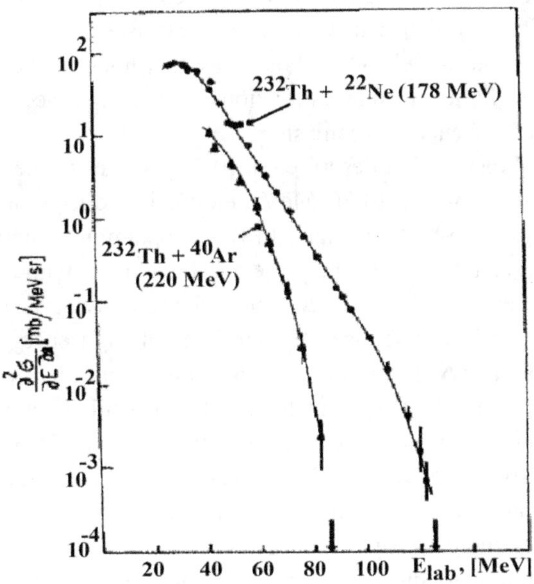

Fig. 1. Differential cross section as function of α-particle energy for the cases of ^{40}Ar+^{232}Th at 220 MeV and ^{181}Ta+^{22}Ne at 178 MeV.

^{268}Sg and ^{199}Tl. The arrows indicate the kinetic energy of α-particles for the case of 2-body kinematics and ***zero excitation energy*** of the residue. Such calculations imply only the masses of the nuclei in the incoming and outgoing channels, beam energy and observation angle. The mass of ^{268}Sg (unknown nucleus) is not known, but the mass extrapolation used in the calculation is given with a precision better than 500 keV.

In [9-11] a number of other examples are given and one can see that depending on the target-projectile combination and the beam energy, the kinematic limit is reached (or not) at values of the cross-sections that are in some cases 7 orders of magnitude below the cross section values in the maximum of the spectrum (case of the ^{181}Ta+^{22}Ne reaction at 178 MeV [11]). These very small values of the cross sections at the limit reflect the very unique character of this phenomenon in which most of the available kinetic energy is concentrated in that carried away by the α-particle. Many hypotheses were elaborated concerning the light particle emission (protons also follow the trend of the α-particles with, however, a sensibly smaller cross-section) see Refs in [9-11], while the heavy residue did not receive much attention. It was generally recognized that the spectra of α-particles present three regions (though not precisely defined): the region of maximum cross section which is attributed to an evaporation process, followed by an exponential fall-off for which pre-equilibrium emission is claimed and an even steeper descent close to the kinematic limit for which a fast, direct process is supposed in practically all proposed theoretical models. One should say that none of these refer explicitly to the kinematic limit and that estimations of the cross-section depending on projectile, target and energy are missing.

Another interesting fact described in [9-11] is that, at energies smaller than the kinematic limit (up to 20-30 MeV below), if the α-particle is detected, it will belong to the two-body channel. In Ref [11], among others, the three body channels with α resulting from the projectile breakup or with α kicked-out from the target were examined. These channels, though having a more favorable *Q* value and apparently giving rise to α particles of high energy (from a simple 3 body calculation: **www.nrv.jinr.ru**), in reality their energies are diminished by the necessary energy to account for Coulomb repulsion among the 3 bodies in the exit channel. All these facts provide already some information about the undetected reaction product: the heavy residue. If at the kinematic limit the heavy residue has zero excitation energy, for lower energies, the cross section rapidly increases and the resulted energy goes almost entirely to the excitation of the heavy residue (the site www.nrv.jinr.ru provides a rapid possibility for a kinematic calculation to see indeed that the kinetic energy of the heavy product is only very little affected).

One can now seriously question the fate of the heavy residue. The question becomes of utmost importance when the heavy residue is a rare, heavy nucleus (e.g. ^{268}Sg). Although the cross sections observed for α-particles are of the order of 20 μb in this case, ***this value is about 6 orders of magnitude higher*** than the typical cross sections for CN synthesis of SHE (few tens of pb). Therefore, the question of survival of the heavy residue is essential. Will an excitation energy of few MeV (below the one neutron separation energy, S_n, or the height of the fission barrier) be sufficient to destroy it? The energy is calculated with the accuracy of the known or the extrapolated values for the masses. To his uncertainty one may add the energy determination for the α-particle (given by the spectrometer and the opening of the telescope), in all a value below 2 MeV. It is difficult to imagine how a nucleus left unexcited could not survive when the α-particle is detected close to the kinematic limit. Table 1 offers some information about the Ar + Th reaction. An important observation in this table refers to the velocity of the heavy residue, ^{268}Sg, which is substantially lower than the velocity of the compound nucleus resulted from the same reaction.

Table 1. Reaction consideration for the case of ^{40}Ar + ^{232}Th at 220 MeV.

Products	Velocity (cm/ns)	TKE (MeV)	q	B (Tm)
^{40}Ar	3.24	220	15 16	0.901 0.845
^{272}Hs	0.48	32.49	15	0.903
^{268}Sg+α	0.39	21.53	15	0.729
	6.1	79.65	2	1.292

This difference can answer why Sg could not be observed in the same reaction in which the compound nucleus ^{272}Hs and its evaporation products (channels 1n and 2n) are searched for: using a velocity filter or a gas-filled spectrometer, Sg simply does not enter in the acceptance window of such a device, not to speak about the incident energy, which is different (generally higher) than those for compound nucleus synthesis.

Another important aspect concerning the heavy residue is that if its excitation energy becomes higher than the neutron separation energy, the residue can evaporate a neutron (if the evaporation is faster than fission, which is to be presumed). The cross section in this case will increase (see Fig. 1) by about an order of magnitude. On the other hand, most of the discovered SHE were obtained by evaporating one or more neutrons from the formed compound nucleus, usually the bombarding energy being of 30-40 MeV above the Bass barrier and leading to an excitation energy allowing for 1 to 4n evaporation.

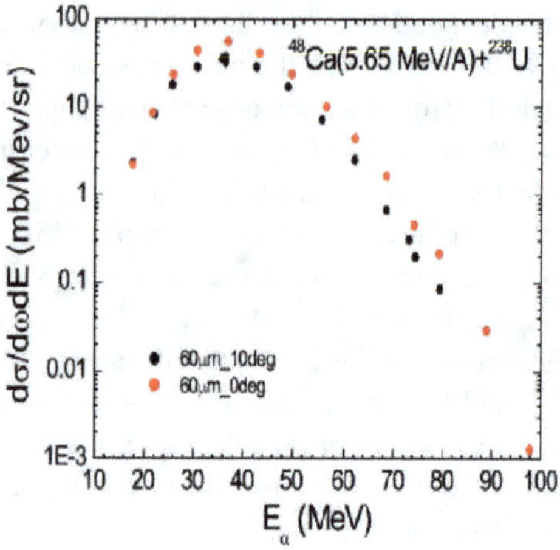

Fig. 2. Differential cross section as function of α-particle energy for the case of $^{48}Ca+^{238}U$ at 270 MeV.

Table 2. Reaction consideration for the case of $^{48}Ca+ ^{238}U$ at 270 MeV.

Products	Velocity (cm/ns)	TKE (MeV)	q	B (Tm)
^{48}Ca	3.28	270	17	0.965
^{286}Cn	0.55	44.9	18	0.906
$^{282}Ds+\alpha$	0.47	32	18	0.76
	6.4	88	2	1.358

Recently, in [13] the reaction $^{48}Ca+^{238}U$ (271 MeV) → α + ^{282}Ds was studied at the MSP144 magnetic spectrometer of FLNR-JINR. The results are shown in Fig. 2 and Table 2 gives the essentials for this reaction. This same reaction was used by Oganessian et al. for the synthesis of Z=112 by fusion-evaporation [14].

The difference is that the bombarding energy was lower by 40 MeV and the excitation energy of the compound is of the order of 40 MeV also, allowing the observation of 2n to 5n channels. The observed cross section for 3n evaporation is of 3 pb. The higher energy in our case, despite the *firm extra-push offered*, leaves the residue with a minimal excitation energy at the kinematic limit and with a higher cross section close to the neutron separation limit (1n evaporation channel).

The problem of survival of the heavy residue in these reactions with α-particle emission in a collinear geometry can *only be settled by observing it* in a dedicated experiment using the same tools as in a synthesis experiment (velocity filter, gas-filled spectrometer). Moreover, in order to be even more convincing, it is necessary that the first experimental confirmation aims at observing an already known/discovered isotope, necessarily with a large α-decay branch such as to render its identification totally unambiguous.

The experiment will be a proof of principle; if successful, it will may a new horizon for the synthesis, in the sense of a more accessible cross sections.

The α-particles recorded in coincidence could become a tool as powerful as present identification of SHE. An additional advantage of this method could be the access to the more neutron-rich species, in particular the region of the island of stability. Nuclei undergoing predominantly SF could be identified only based on the coincidentally recorded α, with its energy.

A thorough work will be needed as a preliminary, to find the most productive bombarding energies for a given target-projectile combination. Irrespective of the result of the experiment aimed at recording the heavy residue, theoretical work will be needed to explain the phenomenon in detail. To our knowledge, one of the last important contributions in this sense is the paper by Knoll et al: [15] which presents a co-operative mechanism for energetic particle emission. Another paper by Bondorf [16] refers to the sound speed velocity in nuclei compared to Fermi velocity, though the implied energies are much higher.

Such fast mechanisms can be imagined also for the cases presented in this paper, keeping in mind the non-negligible fact that the targets are α-emitters and therefore there is a pre-formation probability associated [17].

In conclusion, a phenomenology was presented, observed experimentally, that suggests the possibility to synthesize SHE in a two-body process with the emission of a high-energy α-particle at forward angles. The method is part of a more general effort to overcome the presently encountered difficulties in the synthesis of SHE due to the prohibitively small cross sections and the lack suitable target-projectile combinations.

References

1. S. Hofmann et al., *Eur. Phys. J. A* **52**, 180 (2016).
2. S. Heinz et al., *Eur. Phys. J. A* **51**, 140 (2015).
3. V. Zagrebaev et al., *Phys. Rev. C* **65**, 014607 (2001).
4. R. de Rietz et al., *Phys. Rev. C* **88**, 054618 (2013).
5. E. Kozulin et al., *Eur. Phys. J A* **52**, 293 (2016).
6. S. Kalandarov et al., *Phys. Part. Nucl.* **43**, 85 (2012).
7. S. Wuensch et al., *Phys. Rev. C* **97**, 064602 (2018).
8. N. Morjean et al., *Phys. Rev. Lett.* **101**, 072701 (2008).
9. C. Borcea et al., *Nucl. Phys. A* **351**, 312 (1981).

10. C. Borcea et al., *Nucl. Phys. A* **391**, 520 (1982).
11. C. Borcea et al., *Nucl. Phys. A* **415**, 169 (1984).
12. K. Mendibaev et al., Preprint E7-2017-66 JINR-Dubna.
13. C. Borcea et al., Proceedings of the International Symposium on Exotic Nuclei, 132 (2016).
14. Y. Oganessian et al., *Phys. Rev. C* **70**, 064609 (2004).
15. J. Knoll et al., *Nucl. Phys. A* **483**, 711 (1981).
16. J. P. Bondorf et al., *Nucl. Phys. A* **488**, 31 (1988).
17. A. Andreyev et al., *Phys. Rev. Lett.* **110**, 242502 (2013).

Fission of Super-Heavy Nuclei: Fragment Mass Distributions and their Dependence on Excitation Energy

N. Carjan[1,2], F. A. Ivanyuk[3] and Yu. Ts. Oganessian[1]

[1]Joint Institute for Nuclear Research, 141980 Dubna, Moscow Region, Russia
[2]Horia Hulubei - National Institute for Nuclear Physics and Engineering, P.O.Box MG-6, RO-76900, Bucharest, Romania
[3]Institute for Nuclear Research, Kiev, Ukraine

The mass and total kinetic energy distributions of the fission fragments in the fission of even-even isotopes of superheavy elements from Hs (Z=108) to Og (Z=118) are estimated using a pre-scission point model. We restrict to nuclei for which spontaneous fission has been experimentally observed. The potential energy surfaces are calculated with Strutinsky's shell correction procedure. The parametrization of the nuclear shapes is based on Cassini ovals. For the just before scission configuration we fix α=0.98 [12], what corresponds to $r_{neck} \approx 2$ fm, and take into account another four deformation parameters: α_1, α_3, α_4, α_6. The fragment-mass distributions are estimated supposing they are due to thermal fluctuations in the mass asymmetry degree of freedom just before scission. The influence of the excitation energy of the fissioning system on these distributions is studied. The distributions of the total kinetic energy (TKE) of the fragments are also calculated (in the point-charge approximation). In Hs, Ds and Cn isotopes a transition from symmetric to asymmetric fission is predicted with increasing neutron number N (at N≈168). Super-symmetric fission occurs at N≈160. When the excitation energy increases from 0 to 30 MeV, the peaks (one or two) of the mass distributions become only slightly wider. The first two moments of the TKE distributions are displayed as a function of the mass number A of the fissioning nucleus. A slow decrease of the average energy and a minimum of the width (at N≈162) is found.

PACS numbers: 02.60.Lj, 02.70.Bf, 25.85.Ec

Keywords: SHE; nuclear fission; Hs; Ds; Cn; Fl; Lv; Og isotopes; symmetric vs asymmetric fission; excitation energy dependence.

1. Introduction

Projects to measure fission fragment properties for the heaviest elements (Z > 106) are underway at several heavy-ion facilities around the world. The SHE Factory of JINR-Dubna [1], which will produce its first intense beam in 2019, the HIAF+CUBE of ANU-Canberra [2] which will produce such nuclei through quasifission reactions and the tandem accelerator of the JAEA-Tokai which will use fusion-fission reactions [3] are just few examples. In this context, theoretical calculations are very useful. On one side they can improve the design of such experiments and on the other side they provide predictions to compare with data allowing to test various theoretical assumptions.

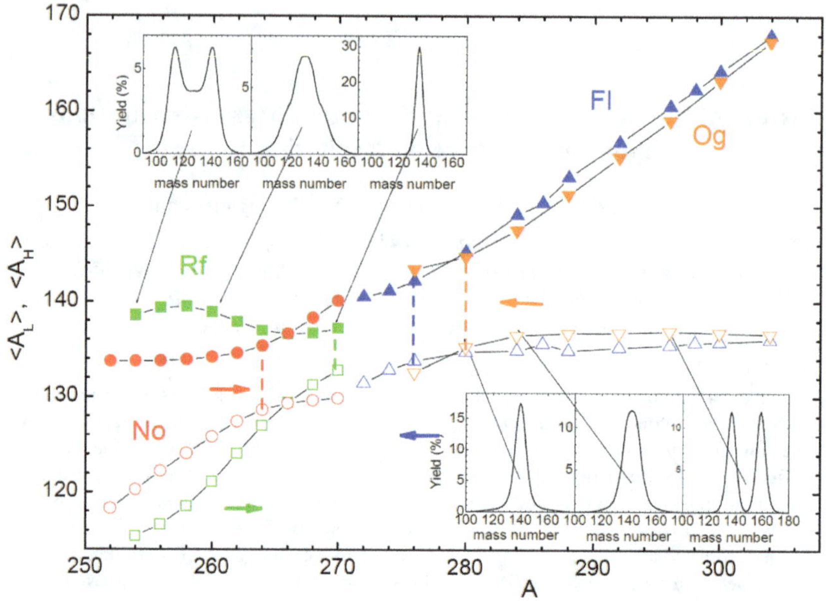

Fig. 1. Average masses of the light and heavy fragments for isotopes of No, Rf, Fl and Og as a function of the mass number A of the fissioning nucleus. The start points of the arrows mark the transition from asymmetric to symmetric mass division and dashed verticals mark the narrowest symmetric mass distribution in each series of isotopes.

In the present study, an improved version of the scission point model [4] that has confirmed its ability to describe the mass and kinetic energy distributions of the fission fragments from the spontaneous fission of the heaviest actinides for which such distributions have been measured [5], is used. More recently, the same model was applied to long series of isotopes with atomic number Z from 110 to 126 in order to study general trends [6]. Thus, if one decreases the number of neutrons N and includes the octupole deformation α_3, a transition from asymmetric to symmetric mass division takes place in Fl, Lv, Og and Z=126 isotopes. It is the mirror of the behaviour in Fm, No, Rf and Sg isotopes where the same transition occurs with increasing N. In this way the fragment mass systematics of the SHE and of the heavy actinides join smoothly together as shown in Fig. 1. It is a test of the reliability of the present approach. It is interesting to notice that each series of isotopes has its narrowest symmetric mass distribution with a full width at half maximum between 5 and 8 amu; hence extremely small. To distinguish it from the regular symmetric fission, we call this type of fission "super-symmetric".

The influence of the double magic ^{132}Sn is clearly seen by the almost constant mass (≈136) of the heavy fragment in actinides and of the light fragment in superheavies. The masses of the complementary fragment lie on a straight line, simply reflecting the conservation of the total mass number A.

However, the fission of many of these nuclei will never be observed and their study is only academic. Here we concentrate on nuclei with Z > 106 for which spontaneous fission has been already detected although the statistics was not enough to build distributions. As mentioned earlier, some of these nuclei will soon be remeasured in better conditions. Our goal is to anticipate such experiments through a detailed theoretical description of their fission fragments' properties.

Although the total kinetic energy of the fragments is also calculated, the accent is put on their masses. Is the mass division expected to be symmetric or asymmetric? How does the excitation energy of the fissioning nucleus influences this mass division? To answer such questions, we use a pre-scission point model and therefore calculate the potential energy surface (PES) of deformation, with Strutinsky's macroscopic-microscopic method [7], for a fissioning nucleus just-before its separation into two fragments. These last mono-nuclear shapes are described by generalized Cassini ovals with up to five deformation parameters $\{\alpha_i\}$. The corresponding collective degrees of freedom (normal to the fission direction) are supposed to be in statistical equilibrium. We therefore estimate the mass and TKE distributions using Boltzman factors for the probability to populate the points $\{\alpha_i\}$ on the PES. As for the excitation energy dependence of PES, the Strutinsky shell correction in excited nucleus is calculated keeping the same entropy (not temperature) in the original and averaged quantities, so that it decreases monotonically, more or less exponentially, with excitation energy.

The computational approach is explained in Sec. II. In Sec. III the predicted fission fragment mass and TKE distributions are presented for the Hs, Ds, Cn, Fl, Lv and Og isotopes for which spontaneous fission has been detected. The effect of the excitation energy on the mass distributions is estimated. A summary and conclusions can be found in Sec. IV.

2. Computational Details

Our model is a "just-before scission" model [6] that uses generalized Cassinian ovals to describe the ensemble of nuclear shapes involved.

$$R(x) = R_0\left[1+\sum_n \alpha_n P_n(x)\right], \tag{1}$$

R_0 is the radius of the spherical nucleus, $P_n(x)$ are Legendre polynomials and α_n are the shape (deformation) parameters. These shapes have in common a parameter α chosen such that at α=1.0 the neck radius is equal to zero irrespective of the values of α_n. For the just before scission configuration we fix α=0.98 [12], and take into account another four deformation parameters α_1, α_3, α_4, α_6.

With the shape parametrization (1) we calculate the potential energy of deformation using the microscopic - macroscopic approach [7]:

$$E_{def} = E_{def}^{LD} + \delta E, \tag{2}$$

where $E^{LD}{}_{def}$ is the macroscopic liquid-drop energy including surface and Coulomb energies and δE contains the microscopic shell and pairing corrections.

Supposing statistical equilibrium for the collective degrees of freedom normal to the fission direction [8], the distribution of these probabilities is

$$P(\alpha_1, \alpha_3, \alpha_4, \alpha_6) \propto e^{-E_{def}(\alpha_1,\alpha_3,\alpha_4,\alpha_6)/T_{coll}}. \tag{3}$$

Projecting $P(\alpha_1, \alpha_3, \alpha_4, \alpha_6)$ on the fixed value of mass asymmetry $\eta = (A_H - A_L)/A$ one obtains the fission fragment mass distribution $Y(\eta)$,

$$Y(\eta) = \frac{\sum_{ijk} P(\alpha_1(\eta), \alpha_{3i}, \alpha_{4j}, \alpha_{6k})}{\int d\eta \sum_{ijk} P(\alpha_1(\eta), \alpha_{3i}, \alpha_{4j}, \alpha_{6k})}. \tag{4}$$

T_{coll} is an unknown parameter that controls the overall width of the distribution. In a sense it takes partially into account the dynamics. In a similar way one obtains the fission fragment total kinetic energy (TKE) distribution. For each point (α_1, α_3, α_4, α_6) one calculates the Coulomb interaction of the fragments

$$E_{coul}^{int} = e^2 Z_L Z_H / D_{cm} = TKE \tag{5}$$

Within a quasi-static approach one cannot take into account the pre-scission kinetic energy. So, we include into TKE only the Coulomb repulsion energy. The TKE distribution

$$Y(TKE) = \sum_{ijkl} P(\alpha_{1i}, \alpha_{3j}, \alpha_{4k}, \alpha_{6l}) \frac{e^{-\left(\frac{TKE_{ijkl}-TKE}{\Delta E}\right)}}{\sqrt{\pi}\Delta E}. \tag{6}$$

accounts for the finite energy resolution through the parameter ΔE. The first two moments are calculated with the following equations:

$$\langle TKE \rangle = \frac{\int TKE\, Y(TKE)\, dTKE}{\int Y(TKE)\, dTKE}, \tag{7}$$

and

$$\sigma_{TKE}^2 = \frac{\int (TKE - \langle TKE \rangle)^2\, Y(TKE)\, dTKE}{\int Y(TKE)\, dTKE}. \tag{8}$$

3. Predicted Fission Fragment Mass and Kinetic Energy Distributions

The model described in the previous section is now applied to fragment properties for superheavy nuclei for which spontaneous fission has been already detected and could therefore be re-measured with better statistics in the near future.

These are $^{264-278}$Hs, $^{268-280}$Ds, $^{276-286}$Cn, $^{285-287}$Fl, $^{290-293}$Lv and ^{294}Og [9–11]. The calculated mass distributions corresponding to Hs, Ds and Cn isotopes are presented in Fig. 2. In all these three series, a transition from symmetric to asymmetric fission is predicted with increasing neutron number N. The transition point is N ≈ 168. The "super-symmetric" fission occurs at N ≈ 160.

The dependence on the excitation energy (under assumption of constant entropy) is also shown. There is no noticeable difference between E^*= 0 and 15 MeV and very small between E^*=15 and 30 MeV. It is good news since the

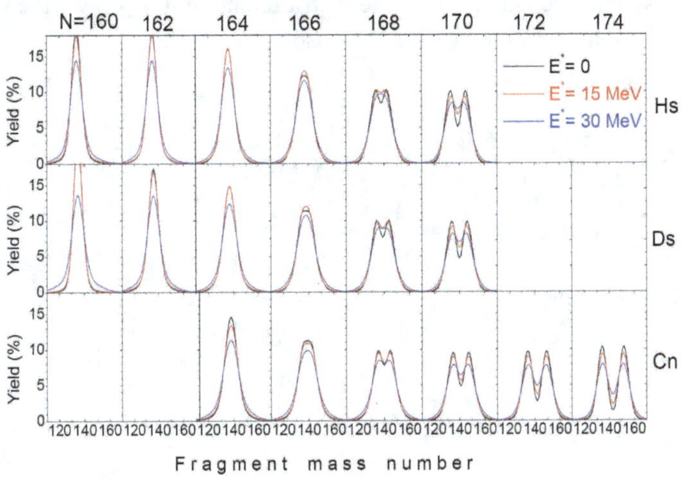

Fig. 2. The calculated fragment mass distributions for isotopes of Hs, Ds and Cn for which spontaneous fission has been detected. Three values of the excitation energy E^* have been considered. T_{coll} = 2 MeV, E_d = 40 MeV.

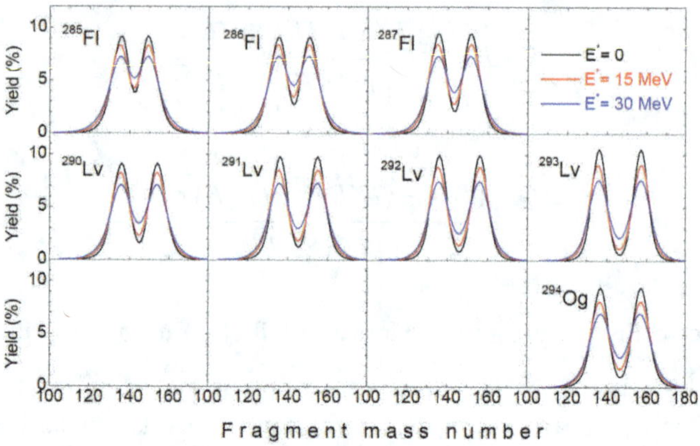

Fig. 3. The same as in Fig. 2 but for isotopes of Fl, Lv and Og.

features discussed above are not expected to be washed out if these nuclei are produced with moderate excitation. So the mass distributions in the SHE region is quite robust with respect to the excitation energy of the fissioning nucleus. The case of the heaviest nuclei ever produced, Fl, Lv and Og, is presented in Fig. 3. All these nuclei are predicted to fission into fragments with unequal masses.

The dependence on the excitation energy is slightly stronger this time. As expected, the mass-symmetric yield increases with E^* but not enough to overturn the mass-asymmetric character of the distributions. Fig. 4 shows the calculated mass distributions for a long series of Ds isotopes.

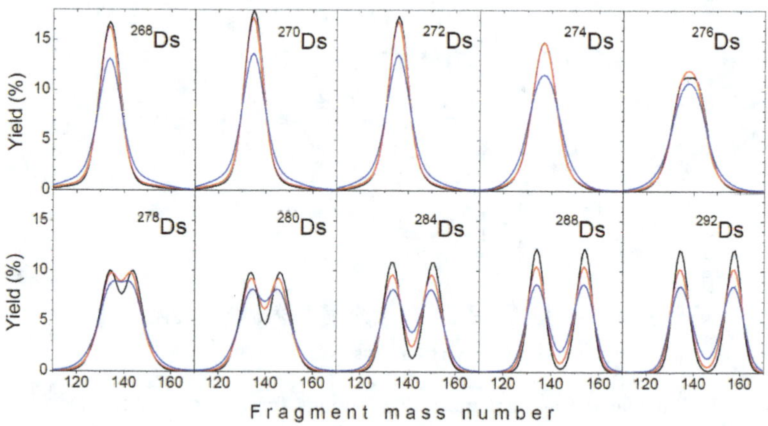

Fig. 4. The same as in Fig. 2 but for a larger series of Ds isotopes.

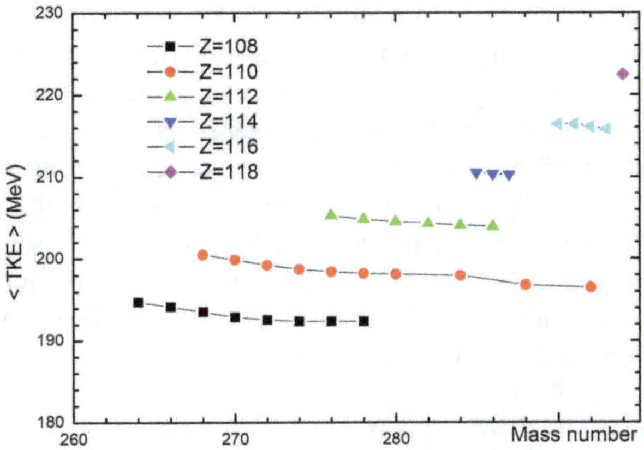

Fig. 5. The average total kinetic energy of the fragments for the fission of isotopes of Hs, Ds, Cn, Fl, Lv and Og for which spontaneous fission has been detected. $T_{\mathrm{coll}} = 2$ MeV, $\Delta E = 10$ MeV.

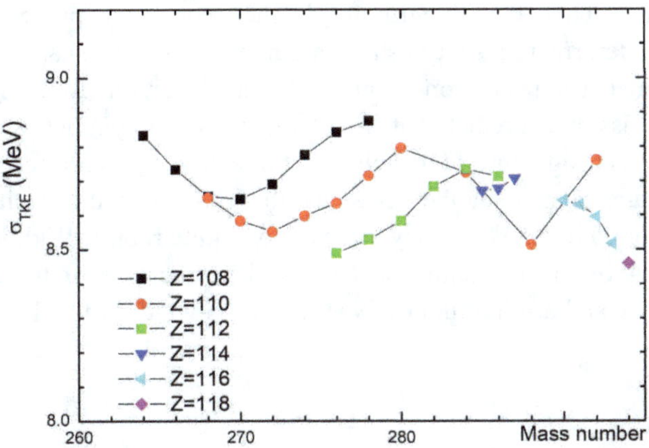

Fig. 6. Second moments of the total kinetic energy distributions for the fission of isotopes of Hs, Ds, Cn, Fl, Lv and Og for which spontaneous fission has been detected.

The heaviest three isotopes have not been detected but they are candidates to be found together with Platinum in cosmic rays or in ores (as eka-Pt)[13]. In agreement with the trends observed previously, they are predicted to fission asymmetrically.

Let us now move to the total kinetic energy distributions. For all nuclei studied here their shapes are identical (quasi-gaussian) and therefore a similar presentation as for the mass distributions is not appropriate. Instead we show the first two moments of these distributions as a function of the mass A of the fissioning isotope in Figs. 5 and 6 respectively. The prescssion contribution to the average total kinetic energy is neglected meaning that the values given are lower limits. For each series of isotopes there is a slight decrease with A due to the increase of the radius ($A^{1/3}$) and a decrease of the product $Z_L \times Z_H$. Concerning the width of the TKE distributions, they exhibit a more complex variation with the mass of the isotope. For Hs and Ds isotopes there is a minimum at N≈162 close to the neutron number where the mass distributions are also the narrowest. It is a sign of a strong nuclear structure effect in extremely deformed nuclei (pre-scission shapes).

4. Summary and Conclusions

The mass and TKE distributions of the fission fragments for the fission of selected even-even isotopes of Hs, Ds, Cn, Fl, Lv and Og are estimated using a pre-scission point model. The influence of the excitation energy of the fissioning system on these distributions is studied. The underlying potential energy surfaces are calculated with Strutinsky's shell correction procedure in a four dimensional deformation space using Cassini ovals as basic shapes.

With increasing neutron number N, a transition from symmetric to asymmetric fission is predicted at N≈168. Very narrow symmetric fission occurs at N≈160. It is a signature of a shell effect in extremely deformed nuclei. There is no dramatic change in the mass distributions when the excitation energy increases from 0 to 30 MeV; they are therefore quite robust. With increasing the total mass A of the fissioning nucleus, a slow decrease of the average total kinetic energy and a minimum (at N≈162) of the width of the TKE distribution is found.

References

1. S. Dmitriev, M. Itkis, and Yu. Oganessian, *EPJ Web of Conferences* **131**, 08001 (2016).
2. D. J. Hinde *et al.*, *EPJ Web of Conferences* **66**, 03037 (2014).
3. K. Nishio, *EPJ Web of Conferences* **62**, 03001 (2013).
4. N. Carjan, F. Ivanyuk, Yu. Oganessian, and G. Ter-Akopian, *Nucl. Phys. A* **942**, 97 (2015).
5. N. Carjan, F. Ivanyuk, and Yu. Oganessian, *J. Phys.: Conf. Ser.* **863**, 012044 (2017).
6. N. Carjan, F. Ivanyuk, and Yu. Oganessian, *Nucl. Phys. A* **968** (2017) 453.
7. M. Brack, J. Damgaard, A. S. Jensen, H. C. Pauli, V. M. Strutinsky, and C. Y. Wong, *Rev. Mod. Phys.* **44**, 320 (1972).

8. W. Norenberg, in Second Symposium on Physics and Chemistry of Fission, Vienna, 1969, IAEA, Vienna, 1969, p. 51.
9. Yu. Oganessian, and V. Utyonkov, *Rep. Prog. Phys*. **78**, 036301 (2015).
10. Yu. Oganessian, A. Sobiczewski, and G. Ter-Akopian, *Phys. Scr*. **92**, 023003 (2017).
11. F.-P. Hessberger, *Eur. Phys. A* **53**, 75 (2017).
12. V.V. Pashkevich, *Nucl. Phys. A* **477**, 1 (1988).
13. A. Svirikhin *et al*., *AIP Conference Proceedings* **1175**, 297 (2009).

Present Status and Perspectives of SHE Researches at RIKEN

H. Haba* for RIKEN SHE Collaboration
Nishina Center for Accelerator-Based Science, RIKEN,
Wako, Saitama 351-0198, Japan
**E-mail: haba@riken.jp*
www.riken.jp/en

A gas-jet transport system has been installed to the RIKEN GAs-filled Recoil Ion Separator (GARIS) at the RIKEN linear accelerator to start up superheavy element (SHE) chemistry. Isotopes of ^{261}Rf^{a,b}, ^{262}Db, and ^{265}Sg^{a,b} useful for chemistry studies were produced in the ^{248}Cm(^{18}O,5n)^{261}Rf^{a,b}, ^{248}Cm(^{19}F,5n)^{262}Db, and ^{248}Cm(^{22}Ne,5n)^{265}Sg^{a,b} reactions, respectively, and were extracted to a chemistry laboratory by the gas-jet method. Production and decay properties of ^{261}Rf^{a,b}, ^{262}Db, and ^{265}Sg^{a,b} were investigated with a rotating wheel apparatus for α and spontaneous fission spectrometry. Chemistry experiments on Rf, Db, and Sg are underway with the GARIS gas-jet system as well as the conventional gas-jet system at the AVF cyclotron. Present status and perspectives of new element syntheses at RIKEN are also presented.

Keywords: Superheavy elements; RIKEN gas-filled recoil ion separator (GARIS); ^{261}Rf^{a,b}; ^{262}Db; ^{265}Sg^{a,b}.

1. Introduction

Chemistry studies of superheavy elements (SHEs, atomic number $Z \geq 104$) are extremely interesting and challenging research subjects in modern nuclear and radiochemistry [1,2]. SHEs are produced at accelerators in heavy-ion-induced nuclear fusion reactions. Due to short half-lives and low production rates of SHEs, any chemistry experiments must be done on a one-atom-at-a-time basis. So far, by coupling a gas- or liquid-chromatography apparatus to a target/recoil-catcher chamber through a gas-jet transport line, chemical properties of SHEs have been investigated for element 104 (Rf) to 108 (Hs) and element 112 (Cn) to 114 (Fl) [1–3]. At the same time, these successful experiments have demonstrated the limitations of the conventional gas-jet technique just behind the target. Large amounts of background radioactivities from unwanted reaction products become unavoidable for identifications of SHEs with larger Z. High-intensity beams from advanced heavy-ion accelerators give rise to a problem in that the plasma formed by the beam in the target/recoil-catcher chamber significantly reduces gas-jet transport yields. To overcome these limitations, a coupling of SHE chemistry to a recoil separator for nuclear physics studies has been considered [4].

At RIKEN, a gas-jet transport system was installed to the GAs-filled Recoil Ion Separator (GARIS) at the RIKEN Linear ACcelerator (RILAC). The performance of the system was first investigated using ^{88}Nb, ^{170}Ta, ^{206}Fr, ^{245}Fm, and ^{255}No produced in the ^{nat}Ge(^{19}F,xn)^{88}Nb, ^{nat}Gd(^{19}F,xn)^{170}Ta, ^{169}Tm(^{40}Ar,3n)^{206}Fr, ^{208}Pb(^{40}Ar,3n)^{245}Fm, and ^{238}U(^{22}Ne,5n)^{255}No reactions, respectively [5–7]. ^{261}Rf^{a,b}, ^{262}Db, and ^{265}Sg^{a,b} useful for chemistry studies were then produced in the ^{248}Cm(^{18}O,5n)^{261}Rf^{a,b}, ^{248}Cm(^{19}F,5n)^{262}Db, and ^{248}Cm(^{22}Ne,5n)^{265}Sg^{a,b} reactions, respectively, and production and decay properties of those isotopes were investigated in detail [8–11]. In this paper, the recent studies on the production and decay properties of ^{261}Rf^{a,b}, ^{262}Db, and ^{265}Sg^{a,b} and on chemical properties of Rf, Db, and Sg are summarized. Present status and perspectives of new element syntheses at RIKEN are also presented.

2. SHE Chemistry at RIKEN

2.1. *GARIS gas-jet system*

A schematic of the GARIS gas-jet system is shown in Fig. 1. The ^{18}O, ^{19}F, and ^{22}Ne beams were extracted from RILAC for the production of ^{261}Rf^{a,b}, ^{262}Db, and ^{265}Sg^{a,b}, respectively. The $^{248}Cm_2O_3$ targets of about 300 μg cm^{-2} thickness were prepared by a molecular-plating method on a 2-μm Ti backing foil. The evaporation residues (ERs) of interest were separated in-flight from the beam particles and the majority of the nuclear transfer products with GARIS. GARIS was filled with He gas at a pressure of 33 Pa. At the focal plane, the ERs passed through a Mylar vacuum window foil of 0.5-μm (^{261}Rf^{a,b} and ^{262}Db) or 0.7-μm (^{265}Sg^{a,b}) thickness which was supported by a honeycomb-hole grid with 84%

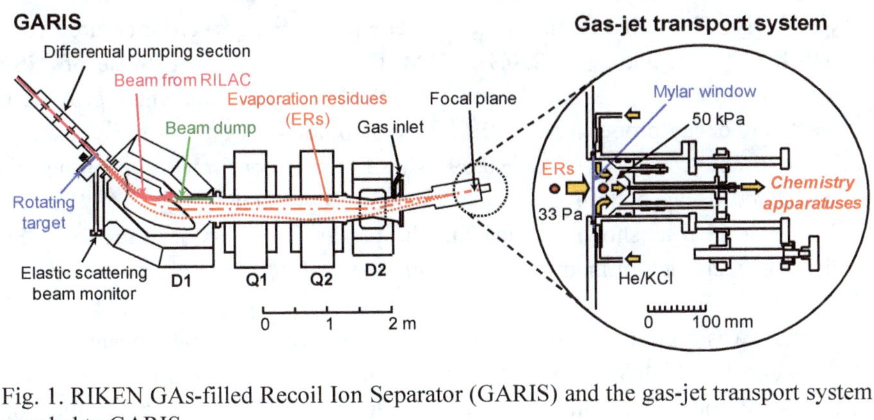

Fig. 1. RIKEN GAs-filled Recoil Ion Separator (GARIS) and the gas-jet transport system coupled to GARIS.

transparency. In the gas-jet chamber (100-mm inner diameter × 20-mm depth), the ERs were thermalized in He gas, attached to aerosol particles of KCl, and were continuously transported through a Teflon capillary (2.0-mm inner diameter × 10-m length) to a chemistry laboratory, where the Measurement system for Alpha-particle and spontaneous fissioN events ON line (MANON) was equipped. The flow rate of the He gas was 2.0 L min^{-1}, and the inner pressure of the gas-jet chamber was about 50 kPa.

2.2. *Production and decay properties of $^{261}Rf^{a,b}$, ^{262}Db, and $^{265}Sg^{a,b}$*

^{261}Rf with a half-life of $T_{1/2}$ = 68 s was produced in the ^{248}Cm(^{18}O,5n)^{261}Rf reaction [8]. The beam energy was 95.5 MeV in the middle of the $^{248}Cm_2O_3$ target, and the average beam intensity was 5 particle μA (pμA). In the α-particle spectrum measured with MANON, α-peaks of ^{261}Rf and its daughter nuclide ^{257}No were clearly identified under low background conditions. The 3-s isomeric state $^{261}Rf^b$ found on the α-decay chains from ^{277}Cn and ^{269}Hs had not been observed as an evaporation residue. Thanks to a low background condition attained with the GARIS gas-jet system, $^{261}Rf^b$ was successfully identified based on α-α correlations linking α decays of $^{261}Rf^b$ and its daughter ^{257}No [9]. The α-particle energy of $^{261}Rf^b$ was measured to be 8.52±0.05 MeV. The half-life was determined to be 1.9±0.4 s based on both 8.52-MeV α and SF decays. The α and SF branches are b_α = 0.27±0.06 and b_{SF} = 0.73±0.06, respectively. The cross section for the ^{248}Cm(^{18}O,5n)$^{261}Rf^b$ reaction is $\sigma(^{261}Rf^b)$ = 11±2 nb at 95.1 MeV, which gives a cross-section ratio of $\sigma(^{261}Rf^a)/\sigma(^{261}Rf^b) = 1.1\pm0.2$.

Two isomeric states of ^{265}Sg, i.e., $^{265}Sg^a$ and $^{265}Sg^b$, proposed in [12] were produced in the ^{248}Cm(^{22}Ne,5n)$^{265}Sg^{a,b}$ reaction at the beam energy of 117.8 MeV [10]. The beam intensity was 3 pμA. Based on the correlated α-α(-α) and α-SF decay chains, 18 and 24 events were assigned to $^{265}Sg^a$ and $^{265}Sg^b$, respectively. The half-life and α-particle energy of $^{265}Sg^a$ were measured to be $T_{1/2} = 8.5^{+2.6}_{-1.6}$ s and E_α = 8.84±0.05 MeV, respectively, and those of $^{265}Sg^b$ were $T_{1/2} = 14.4^{+3.7}_{-2.5}$ s and E_α = 8.69±0.05 MeV. As a daughter product of $^{265}Sg^{a,b}$, the decay properties of $^{261}Rf^b$ were also derived: $T_{1/2} = 2.6^{+0.7}_{-0.5}$ s, E_α = 8.51±0.06 MeV, and b_{SF} = 0.82±0.09. These results confirmed and refined the complicated decay pattern suggested for the decay chain $^{265}Sg^{a,b} \rightarrow {}^{261}Rf^{a,b} \rightarrow {}^{257}No \rightarrow$ in [12] as shown in Fig. 2a. The production cross sections for $^{265}Sg^a$ and $^{265}Sg^b$ were determined to be $\sigma = 180^{+80}_{-60}$ and 200^{+60}_{-50} pb, respectively at 117.8 MeV.

^{262}Db was produced in the ^{248}Cm(^{19}F,5n)^{262}Db reaction at the beam energies of 97.4 and 103.1 MeV: at the latter energy, the excitation function calculated by the HIVAP code [13] exhibits the maximum cross section. The beam intensity

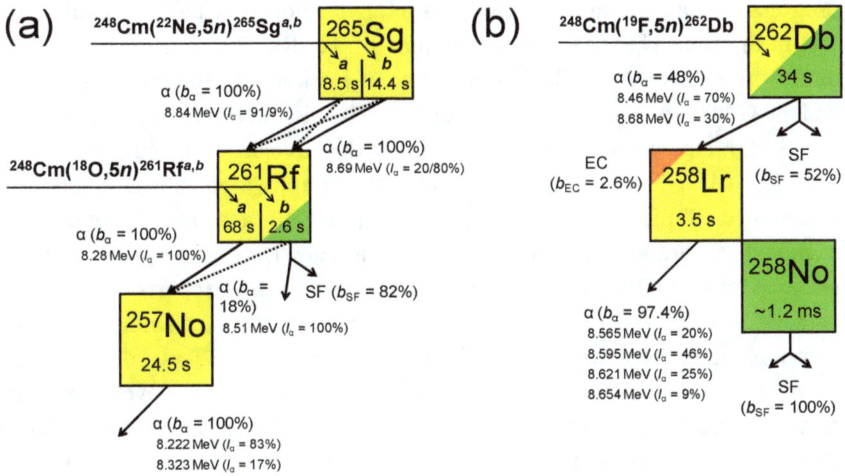

Fig. 2. (a) Decay patterns for the chain (a) ^{265}Sg^{a,b} → ^{261}Rf^{a,b} → ^{257}No → and (b) ^{262}Db → ^{258}Lr → ^{258}No →.

was 4 pμA. Based on the genetically correlated 73 α-α and 2 α-SF decay chains of ^{262}Db and its daughter ^{258}Lr, two α lines at $E_α$ = 8.46±0.04 and 8.68±0.03 MeV with α intensities $I_α$ = 70±5% and 30±5%, respectively, were assigned to ^{262}Db, though three energies of $E_α$ = 8.45 ($I_α$ = 75%), 8.53 (16%), and 8.67 MeV (9%) had been adopted for ^{262}Db in [14]. The half-life of $T_{1/2} = 33.8^{+4.4}_{-3.5}$ s obtained for ^{262}Db confirmed the literature data [14]. The SF branch of b_{SF} = 52±4% is fairly larger than that adopted in [14] (b_{SF} = ~33%). The average α-particle energy of $E_α$ = 8.61 MeV and the half-life of $T_{1/2} = 3.54^{+0.46}_{-0.36}$ s were obtained for ^{258}Lr, and they agree with the literature values [14]. The EC decay was observed in ^{258}Lr for the first time with b_{EC} = 0.026±0.018. The decay pattern for the chain ^{262}Db → ^{258}Lr → ^{258}No → is shown in Fig. 2b. The production cross section for the ^{248}Cm(^{19}F,5n)^{262}Db reaction was $σ$ = 2.1±0.7 nb at 103.1 MeV.

2.3. *SHE chemistry behind GARIS*

The GARIS gas-jet system is expected to open new chemistry studies that were not accessible before. In the gas-phase chemistry, the beam free condition in the gas-jet chamber makes it possible to synthesize a large variety of chemical compounds. The chemical synthesis and gas-chromatographic analysis of the first organometallic compound of SHEs, Sg(CO)$_6$ were successfully conducted under a large international collaboration lead by the Univ. Mainz and GSI (GSI Helmholtzzentrum für Schwerionenforschung GmbH) groups [15]. A detailed experiment to investigate the stability of the metal carbon bond in Sg(CO)$_6$ is in progress with a thermal decomposition setup developed by the PSI (Paul Scherrer Institute) group [16]. Also investigations of Tc, Ru, Rh, and Re carbonyls are under study using the ^{252}Cf fission source of Institute of Modern Physics [17,18] and the GARIS gas-jet system [19] for future studies on Bh, Hs, and Mt carbonyls.

So far, aqueous-chemistry studies of SHEs have been carried out for Rf, Db, and Sg using batch-wise column chromatography apparatuses such as the Automated Rapid Chemistry Apparatus (ARCA) and the Automated Ion-exchange separation apparatus coupled with the Detection system for Alpha spectroscopy (AIDA) [1,2]. However, there have been no reports on the aqueous chemistry of Sg and the heavier SHEs after the pioneering studies on fluoride complexation and hydrolysis of Sg in the 1990s [1,2]. Toward the aqueous-chemistry studies of Sg and element 107, Bh, we have been developing a rapid solvent extraction apparatus coupled to the GARIS gas-jet system. This new chemistry system consists of a continuous dissolution apparatus, the flow solvent extractor consisting of a Teflon capillary and a phase separator, and a flow liquid scintillation detector for α/SF-spectrometry [20].

2.4. *SHE chemistry using the conventional gas-jet transport system*

A conventional target/gas-jet system for the production of SHEs was installed on the beam line of the RIKEN AVF cyclotron [21]. Recently, AutoMated Batch-type solid-liquid Extraction apparatus for Repetitive experiments of transactinides (AMBER) [22] was developed by the Osaka Univ. group and was coupled to the automated rapid α/SF detection system for aqueous chemistry of SHEs [23]. The distribution coefficients (K_d) of Rf were determined in HCl by observing extraction equilibrium [24]: the K_d values follow the order of Rf > Zr > Hf in 9 M HCl. Reversed-phase extraction behavior of Rf with 2-thenoyltrifluoroacetone (TTA) in HF/HNO_3 was investigated with ARCA [25]: the K_d order of Zr ≥ Hf > Rf suggests that the fluoride complexation of Rf is weaker than that of either Zr or Hf. The reversed-phase extraction behavior of Db with tributyl phosphate (TBP) in HF was also investigated using ARCA [26]: Db does not form a Ta-like fluoride complex $[DbF_6]^-$ in 1.0 M HF, where Ta forms an extractable species, $[HTaF_6(TBP)_3]$.

3. SHE Synthesis at RIKEN

Element 113 was synthesized in the cold fusion reaction of ^{209}Bi(^{70}Zn,*n*)^{278}113 at RIKEN [27]. In 2016, the name "nihonium" and the symbol "Nh" was approved by the International Union of Pure and Applied Chemistry together with moscovium (Mc, $Z = 115$), tennessine (Ts, $Z = 117$), and oganesson (Og, $Z = 118$) [28]: the 7th period of the periodic table has been complete.

The search for SHEs beyond 118 is proving to be a great challenge. Man-made actinide targets such as the ones used to synthesize elements 114–118 in the ^{48}Ca-induced fusion reactions can only be produced at high-flux nuclear reactors such as that in Oak Ridge National Laboratory (ORNL) [29]. However, those target materials can only be prepared in sufficient quantities (> ~1 mg) up to Cf isotopes (Z = 98), which means that in combination with ^{48}Ca (Z = 20) these reactions can only be used to synthesize elements up to Z = 118. Attempts to synthesize elements 119 and 120 are underway using the heavier ions than ^{48}Ca: the synthesis of element 120 was attempted through the reactions ^{244}Pu + ^{58}Fe at Joint Institute for Nuclear Research in Dubna [30] and ^{238}U + ^{64}Ni [31], ^{248}Cm + ^{54}Cr [32], and ^{249}Cf + ^{50}Ti [33] at GSI, while the synthesis of element 119 was attempted through ^{249}Bk + ^{50}Ti at GSI [33], but those have remained unsuccessful.

At RIKEN, the gas-filled recoil ion separator GARIS II, which was newly designed for hot fusion reactions [34], was moved from the RILAC facility to the RIKEN Ring Cyclotron (RRC) in 2017. The ^{248}Cm target material was shipped from ORNL to RIKEN. Since 2018, an experiment to synthesize element 119 is underway with RRC + GARIS II through the ^{248}Cm + ^{51}V reaction [35]. Since 2016, RILAC has been under upgrade to increase its beam intensity by 5–10 times, with a 28 GHz superconducting ECR ion source, a super-conducting quarter-wavelength resonator, and the new separator (GARIS III) being installed. A new chemistry building was constructed next to the room of GARIS and GARIS III. The upgraded RILAC is expected to be a powerful facility for future SHE synthesis and chemistry at RIKEN.

References

1. M. Schädel and D. Shaughnessy (Eds.), *The Chemistry of Superheavy Elements, 2nd ed.* (Springer, Heidelberg, New York, Dordrecht, and London, 2013).
2. A. Türler and V. Pershina, Advances in the production and chemistry of the heaviest elements, *Chem. Rev.* **113**, 1237 (2013).
3. S. N. Dmitriev et al., Pioneering experiments on the chemical properties of element 113, Mendeleev Commun. **24**, 253 (2014).
4. M. Schädel and A. Türler (Eds.), *Proc. ChemSep 02, The 1st Workshop on Recoil Separator for Superheavy Element Chemistry* (Darmstadt, Germany, March 20–21, 2002), https://www-win.gsi.de/tasca/workshops/chemsep02/workshops_chemsep02.html
5. H. Haba et al., Development of gas-jet transport system coupled to the RIKEN gas-filled recoil ion separator GARIS for superheavy element chemistry, *J. Nucl. Radiochem. Sci.* **8**, 55 (2007).

6. H. Haba et al., Performance of the gas-jet transport system coupled to the RIKEN gas-filled recoil ion separator GARIS for the ^{238}U(^{22}Ne,5n)^{255}No reaction, *J. Nucl. Radiochem. Sci.* **9**, 27 (2008).
7. M. Huang et al., Production of ^{88}Nb and ^{170}Ta for chemical studies of element 105, Db, using the GARIS gas-jet system, *J. Radioanal. Nucl. Chem.* **304**, 845 (2015).
8. H. Haba et al., RIKEN gas-filled recoil ion separator GARIS as a promising interface for superheavy element chemistry –Production of element 104, ^{261}Rf, using the GARIS/gas-jet system–, *Chem. Lett.* **38**, 426 (2009).
9. H. Haba et al, Production and decay properties of the 1.9-s isomeric state in ^{261}Rf, *Phys. Rev. C* **83**, 034602 (2011).
10. H. Haba et al., Production of ^{265}Sg in the ^{248}Cm(^{22}Ne,5n)^{265}Sg reaction and decay properties of two isomeric states in ^{265}Sg, *Phys. Rev. C* **85**, 024611 (2012).
11. H. Haba et al., Production of ^{262}Db in the ^{248}Cm(^{19}F,5n)^{262}Db reaction and decay properties of ^{262}Db and ^{258}Lr, *Phys. Rev. C* **89**, 024618 (2014).
12. Ch. E. Düllmann and A. Türler, ^{248}Cm(^{22}Ne,xn)$^{270-x}$Sg reaction and the decay properties of ^{265}Sg reexamined, *Phys. Rev. C* **77**, 064320 (2008).
13. W. Reisdorf and M. Schädel, How well do we understand the synthesis of heavy elements by heavy-ion induced fusion?, *Z. Phys. A* **343**, 47 (1992).
14. R. B. Firestone and V. S. Shirley, *Table of Isotopes, 8th ed.* (John Wiley & Sons, New York, 1996).
15. J. Even et al., Synthesis and detection of a seaborgium carbonyl complex, *Science* **345**, 1491 (2014).
16. I. Usoltsev et al., Decomposition studies of group 6 hexacarbonyl complexes. Part 1: Production and decomposition of $Mo(CO)_6$ and $W(CO)_6$, *Radiochim. Acta* **104**, 141 (2015).
17. Y. Wang et al., Gas-phase chemistry of technetium carbonyl complexes, *Phys. Chem. Chem. Phys.* **17**, 13228 (2015).
18. S. Cao et al., Gas-phase chemistry of ruthenium and rhodium carbonyl complexes, *Phys. Chem. Chem. Phys.* **18**, 119 (2016).
19. Y. Wang et al., Gas-phase chemistry of Re carbonyl complexes with short-lived isotopes, *RIKEN Accel. Prog. Rep.* **50**, 25 (2017).
20. H. Haba, Superheavy element chemistry at GARIS, *EPJ Web Conf.* **131**, 07006 (2016).
21. H. Haba et al., Startup of superheavy element chemistry at RIKEN, *Eur. Phys. J. D* **45**, 81 (2007).
22. Y. Kasamatsu et al., Development of an automated batch-type solid-liquid extraction apparatus and extraction of Zr, Hf, and Th by triisooctylamine

from HCl solutions for chemistry of element 104, Rf, *Radiochim. Acta* **103**, 513 (2015).

23. H. Haba et al., Automated rapid α/SF detection system for studying aqueous chemistry of superheavy elements at RIKEN, *RIKEN Accel. Prog. Rep.* **45**, 204 (2012).
24. T. Yokokita et al., Observation of the chemical reaction equilibria of element 104, rutherfordium: Solid-liquid extraction of Rf, Zr, Hf and Th with Aliquat 336 resin from HCl, *Dalton Trans.* **45**, 18827 (2016).
25. A. Yokoyama et al., Extraction behavior of rutherfordium as a cationic fluoride complex with a TTA chelate extractant from HF/HNO_3 acidic solutions, *Radiochim. Acta* **107**, 29 (2019).
26. M. Murakami et al., Reversed-phase extraction behavior of the 105th element, Db, with tributyl phosphate, *RIKEN Accel. Prog. Rep.* **48**, 279 (2015).
27. K. Morita et al., New result in the production and decay of an isotope, $^{278}113$, of the 113th element, *J. Phys. Soc. Jpn.* **81**, 103201 (2012).
28. L. Öhrström and J. Reedijk, Names and symbols of the elements with atomic numbers 113, 115, 117 and 118 (IUPAC Recommendations 2016), *Pure Appl. Chem.* **88**, 1225 (2016).
29. J. B. Roberto et al., Actinide targets for the synthesis of super-heavy elements, *Nucl. Phys. A* **944**, 99 (2015).
30. Yu. Ts. Oganessian et al., Attempt to produce element 120 in the $^{244}Pu + ^{58}Fe$ reaction, *Phys. Rev. C* **79**, 024603 (2009).
31. S. Hofmann, Super-heavy nuclei, *J. Phys. G* **42**, 114001 (2015).
32. S. Hofmann et al., Review of even element super-heavy nuclei and search for element 120, *Eur. Phys. J. A* **52**, 180 (2016).
33. Ch. E. Düllmann, On the search for elements beyond Z =118. An outlook based on lessons from the heaviest known elements, *EPJ Web Conf.* **131**, 08004 (2016).
34. K. Morita et al., Discovery of new element, nihonium, and perspectives, *Prog. Nucl. Sci. Technol.* **5**, 8 (2018).
35. H. Haba, A new period in superheavy-element hunting, *Nat. Chem.* **11**, 10 (2019).

Investigation of the Spontaneous Fission Properties of Neutron-Deficient Nobelium Isotopes*

A. V. Isaev†, A. I. Svirikhin, A. V. Andreev, A. V. Yeremin, I. N. Izosimov, A. N. Kuznetsov, A. A. Kuznetsova, O. N. Malyshev, A. G. Popeko, Yu. A. Popov, E. A. Sokol, M. L. Chelnokov, V. I. Chepigin and T. M. Schneidman

Joint Institute for Nuclear Research,
Dubna, 141980, Russia
†E-mail: isaev@jinr.ru
www.jinr.ru

B. Gall, O. Dorvaux, P. Brionnet and A. Sellam

IPHC, IN2P3-CNRS,
Strasbourg, 67037, France

K. Hauschild, A. Lopez-Martenz and K. Rezynkina

CSNSM, IN2P3-CNRS,
Orsay, 91405, France

S. Mullins

iThemba LABS,
Cape Town, South Africa

P. Mosat

Comenius University,
Bratislava, 81499, Slovakia

In the last years we carried out several experiments aimed to investigate properties of short-lived SF isotopes. The neutron-deficient isotopes of nobelium were produced in fusion-evaporation reactions using ^{206,208}Pb targets and an intense ^{48}Ca-beam. Fusion-evaporation residues were separated by the SHELS separator and implanted into a large-area double-sided 48×48 strip silicon detector surrounded by ^{3}He-based neutron counters. Half-lives and decay branching ratios for ^{252,254}No isotopes were measured. The average number of neutrons per spontaneous fission of ^{254}No determined for the first time.

Keywords: Neutron multiplicity; spontaneous fission; fission fragment.

* This work is supported by grants RFBR 17-02-00867 and RFBR 18-52-15004.

1. Separator and Detectors

Main purpose of our experiments is collection of data about characteristics of spontaneous fission of short-lived nuclei from transfermium region. Separator for Heavy ELement Spectroscopy (SHELS) was used, see Figure 1. SHELS consists of two triplets of magnetic quadrupole lenses and velocity filter that are formed by two dipole magnets and two electrostatic deflectors [1].

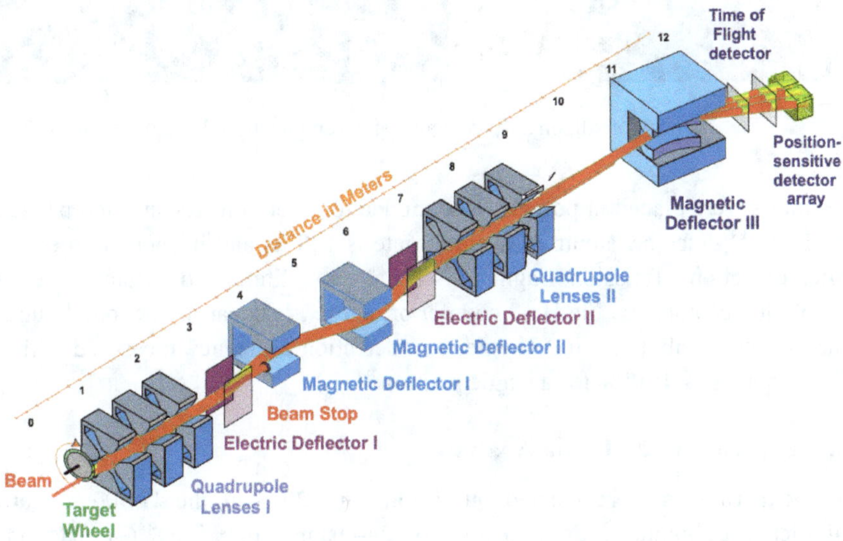

Fig. 1. SHELS separator.

At the focal plane of separator, the combined detection system placed. After MCP-based time of flight detectors, reaction products from separator implanting to 48×48 strip silicon detector [2], see Figure 2.

The detector has 58×58 mm^2 active area and 300 μm thickness. Energy resolution is 20 keV for α-particles with 6-8 MeV. Detection efficiencies of focal detector are 50% for α-particles and 100% for at least one fission fragment. For the total fission fragment energy calculation, we using summarization of focal plane and side detectors signals, detection efficiency in this case is about 70%.

Neutron detector, surrounding silicon detectors array, consist of 54 ^3He neutron counters, see Figure 2. Pressure in counter is 7 atm. Each counter has length 500 mm and diameter 32 mm. Neutron detector allows us to detect multiple prompt neutrons from the spontaneous fission of registering nuclei [3].

Fig. 2. Array of silicon strip detectors (left) and neutron detector (right).

Counters were placed in polyethylene moderator. Neutrons registration triggered by DSSSD detector. Neutrons waiting gate is 128 μs and it opens immediately after detection of fission fragment. The average lifetime of a single neutron before it's captures by the He-3 counter or before it's departure beyond bound of the detector is about 25 μs. The neutron detection efficiency measured with the ^{248}Cm source is 45±1% for a single neutron.

2. Experiments and Data Analysis

The experiments were carried out in January 2015 at the U-400 cyclotron of Flerov Laboratory. Cross-section of 2n-channel of ^{208}Pb+^{48}Ca reaction is about 1800 nanobarns. The cross-section for the same evaporation channel of ^{206}Pb+^{48}Ca reaction is about 800 nanobarns. Beam intensity was 0.5 pμA and beam energy was 215±2 MeV.

The main interest for us was related to the first reaction. The second reaction was used for calibration purpose mostly.

A highly enriched ^{208}PbS target (370 μg/cm^2) was used for ^{254}No synthesis. Enrichment by ^{208}Pb was about 99.6%. But despite such enrichment the target still contained admixtures: ^{206}Pb ~ 0.14% and ^{207}Pb ~ 0.29%, this leads to some difficulties in the data analysis. For ^{252}No synthesis ^{206}PbS target (350 μg/cm^2) was used.

The reaction cross-section for ^{254}No synthesis is huge, but the spontaneous fission branching ratio for this isotope is very small, it's only about 0.17% [4]. The calculated value of the number of spontaneous fissions of ^{254}No should have been about 175 events. We calculate this value based on the number of registered α-particles with detection efficiency about 50% and we took into account that detection efficiency for two fission fragments is about 70%.

With correlation analysis, we got 115 fission events of ^{254}No. In order to get rid of the high-energy background we did a search only of fission fragments – recoil correlations in the time interval from 15 to 200 seconds. For reliability, we filter out events with only one fission fragment detected (both signals from side and focal detectors must be found). A fission fragments spectra was obtained using summarization of energy from focal and side detectors, see Figure 3. TKE data for ^{254}No was approximated based on the known value of TKE for ^{252}No. Received value of TKE for ^{254}No is about 184 MeV. Measured multiplicity of neutrons (these values do not include detector efficiency) was 2.2 neutrons per act for ^{254}No and 1.89 neutrons per act for ^{252}No. Thus, we see that the total kinetic energy decreases in inverse proportion to the average number of neutrons per SF.

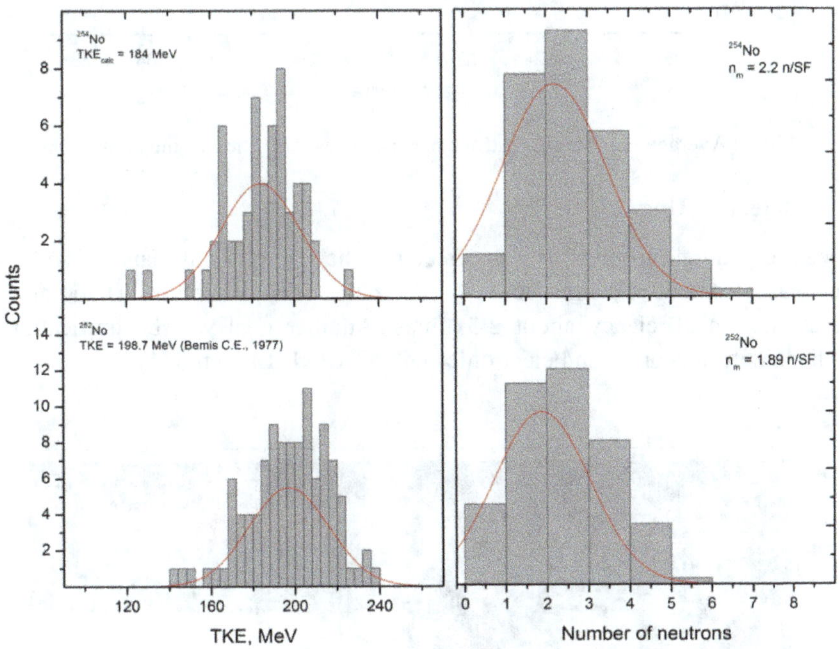

Fig. 3. Measured SF-properties of ^{252,254}No isotopes.

Half-lives received in discussed experiments were 2.76±0.18s for ^{252}No and 55±5s for ^{254}No. Average number of neutrons per spontaneous fission act were determined (the efficiency of neutron detector taking in to the account) as 4.33±0.17 for ^{252}No and 5.07±0.27 for ^{254}No, see Figure 4.

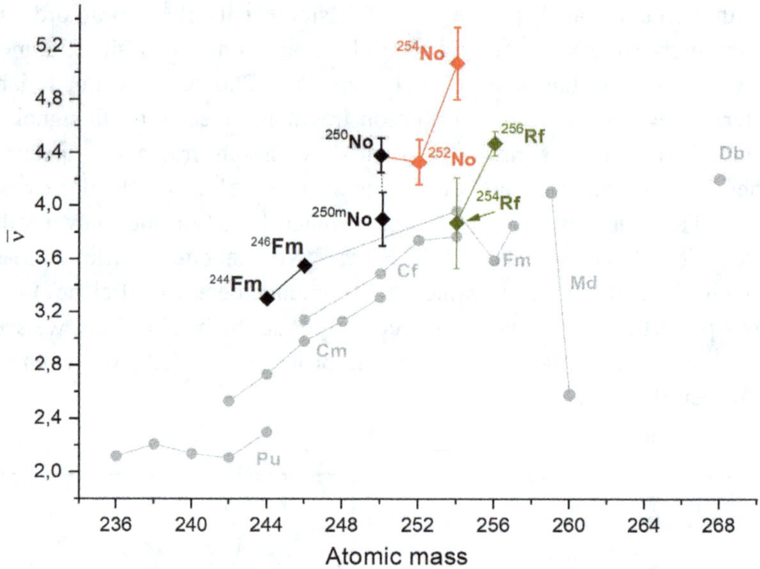

Fig. 4. Average number of neutrons per fission as a function of the atomic mass.

3. Detectors Upgrade

We are planning to upgrade our detectors. The size of focal plane DSSSD will increase from 60×60 mm^2 to 100×100 mm^2. This will allow us to improve transmission efficiency about 2-3 times. Another goal will be unification of silicon detectors array and electronics with GABRIELA setup [1].

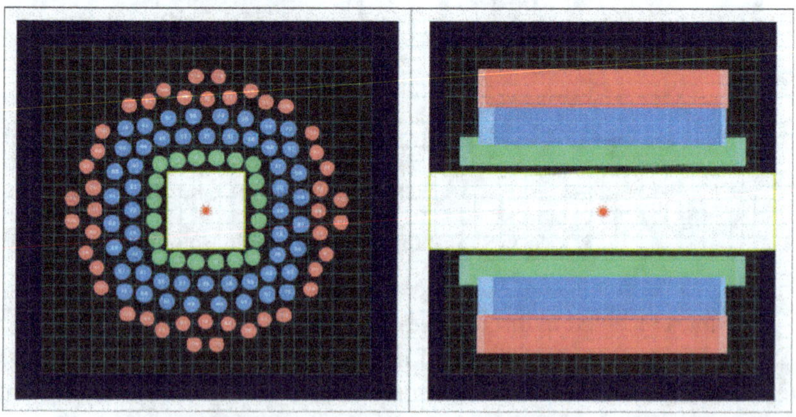

Fig. 5. The optimal geometry. A grid step is 25 mm.

Current vacuum chamber with DSSSD is cylinder with diameter 130 mm. It will be replaced by rectangular cuboid with focal size 155×155 mm^2. The NeutronGeometry application was created to automate the process of new geometry definition [5]. When the new geometry constructed user will be able to export it to MCNPX input file format [6].

The best result we found was for geometry with square layers, see Figure 5. We got neutrons registration efficiency about 57% using 116 counters.

4. Results

Various characteristics of spontaneous fission of ^{252,254}No isotopes were measured. The average number of neutrons per spontaneous fission of ^{254}No is 5.07±0.27. This value determined for the first time.

The new neutron detector configuration found. Our efforts to reach the maximum efficiency to be continue. The NeutronGeometry application solution can be reused for similar tasks.

References

1. A. G. Popeko, A. V. Yeremin, O. N. Malyshev, V. I. Chepigin, A. V. Isaev, Yu. A. Popov, A. I. Svirikhin, K. Haushild, A. Lopez-Martens, K. Rezynkina and O. Dorvaux, *Separator for Heavy ELement Spectroscopy – velocity filter SHELS*, Nucl. Instrum. Methods Phys. Res. B **376**, 140 (2016).
2. A. V. Isaev, A. V. Yeremin, N. I. Zamyatin, A. N. Kuznetsov, O. N. Malyshev, A. I. Svirikhin, M. L. Chelnokov, V. I. Chepigin, K. Hauschild, A. Lopez-Martens and O. Dorvaux, *Application of a double-sided stripped Si detector in the focal plane of the VASILISSA separator*, Instrum. Exp. Tech. **54**, 37 (2011).
3. A. I. Svirikhin, A. V. Isaev, A. V. Yeremin, A. N. Kuznetsov, O. N. Malyshev, A. G. Popeko, E. A. Sokol, M. L. Chelnokov and V. I. Chepigin, *A detector for studying the characteristics of spontaneous fission of short-lived heavy nuclei*, Instrum. Exp. Tech. **54**, 644 (2011).
4. J. Magill, G. Pfennig, R. Dreher and Z. Sóti, *Karlsruher nuklidkarte* (Nucleonica GmbH, 2015).
5. A. V. Isaev, NeutronGeometry (2018), https://github.com/siberianisaev/NeutronGeometry.
6. D. B. Pelowitz, J. W. Durkee, J. S. Elson, M. L. Fensin, J. S. Hendricks, M. R. James, R. C. Johns, G. W. McKinney, S. G. Mashnik, J. M. Verbeke, L. S. Waters and T. A. Wilcox, *MCNPX 2.7.0 Extensions*, LA-UR-05-2675, Los Alamos National Laboratory (2005), https://mcnp.lanl.gov.

Structure of Superheavy Nuclei

R. V. Jolos[1,2,*], N. V. Antonenko[1], G. G. Adamian[1], A. N. Bezbakh[1], V. G. Kartavenko[1], L. A. Malov[1], T. M. Shneidman[1], N. Yu. Shirikova[1] and A. V. Sushkov[1]

[1]*Bogoliubov Laboratory of Theoretical Physics, Joint Institute for Nuclear Research, Dubna, 141980, Russia*

[2]*Dubna State University, Dubna, 141980, Russia*

[*]*E-mail: jolos@theor.jinr.ru*

A short review of the results on the structure of superheavy nuclei is presented.

Keywords: Superheavy; nuclear mean field; alpha-decay.

1. Introduction

Superheavy nuclei form a new region of the nuclide chart whose structure is only started to be investigated. It is interesting if their structure is similar to that of nuclei which have been already studied or unexpected features will be discovered. The specific feature of atomic nuclei is a selfconsistent mean field whose characteristics determine a shape of atomic nucleus, its excitation spectra and decay modes. The following theoretical approaches are used to determine the properties of selfconsistent mean field: the microscopic-macroscopic method and the selfconsistent mean field models (relativistic or nonrelativistic) based on nuclear energy density functional. Both approaches are phenomenological, however, in the second one the nuclear mean field is consistently determined with a good description of the global nuclear properties.

2. Shell Effects

We use in our calculations the two-center shell model potential (TCSM) which is useful for the description of nuclear structure and reactions. The parameters of the nuclear average mean field potential were set to describe the spins and parities of known rare earth, actinides and superheavy nuclei [1].

Weak dependence on $(N-Z)$ was incorporated in the momentum dependent part of the single particle Hamiltonian. The ground state spins of nuclei with $N = 147 \div 161$ are presented in Tables I and II. One can see a good description up to Sg isotopes.

The structure of superheavy nuclei crucially influences the evaporation residue cross sections in actinide-based reactions. The results of calculations of $(Bf - Bn)$ provides the shell at Z=114. However, the shell effects at Z=120-126

Table 1. Calculated and experimental ground state spins of the indicated nuclei with N=147-161. The tentative assignments of spins are in brackets.

N	Cm		Cf		Fm	
	Exp	Cal	Exp	Cal	Exp	Cal
147	$5/2^+$	$5/2^+$	$(1/2^+)$	$5/2^+$		
149	$7/2^+$	$7/2^+$	$(7/2^+)$	$7/2^+$	$(7/2^+)$	$7/2^+$
151	$9/2^-$	$9/2^-$	$9/2^-$	$9/2^-$	$(9/2^-)$	$9/2^-$
153	$1/2^+$	$1/2^+$	$1/2^+$	$1/2^+$	$1/2^+$	$1/2^+$
155	$(1/2^+)$	$1/2^+$	$(7/2^+)$	$7/2^+$	$7/2^+$	$7/2^+$
157			$(7/2^+)$	$7/2^+$	$(9/2^+)$	$9/2^+$

Table 2. Calculated and experimental ground state spins of the indicated nuclei with N=147-161. The tentative assignments of spins are in brackets.

N	No		Rf		Sg	
	Exp	Cal	Exp	Cal	Exp	Cal
149	$(7/2^+)$	$7/2^+$				
151	$(9/2^-)$	$9/2^-$	$(9/2^-)$	$9/2^-$		
153	$(1/2^+)$	$1/2^+$	$(1/2^+)$	$1/2^+$	$(1/2^+)$	$1/2^+$
155	$(3/2^+)$	$1/2^+$		$1/2^+$		$1/2^+$
157	$(9/2^+)$	$9/2^+$		$11/2^-$		$11/2^-$
159					$(9/2^+)$	$9/2^+$

are rather strong. Since for nuclei with Z=120-126 the Q_α is minimal at Z=120 where the fission barrier is rather high, the nuclei with Z=120, N=180-184 are expected to be the most stable beyond those with Z=114, N=176-178.

The information on the proton magic number after Z=82 is contained also in the spectra of the proton two-quasiparticle states. While for nuclei with $Z \leq 118$ the calculated energies of the first proton two-quasiparticle states are smaller than 1.2 MeV, in ^{296,298}120 the calculated energies of the first proton two-quasiparticle states are above 1.9 MeV (Fig. 1) [2]. This indicates a large gap in the proton single-particle spectrum.

The measurements of the α-decay spectra of superheavy nuclei can give us an information on the isomers. Around ^{250}Fm one can find both protons and neutrons pairs of levels close to the Fermi level that can be coupled to low-lying states with high K, such as $(9/2^-[734] \times 7/2^+[624])_n 8^-$ for neutrons, or $(9/2^+[624] \times 7/2^-[514])_p 8^-$ and $(7/2^-[514] \times 7/2^+[633])_p 8^-$ for proton pairs.

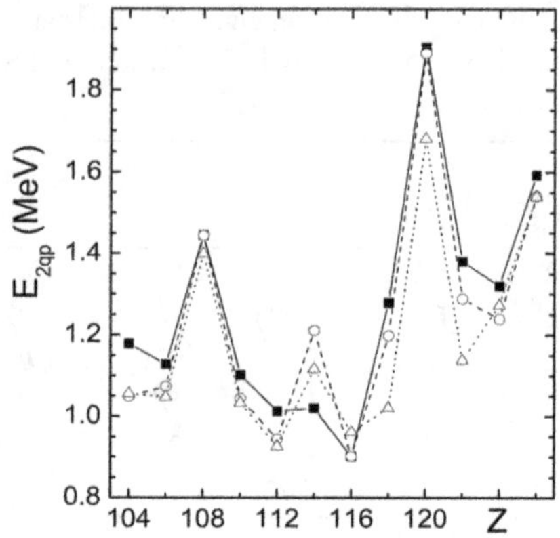

Fig. 1. Energies of the two-quasiparticle proton states in alpha-decay chains containing $^{296,298,300}120$ nuclei.

The lowest calculated states in the N=155 isotonic chain have $\Delta K \geq 3$. Therefore, the long-living isomeric state is expected in N=155 isotones (Fig. 2) [3].

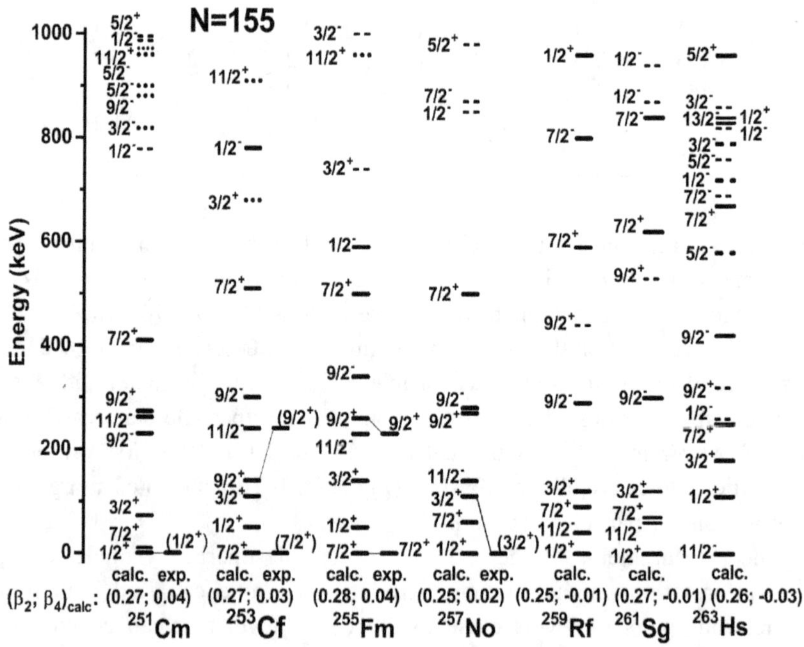

Fig. 2. The lowest calculated states in the N=155 isotonic chain.

If the energy of this isomer is small, its lifetime is long enough and the α-decay could occur from this isomer. The calculated ground state spin changes in ^{257}No when the value of β_4 decreases. Therefore, a small change of deformation could cause the inversion of the levels $1/2^+$ and $7/2^+$ which are close in energy in dense spectrum.

The lowest calculated states in ^{263}Sg and ^{265}Hs have $\Delta K \geq 4$. So, the first excited states in ^{263}Sg and ^{265}Hs can be related to the observed states at 130 keV and above 300 keV, respectively. The lowest $3/2^+$ state can be isomeric in ^{257}Fm, ^{259}No, and ^{261}Rf (Fig. 3) [3].

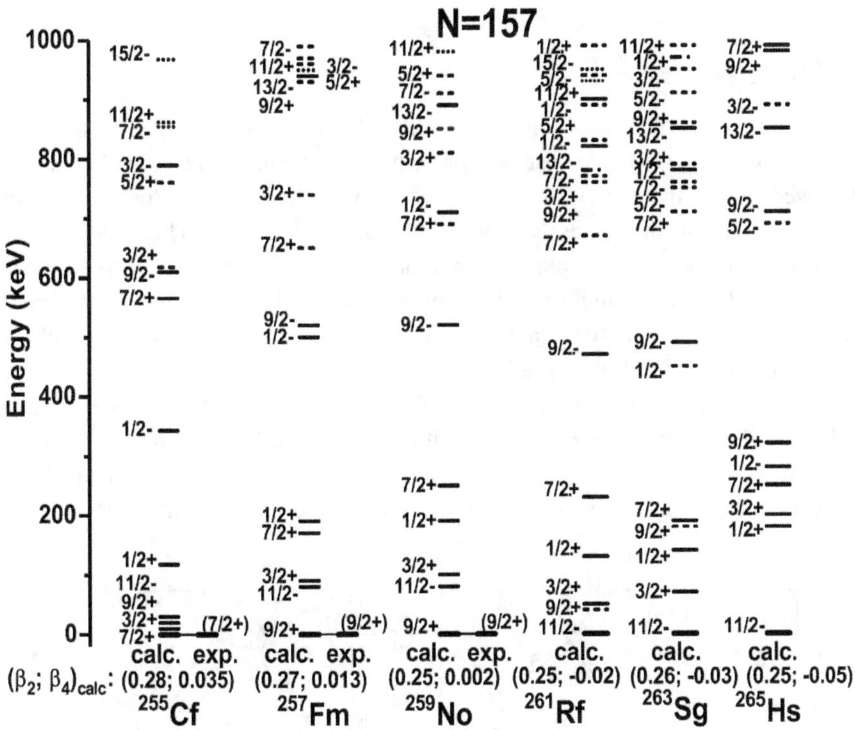

Fig. 3. The lowest calculated states in the N=157 isotonic chain.

3. α-Decay

In ^{285}Fl the single-particle state 9/2[604] could be the isomeric one from which α-decay can occur with $T_\alpha \approx 3$ ms.

The α-decay of ^{277}Ds can occur from the isomeric 9/2[604] and ground state 3/2[611]. The ground 3/2[611] and isomeric 13/2[716] states of ^{273}Hs are populated in the α-decay chain of ^{285}Fl. The α-decay from these states populates the corresponding states in ^{269}Sg from which the α-decays with $T_\alpha \geq 410$ s occur and compete with spontaneous fission. In ^{293}Lv the 7/2[604] state is close in energy to the ground state.

The Skyrme-Hartree-Fock calculations indicate possibility of the two close lines in α-decay spectrum of ^{293}Lv. After the γ-transitions the α- decay of ^{289}Fl likely occurs from the ground state. So, in the α-decay chain of ^{293}Lv there are single lines in the α-decay spectra of ^{289}Fl, ^{285}Cn, ^{281}Ds, and ^{277}Hs. The isomeric states 15/2[707] (^{289}Fl, ^{285}Cn), 11/2[606] (^{281}Ds) and 9/2[604] (^{277}Hs) do not affect the α-decay chain and can be explored in the direct production of these nuclei.

4. Phase Transition

Phase transition phenomenon, especially, shape phase transition is a very interesting subject of investigations which became widespread in nuclei removed out of the valley of stability. It is interesting to look for the possible manifestation of this phenomenon in superheavy nuclei. There are several known indicators of the phase transitions. One of them is a level density parameter. In Fig. 4 the calculated proton a_Z and neutron a_N level density parameters are shown for superheavy nuclei belonging to the α-decay chains containing ^{296,298,300}120 isotopes [4].

As seen, a_Z has a minimum at Z=120 and a_N has a minimum at N=184. This indicates Z=120 and N=184 as a possible proton and neutron magic numbers, respectively. The proton level density parameter has a maximum at Z=112 indicating Z=112 as a transition point from deformed to spherical nuclei.

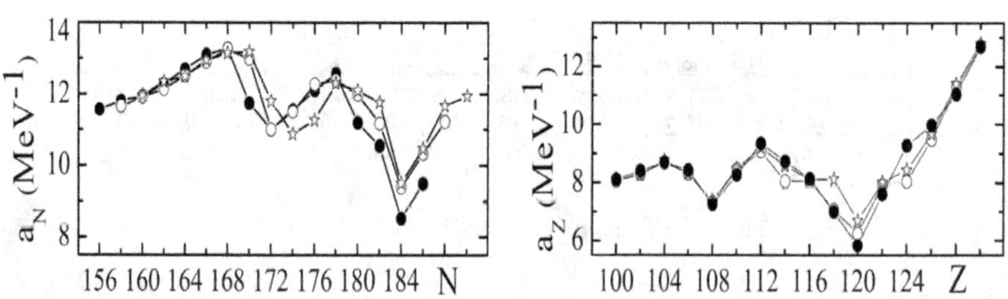

Fig. 4. Calculated neutron (a_N) and proton (a_Z) level density parameters as a function of neutron number N (left) and proton number Z (right). The nuclei from α-decay chains containing ^{296}120, ^{298}120, ^{300}120 are marked by closed circles, open circles, and stars, respectively.

References

1. A. N. Kuzmina, G. G. Adamian, N. V. Antonenko, and W. Scheid, Influence of proton shell closure on production and identification of new superheavy nuclei, *Phys. Rev. C* **85**, 014319 (2012).
2. A. N. Kuzmina, G. G. Adamian, and N. V. Antonenko, Impact of nuclear structure on production and identification of new superheavy nuclei, *Eur. Phys. J. A* **47**, 145 (2011).
3. G. G. Adamian, L. A. Malov, N. V. Antonenko, and R. V. Jolos, Nonrotational states in isotonic chains of heavy nuclei, *Phys. Rev. C* **97**, 034308 (2018)
4. A. N. Bezbakh, T. M. Shneidman, G. G. Adamian, and N. V. Antonenko, Level densities of heaviest nuclei, *Eur. Phys. J. A* **50**, 97 (2014).

Fission Timescale of Superheavy Element Z = 120 from the Langevin Dynamical Model

M. T. Senthil Kannan

Department of Physics, Bharathiar University, Coimbatore-641046, India
E-mail: senthilthulasiram@gmail.com

Jhilam Sadhukhan

Physics Group, Variable Energy Cyclotron Centre, Kolkata-700064, India
Homi Bhabha National Institute, Mumbai-400094, India
E-mail: jhilam@vecc.gov.in

B. K. Agrawal

Saha Institute of Nuclear Physics, Kolkata-700064, India
Homi Bhabha National Institute, Mumbai-400094, India
E-mail: bijay.agrawal@saha.ac.in

M. Balasubramaniam

Department of Physics, Bharathiar University, Coimbatore-641046, India
E-mail: m.balou@gmail.com

Santanu Pal*

CS-6/1, Golf Green, Kolkata-700095, India
E-mail: santanupal1950@gmail.com

The measurement of fission lifetime using nuclear and atomic probes strongly disagree with each other. To examine this discrepancy in measurements, we applied a state-of-the-art Langevin dynamical model including the deformation and energy dependent quantum corrections in nuclear potential energy. For a realistic calculation, light particles and γ-ray evaporations are coupled with the dynamical evolution of fissioning system. We show that the pre-scission neutron multiplicity is an inappropriate probe to predict fission lifetime. Moreover, our calculations on fission lifetime are in agreement with the atomic measurements for a wide range of compound nuclear masses. Specifically, the calculated lifetime of superheavy element Z = 120 is found to support the atomic probe.

Keywords: Fission timescale; Langevin dynamical model; superheavy nuclei.

---

* Formerly with Variable Energy Cyclotron Centre, Kolkata-700064, India.

1. Introduction

The existence and stability of superheavy elements (SHE) rely on nuclear shell effect which depends on the temperature and deformation of the nucleus [1, 2]. Over the years, vast experimental studies have been devoted to synthesize the SHE [3-6]. Fission lifetime is one of the important observable that probes the stability of SHE.

Fission lifetime can be inferred from the pre-scission neutron [7-17], charged particle [10, 14, 17-19], and γ ray [20-23] multiplicities. Also, the concept of transient time is introduced [24] to explain the measured large pre-scission multiplicities by Gavron et al. [25, 26]. Majority of these nuclear measurements assume a statistical model for fission to reproduce the measured multiplicities and, then, the fission lifetime is predicted from this model. In general, a statistical model encompasses different combinations of free parameters such as fission delay time, nuclear dissipation strength, nuclear level density parameter, and fission barrier. Therefore, the prediction of fission lifetime depends on the underlying assumptions of the theoretical model.

In contrast, atomic techniques like the crystal blocking method [1, 27-31] and x-ray method [32, 33] are considered to be more direct probes for fission lifetime measurements [15]. The fission lifetime measured with different atomic probes are found to be $\sim 10^{-18}$s which is, in general, almost two orders slower than nuclear measurements. Further, the long fission timescale of superheavy element Z = 120 is reported using the K x-ray method [33] in agreement with an earlier crystal blocking measurement [1].

Important theoretical efforts are pursued to understand fission timescales [34-36]. The measured attosecond (10^{-18}s) timescale in the actinide region has been reproduced using a combined dynamical and statistical model (CDSM) code [34]. However, a large dissipation strength was required to reach the experimental values [34]. Further, a two dimensional variant of CDSM is developed in combination with one-body dissipation model to reproduce the prescission neutron multiplicities in coincidence with fission lifetime [36]. However, CDSMs are limited to small dynamical time to track the initial buildup of fission probability and the rest of the evolution is followed with suitable statistical model where the appropriate shape-dependence of all the input quantities are sacrificed.

2. The Model

We have applied the stochastic Langevin dynamical model to study the full dynamical evolution of an excited compound nucleus starting from the ground-state configuration up to scission. Here, the collective motion of the elongation coordinate *c*, defined from the Funny-Hills shape parameterization [37], is governed by the one dimensional Langevin equations [35, 38, 39],

$$\frac{dp}{dt} = -\frac{p^2}{2}\frac{d}{dc}\left(\frac{1}{M(c)}\right) - \frac{dF}{dc} - \eta\frac{p}{M(c)} + g\Gamma(t),$$

$$\frac{dc}{dt} = \frac{p}{M(c)}, \tag{1}$$

where M(c) is the shape dependent collective inertia tensor calculated within the Werner-Wheller approximation [40, 41] for incompressible and irrotational flow of nuclear fluid; η is the one-body dissipation strength weighted with the chaos factor [41, 42]. The $\Gamma(t)$ is the fluctuating force with the same time-correlation property as a Gaussian random number. The strength of random force g obeys the Einstein dissipation-fluctuation theorem: $g^2 = \eta T$, where T is the compound nucleus temperature. The T is calculated from the ground state excitation energy using Fermi gas model: $T = \sqrt{E^*/a_0}$; α_0 being the level density parameter at the ground state deformation. The excitation energy E^* of the CN is calculated using the energy conservation $E^* = E_{c.m.} - V_Q - V(c)$, where $E_{c.m.}$ is the center-of-mass energy, V_Q is the experimental *Q*-value of the fusion-fission reaction, and *V(c)*, is the shape dependent potential energy.

In the present work, the driving potential *F* is calculated using the Helmholtz free energy expression $F(c, T) = V(c)-(a(c)-a0)\,T^2$. The shape and energy dependent level density parameter *a*(*c*) is calculated within the modified Ignatyuk prescription [43]. *V*(*c*) is evaluated from the double folding Yukawa-plus-exponential form [44] of liquid drop model. The associated shell correction energy is calculated following the Strutinsky's method [37, 45] of shell correction to the nucleonic levels generated with the two centered Woods-Saxon mean field [46]. We use the BCS pairing to account for super fluidity [37, 46]. The calculated potential barriers are found to be in good agreement with the existing results [47].

Although the pre-scission neutron emission is the major competitive process with the fission decay, all the possible particle emissions (n, p, α and statistical γ-rays) are included in the calculation. The statistical theory of Weisskopf [35, 48] is employed to calculate the particle evaporation widths.

During the time evolution, changes in the compound nuclear properties are updated following the associated evaporation channel. We have considered an ensemble of 10^6 Langevin events to generate the result for a single macrostate. The time evolution for each Langevin trajectory was considered up to 10^{-15}s with a time-step of 10^{-25}s. The initial angular momentum of the compound nucleus is sampled from the corresponding fusion spin distribution. We consider the scission to occur when the neck radius of the compound nucleus becomes equal to $0.3R_0$ [37, 40]; R_0 being the spherical radius of the nucleus. The time τ_{nl} when the last neutron is evaporated and the scission time τ_f are recorded for each fission event. Here, τ_{nl} defines the neutron-clock for fission with the underlying assumption that scission occurs immediately after the last neutron is evaporated.

3. Results and Discussions

First, we consider the pre-actinide compound nucleus ^{200}Pb formed in the reaction $^{19}F + ^{181}Ta$. This system has been studied experimentally [49] around $E^* = 100$ MeV and, subsequently, analyzed in Ref. [34]. In this analysis, a comparatively large dissipation strength was required within the dynamical model. The average lifetimes $<\tau_{nl}>$ and $<\tau_f>$ from the present calculation are depicted in Fig. 1. At all the excitation energies $<\tau_f>$ is found to be $\geq 10^{-17}$s and it is in agreement with the experimental findings of Ref. [49]. However, the lifetime inferred from the last neutron ($<\tau_{nl}>$) is $\sim 10^{-19}$s.

The ratio $<\tau_f> / <\tau_{nl}>$ is plotted in Fig. 1 for a better visualization of this difference in timescales.

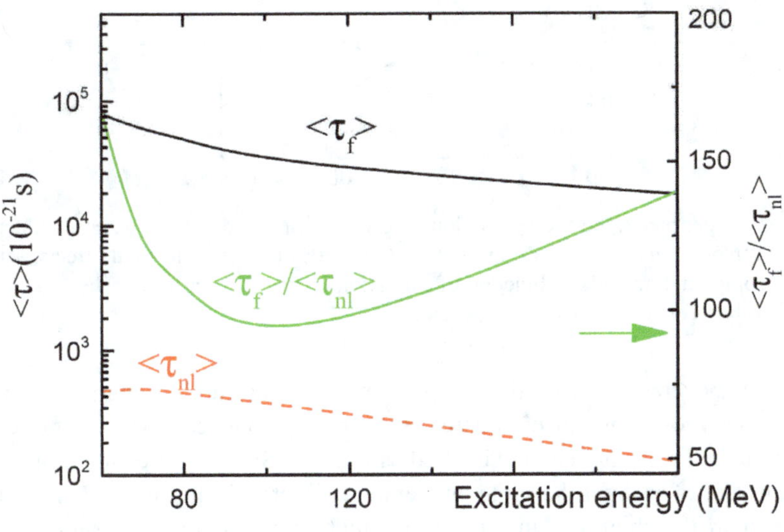

Fig. 1. The average fission timescale of ^{200}Pb compound nuclei as a function of excitation energy. The right scale corresponds $<\tau_f>/<\tau_{nl}>$.

Next, we consider the reaction $^4He + ^{232}Th \rightarrow ^{236}U$. The averages $<\tau_{f1}>$ and $<\tau_{nl1}>$, corresponding to excitation-energy dependent shell-damping within Fermi gas model, are shown in Fig. 2. Also, the fission lifetime of U-like nuclei, extracted from the crystal blocking experiment [50], is shown in the same figure. As evident, $<\tau_f>$ fails to explain the experimental fission time. At low energies $E^* < 60$ MeV, our results agree well with the experimental data. Further, we performed another set of calculation by reducing the rate of shell damping with E^* to 50% of its original value derived from Fermi gas model [51]. In other words, the rate of decrease in shell effects due to temperature effects is 50% reduced. The corresponding averages ($<\tau_{f2}>$ and $<\tau_{nl2}>$) are plotted in Fig. 2.

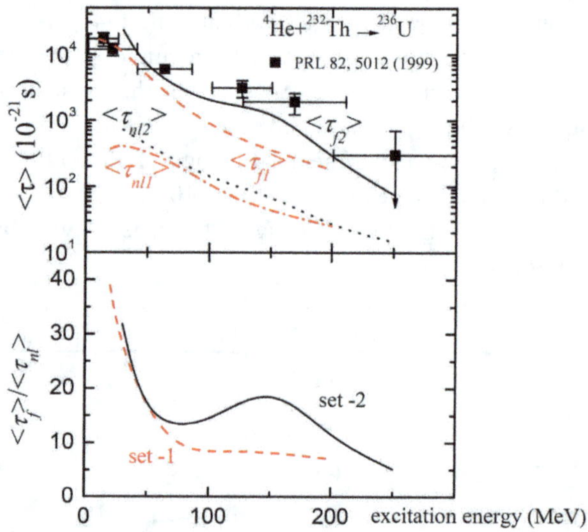

Fig. 2. Upper panel: the average fission timescale $< \tau_{f1} >$ and $< \tau_{nl1}>$ (without reduction in shell effects, set-1) and $< \tau_{f2}>$ and $< \tau_{nl2}>$ (with 50% reduction in shell effects, set-2) of ^{236}U compound nuclei as a function of excitation energy. Lower panel: The $< \tau_f > / < \tau_{nl}>$ ratio values for set-1 and set-2 calculations.

Now, the average scission time ($<\tau_{f2}>$) becomes close to the experimental values over the whole range of E^*, whereas $<\tau_{nl}>$ hardly changes. The use of a reduced rate of shell damping can be justified as follows. Since we are employing Fermi gas model to account the finite temperature effects, all the correlations that help to sustain the shell structure even at a high temperature are absent. A finite-temperature shell model calculation would have been more appropriate. Instead, we are slowing down the disappearance of shell effect to mimic the actual scenario.

Finally, we address the stability of superheavy element Z = 120 through fission timescale calculation. We have selected the $^{238}U + ^{64}Ni \rightarrow ^{302}120$ reaction as it is preferred to be the ideal reaction to reach Z = 120. Accurate calculation of potential energy surface is crucial for this compound system. Specifically, there is a large ambiguity in the determination of fission barrier of SHE [52, 53]. As in case of lighter systems, we have used the same interaction to calculate the potential profiles for the Z = 120 and its daughter systems. We found that the calculated fission barriers are within the theoretical uncertainties.

Calculations are performed for two sets of shell damping as described above. The distributions of τ_f for different E^* are shown in Fig. 3. Clearly the rate of shell damping plays a strong role to enhance the fission lifetime. The average fission lifetime ($<\tau_f>$) for a range of excitation energy relevant to the experimental data is shown in Fig. 4. For the full shell damping, the Z = 120 system survived more than attosecond only if $E^* \leq 10$ MeV. However, when shell effect survives at relatively higher excitation energy, the timescale more than 10^{-18}s can be reached even at $E^* \approx 30$ MeV. From this study the importance of the proper estimation of temperature and shape dependence in shell effect is highlighted. Further, this study suggests the importance of advanced microscopic calculation in the superheavy region [2, 51].

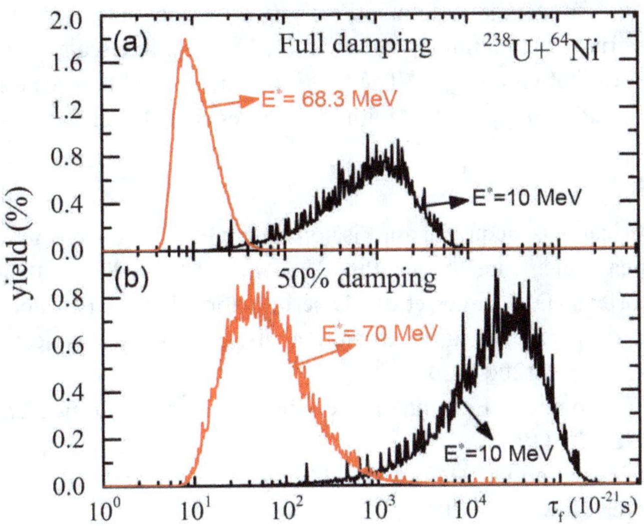

Fig. 3. (a) Distributions of τ_f for full shell damping. (b) Distributions of τ_f for reduced rate of shell damping to 50%.

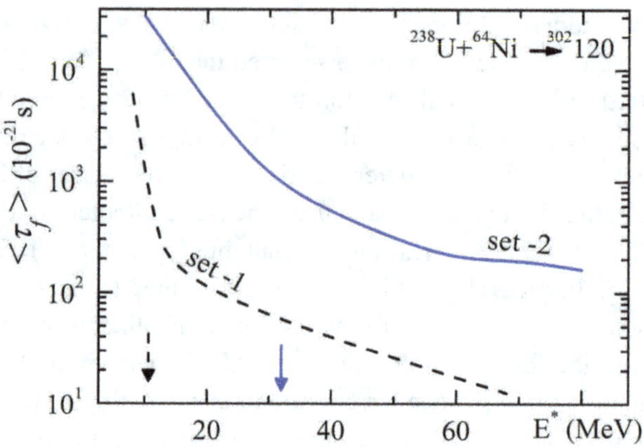

Fig. 4. The fission timescale of superheavy element Z = 120 for set-1 and set-2 calculations. This figure is taken from Ref. [51].

4. Summary

The state-of-the-art Langevin dynamical model, including the shape and temperature dependent shell effects is applied to the study of fission timescale calculations of heavy and superheavy nuclei. The scission timescales are calculated in agreement with the atomic clock measurements. The neutron-clock devised to measure fission lifetime from the last prescission neutron is found to be inappropriate. Pre-actinide region nucleus ^{200}Pb timescale is greater than 10^{-17}s at all excitation energies. The measured attosecond timescale for Z = 120 nucleus is found to be consistent with our theoretical calculations.

References

1. M. Morjean, D. Jacquet et al., Fission time measurements: A new probe into superheavy element stability, *Phys. Rev. Lett.* **101**, 072701 (2008).
2. M. Morjean, D. Jacquet et al., Discrimination between fission and quasi-fission from reaction time measurements, Journal of Physics: Conference Series 282, p. 012009 (2011).
3. S. Bjrnholm and J. E. Lynn, The double-humped fission barrier, *Rev. Mod. Phys.* **52**, 725 (1980).
4. H. J. Krappe and K. Pomorski, Theory of Nuclear Fission: A Textbook (Springer, New York, 2012).
5. Y. T. Oganessian, V. K. Utyonkov et al., Synthesis of superheavy nuclei in the ^{48}Ca + ^{244}Pu reaction, *Phys. Rev. Lett.* **83**, 3154 (1999).
6. Y. T. Oganessian and V. K. Utyonkov, Super-heavy element research, *Rep. Prog. Phys.* **78**, 036301 (2015).

7. D. J. Hinde, R. J. Charity et al., Neutron multiplicities in heavy-ion induced fission: Timescale of fusion-fission, *Nucl. Phys. A* **452**, 550 (1986).
8. D. J. Hinde, D. Hilscher and H. Rossner, Fission timescales of excited nuclei, *Nucl. Phys. A* **502**, 497 (1989).
9. D. J. Hinde, D. Hilscher et al., Neutron emission as a probe of fusion-fission and quasifission dynamics, *Phys. Rev. C* **45**, 1229 (1992).
10. J. P. Lestone, Determination of the time evolution of fission from particle emission, *Phys. Rev. Lett.* **70**, 2245 (1993).
11. D. J. Hinde, Neutron emission as a clock and thermometer to probe the dynamics of fusion-fission and quasi-fission, *Nucl. Phys. A* **553**, 255 (1993).
12. K. Siwek-Wilczynski, J. Wilczynski et al., Time scales of fusion-fission reactions calculated from prescission neutron multiplicities, *Nucl. Phys. A* **583**, 141 (1995).
13. J. Cabrera, T. Keutgen et al., Fusion-fission and fusion-evaporation processes in ^{20}Ne+^{159}Tb and ^{20}Ne+^{169}Tm interactions between e/a = 8 and 16 MeV, *Phys. Rev. C* **68**, 034613 (2003).
14. K. Ramachandran, A. Chatterjee et al., Fission time scale from prescission neutron, proton, and α particle multiplicities in ^{28}Si+^{175}Lu, *Phys. Rev. C* **73**, 064609 (2006).
15. D. Jacquet and M. Morjean, Fission times of excited nuclei: An experimental overview, *Prog. Part. Nucl. Phys.* **63**, 155 (2009).
16. S. Appannababu, M. Cinausero et al., Mass, total kinetic energy, and neutron multiplicity correlations in the binary fragmentation of ^{50}Ti + ^{208}Pb at 294 MeV bombarding energy, *Phys. Rev. C* **94**, 044618 (2016).
17. K. Kapoor, S. Verma et al., Fission time scale from pre-scission neutron and α multiplicities in the ^{16}O + ^{194}Pt reaction, *Phys. Rev. C* **96**, 054605 (2017).
18. J. P. Lestone, J. R. Leigh et al., Fission time scales from prescission charged-particle multiplicities, *Phys. Rev. Lett.* **67**, 1078 (1991).
19. J. Boger, J. M. Alexander et al., Tests for equilibration of ^{149}Tb* composite nuclei produced in the reactions 337 MeV ^{40}Ar+^{nat}Ag and 640 MeV ^{86}Kr+^{63}Cu, *Phys. Rev. C* **49**, 1587 (1994).
20. M. Thoennessen, D. R. Chakrabarty et al., Giant dipole resonance in highly excited thorium: Evidence for strong fission hindrance, *Phys. Rev. Lett.* **59**, 2860 (1987).
21. P. Paul, The GDR as clock for fission dynamics in hot actinide nuclei, *Nucl. Phys. A* **569**, 73 (1994).
22. G. van 't Hof, J. C. S. Bacelar et al., Viscosity, fission time scale and deformation of ^{156}Dy, *Nucl. Phys. A* **638**, 613 (1998).

23. D. J. Hofman, B. B. Back et al., Viscosity of saddle-to-scission motion in hot ^{240}Cf from Giant Dipole Resonance Yield, *Phys. Rev. Lett.* **72**, 470 (1994).
24. P. Grange, L. Jun-Qing and H. A. Weidenmuller, Induced nuclear fission viewed as a diffusion process: Transients, *Phys. Rev. C* **27**, 2063 (1983).
25. A. Gavron, J. R. Beene et al., Time scale of fission at high angular momentum, *Phys. Rev. Lett.* **47**, 1255 (1981).
26. A. Gavron, J. R. Beene et al., Time scale of fission at high angular momentum, *Phys. Rev. Lett.* **48**, 835 (1982).
27. F. Brown, D. A. Marsden and R. D. Werner, Use of blocking in crystals to study the lifetime for fission of ^{238}U by 12-MeV protons, *Phys. Rev. Lett.* **20**, 1449 (1968).
28. J. U. Andersen, E. Laegsgaard et al., Time evolution of heavy-ion induced fission studied by crystal blocking, *Phys. Rev. Lett.* **36**, 1539 (1976).
29. M. Morjean, M. Chevallier et al., Fission lifetime measured by the blocking technique as a function of excitation energy in the 24 A MeV ^{238}U+^{28}Si reaction, *Nucl. Phys. A* **630**, 200 (1998).
30. J. U. Andersen, J. Chevallier et al., Crystal blocking measurements of the time delay of fission induced by ^{32}S, ^{48}Ti, and ^{58}Ni bombardment of W, *Phys. Rev. Lett.* **99**, 162502 (2007).
31. J. U. Andersen, J. Chevallier et al., Attosecond time delays in heavy ion induced fission measured by crystal blocking, *Phys. Rev. C* **78**, 064609 (2008).

A Systematic TDDFT Data for Nuclear Fission Analysis – Z = 120 Isotopes –

Takashi Nishikawa[1*], Yoritaka Iwata[2] and Satoshi Chiba[2]

[1]Nuclear Engineering Ltd, Osaka 550-0001, Japan
[2]Tokyo Institute of Technology, Tokyo 152-8550, Japan
**E-mail: rg12f090@gmail.com*

Dissipation of superheavy nuclei is studied by means of TDDFT+ Langevin model. Much attention is paid to the energy dependence of the friction coefficient, identifying the thermodynamic property of nuclear medium for given nucleon numbers, neutron-richness, and energies. In this article, following the preceding work showing a systematics on Z = 92 to 100 nuclei, macroscopic friction coefficients for Z = 120 isotopes are derived from a microscopic framework. The comparison between Z = 92 and Z = 120 cases clarifies the similarity and the difference of fissions between heavy and superheavy nuclei. That is not only the completion of a systematic theoretical database, but also discovering the existence limit of chemical elements. In addition, the knowledge about the obtained fission probability and fission fragment yield are expected to be useful for superheavy synthesis taking place in accelerators.

Keywords: Superheavy nuclei; nuclear fission; TDDFT; Langevin model.

1. Introduction

The synthesis of superheavy nuclei is a popular issue under the world wide competition for the naming rights of new chemical elements. The nuclei with their proton number Z = 120 collect a special attention, since they are the next unnamed element at this point. Since the fission is highly affected by the dissipation, the dissipation is associated with a certain kind of the stability of heavy nuclei.

In this article, following the preceding work [1] showing a systematics on Z = 92 region (Uranium, Plutonium isotopes and so on), the dissipation in many-nucleon systems is studied by the collective friction effects. A systematics of friction coefficients for Z = 120 isotopes is presented.

2. Proposed Method

The analysis of quasi-fission events is made based on the TDDFT + Langevin model being introduced by Nishikawa et al. [1] after our preparatory works on nuclear fission [2-5], where the terminology "TDDFT" stands for the time dependent density functional theory. The TDDFT + Langevin model is explained in Ref. [1]. The numerical code Sky3D [6] is employed for the

TDDFT calculations, and fission dynamics is calculated by 4D Langevin code [7]. For the TDDFT calculations, we employ three different effective nuclear interactions: SV-bas [8], SLy4 [9], and SkM* [1]0. We carried out the TDDFT calculations of quasi-fission processes by symmetric central collision reactions

$$^{A}Z + {}^{A}Z \rightarrow {}^{2A}2Z \rightarrow {}^{A}Z + {}^{A}Z \tag{1}$$

where A and Z are a mass number and a proton number respectively. In particular, Z is fixed to Z = 120 in this research, and several excitation energies for each choice of A = 276, 284 . . . and 308 are examined. Here the energy E is taken as the initial energy of the collision, and E = A is from 3 MeV to 7 MeV. The distance $R(t)$ between the center-of-mass of two colliding nuclei is extracted from the TDDFT wave function. For the details of $R(t)$, see Ref. [11]. The averaged friction coefficient is calculated by the $R(t)$

$$\underline{\gamma}(E) = \frac{\left\{\frac{1}{2}\mu\dot{R}^2(t_i) + V(t_i)\right\} - \left\{\frac{1}{2}\mu\dot{R}^2(t_f) + V(t_f)\right\}}{\int_{t_i}^{t_f}\left\{\dot{R}(t)\right\}^2 dt}. \tag{2}$$

where $\underline{\gamma}(E)$ is averaged for time t. t_i and t_f are usually taken as the initial and the final time of reaction. Note that $\underline{\gamma}(E)$ values at extra-ordinary energies, which are too low or too high to be treated by the TDDFT, are also obtained using the extrapolation method being explained in Ref. [1].

The $\underline{\gamma}(E)$ at the extrapolated lower energy is substituted to the Langevin coefficient γ_{11}, and fission dynamics is obtained by Langevin calculation. γ_{11} is the friction coefficient of the motion along $R(t)$. Detailed explanation of Langevin calculation is shown in Ref. [7]. By this procedure, both stochastic aspect and microscopic aspect of fission dynamics are incorporated.

3. Result

Figure 1 shows the fission fragment yields (FFYs) for ^{276}Ubn and ^{300}Ubn, where unnamed Z = 120 element is generally denoted by Ubn. The transitions from symmetric fission to asymmetric fission are well described. The mass number is large enough for the sizable effect of fission in which more than 10% of excited Z = 120 nuclei result in fission. In this case, we found no corresponding experimental fission data to compare, so that the calculation is carried out as the prediction for future research in the context of figuring out the role of dissipation in superheavy synthesis. By comparing low and high energy cases, the energy dependence is remarkable in which asymmetric fissions are favored for lower energies and symmetric fissions are favored for higher energies. Indeed, for low energy fission, A = 126 and 150 nuclei are more produced in

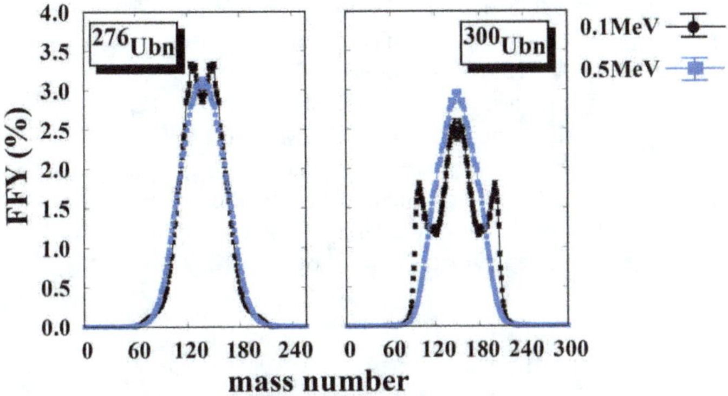

Fig. 1. Fission fragment yields (FFYs) of Z = 120 isotopes calculated by the TDDFT + Langevin model (SV-bas). The FFYs at 0.1 MeV and 0.5 MeV per nucleon are shown for ^{276}Ubn and ^{300}Ubn. The corresponding fission probability is 23% for 0.1 MeV and 41% for 0.5 MeV in case of ^{276}Ubn, while 12% for 0.1 MeV and 39% for 0.5 MeV in case of ^{300}Ubn.

the fission of ^{276}Ubn, while A = 98, 147, 153 and 202 nuclei are obtained in the fission of ^{300}Ubn.

Fission events with non-negligible fission probability are calculated to be roughly equal to 15% for 0.1 MeV and 40% for 0.5 MeV. The present prediction is worth believing by accounting for the agreement between the TDDFT + Langevin result and the experimental FFYs in Uranium and Plutonium isotopes [1].

The friction coefficient $\gamma(E)$ at the energy closed to the Coulomb barrier is shown in Table 1. Fission at or around the barrier energy is useful to analyze the low-energy fission events such as quasi-fission, photo-fission, neutron-induced fission and so on.

Energy and mass dependence of γ can be found in Fig. 2. As a trend, the amplitude of γ becomes smaller for higher energies.

Such a trend is reasonable, because the shorter duration time in higher energy collisions should lead to the smaller amount of dissipated energy in reaction processes. The neutron number dependence is not definitely monotonous. Although the minimal value can be found in Fig. 2 (noted only for SV-bas): e.g., the minimum at N = 180, its origin is not easy to be identified, because it is highly affected by the choice of interaction with many factors such as shell effect and/or Pauli effect. The detail investigation on the energy and mass dependence of $\gamma(E)$ is a future problem in which the different properties of effective nuclear interaction should be investigated more carefully.

Table 1. The friction coefficients $\gamma(E)$ of unit $\hbar$ calculated for Ubn isotopes (Z=120) by three effective nuclear interactions (SV-bas, SLy4 and SkM*). The energy is E = 20 MeV (E/A~0.17 MeV) being calculated by the extrapolation method [1]. In the column "Average", the averaged values for these three interactions are shown.

N	SV-bas	SLy4	SkM*	Average
156	171.08	159.61	165.93	165.54
164	170.57	140.38	163.29	158.08
172	165.52	143.49	162.68	157.23
180	148.33	138.14	170.03	152.17
188	183.14	146.06	143.50	157.57

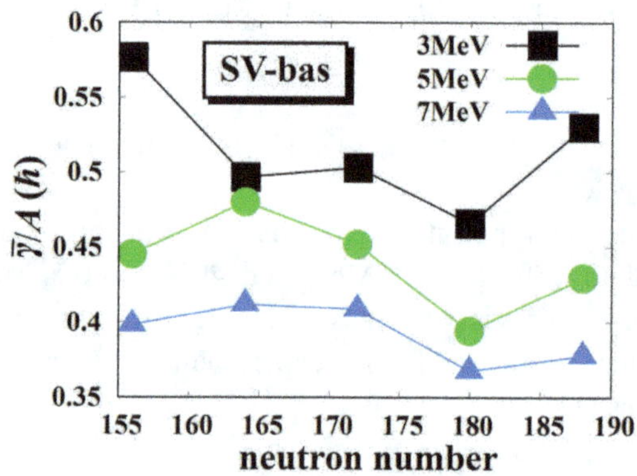

Fig. 2. Energy and mass dependence of friction coefficient γ per nucleon, where A denotes the mass number. The amplitude of γ stands for the amount of dissipated energy at a given excitation energy.

Consequently, for Ubn isotopes, the energy dependence for E/A = 5 MeV and 7 MeV are quite similar, then it suggests the dissipation mechanism of one kind is activated between 3 MeV and 5 MeV per nucleon. Since this effect depends on the neutron number, the dissipation of superheavy nuclei is suggested to be sensitive to the symmetry energy.

4. Summary

Ubn isotopes are regarded as the next unnamed elements at this point. The present research will contribute to identify rather stable nuclei among Ubn isotopes from the point of view of fission process. The friction coefficients being utilizable to the Langevin calculations have been obtained based on the TDDFT framework. In a more microscopic point of view, possible dissipation for a given excitation energy is systematically obtained for Z = 120 isotopes. Since the low-energy heavy-ion reactions (~E/A < 10 MeV) have known to be well described by the TDDFT (for a review, see Ref.[15]), the reliability of proposed method is expected to be sufficiently high.

The sizable effect of binary fission on the stability of Ubn isotopes is presented; the fission probability is almost 10% for E/A = 0.1 MeV, and almost 40% for E/A = 0.5 MeV. Asymmetric fission occurs for low energy (E/A = 0.1 MeV), while symmetric fission occurs for high energy (E/A = 0.5 MeV).

Table 2. The friction coefficients $\gamma(E)$ of unit $\hbar$ calculated for Ubn isotopes (Z = 120) by three effective nuclear interactions (SV-bas, SLy4 and SkM*). The energy is taken for 3 MeV, 5 MeV and 7 MeV per nucleon. In the column "Average", the averaged values for these three interactions are shown.

a) $E/A = 3$ MeV

N	SV-bas	SLy4	SkM*	Average
156	159.13	139.87	156.61	151.87
164	141.06	123.64	146.87	137.19
172	146.88	122.21	134.68	134.59
180	139.78	113.53	137.84	130.39
188	163.28	130.63	127.33	140.41

b) $E/A = 5$ MeV

N	SV-bas	SLy4	SkM*	Average
156	122.86	103.60	109.78	112.08
164	136.35	99.38	120.62	118.78
172	131.97	105.25	117.55	118.26
180	118.24	96.37	115.17	109.93
188	132.36	100.58	110.01	114.32

c) $E/A = 7$ MeV

N	SV-bas	SLy4	SkM*	Average
156	109.85	78.00	90.21	92.69
164	116.98	79.18	86.08	94.08
172	119.30	78.34	89.29	95.64
180	110.28	84.01	102.69	98.99
188	116.35	80.87	97.70	98.30

In terms of the future utility, the obtained coefficients are summarized in Table 2. The γ values depend more prominently on the energy than on the neutron-richness. Although smaller γ is calculated for heavier cases in Z = 50 and Z = 92 regions, such an effect is weaken in the superheavy cases (Z = 120). For E = 20 MeV roughly comparable to Table 1, γ values distribute from 110 to 120 for the uranium isotopes (N = 142 to 150). It also tells us another aspect of the mass dependence; i.e., Z dependence. Table 2 provides a part of theoretical database of systematic friction coefficients. The presented result should be useful to superheavy nuclei in accelerators, natural reactors, and artificial reactors.

Acknowledgement

Numerical computation was carried out at a workstation system at Tokyo Institute of Technology (AEGIS system). This work was supported by JSPS KAKENHI Grant No. 17K05440. One of the authors (Y. I.) is grateful to Prof. Dr. Hans Feldmeier for presenting an essential idea at the preparatory stage of this research.

References

1. T. Nishikawa, Y. Iwata, and S. Chiba, in preparation.
2. T. Nishikawa, Y. Iwata, and S. Chiba, to appear in JAEA-Conf (2018).
3. Y. Iwata and S. Heinz, Exotic Nuclei, EXON-2012, World Scientific, 153 (2013).
4. Y. Iwata and S. Heinz, CERN Proceedings 2012 - 002, 241 (2013).
5. Y. Iwata, JAEA-Conf 2015-003, INDC(JPN)-201, 93 (2015).
6. J. A. Maruhn, P.-G. Reinhard et al., Comput. Phys. Commun. **185**, 2195 (2014).
7. M. D. Usang, F. A. Ivanyuk, C. Ishizuka, and S. Chiba, *Phys. Rev. C* **94** 044602 (2016).
8. P. Klupfel, P.-G. Reinhard, J. Burvenich, and J. A. Maruhn, *Phys. Rev. C* **79** 034310 (2009).
9. E. Chabanat, P. Bonche, P. Haensel, J. Meyer, and F. Schaeffer, *Nucl. Phys. A* **635**, 231 (1998).
10. J. Bartel, P. Quentin, M. Brack, C. Guet, and H. B. Hakansson, *Nucl. Phys. A* **386**, 79 (1982).
11. Y. Iwata, H. Feldmeier, and J. A. Maruhn, *AIP Conf. Proc.* **1377**, 365 (2011).
12. K. Washiyama and D. Lacroix, *Phys. Rev. C* **78** 024610 (2008).
13. K. Washiyama, D. Lacroix, and S. Ayik, *Phys. Rev. C* **79**, 024609 (2009).
14. T. Dossing and J. Randrup, *Nucl. Phys. A* **475**. 557 (1987),
15. C. Simenel, and A. S. Umar, *Prog. Part. Nucl. Phys.* in press (2018).

Heavy-Ion Reaction and Fission Studies at the JAEA Tandem Accelerator Facility

K. Nishio
Advanced Science Research Center, Japan Atomic Energy Agency, Tokai, Ibaraki, 319-1195, Japan
E-mail: nishio.katsuhisa@jaea.go.jp

The tandem accelerator facility of Japan Atomic Energy Agency now gives the highest terminal voltage over the world, and offers unique opportunities for nuclear physics, chemistry, and material studies, especially by the availability of various radioactive target materials. Concerning the nuclear reaction and fission studies, we started to promote programs to use multi-nucleon transfer (MNT) reactions for the fission and nuclear structure studies as well as the precise mass measurement, particularly for neutron-rich and heavy-element nuclides. In this contribution, recent results on MNT induced fissions will be presented.

Keywords: Tandem accelerator; multi-nucleon transfer reaction; fission.

1. Introduction

Production of the heaviest elements/nuclei is one of the biggest challenges in nuclear physics, and nuclides with elements up to 118 (Og) are produced so far [1, 2]. They are produced by heavy-ion fusion and evaporation reactions using stable isotope beams and available target materials. The island of stability, created by the shells of neutron number N=184 and proton number Z=114 or 126, cannot be accessed in such reactions, however, due to the lack of neutrons given to the compound nucleus. In comparison to the heavy-ion fusion reaction, multi-nucleon transfer (MNT) reaction has an unique feature that relatively neutron-rich nuclei can be produced by opening the channels transferring larger number of neutrons than protons. For example production of neutron-rich super heavy nuclei is discussed in the reaction $^{238}U+^{248}Cm$ [3].

We report the application of MNT reaction in fission studies [4-6]. Another unique feature of the MNT reaction for fission study is that excited states in compound nucleus distribute widely from nearly the ground state to several tens MeV, depending on the choice of the reaction and reaction energy, thus allowing us to study evolution of fission properties, such as fission fragment mass distributions (FFMDs), with excitation energies. Threshold for fission, appearing in the excitation function of fission probability, gives the fission barrier height, important quantity to describe fission. By using a relatively heavy ion beam such as ^{18}O, as shown in this report, new fission barrier data for larger number nuclides can be obtained.

Figure 1 is the schematic explanation of the experiment to study MNT induced fissions. We are systematically taking data using ^{18}O projectile, bombarded to several radioactive targets, including Np, Pu, Am, Cm, Bk, Cf isotopes. We also plan to carry out the experiment using ^{254}Es (Z=99). Taking advantage of many transfer channels yielding different compound nuclei in one reaction, significantly wide range of nuclides can be investigate in this campaign.

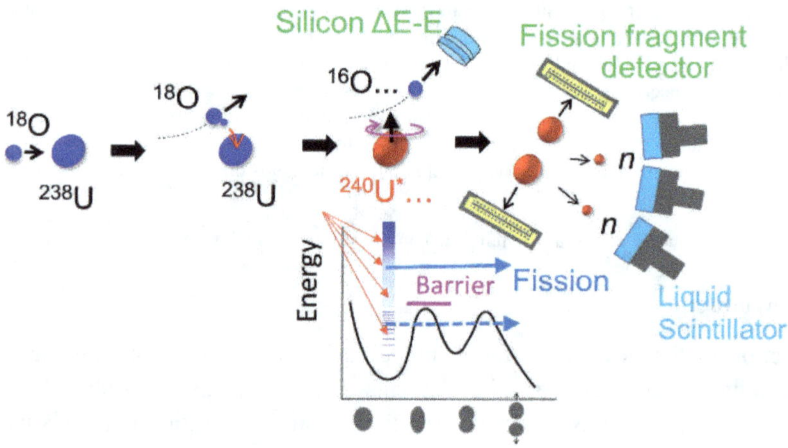

Fig. 1. Principle to study fission using multi-nucleon transfer channels.

2. Experimental Setup

Figure 2 shows the experimental setup. The target (typically ~50-100 μg/cm2), made by electrically depositing the material on a nickel foil (~300 μg/cm2), are irradiated by the ^{18}O beam at energy of about 9 MeV/u. Transfer channels were identified by detecting the projectile-like nucleus using a silicon ΔE-E telescope, mounted to the forward direction of the target. The thickness of the ΔE layer is 75 μm. Twelve pieces of the ΔE detectors were mounted conically around the beam axis to make an efficient collection of the projectile-like nuclei. Particles passing through the ΔE detector were detected by a silicon strip E detector (SSD, 300 μm) to measure the residual energy (E_{res}). The E detector is the annular type strip detector, which can accept scattered particle at angles from ~17° to 31° relative to the beam direction. Figure 3 shows an example of the projectile-like nuclei plotted on the (E_{res}, ΔE) plane. Oxygen isotopes are clearly

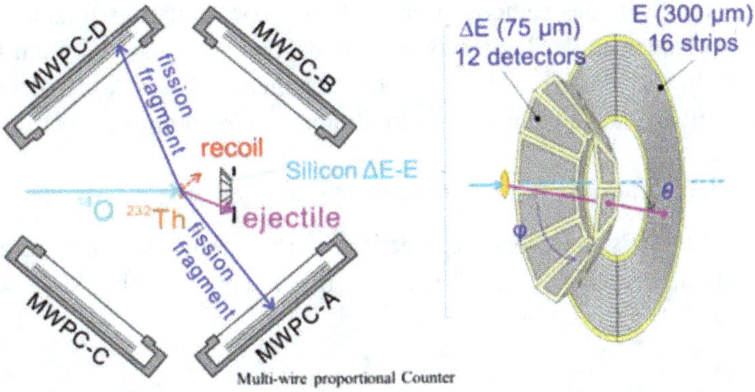

Fig. 2. Experimental setup for the MNT induced fission studies. Neutron detectors are not shown.

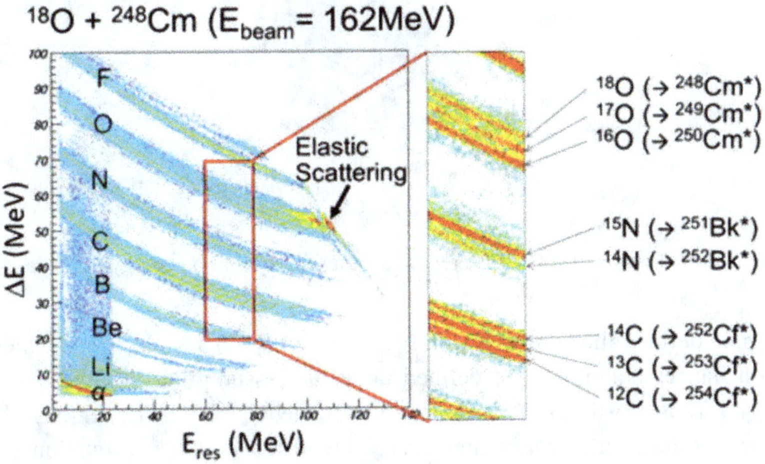

Fig. 3. Particle identification with the silicon ΔE-E telescope in the reaction of $^{18}O+^{248}Cm$.

separated as well as those of lighter-element isotopes. By choosing a specific transfer channel line, we can assign the corresponding compound nucleus by assuming a binary reaction process.

Fission fragments were detected using four multi-wire proportional counters (MWPCs). Each MWPC has an active area of 200×200 mm^2. The MWPC consists of the central cathode which is sandwiched by two wire planes. The wire planes were designed to detect the incident position of a fission fragment.

Induced charge in the cathode was used to separate fission fragments from scattered particles and/or lighter ions. Time difference signal, ΔT, from the two facing MWPCs was recorded. Fission fragment masses were determined cinematically using ΔT value and fission fragment directions.

3. Selected Experimental Results and Discussions

Figure 4 shows fission events recorded on the plane on the fragment mass vs total excitation energy E^*_{tot} of the system, obtained in the MNT channel ^{237}Np (^{18}O, ^{19}O) ^{236}Np*.

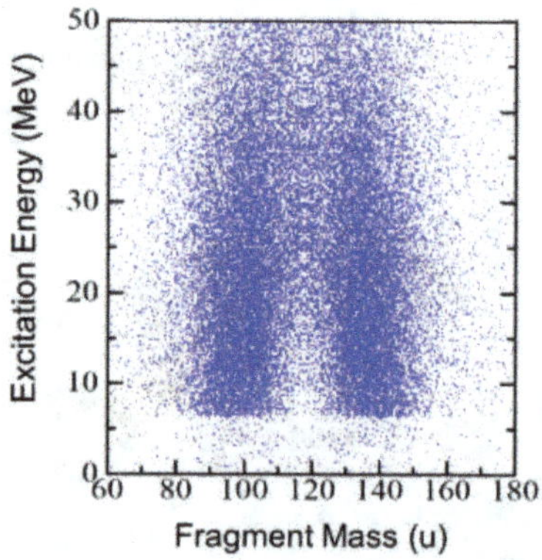

Fig. 4. Fission events registered on the mass vs excitation energy of ^{236}Np.

The plot defines the threshold for fission at $E^*_{tot} = 6.3$ MeV, from which the fission barrier height can be defined under the assumption that the excitation energy E^*_{tot} is mostly given to the recoiled nucleus, ^{236}Np*. The regions of light and heavy fragment groups are clearly identified at lower-excitation energy region, which gradually ambiguous toward the higher energies due to the smearing of the shells responsible for mass asymmetric fission.

Figure 5 shows the FFMDs of 23 nuclides and their E^*_{tot} dependence, obtained in the MNT channels of thc ^{18}O+^{237}Np reaction.

At the lowest energy region, all the studied nuclei show the prominent mass asymmetric fissions. The structure tends to have a broader symmetric fission at higher excitation energies. It is also found by inspecting the spectra of $E^*_{tot} = 20 - 30$ MeV that the peak-to-valley ratio (P/V) of the FFMD gradually decreases with proton number of fissioning nucleus. In accordance with the number of transferred protons from the projectile to the target nucleus, it is evident that population of low excited states is hindered. One of the plausible

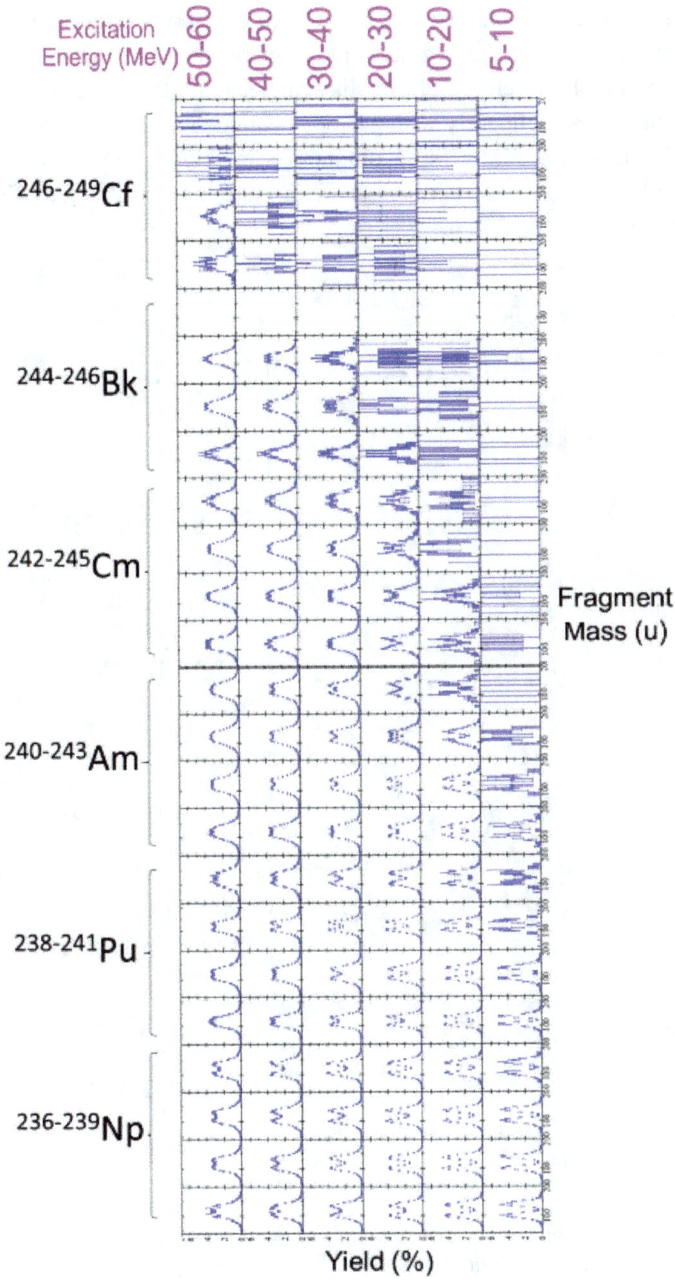

Fig. 5. Fission fragment mass distribution obtained in the MNT channels of the $^{18}O+^{237}Np$ reaction. Excitation energy of the system is shown in the right side.

interpretation would be that the several nucleons are transferred with the form of cluster, instead of transferring individual nucleons sequentially, so that the Coulomb barrier between the cluster-like nucleus and the target nucleus can be defined and enhanced linearly with protons contained in the cluster.

The FFMDs of ^{236}Np obtained in the MNT channels of ^{237}Np (^{18}O,^{19}O)^{236}Np* are compared with the proton-induced fissions of ^{235}U (^{236}Np*) [7, 8], see Fig. 6.

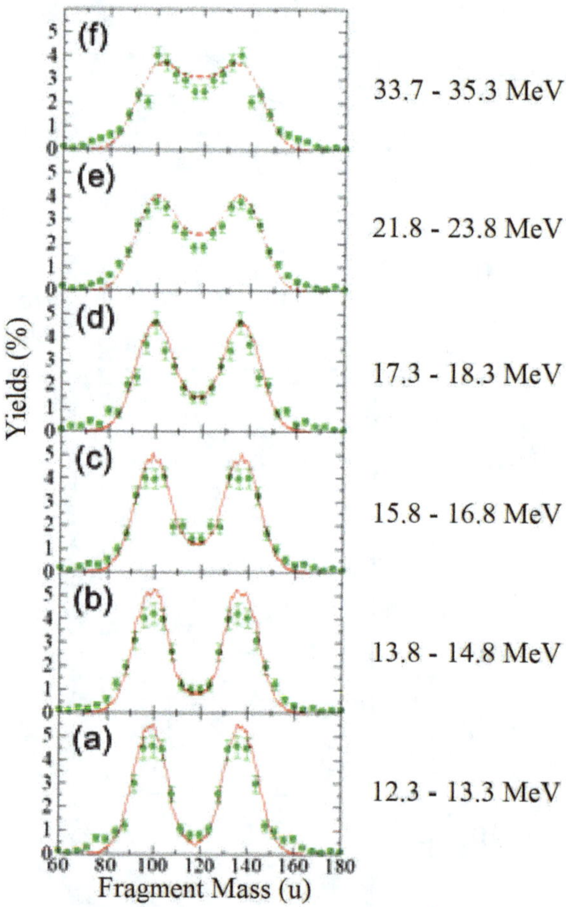

Fig. 6. Fission fragment mass distributions of ^{236}Np* obtained in the MNT reactions (solid circles) are compared with the data $p + {}^{235}$U of [7] (a)-(d) (solid curves) and [8] (e)-(f) (dashed curves).

To make a comparison, center for the total-excitation energy (E^*tot) gate of the MNT channel is adjusted to fit the excitation energy of compound nucleus populated by p + ^{235}U at each incident proton energy [7, 8]. Apart from the small and minor disagreement at the highest energy of E^*_{tot} = 33.7 - 35.7 MeV, the data from MNT reaction agree with the proton-induced fissions. The results support that the total excitation energy after MNT process is mostly given to the recoiled nucleus (fissioning nucleus) at least up to about E^*_{tot} = 35 MeV, the E^*_{tot} is reasonably approximated to the excitation energy of compound nucleus E^*.

Extended data for FFMDs over the isotopes and excitation energies are useful to understand the fission process. One of the finding from the data is the effects of neutron-evaporation prior to fission, called as "multi-chance fission".

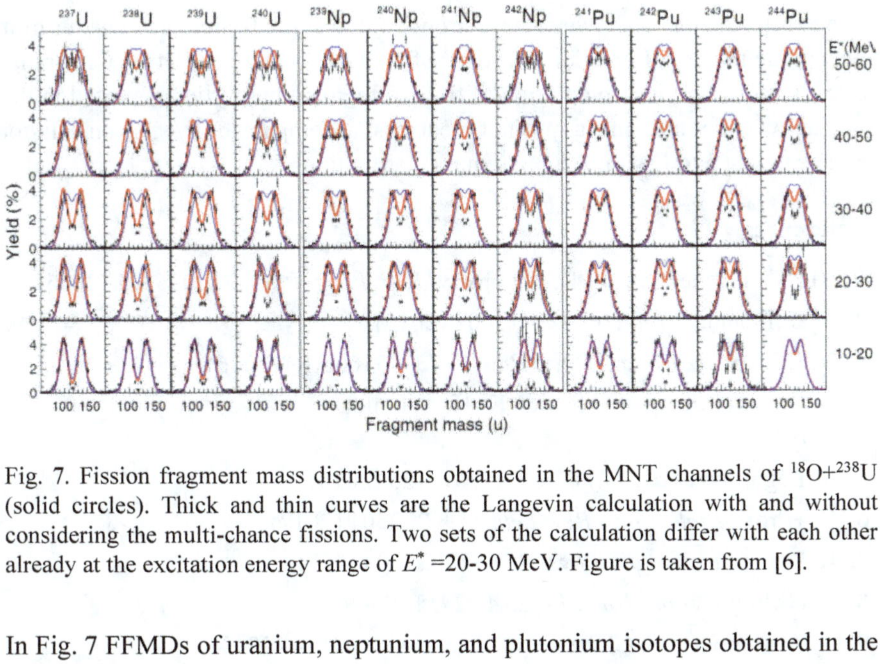

Fig. 7. Fission fragment mass distributions obtained in the MNT channels of ^{18}O+^{238}U (solid circles). Thick and thin curves are the Langevin calculation with and without considering the multi-chance fissions. Two sets of the calculation differ with each other already at the excitation energy range of E^* =20-30 MeV. Figure is taken from [6].

In Fig. 7 FFMDs of uranium, neptunium, and plutonium isotopes obtained in the ^{18}O + ^{238}U reactions are shown [7]. The data are compared with the Langevin calculation [9], where thick and thin curves are the results with and without taking into account the concept of multi-chance fission. By evaporating neutrons before fission, fission fragments from less excited and lighter compound nucleus are generated and detected in the fission setup. As the shell-structure responsible for mass-asymmetric fission revives in proportion to pre-fission neutrons, the resultant shape of FFMD tends to have pronounced double-peak structure with larger P/V ratio. Comparing two types of calculations, with and without multi-chance fission, it is evident that the FFMDs of the excitation energy E^*_{tot} = 20-30 MeV already require the process of multi-chance fission, to give better agreement. Reduced P/V ratio of FFMDs at heavier element and enhanced P/V ratio toward neutron-rich isotopes in Fig. 5 and Fig. 7, revealed in the JAEA setup, can be explained only by invoking the concept of multi-chance fission. Alternatively, fission from the particular excited state is clearly revealed in this study, shown as the thin curves in Fig.7. At the excitation energy of ~40 MeV, the FFMD shows almost single-humped Gaussian-like structure, and the nucleus ends to behave as a charged liquid of nuclear matter, due to the smearing of the shells.

Conclusion

The MNT reaction is shown to be a strong tool to study fission, originating from the unique feature of opening many transfer channels and populating wide range of excitation energies. In addition to fission fragment properties, reported in this article, we are measuring prompt neutrons accompanied by fission, fission barrier height and fission fragment mass distributions.

References

1. Yu. Ts. Oganessian, V. K. Utyonkov, *Rep. Prog. Phys.* **78**, 036301 (2015).
2. S. Hofmann, *J. Phys. G* **42**, 114001 (2015).
3. V. I. Zagrebaev *et al.*, *Phys. Rev. C,* **73**, 031602(R) (2006).
4. A. N. Andreyev. K. Nishio, K.-H. Schmidt, *Rep. Prog. Phys.* **81**, 016301 (2018).
5. R. Léguillon *et al.*, *Phys. Lett. B* **761**, 125 (2016).
6. K. Hirose *et al.*, *Phys. Rev. Lett.* **119**, 222501 (2017).
7. R. L. Ferguson *et al.*, *Phys. Rev. C* **7**, 2510 (1973).
8. S. I. Mulgin *et al.*, *Nucl. Phys. A* **824**, 1 (2009).
9. Y. Aritomo and S. Chiba, *Phys. Rev. C* **88**, 044614 (2013).

Evaluation for Possibility of Synthesizing New Superheavy Elements

Nur Liyana Mohd Anuar,* Yoshihiro Aritomo, Shoya Tanaka and Baku Yanagi

Faculty of Science and Engineering, Kindai University,
3-4-1 Kowakae, Higashiosaka City, Osaka 577-8502, Japan
**E-mail: yaya940817@gmail.com*

The evaporation residue cross section of synthesizing superheavy elements requires touching probability, formation probability and survival probability. Survival probability and evaporation residue cross section appear in the last stage of synthesizing superheavy element. There are many uncertain parameters in each stage, here we focused on uncertain parameter in the survival probability that has also affect the evaporation residue cross section. As the parameters change, the calculation results also change. To adjust these parameters to fit with experiment value, we adopt the covariance method to arrange it more mathematically. There are also many possible systems for the same element especially the element that has not been discovered yet.

Keywords: Evaporation residue; parameters; superheavy element.

1. Introduction

On 28 November 2016, IUPAC announced some new superheavy elements. However, there are still many elements that has not be discovered yet and the periodic table will be continued to the 8th period.

Accurate prediction is important to synthesize new nuclei and superheavy elements. Here, we focused on the last stage of synthesis of superheavy element which includes survival probability and evaporation residue cross section. Since there are uncertainties of some parameters, we discuss the parameters and look how these uncertainties affect the calculation results. We calculated the superheavy element from element 104 Rutherfordium till superheavy element 124 and more focusing on atomic number 119 and above which are the elements that has not been discovered yet.

2. Synthesis of Superheavy Element

The synthesis of superheavy element are divided into 3 stages by different of reaction time. The first stage is touching stage, where the projectile and target nucleus are approaching. The second stage is formation or fusion stage where the fusion and quasi-fission processes are competing. Fusion probability P_{fus} can be calculated using Langevin equation, which is a dynamical model [1,2].

$$P_{fus}(E,l)=\int_0^1 d(\cos\theta)T_l(E;\theta)P_{CN}(E,l,\theta) \tag{1}$$

The third stage is a decay process which in this stage the evaporation residue (ER) and fission process will be occur. It is important to know the survival probability P_{sur} and it can be calculated by statistical model [3]. From all the probabilities in these 3 stages, we calculated the evaporation residue cross sections σ_{ER}

$$P_{sur}=\prod_{i=1}^{N}\frac{\Gamma_n^{(i)}}{\Gamma_n^{(i)}+\Gamma_f^{(i)}} \tag{2}$$

$$\sigma_{ER}\frac{\pi h}{2\mu_0 E_{cm}}\sum_{l=0}^{l_{max}}(2l+1)T_l(E_{cm},l)P_{fus}(E^*,l)P_{suv}(E^*,l) \tag{3}$$

In survival probability equation (2), there are uncertain parameters include fission barrier height parameter δB_f, friction parameter γ and level density parameter a_f/a_n [4,5]. In the superheavy element region, it is no determined which values to be used. The survival probability can be expresses as

$$\frac{\Gamma_n}{\Gamma_f}=\frac{k_{coll}(gr.st.)}{k_{coll}(saddle)\cdot k_{kramers}}A_0\exp\left[2\sqrt{a_nE_n^*}-2\sqrt{a_fE_f^*}\right]$$

$$E_n^*=E_{int}-B_n,\quad E_f^*=E_{int}-B_f$$

$$k_{kramers}=\frac{\hbar\omega_1}{\sqrt{E_{int}}}\left(\sqrt{1+x^2}-x\right),\ x\equiv\gamma/2\omega_1 \tag{4}$$

We need to know the parameter dependence of the survival probability and to see how these uncertainties influence the survival probability and evaporation residue cross section.

3. Results

We change these three parameters; fission barrier height parameter δB_f (MeV), friction parameter γ ($\times 10^{21}$ s^{-1}) and level density parameter a_f/a_n, and see the differences in each parameter. We also compare the calculation result for different type of reaction system that suggested by JINR(Dubna), RIKEN, GSI and others. The used parameters are standard parameters ($\delta Bf=0$, $\gamma=1$, $af/an=1.02$, $\Delta Q=0$) as a constant parameter. (Figure 1)

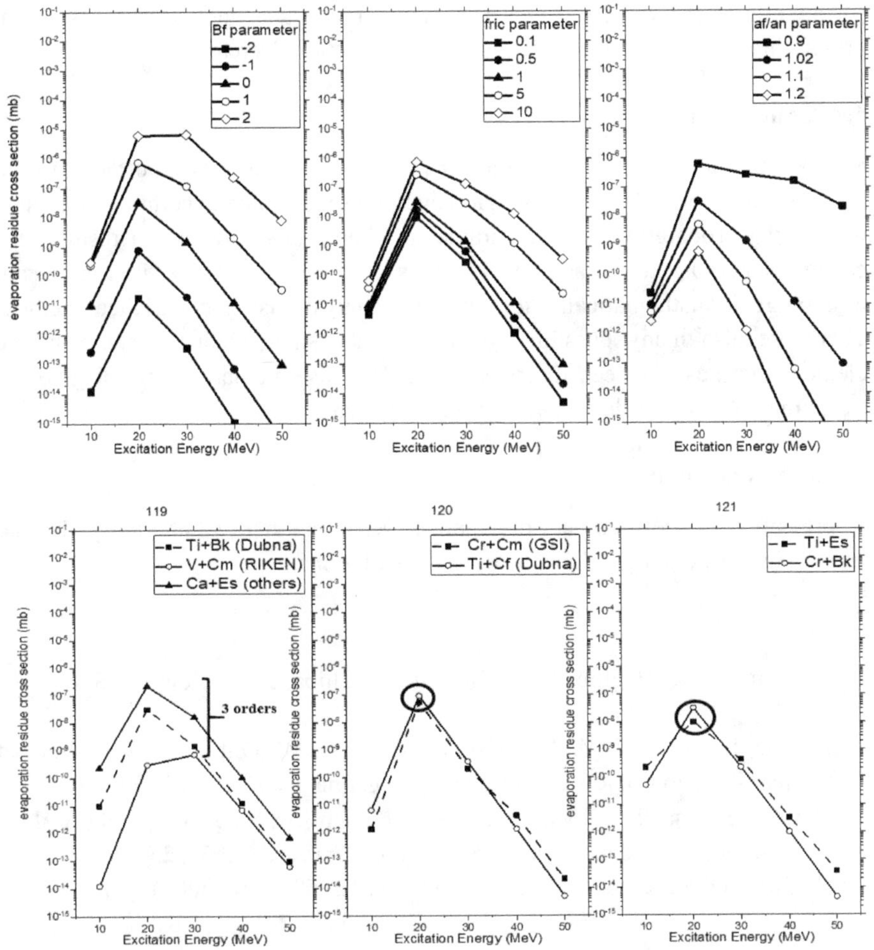

Fig. 1. The calculation results of ER cross section for $^{50}Ti+^{249}Bk \rightarrow ^{299}119$ (above) and the comparison results for different type of reaction systems but the same nuclei number (below). The calculated systems are; $^{50}Ti+^{249}Bk \rightarrow ^{299}119$, $^{51}V+^{248}Cm \rightarrow ^{299}119$, $^{48}Ca+^{254}Es \rightarrow ^{302}119$, $^{54}Cr+^{248}Cm \rightarrow ^{302}120$, $^{50}Ti+^{250}Cf \rightarrow ^{300}120$, $^{50}Ti+^{254}Es \rightarrow ^{304}121$ and $^{54}Cr+^{249}Bk \rightarrow ^{303}121$.

4. Discussion

The patterns of the changes in each element are uniform. As the fission barrier height parameter δB_f and friction γ parameter increases, the fission process becomes hard and the fusion process will become easier. The survival probability will increase thus cause the increase of the ER cross section.

Meanwhile as the value of level density parameter a_f/a_n increases, the survival probability decreases thus cause the ER cross section decrease.

5. Conclusion

Survival probability and evaporation residue are some of the important parts in synthesizing superheavy element. However, there are some uncertain parameters in survival probability. As the parameters change, the calculation results also change. To adjust these parameters to fit with experiment value, it is important to arrange it mathematically for example analysis using covariance method. There are also many possible systems for the same element especially the element that has not been discovered yet. By this calculation, it will give the experiment team a guide to discover the new elements.

Acknowledgements

Appreciation to the members of Nuclear Reaction Energy Laboratory, Kindai University for helping to do some of the calculations.

References

1. Y. Aritomo, K. Hagino, K. Nishio and S. Chiba, *Phys. Rev. C* **85**, 044614 (2012).
2. Y. Abe, K. Okazaki, Y. Aritomo, T. Wada and M. Ohta, Int. Workshop on Nuclear Reactions and Beyond, Lanzhou, China, 24-27 Aug, 1999.
3. M. Ohta, Dubna, 25-27 May 2000, World Scientific, Singapore, 110 (2001).
4. Y. Aritomo, T. Wada, M. Ohta and Y. Abe, *Phys. Rev. C* **59**, 2 (1999).
5. M. Ohta and Y. Aritomo, JNST, Supplement 2, 586-591 (2002).

Achievements and Challenges in Theoretical Predictions of Chemical Properties of Superheavy Elements

V. Pershina

Helmholtzzentrum für Schwerionenforschung GmbH
Darmstadt, Planckstr. 1, D-64291 Darmstadt, Germany
E-mail: V.Pershina@gsi.de

Adsorption energies, E_{ads}, of superheavy elements Cn, Nh, and Fl and their homologs Hg, Tl, and Pb, respectively, on surfaces of gold and quartz were predicted using periodic ADF BAND code. Accurate molecular calculations were performed for hydroxides of Nh and its lighter homologs In and Tl using molecular ADF package. The calculated properties of MOH (M = In, Tl and Nh) were used to predict E_{ads} of these molecules on the surface of Teflon to assure their transportation through the capillaries to the chemistry set up. It was shown that a trend in E_{ads} of group-13 atoms on gold, decreasing with Z in the group, is opposite to the trend in E_{ads} of the hydroxides, that should increase in the group. Therefore, it will be difficult to differentiate between Nh and NhOH by their adsorption on gold, because their E_{ads} should be similar.

Keywords: Superheavy elements; relativistic calculations; adsorption.

1. Introduction

Presently, superheavy elements (SHEs) Cn, Nh and Fl are in the focus of experimental and theoretical studies [1-7]. These are the elements where relativistic effects – strong relativistic stabilization of the $7s_{1/2}$- and $7p_{1/2}$ AOs and large spin-orbit (SO) splitting of the 7p-AOs - are expected to profundly influence their chemical properties. As a consequence, an unusually low reactivity, or high volatility as compared to the lighter homologs in the groups are expected [8]. Experimentally, volatility of single species is studied by gas-phase chromatography technique providing adsorption enthalpies, ΔH_{ads}, of the SHE species on some surfaces. Usually, gold or quartz are used. Those ΔH_{ads} are compared then with ΔH_{ads} of the lighter homologs in the chemical groups, which allows for assessing their similarity.

Adsorption behaviour of Hg, Tl, and Pb on gold and quartz has meanwhile been studied [9-14]. It is known that Hg interacts strongly with Au, but weakly with quartz. Tl reveals strong interaction both with gold and quartz [10,13]. Pb adsorbs strongly both on quartz and on gold either, even stronger than Tl. For SHEs, Cn was shown to adsorb on gold with $-\Delta H_{ads}$ of 34^{+54}_{-11} kJ/mol [3], but no interaction with quartz is expected. For Fl, there are two types of measurements showing contradictive results: according to one of them [4], Fl adsorbs on the gold surface at very low temperatures, which means it is very volatile, while according to other [5], Fl should be more metallic with $-\Delta H_{ads} >$ 48 kJ/mol. No adsorption of Fl on quartz is expected.

Our earlier works were devoted to predictions of E_{ads} of SHEs and their homologs on gold using fully relativistic Density Functional Theory (4c-DFT) calculations and a cluster approach [1,2,15]. Due to the recent development of periodic codes, i.e., of computational algorithms and basis set techniques, calculations of E_{ads} of SHEs using those periodic codes became possible. Some results are presented in this work. Accurate molecular calculation were also used in predicting adsorption of SHE on inert materials to assure their transport through Teflon capillaries.

2. Details of the Calculations

The computational software used in this work is the Amsterdam Density Functional Theory (ADF) suite [16]. The method lying in the basis of it uses the zeroth order regular approximation (ZORA) two-component (2c) Hamiltonian incorporating spin-orbital (SO) effects (SO-ZORA) [17]. For molecular calculations, the exchange-correlation potential, V_{xc}, was chosen as BP86 in order to facilitate comparison of results with the previous 4c-DFT ones [9] using B88/P86. Also revPBE and dispersion-corrected V^{xc}, revPBE-D3(BJ), were applied. The TZ2P basis sets as a combination of the Slater Types of Orbitals (STO) with numerical functions were found to be sufficiently good for the required accuracy. Molecular calculations were performed with all electrons.

The ADF BAND suite was used for periodic calculations of adsorption of these elements on quartz and gold [18-20]. For the gold-slabs, the PW91 functional proved to be most useful, while for quartz, revPBE was found to be sufficiently good. For calculations of E_{ads} of molecules, the usage of revPBE-D3(BJ) was essential.

The E_{ads} of M on such a slab/supercell is obtained using the known expression

$$E_{ads}(\text{M-slab}) = -[E_f(\text{M-slab}) - E_f(\text{M}) - E_f(\text{slab})], \tag{1}$$

where E_f(M) is a correction for the atomic ground state within the ZORA approximation. As a start, structures of bulk quartz and gold were optimized. Then a slab (for a regular adsorbate coverage), or a supercell (for a lower adsorbate coverage) were cut out of the optimized bulk structure. An adsorbate atom was then placed on the surface of such a slab, or a supercell and the formation energy, E_f, of the whole system was calculated.

Various positions of ad-atoms M on a slab, or a supercell were considered. For geometry optimization, scalar-relativistic (SR) non-spin polarized and spin-polarized non-collinear approaches were used. The SO corrections were taken into account as single point calculations for the optimized structures. A frozen core approximation was used with the smallest frozen core. This means that for Cn and Fl, orbitals from 1s through 5d were frozen, while those from 6s through 5f were explicitly included in the calculations of integrals.

3. Results and Discussion

As is known, the most stable gold surface is Au(111). Adsorption of the single M atoms (M = Hg/Cn, Tl/Nh and Pb/Fl) on it was modeled by the (4 x 4)M/Au(111) supercells [20].

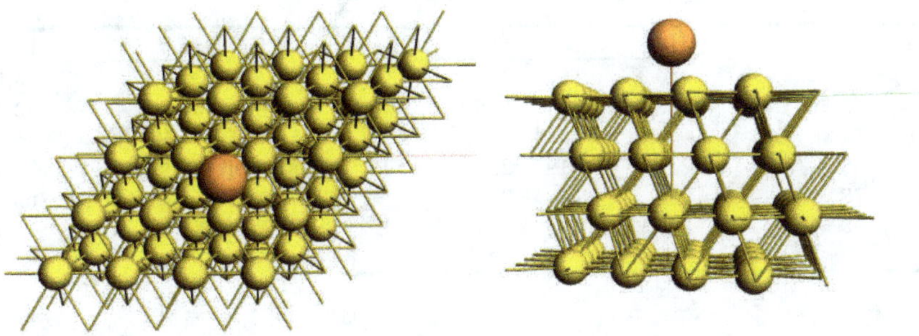

Fig. 1. A (4 x 4)M/Au(111) unit supercell modeling adsorption of atom M on the Au(111) surface.

Four possible adsorption positions were considered: (1) on top of the upper Au atom, (2) on the bridge between two gold atoms, (3) a "hollow-1" position, on top of the Au atom in the second row from the top and (4) a "hollow-2" one, on top of the Au atoms in the 3d row from the top. The "hollow-2" position was found to be preferential for all the atoms of interest. The obtained results along with those from the earlier 4c-DFT cluster calculations [15] for gold and quartz are summarized in Table 1. They show that PW91 relatively well (slightly underestimates) reproduces E_{ads} on gold, much better than revPBE giving even negative E_{ads} for Cn and Fl. For Hg, as expected, PW91 E_{ads} is significantly underestimated with respect to the experimental value of almost 1 eV, as was also found in all other, both cluster and periodic, calculations. This is probably due to the fact, that Hg dissolves in gold, which needs a special simulation. Cn turned out to be non-interactive with the Au(111) surface at the PW91 level of theory having a zero adsorption energy and a very large separation from the surface. The revPBE-D3(BJ) functional gives better agreement with experiment for Hg and Tl, however, it is not available for elements heavier than Z=92. Estimates of dispersion-corrected V^{xc}, revPBE-D3(BJ), as those for the 6th row elements were introduced in the ADF code only very recently and were used in the present calculations.

For SHEs, earlier 4c-DFT results [15] are in consonance with experiment for Cn [3]. Experiments on adsorption of Fl on gold [4,5] did not agree on the final E_{ads} value and, therefore, new measurements are on the way [7]. First experiments on volatility of Nh indicated its high volatility and reactivity towards gold [6], in agreement with the predictions [20].

Table 1. Summary of predictions of $-\Delta H_{ads}$ (in kJ/mol) of Hg, Cn, Tl, Nh, Pb, and Fl on Au(111) and hydroxylated quartz surfaces from ADF BAND calculations in comparison with experiment and earlier 4c-DFT cluster calculations.

Surf.	Hg	Cn	Tl	Nh	Pb	Fl	V^{xc}	Ref.
SiO_2	54	-38	150	58	152	-22	ADF (revPBE)	18,19
	42 ± 2	-	158 ± 3	-	165 ± 4	-	Exp.	12-14
Au	54	45	-	-	232	68	4c-DFT (B88/P86)	15
	45	0	213	119	230	57	ADF (PW91)	20
	78	-	254	-	268	-	ADF (revPBE-D3BJ)	this
	98 ± 3	52^{+4}_{-3}	270 ± 10	-	234	34^{+54}_{-11} ≥ 48	Exp.	14-16, 45

Nh, as Tl can, however, form hydroxides, MOH, because humidity is always present in the set up. To estimate volatility of the MOH (M = In, Tl and Nh) molecules as E_{ads} on Teflon, accurate molecular calculations were performed for these species using the ADF (SO-ZORA) code [21].

Table 2. Calculated properties of InOH, TlOH, and NhOH: bond lengths, R_e, (in Å), dipole moments, μ (in D), average (α_{av}) and anisotropic (α_{anis}) polarizabilities (in a.u.), and ionization potentials, IP (in eV).

Molecule	R_e(M-H)	μ	α_{av}	α_{anis}	IP
InOH	3.037	2.13	48.91	43.31	8.89
TlOH	3.129	2.72	41.76	115.11	9.25
NhOH	3.258	3.03	39.60	201.90	9.39

The molecule-surface E_{ads} should be composed of the molecular polarizability-surface and molecular dipole moment-surface terms. Table 2 contains results of the calculations: dipole moment (μ), polarizabilities (α), average (α_{av}) and anisotropic (α_{anis}), and ionization potentials (IP) of MOH. Table 3 contains calculated components of α and μ along the x, y and z axes. One can see that μ(MOH) increase in the row In < Tl < Nh, as do the anisotropic α, while α_{av} slightly decrease in this direction. The increase in μ(MOH) is due to an increase in the M-OH distances in this row caused by expansion of the valence d-orbitals with Z.

Table 3. Calculated properties of InOH, TlOH, and NhOH: x, y and z components of polarizabilities, α (in a.u.), and dipole moments, μ (in D).

	α			μ		
Molecule	x	y	z	x	y	z
InOH	49.3	54.2	43.3	2.07	0.46	0
TlOH	43.9	50.4	34.2	2.64	1.04	0
NhOH	42.5	49.6	29.3	2.86	1.31	0

Due to the strong anisotropicity and a low symmetry of the molecule, it is very difficult to estimate its E_{ads} on Teflon, since its van der Waals radius cannot be defined. E_{ads} also depends very much on the orientation of the MOH molecule. Thus, if the molecule is parallel to the surface, the molecular dipole-surface interaction energy is zero and a pure van der Waals E_{ads} should be of 25 kJ/mol, 18 kJ/mol and 14 kJ/mol for the In, Tl, and Nh hydroxides at the molecule-surface separation of 2.501 Å, 2.790 Å and 2.935 Å, respectively. However, if the molecule has a non-zero angle with respect to the surface, then a contribution of μ(MOH)-surface to E_{ads} might be significant and it should increase in the row In < Tl < Nh. It should, however, be still smaller than the dispersion part of E_{ads}. Thus, according to the molecular data, we believe that NhOH should be transported along the Teflon capillaries similarly to TlOH.

The calculated molecular properties listed in Table 2 are evident that E_{ads} of the molecular species on the gold surface should increase in the direction InOH < TlOH < NhOH due to an increase in the dipole moments [21]. Taking into account the reversed trend in E_{ads} of the atoms, i.e., In > Tl > Nh, it may come to the situation when Nh and NhOH will adsorb on the gold surface at about the same temperatures, whereby their differentiation will be difficult.

4. Conclusions

Calculations of adsorption energies of Cn, Nh, and Fl and of their homologs Hg, Tl, and Pb on quartz and gold were performed using a periodic ADF BAND program package. According to the results, Cn and Fl should be very volatile, while Nh should adsorb rather well both on quartz and gold, however, less strongly than Tl. Adsorption of group-13 atoms should show the following trend In > Tl > Nh, while for molecular species, it should be the other way around, i.e., InOH < TlOH < NhOH. This means that adsorption of NhOH on gold should take place in the area of adsorption temperature of Nh, which will be a problem for their chemical identification.

Acknowledgments

The author is grateful to A. Yakushev and R. Eichler for fruitful discussions of experimental results.

References

1. A. Türler and V. Pershina, Chem. Rev. **113**, 1237 (2013).
2. V. Pershina, Nucl. Phys. A **944**, 578 (2015).
3. R. Eichler, N. V. Aksenov, A. V. Belozerov *et al.*, Nature **447**, 72 (2007).
4. R. Eichler, N. V. Aksenov, Yu. V. Albin *et al.*, Radiochim. Acta **98**, 133 (2010).
5. A. Yakushev, J. M. Gates, A. Türler *et al.*, Inorg. Chem. **53**, 1624 (2014).
6. S. N. Dmitriev, N. V. Aksenov, and Y. V. Albin, Mendeleev Commun. **24**, 253 (2014).
7. A. Yakushev, private communication, 2016.
8. K. S. Pitzer, J. Chem. Phys. **63**, 1032 (1975).
9. S. Soverna, R. Dressler, C. E. Düllmann *et al.*, Radiochim. Acta **93**, 1 (2005); J. P. Niklaus, R. Eichler, S. Soverna *et. al.*, PSI Annual Report, 2000.
10. A. Serov, R. Eichler, R. Dressler *et al.*, Radiochim. Acta **101**, 421 (2013).
11. F. Haenssler, R. Eichler, H. W. Gäggeler *et al.*, PSI Annual Report 2005, 2006; R. Eichler, private communication, 2008.
12. S. Soverna, Doctoral Thesis, Universität Bern, 2004.
13. P. Steinegger, M Asai, R. Dressler *et al.*, J. Phys. Chem. C **120**, 7122 (2016).
14. R. Eichler, private communication 2017.
15. V. Pershina, J. Anton, and T. Jacob, J. Chem. Phys. **131**, 084713 (2009).
16. ADF BAND program package, see `http://www.scm.com`.
17. R. A. Olsen, P. H. T. Philipsen, and E. J. Baerends, J. Chem. Phys. **119**, 4522 (2003).
18. V. Pershina, Phys. Chem. Chem. Phys. **18**, 17750 (2016).
19. V. Pershina, J. Phys. Chem. C **120**, 20232 (2016).
20. V. Pershina, Inorg. Chem. **57**, 3948 (2018).
21. V. Pershina and M. Ilias, Chem. Phys. Lett. **694**, 107 (2018).

$^{136}Xe/^{238}U + ^{251}Cf$ Collisions at Near-Barrier Energies as Method for Production of Heavy Nuclides

V. V. Saiko and A. V. Karpov*

Flerov Laboratory of Nuclear Reactions, Joint Institute for Nuclear Research, Dubna, 141980, Russia
and
Dubna State University, Dubna, 141982, Russia
**E-mail: karpov@jinr.ru*
www.jinr.ru

In the paper we analyze production yields of various isotopes of heavy elements in the MNT reactions of Xe/U + Cf within a Langevin-type dynamical model. The influence of the projectile nucleus on formation of reaction products has been discussed.

Keywords: Multi-nucleon transfer reactions; neutron-enriched nuclei; Langevin equations.

1. Introduction

The process of multi-nucleon transfer (MNT) is an important feature of deep inelastic collisions of heavy nuclei, because it results in a broad mass distribution of reaction products. Particularly, the final fragments can cover the area of unknown neutron-enriched isotopes of heavy nuclei hardly achievable by other reaction mechanisms, e.g. fusion or fragmentation. It makes MNT to be a promising method of production of these nuclides and stimulates both experimental and theoretical investigations of these reactions.

Langevin-type dynamical models are a powerful tool for theoretical analysis of MNT reactions. They can provide a rather complete information about collision dynamics [1,2]. A multidimensional dynamical model [3] is able to describe various observable characteristics, such as mass, energy and angular distributions of reaction products.

It is well known that the yields of products, which are heavier than the target drop exponentially with increasing atomic number of the product in collisions of heavy actinides. We aimed to analyze the influence of the mass of projectile nuclei on the final cross sections of the nuclides heavier than the actinide target.

2. Model

The complete description of the model as well as peculiarities of accounting the mutual orientations of the statically deformed nuclei can be found in [3,4]. Only the basics of the model are discussed here. We consider a collision of two heavy nuclei consisting of three main stages: approaching stage, formation of the dinuclear system, and its decay into two fragments. The model is based on the system of the Langevin equations:

$$\begin{aligned} \dot{q}_i &= \sum_j \mu_{ij} p_j, \\ \dot{p}_i &= F^{\text{driving}} + F^{\text{friction}} + F^{\text{random}}, \end{aligned} \tag{1}$$

where q_i and p_i are the collective degrees of freedom considered in the model and their conjugate momenta. $\mu_{ij} = m_{ij}^{-1}$ is the inverse of the inertia tensor. The friction and the random forces describe energy dissipation and fluctuations in the system of two heavy nuclei.

The set of equation (1) is solved numerically and a solution is a trajectory in multidimensional space of the collective degrees of freedom. One should simulate a large number of trajectories in order to obtain distributions of various characteristics of the reaction. Each trajectory contains a complete information about a single collision during all the stages of the reaction. In particular, it includes characteristics of the excited fragments formed in the exit channel, e.g. nucleon numbers, kinetic energies, angles, etc.

The final reaction products are obtained from the primary excited ones by simulating the particle evaporation cascade within the statistical model [5]. Finally, the differential cross section of the reaction products is calculated as follows:

$$\frac{d^2\sigma}{dZdAdEd\Omega}(Z,A,E,\theta) = \int_0^{b_{\max}} \frac{\Delta N(b,Z,A,E,\theta)}{N_{\text{tot}}(b)} \frac{bdb}{\Delta Z \Delta A \Delta E \sin\theta \Delta\theta} \tag{2}$$

3. Results

Recent results of calculations performed in the framework of the model show a reasonable description of the yields of reaction products heavier than target nuclei produced in collisions of heavy actinides ^{238}U + ^{238}U/^{248}Cm [6]. It gives one a confidence that the model has some predictive power in analysis of reactions with actinide targets.

It was concluded in Refs [4,6] that the use of the heaviest available target allows production of heaviest nuclides. In this respect the ^{251}Cf target is rather interesting. We have performed modeling the collisions induced by ^{136}Xe and ^{238}U nuclei with ^{251}Cf target. The set of the model parameters was the same as in Ref. [6]. Collision energies were slightly above the Bass barrier heights $E=1.2V_B$: $E_{c.m.}$=620 MeV and 1080 MeV for the ^{136}Xe + ^{251}Cf and the ^{238}U + ^{251}Cf reactions, respectively. The cross sections of final (cold) nuclides with Z>91 produced in the both reactions superimposed on the chart of known nuclides are shown in Fig. 1.

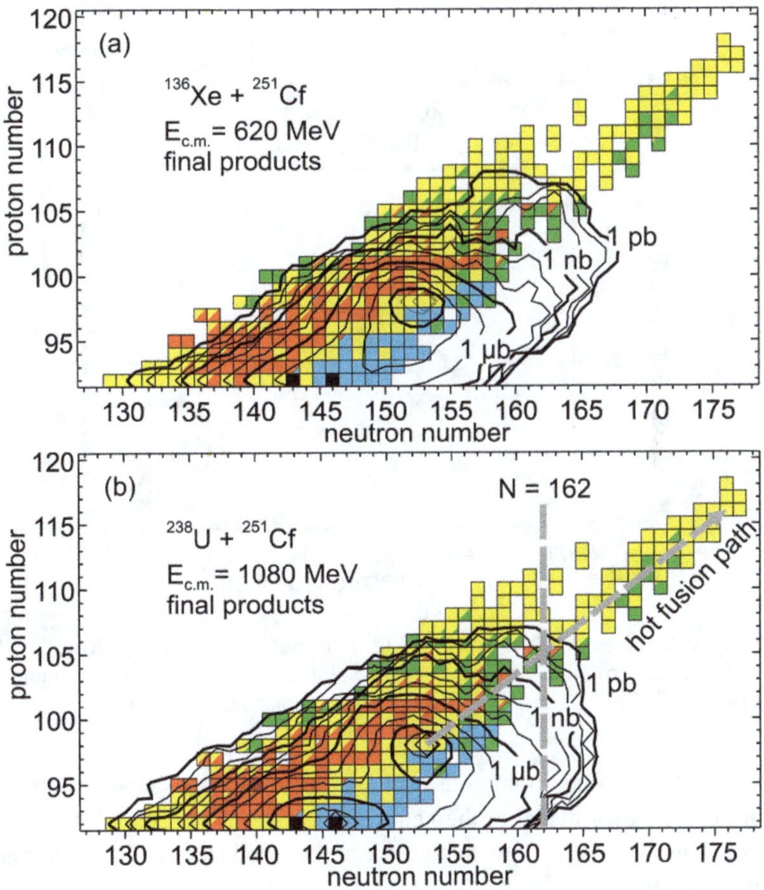

Fig. 1. Production cross sections of transuranium nuclides in the ^{136}Xe/^{238}U + ^{251}Cf reactions superimposed on the chart of known nuclides. Collision energies are 20% above the Bass barrier heights. The contour lines are drawn over an order of magnitude down to pb, the labels correspond to the thick lines. The dashed lines show the nuclides analyzed in Fig. 2.

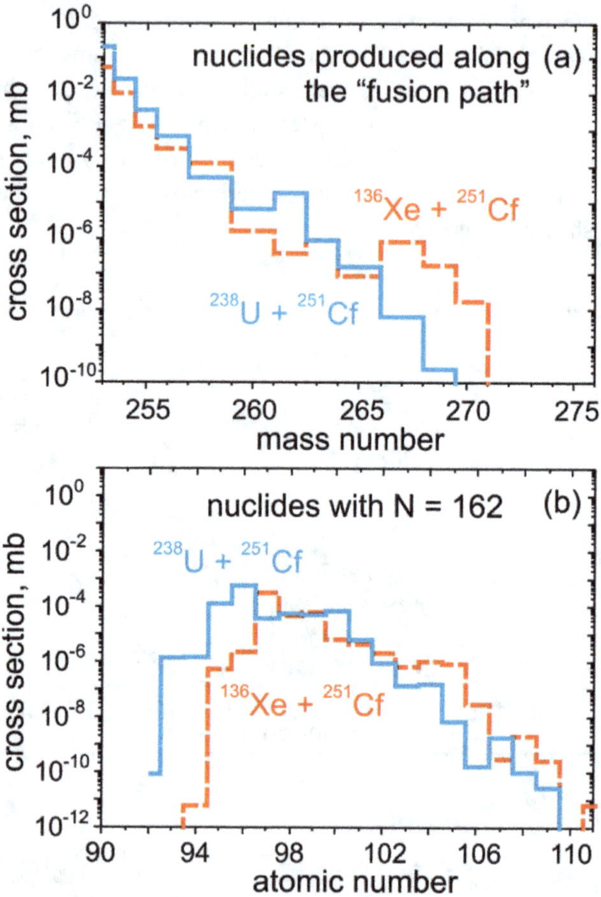

Fig. 2. The cuts of the data presented in Fig. 1: a) nuclides produced along the "fusion path" (see Fig. 1), b) isotones with N=162. Solid and dashed histograms indicate the results for the ^{238}U + ^{251}Cf and the ^{136}Xe + ^{251}Cf reactions, respectively.

The first thing one might note is that the cross sections cover a rather large area of yet-unknown nuclei in the neutron-rich side of the chart. Thus, the both reactions can be used. New nuclides can be produced with experimentally achievable cross sections exceeding 1 μb.

An obvious difference between considered systems is that the shape of the product distributions is broader in the ^{238}U + ^{251}Cf reaction than in the ^{136}Xe + ^{251}Cf one. It makes heavier projectiles more preferable for production of the heaviest neutron-enriched nuclides. Therefore, the choice of the projectile nucleus is an important problem at planning the experiments aimed at synthesis of unknown nuclides in MNT reactions with actinides.

A more quantitative analysis was done by comparing the different sections of the two-dimensional plots in Fig. 1. The directions of the sections are indicated by the dashed lines in panel b) of Fig. 1. They correspond to the isotones with N=162 and heavy nuclides, which can be produced in the hot fusion reactions. The corresponding distributions are shown in Fig. 2.

The cross sections of nuclei along the decay chains of the products of the hot fusion reaction are comparable for the two reactions and drop rapidly with number of transferred nucleons. Moreover, in the region of mass numbers ~270 the MNT reaction cross sections become lower than the hot fusion cross sections. However, the gap between the products of the hot fusion reactions and the "continent" can be filled by the MNT products synthesized with the cross sections ~1 nb. The plateau in the distribution for the $^{136}Xe + ^{251}Cf$ reaction is due to the neutron sub-shell closure N=162.

The distributions of isotones with N=162 reveal the fact that the most neutron-enriched nuclides can be produced with larger cross sections in reactions induced by the heaviest available projectile ^{238}U.

Conclusions

In MNT reactions with actinide targets the production cross sections of heavy nuclides drop down very quickly with increasing number of transferred nucleons. However, there is a real chance to produce new isotopes of heavy actinides in these reactions. The heavier projectiles are more preferable to achieve the larger neutron-to-proton ratio of the products.

References

1. P. Fröbrich and I. I. Gontchar, *Physics Reports* **292**, 131 (1998).
2. V. Zagrebaev and W. Greiner, *Journal of Physics G* **31**, 825 (2005).
3. V. Karpov and V. V. Saiko, *Phys. Rev C* **96**, 024618 (2017).
4. V. V. Saiko and A. V. Karpov, http://arxiv.org/pdf/1810.04025.pdf
5. *NRV: Decay of excited nuclei*, http://nrv.jinr.ru/nrv/webnrv/stat_model/
6. V. V. Saiko and A. V. Karpov, accepted in *Acta Phys. Pol. B* **50** (2019).

Study of Neutron-Deficient Nuclei in the $^{239,240}Pu+^{48}Ca$ Reactions

M. V. Shumeiko,[1,*] V. K. Utyonkov,[1] N. T. Brewer,[2] Yu. Ts. Oganessian,[1]
K. P. Rykaczewski,[2] F. Sh. Abdullin,[1] S. N. Dmitriev,[1] R. K. Grzywacz,[2,3]
M. G. Itkis,[1] K. Miernik,[2,4] A. N. Polyakov,[1] J. B. Roberto,[2] R. N. Sagaidak,[1]
I. V. Shirokovsky,[1] Yu. S. Tsyganov,[1] A. A. Voinov,[1] V. G. Subbotin,[1]
A. M. Sukhov,[1] A. V. Karpov,[1] A. G. Popeko,[1] A. V. Sabelnikov,[1] A. I. Svirikhin,[1]
G. K. Vostokin,[1] J. H. Hamilton,[5] N. D. Kovrizhnykh,[1] L. Schlattauer,[1,6]
M. A. Stoyer,[7] Z. Gan,[8] W. X. Huang,[8] and L. Ma,[8]

[1]*Joint Institute for Nuclear Research, RU-141980 Dubna, Russian Federation*
[2]*Oak Ridge National Laboratory, Oak Ridge, Tennessee 37831, USA*
[3]*Department of Physics and Astronomy, University of Tennessee, Knoxville, Tennessee 37996, USA*
[4]*Faculty of Physics, University of Warsaw, PL-02-093 Warsaw, Poland*
[5]*Department of Physics and Astronomy, Vanderbilt University, Nashville, Tennessee 37235, USA*
[6]*Faculty of Science, Palacký University, CZ-77147 Olomouc, Czech Republic*
[7]*Lawrence Livermore National Laboratory, Livermore, California 94551, USA*
[8]*Institute of Modern Physics, Chinese Academy of Sciences, Lanzhou 730000, China*
**E-mail: m.v.shumeyko@gmail.com*

Three new decay chains of ^{285}Fl were observed in the $3n$-evaporation channel of the ^{240}Pu + ^{48}Ca reaction [1]. The irradiation was performed at the Dubna Gas Filled Recoil Separator (DGFRS) and using 250 MeV ^{48}Ca beam accelerated at the U400 cyclotron of the Flerov Laboratory of Nuclear Reactions, JINR. The decay properties of the observed nuclei are mostly in agreement with a chain identified at the BGS in the ^{242}Pu(^{48}Ca,$5n$) reaction [2] and three chain previously observed at the DGFRS in the same ^{240}Pu(^{48}Ca,$3n$) reaction at lower energy of 245 MeV [3]. The lifetime of ^{269}Sg observed in one chain exceeds that derived from other five decays by a factor of 33, which might indicate the observation of transitions through different levels in ^{285}Fl and descendants. The cross section of the reaction was measured to be $0.58^{+0.60}_{-0.33}$ pb, which is lower by a factor of 4-5 than the value measured at 245 MeV ^{48}Ca and is in agreement with expectations for the $3n$-evaporation channel.

Keywords: Superheavy elements; α decay; spontaneous fission.

1. Introduction

The aim of the experiment was to study neutron-deficient Fl isotopes produced in the ^{240}Pu+^{48}Ca reaction. Synthesis of neutron-deficient nuclei and the study of the properties of superheavy nuclei (SHN) could help to clarify the stabilizing effect of the neutron shell closures of N=162 and N=184.

The area of interest includes three isotopes of Fl: ^{285}Fl, ^{284}Fl, and ^{283}Fl. During sequential α decays of Fl nuclei, having neutron numbers N=170-175, descendant nuclei are formed, which are located closer to N=162 and even cross it in some decays. Thus the stability of such nuclei is governed largely by the influence of these shells. In addition, α-decay chains of the odd-N nucleus ^{283}Fl, the product of the 5n-evaporation channel of the reaction, could reach the area of known nuclei connecting the region of SHN to the nuclear mainland.

From the systematics of α-decay energies for isotopes with Z=105–118 and half-life systematics provided by Viola-Seaborg formula one can extract extrapolated value of half-life for ^{284}Fl. According to the events detected in previous experiments which could be assigned to the spontaneous fission (SF) of ^{284}Fl with half-life of $T_{1/2}$=2.8 ms we could expect an α-decay branch of about 20% for ^{284}Fl. Observation of this decay mode would be very important for final identification of the even-even isotope ^{284}Fl through its characteristic α-decay energy. In addition, the registration of its descendant, ^{280}Cn, that presumably undergoes SF, would be important for tracing the properties of SHN in the $N\approx$168–170 region of nuclei that exhibit the lowest stability against SF. For ^{280}Cn (N=168) a further decrease of T_{SF} is predicted as observed for neighboring Cn isotopes with decreasing N, but not as much as in neighboring isotopes ^{282,284}Cn. Experimental verification of this prediction could shed light on stability of neutron-deficient isotopes of Fl, Cn, and other nuclei in this region.

2. Experimental Setup

The experiments were performed employing the Dubna Gas-Filled Recoil Separator (DGFRS) and using ^{48}Ca beams accelerated at the U400 cyclotron of the Flerov Laboratory of Nuclear Reactions, JINR. The maximum beam intensity of ^{48}Ca ions was 1.1 particle μA. The beam energy was measured with a systematic uncertainty of 1 MeV by a time-of-flight system. In the experiment the same ^{240}Pu target was used as in [3]. The target material was provided by Oak Ridge National Laboratory (ORNL) (enrichment 98.97%, impurities of other Pu isotopes: 0.77% ^{239}Pu, 0.09% ^{241}Pu, and 0.17% ^{242}Pu) and JINR (enrichment 88.9%, impurities of other Pu isotopes: 7.7% ^{239}Pu, 1.4% ^{241}Pu, and 2.0% ^{242}Pu). The average thickness of the target for ^{240}Pu was 0.39±0.04 mg/cm^2 for the mixed ORNL/JINR target material combined in the ratio 1/5. The material was electrodeposited as PuO_2 oxide onto 0.72 mg/cm^2 Ti foils. The laboratory-frame beam energy in the middle of the target layer was about 250 MeV. Taking into account the energy spread of the incident cyclotron beam, the small variation of the beam energy during irradiation, and the energy losses in the target, we calculated the resulting ^{288}Fl compound nuclei (CN) to have an excitation energy range of 40.4–45.2 MeV. The total beam dose of ^{48}Ca particles was about 1.4×10^{19}.

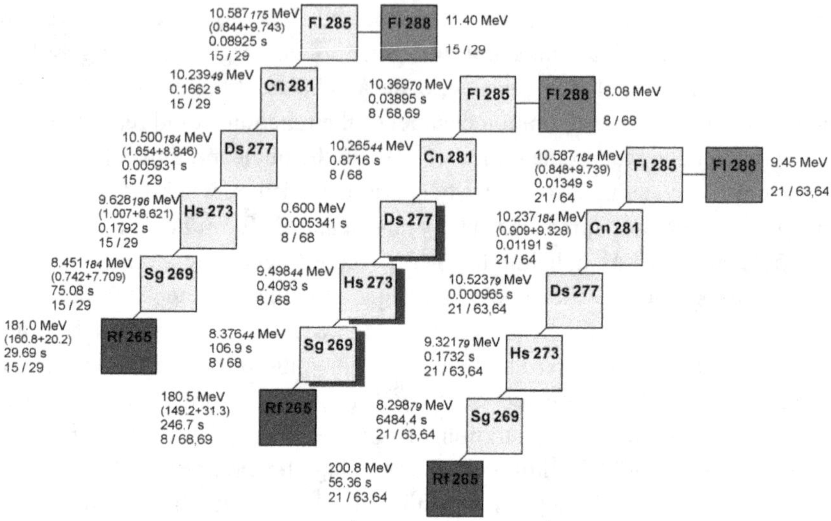

Fig. 1. Decay properties of ^{285}Fl and descendant nuclei observed in the ^{240}Pu+^{48}Ca reaction. The top right rows for each chain show ER energies and strip numbers (front and back). The left rows provide energies, time intervals between events and their strip numbers for α decay and SF. Energies of summed signals are given in parentheses. Three events marked with a shadow were registered during the beam-off period. The FWHM α-particle energy errors are shown by smaller italic numbers.

3. Results

In this experiment, performed at a ^{48}Ca beam energy of 250 MeV, we observed three decay chains of ^{285}Fl (Figure 1). In the first experiment [3], carried out at the same ^{48}Ca energy but at a lower beam dose (4.7×10^{18}), only an upper cross-section limit was determined for the 3*n* channel of the ^{240}Pu+^{48}Ca reaction (≤1.3 pb).

For the first chain, the total number of random ^{285}Fl-like decay chains N_{random} was about 2×10^{-14}. For the second chain, an escape event of ^{277}Ds was not taken into account, $N_{\mathrm{random}}<2\times10^{-17}$. In the last case, the decays of ^{269}Sg and ^{265}Rf with lifetime of about 2 h were not taken into account as well, $N_{\mathrm{random}}<4\times10^{-8}$. Thus, it is very unlikely that any of the three above decay chains of ^{285}Fl are due to random correlation of unrelated events.

4. Discussion

In this experiment, three new decay chains of ^{285}Fl were observed, in addition to one chain identified at Berkeley Gas-filled Separator in the ^{242}Pu(^{48}Ca,5n) reaction [2] and three chains detected at DGFRS in the ^{240}Pu(^{48}Ca,3n) reaction [3].

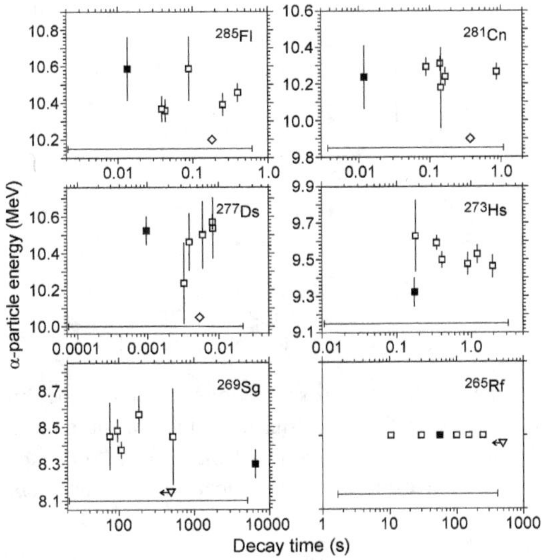

Fig. 2. Measured α-particle energies $E\alpha$ (with error bars) vs. decay times of isotopes assigned to ^{285}Fl, ^{281}Cn, ^{277}Ds, ^{273}Hs, ^{269}Sg, and ^{265}Rf. For spontaneously fissioning ^{265}Rf, only decay times are shown. Decay properties of nuclei observed in the third decay chain in Figure 1 are shown by filled squares. Time ranges for ^{269}Sg and ^{265}Rf in one chain with missing α-decay of ^{269}Sg [3] are determined as intervals between decays of ^{273}Hs and ^{265}Rf and are shown by triangles with arrows (upper limits). Decay times for events with partially measured E_α values (full energy was not registered) are shown by diamonds on the bottom part of panels for ^{285}Fl [2], ^{281}Cn [3], and ^{277}Ds [1]. Time intervals corresponding to a detection probability of 97% of decays are shown by horizontal lines.

The decay properties of most nuclei in the new chains are in agreement with previous observations (Figure 2). However, in one case the decay time of ^{269}Sg exceeds the average lifetime determined for the five other observed events by a factor of 33. Despite the apparent difference in lifetimes for ^{269}Sg, the standard deviation of the logarithm of all the measured decay times is 1.54, which fits into the interval of 0.48–1.89 proposed in [4] for six exponentially decaying events. Thus, the available set of data does not provide a valid reason, which could confidently point out the inconsistency of the results obtained for ^{269}Sg in one of the seven decay chains. This behavior is not unexpected for statistically decaying nuclei. However, this long decay time could also be associated with existence of the isomeric state of the nucleus.

More than 20 SF nuclei with lifetimes of about 10 μs and 1 ms were observed in [1] which could originate from several sources: $4n$-evaporation product, ^{284}Fl, as well as SF isomers of Pu and Am isotopes or product of $\alpha 2n$

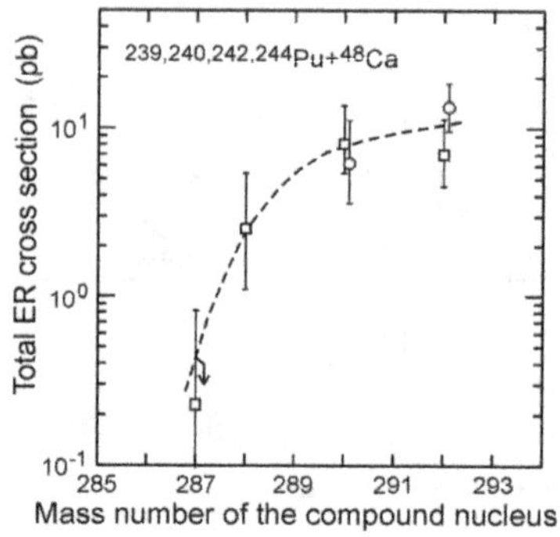

Fig. 3. Maximum of the production cross sections of Fl isotopes in the ^{48}Ca-induced reactions with targets shown in figure vs mass number of the compound nucleus. Data from DGFRS are shown by squares as well as from BGS (^{290}Fl) and TASCA (^{292}Fl) by circles. Vertical error bars correspond to total uncertainties. The dashed line is drawn to guide the eye.

channel, ^{282}Cn. Because of difficulties in identification of SF nuclei, we cannot definitely assign at least part of these events to ^{284}Fl, including events observed in the experiments with ^{239,240}Pu targets in [3] and tentatively attributed to ^{284}Fl. The unambiguous identification of ^{284}Fl still requires further studies.

The unexpected result of these series of experiments is the observation of a strong decrease in the production cross section of the ^{239}Pu+^{48}Ca reaction when compared with heavier target nuclei. (see Figure 3) The cross section of the ^{239}Pu+^{48}Ca reaction decreases by a factor of about 50 in comparison with that for the reaction with the heavier isotope ^{244}Pu. However, the value of about 3 pb was calculated in [5] for the ^{239}Pu+^{48}Ca reaction at excitation energy $E^*\approx32.5$ MeV. According to recent predictions [6], the total ER cross section could reach about 4 pb at E^*=38-39 MeV. Thus, the experimental value was found to be lower by factor of about 20 than the calculated value. Therefore, a considerable difference with theoretical results, which were based on predicted fission barriers, would be caused by a stronger drop of fission barriers than expected. An approximate 1 MeV reduction in fission barrier height would be consistent with the experimental data (Ref. [7]).

Acknowledgments

We would like to express our gratitude to the personnel of the U400 cyclotron and the associates of the ion-source group for obtaining intense ^{48}Ca beams. The ^{240}Pu used in this research was partially supplied by the United States Department of Energy Office of Science by the Isotope Development and

Production for Research and Applications Program in the Office of Nuclear Physics. These studies were supported by the Russian Foundation for Basic Research, including recent Grant No. 16-52-55002. Research at ORNL was supported by the U.S. DOE Office of Nuclear Physics under DOE Contract No.DE-AC05-00OR22725 with UT Battelle, LLC. Research at LLNL was supported by LDRD Program Project No. 08-ERD-030, under DOE Contract No. DE-AC52-07NA27344 with Lawrence Livermore National Security, LLC. This work was also supported by the U.S. DOE through Grant No. DE-FG-05-88ER40407 (Vanderbilt University) and by the National Natural Science Foundation of China (Grant No. 11661131003).

References

1. V.K. Utyonkov, N.T. Brewer, Yu.Ts. Oganessian, K.P. Rykaczewski, F.Sh. Abdullin, S.N. Dmitriev, R.K. Grzywacz, M.G. Itkis, K. Miernik, A.N. Polyakov, J.B. Roberto, R.N. Sagaidak, I.V. Shirokovsky, M.V. Shumeiko, Yu.S. Tsyganov, A.A.Voinov, V.G.Subbotin, A.M. Sukhov, A.V. Karpov, A.G. Popeko, A.V. Sabel'nikov, G.K. Vostokin, J.H. Hamilton, N.D. Kovrizhnykh, L. Schlattauer, M.A. Stoyer, Z. Gan, W.X. Huang, and L. Ma, *Phys. Rev. C* **97**, 014320 (2018).
2. P.A. Ellison, K.E. Gregorich, S. Berryman, D.L. Bleuel, R.M. Clark, I. Dragojević, J. Dvorak, P. Fallon, C. Fineman-Sotomayor, J.M. Gates, O.R. Gothe, I.Y. Lee, W.D. Loveland, J.P. McLaughlin, S. Paschalis, M. Petri, J. Qian, L. Stavsetra, M. Wiedeking, and H. Nitsche, *Phys. Rev. Lett.* **105**, 182701 (2010).
3. V.K. Utyonkov, N.T. Brewer, Yu.Ts. Oganessian, K.P. Rykaczewski, F.Sh. Abdullin, S.N. Dmitriev, R.K. Grzywacz, M.G. Itkis, K. Miernik, A.N. Polyakov, J.B. Roberto, R.N. Sagaidak, I.V. Shirokovsky, M.V. Shumeiko, Yu.S. Tsyganov, A.A. Voinov, V.G. Subbotin, A.M. Sukhov, A.V. Sabel'nikov, G.K. Vostokin, J.H. Hamilton, M.A. Stoyer, and S.Y. Strauss, *Phys. Rev. C* **92**, 034609 (2015).
4. K.-H. Schmidt, *Eur. Phys. J. A* **8**, 141 (2000).
5. G.G. Adamian, N.V. Antonenko, and W. Scheid, *Phys. Rev. C* **69**, 014607 (2004).
6. V.I. Zagrebaev, A.V. Karpov, and W. Greiner, *Phys. Rev. C* **85**, 014608 (2012).
7. V.I. Zagrebaev, Y. Aritomo, M.G. Itkis, Yu.Ts. Oganessian, and M. Ohta, *Phys. Rev. C* **65**, 014607 (2001).

Fission and Quasifission Toward the Superheavy Mass Region*

E. Vardaci[1,2,†], M. G. Itkis[3], I. M. Itkis[3], G. Knyazheva[3] and E. M. Kozulin[3]
[1]*Dipartimento di Fisica "Ettore Pancini", Università degli Studi di Napoli Federico II, Napoli, 80126, Italy*
[2]*Istituto Nazionale Fisica Nucleare, Sezione di Napoli, Napoli, Italy*
[3]*Flerov Laboratory of Nuclear Reactions, JINR, Dubna, Russia*
†*E-mail: vardaci@na.infn.it*

The pathway toward the production of superheavy elements crosses with the one of an in depth study of the fission and quasifission processes. Quasifission affects the probability of forming a compound nucleus after capture and fission plays a decisive role during the cooling of the compound nucleus. To search for optimal reactions to produce superheavy elements in fusion reactions and to provide an estimate of their production cross section measurements of the capture and fusion cross sections along with the survival probabilities are a mandatory task. In this presentation an overview of the impact of fission and quasifission over the synthesis of superheavy elements is discussed along with the need to perform measurements over a wide range of masses and energies.

Keywords: Nuclear reactions; fusion-fission; quasi-fission; capture.

1. Introduction

The major aim of this presentation is to discuss on the importance of studying fission and quasifission processes in connection with the research finalized to the production of superheavy elements (SHE) via heavy ion fusion reactions. This importance is justified by the fact that in the long reaction chain which brings to the formation of SHE two bifurcations must be crossed. Picturing the time evolution of the reaction as shown in Fig. 1 [1], in the first one, after contact or capture, the formation of a compound nucleus (CN) has to overcome the competition with the quasifission process; in the second one, if a compound nucleus is formed, it has to survive its own fission to become an evaporation residues. The fate of each of these two bifurcations relies on the competition with very strong processes. Quasifission (QF) and CN fission (CNF) are indeed so strong that approximately one out of 10^6 decay paths end up in an evaporation residue. The understanding of the features of quasifission and fission is therefore crucial to search for the optimal conditions to ignite SHE production and to estimate fusion cross sections.

* This work was supported by FLNR, JINR and INFN.

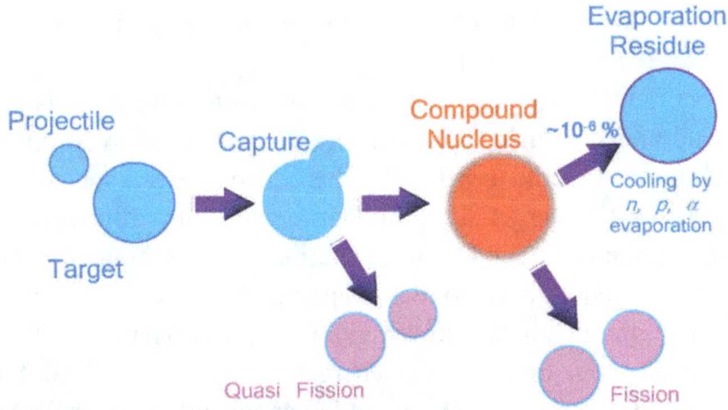

Fig. 1. Schematic picture of the present knowledge of the process chain leading to SHE in heavy ion fusion reactions (from [1]).

In the following, few issues connected to the properties of quasifission and fission will be discussed in some detail along with the strategy adopted to choose the optimal entrance reaction channels to produce elements heavier than *Og*.

2. The SHE Production Cross Section

As schematically shown in Fig. 1, the formation of SHE is pictured to occur in three sequential, independent steps: contact or capture, CN formation, survival of the CN to fission. Consequently, the evaporation residues cross section σ_{ER} is the product of three terms [1,2]: the capture cross section $\sigma_{capture}$, the probability of forming a CN (P_{CN}) after contact, and the probability that it survives against fission ($W_{survival}$), namely:

$$\sigma_{ER} = \sigma_{capture} \times P_{CN} \times W_{survival} \tag{1}$$

Experimentally, $\sigma_{capture}$ is defined as the sum of the QF, CNF and ER cross sections:

$$\sigma_{capture} = \sigma_{QF} + \sigma_{CNF} + \sigma_{ER}$$

Both capture cross section and fusion probability depend on the incident energy and angular momentum. The survival probability is a property of CN and depends on its excitation energy of CN and angular momentum, whereas the fusion probability depends strongly on the reaction entrance channel.

The separation of the SHE reaction chain into three steps is clearly an oversimplification. However, the three-step concept is very helpful from the experimental point of view which only gives observables on the final stage of a reaction. Accurate measurements of the above cross sections are very desirable to estimate the production rate of SHE and are of crucial importance to setup experiments for their production.

At present, both "cold" and "hot" fusion reactions have been used to produce SHE [2]. Cold and hot fusion reactions are characterized by strongly different entrance channels. In cold reactions, where fusion occurs near the reaction threshold, ^{208}Pb or ^{209}Bi target nuclei are bombarded with projectiles heavier than Ca to form compound nuclei with low excitation energy (about 10–20 MeV). One neutron emission is enough to produce evaporation residues. Hot fusion reactions involve actinide target nuclei (lower entrance mass asymmetry) and lead to evaporation residues with excitation energy between 30 and 60 MeV, which results in higher neutron evaporation and lower survival against fission. Besides this, there are profound differences in the production cross section as shown in Fig. 2.

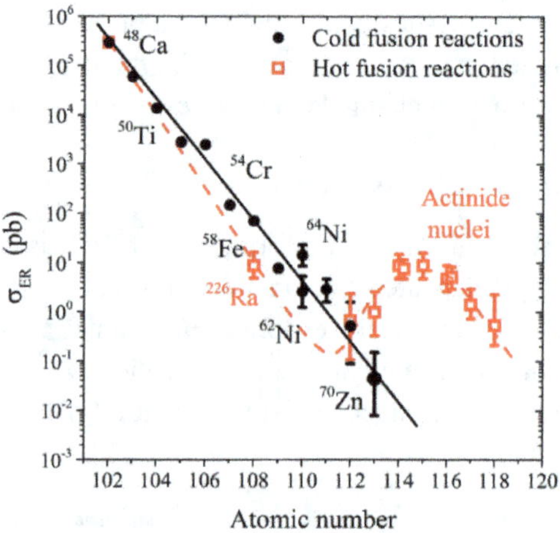

Fig. 2. Maximal cross sections of 1n-evaporation channel in the cold fusion reactions of ^{208}Pb and ^{209}Bi target nuclei with different projectiles (labelled in the figure) and cross sections of 3 and 4n-channels in the hot fusion reactions of ^{48}Ca as a function of compound nucleus atomic number. The lines are drawn as guide the eye (adapted data from [3]).

It is quite evident that hot fusion reactions involving ^{48}Ca are more suitable to produce SHE with $Z > 112$. The fast monotonic decrease of the ER cross sections with increasing charge of SHE synthesized in the "cold" fusion reactions must be compared with the rise of the cross sections measured in hot reactions leading to heavier SHE. The change in the trend occurs at Z=112. There are a number of reasons for this drastic change which spans several orders of magnitude. A comprehensive overview of such reasons is given from the trend of the calculated cross section proposed in Ref. [1] and shown in Fig. 3. The dashed lines describe the trend for the capture, fusion and evaporation residues cross sections in the cold fusion reactions induced by several projectiles on ^{208}Pb. The solid lines refer to the same observables but for the hot fusion reactions induced by ^{48}Ca on several actinides. There are several major differences. Besides the difference of about one order of magnitude in the capture cross sections, the most striking feature is in the gap between σ_{cont} and the compound nucleus cross section σ_{CN} and the evaporation residues cross section σ_{ER}. These remarkably different trends for the cold and hot cases mean a different balance between the probabilities to pass the bifurcation points above.

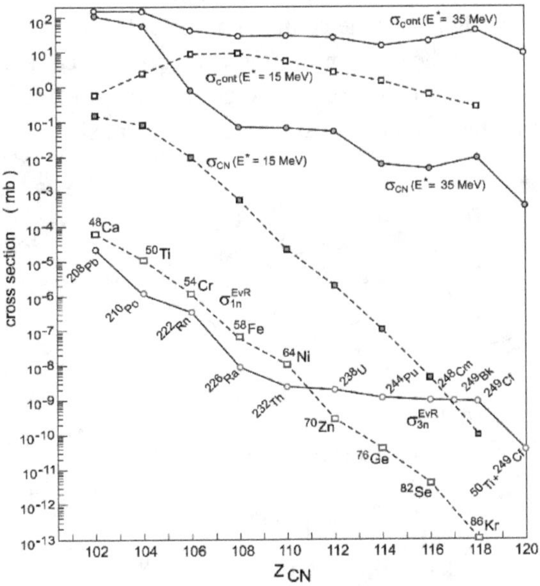

Fig. 3. Calculated contact (capture), fusion and evaporation residue cross sections in the "cold" ^{208}Pb induced (rectangles joined by dashed lines, projectiles are shown) and in "hot" ^{48}Ca induced (circles joined by solid lines, targets are shown) fusion reactions. The cross sections are calculated at beam energies corresponding to 15 MeV ("cold" fusion, 1n channel) and 35 MeV ("hot" fusion, 3n channel) excitation energies of the compound nuclei. From Ref. [1].

The gap between σ_{cont} and σ_{CN} gives the measure of what is the fate of the reaction at the first bifurcation point. In the case of a cold reaction, for instance, to produce element Z=116, one out of 10^8 paths end up into a CN whereas one out of 10^4 in the case of hot reactions. This is a dramatic gain in favor of hot reactions.

This big lost in the flow for cold reactions is due to the competition with the QF channel which is comparatively less strong in the hot case because the entrance channel mass asymmetry is lower and the target nuclei are usually deformed in their ground state.

Continuing in the timeline, the survival of the formed CN comes into play. As expected from the lower excitation energy, the "cold" reaction case gives rise to a larger survival probability, about 10^{-3}, whereas in the "hot" reactions this probability is about 10^{-7} for the 3n channel. Combining these two numbers, the outcome is a total gain of three orders of magnitude in the production cross section in hot reactions. This result is mainly due to the fact that in hot reactions the compound nuclei are located closer to the island of stability, and the N/Z values, even after evaporating 3 neutrons, remains enough large to make the fission barrier sufficiently higher for the compound nuclei to survive fission with larger probability. This is not the case for the nuclei involved in cold reactions which, in spite of the lower excitation energy, undergo fission with larger probability after losing one neutron being on the most neutron deficient side.

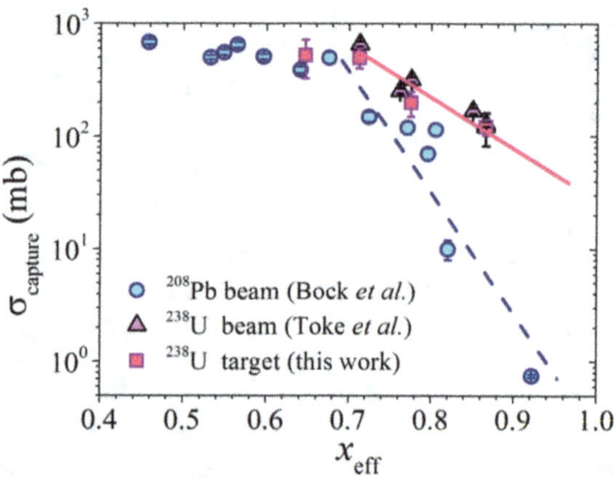

Fig. 4. The capture cross section as a function of the effective fissility parameter for the reactions with ^{208}Pb and ^{238}U nuclei at interaction energies of $E_{c.m.}/E_{Bass}$ = 1.1−1.2. The lines are a guide for the eye. From Ref. [5].

The relative balance between the competition with QF, which affects P_{CN}, and the survival against fission explains why all the elements above Z=113 where synthesized via hot fusion reactions and gives the guidelines on how to search for reactions and mechanisms to produce heavier nuclei. It is therefore clear that in choosing the reactions to produce SHE it is advisable to maximize all the three factors, namely $\sigma_{capture}$, P_{CN} and $W_{survival}$. Consequently, independent measurements of $\sigma_{capture}$, P_{CN} and $W_{survival}$ over a large span of reactions are of paramount importance to understand the individual contributions to σ_{ER}. These measurements are indeed even more relevant considering that the models known to predict these quantities differ dramatically between each other [4] and accurate predictions of σ_{ER} are crucial for the design of new experiments which may last many months.

Many models predict quite well the trend of $\sigma_{capture}$ in hot and fusion reactions. From the experimental point of view, the same trend is also observed as in Fig. 3. One recent compilation [5] of old and new data is shown in Fig. 4. Capture cross sections are reported as a function of the effective fissility parameter $\varkappa_{eff}$, a kind of scaling parameter which gives a more comprehensive overall view. In the case of hot fusion reactions (reactions on ^{238}U) the cause of the less steep dropping of the capture cross section has been searched into the ground state deformation of the target nuclei [2].

One essential step in the search for optimal reactions to synthesize SHE is to measure P_{CN} or fusion cross sections. This task is particularly important because no satisfactory model exist to predict the behavior of P_{CN}. The measurement of P_{CN} is not an easy and fully understood task because QF events overlap with pure CNF events and make the counting of true fusion events blurred. A detailed knowledge of the QF process is therefore mandatory and is still a matter of open questions.

3. The Fusion Cross Section

The basic task in the measurement of the fusion cross section consists in counting the number of events which arise from the complete fusion of the projectile and target nuclei. In the light/medium mass nuclei, evaporation residues and fission fragments are the most copious events. The sum of them gives rise to the fusion cross section:

$$\sigma_{fusion} = \sigma_{CNF} + \sigma_{ER} \tag{2}$$

When more massive nuclei are involved in the fusion process, at energies around the Coulomb barrier, because of the substantial increase of the Coulomb repulsion, the QF arises as the most important process counteracting complete fusion [2]. QF gives rise to a binary reaction without forming a compound nucleus. The mass distribution is predominantly asymmetric but events extend up to the symmetric region of the mass distribution. The counting of pure fusion-fission events, if any, is therefore polluted by the occurrence of non-compound events in the same mass region or at its tails. One such example is given in Fig. 5.

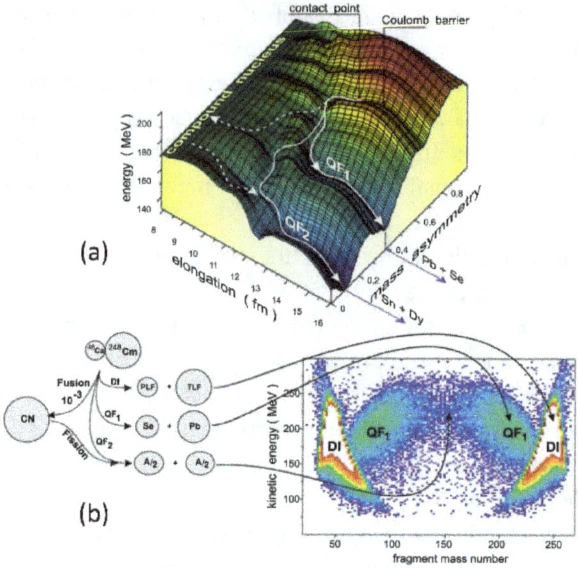

Fig. 5. (a) Potential energy surface for the nuclear system formed in $^{48}Ca + ^{248}Cm$ collision (from [6]). The solid lines with arrows show schematically the quasi-fission trajectories going to the lead and tin valleys. The dashed curves correspond to fusion (CN formation) and fission processes. (b) Mass-energy distribution of reaction fragments in collision of ^{48}Ca with ^{248}Cm at 203 MeV center-of-mass energy (from [7]).

The potential energy surface (PES) for the $^{48}Ca + ^{248}Cm$ reaction leading to the superheavy composite system $^{296}116$ as a function of elongation and mass asymmetry (collective variables) is shown in Fig. 5(a). After just overcoming the Coulomb barrier and reaching the contact point, the intermediate nuclear system can follow different trajectories (drawn schematically, without fluctuations), some of which may end up in the same binary channel, also known as “fission channel”, hollowed by the shell effects (magic numbers) with or without forming a compound nucleus. The symmetric QF2 path merges with the one following CNF and causes the pollution. The corresponding mass-energy distribution in Fig. 5(b) shows loci that can be readily connected to the valleys originated by the shell corrections. This interpretation of the QF process also manifests one of the proof of its strong connection with shell closures. Another very important evidence is the occurrence of the inverse QF [8].

The amount of pollution is rather strongly related to some entrance channel features of the reaction. For instance, QF grows with growing mass symmetry or with the Z_PZ_T products of the charges of the projectile and target, respectively [2].

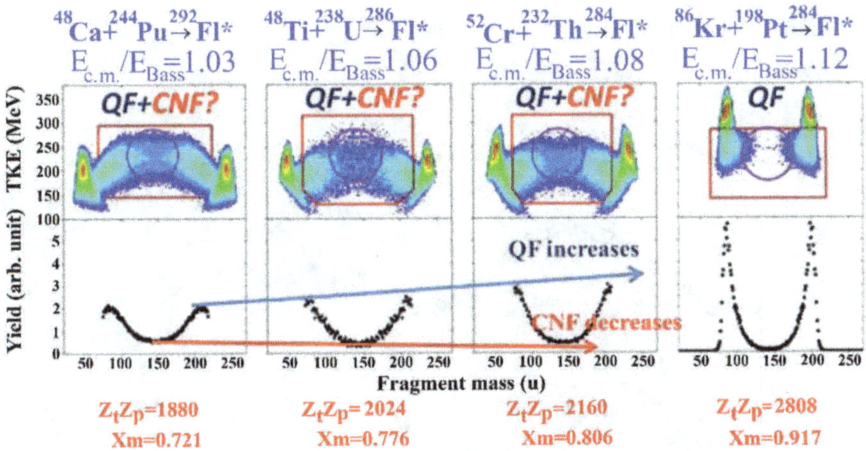

Fig. 6. (top) The mass-energy matrices of binary fragments formed in several reactions candidate to the formation of composite systems with $Z = 114$ at energies above the Bass barrier. (bottom) mass distributions as a function of mass of fissionlike fragments inside the contour lines shown on the mass-energy matrices. From [9].

Such an example is given in Fig. 6. It is fairly evident the dependence on the entrance channel mass asymmetry and on the product $Z_P Z_T$ [9].

Since QF is a multi-facet process, still not completely identified, for which no reliable theory exists yet, semi-empirical methods have been developed on purpose to isolate the CNF component from the QF one. In order to achieve this separation several techniques have been developed based on the analysis of different experimental observables of fissionlike fragments [2], namely, angular distributions, width of mass distributions, mass and energy distributions, mass and angular distributions. One of the main procedure used for decomposing fissionlike fragments into CNF and QF is to compare the experimental properties of fissionlike fragments for reactions leading to the formation of the same composite system but having different entrance channel properties, either with experimental fission systematics, or with theoretical predictions in the frame of the liquid drop model (LDM), or the transition state model or modern multidimensional calculations [2].

In the lack of a comprehensive theoretical model it is necessary to proceed empirically by searching for correlations between observables, and extrapolating the systematics to unknown regions. One such example is given in Fig. 7. On the left, the Total Kinetic Energy (TKE) distributions for the fragments with masses $A_{CN/2} \pm 20$ u for the reactions ^{36}S, ^{48}Ca, ^{48}Ti, and ^{64}Ni+^{238}U are shown. Clearly, the TKE distributions change markedly showing nearly Gaussian shape in the

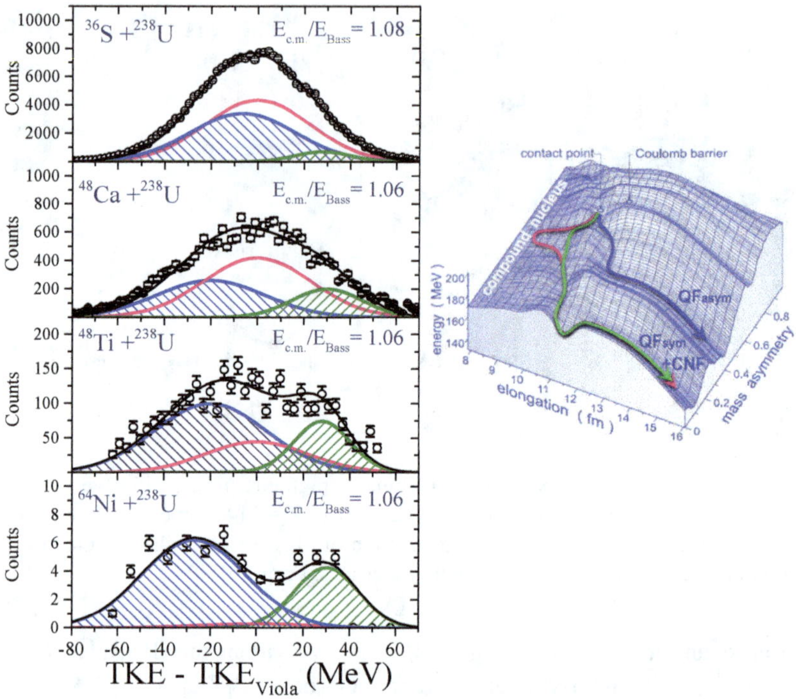

Fig. 7. Total Kinetic Energy distributions of fragments with masses $A_{CN/2} \pm 20$ u for the reactions ^{36}S, ^{48}Ca, ^{48}Ti, and ^{64}Ni + ^{238}U at energies above the Coulomb barrier. The filled region corresponds to the TKE distribution for CN fission. The dashed and dashed-dotted curves are associated with asymmetric and symmetric QF, respectively (from [5]).

case of ^{36}S and ^{48}Ca ions and two-humped shape for the reactions with ^{48}Ti and ^{64}Ni. One may speculate about the presence of other processes together with the CN fission in the symmetric mass region. The disentanglement of possibly three components is based on the fact that different paths in the PES, as shown in Fig. 7 (right), have different TKE. If it is assumed that the mass-symmetric fragments may be formed by three different modes (CN fission, symmetric QF and a tail of the asymmetric QF process), to evaluate the contribution of the CN-fission process in the symmetric mass region the TKE distributions can be decomposed as a sum of three Gaussians. One of them is associated to the CN-fission process (filled region in Fig. 7 (left)). To constraint the fit of three Gaussians to some physical ground, the means and variances can be fix considering some systematics. For the CN component, the fission systematics can fix mean and variance of this component. For the QF components, additional information can be gained from systems where it is fairly known that only QF occurs in the symmetric mass region. For instance, in the case of the ^{58}Fe+^{208}Pb reaction [10] where the asymmetric QF is the main process, even in the symmetric mass region of fragments, it is found that the variance of TKE for QF does not depend on the mass of the QF fragments. Attributing the lower energy component to QFasym and the higher energy one to QFsym, in the fitting procedure the variance of the QFasym components can be kept fix.

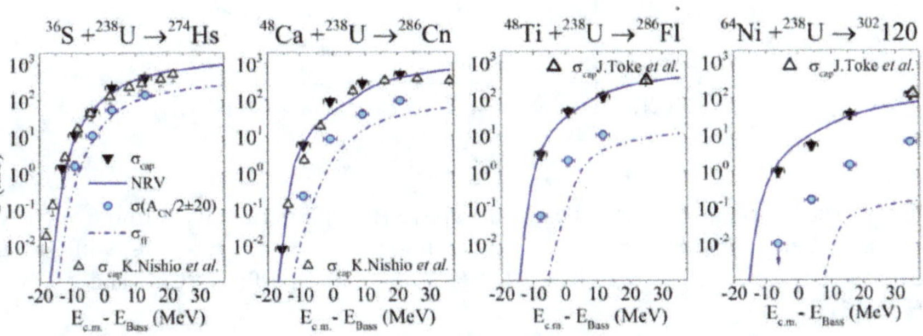

Fig. 8. The solid triangles are the cross sections for fissionlike fragments in the reactions ^{36}S, ^{48}Ca, ^{48}Ti, and ^{64}Ni+^{238}U and the open triangles are the cross sections for fissionlike fragments. The solid lines are the empirical model calculations. The circles are the cross sections for fragments with masses $A_{CN/2}$ ± 20 u, and the dashed-dotted lines are the estimated CN-fission cross sections. E_{Bass} is the Bass barrier. (see Ref. [5] for details).

From the analysis above on the fusion events, the fusion cross section can be extracted by means of an integration over 4π considering an angular distribution of the type $1/\sin \theta_{c.m.}$. This is clearly an important approximation, but the only hypothesis possible since no information was obtained for the data in [5] about the angular distribution both in fission and capture events. The fusion cross sections are shown in Fig. 8 along with the capture cross sections. One can notice the drastic drop in the capture cross section going from ^{36}S to ^{64}Ni. This in an important result in view of the search of reactions to produce superheavy elements heavier than *Og* by using heavier projectiles.

4. The Fusion Probability

Once the fusion cross section is extracted, the fusion probability P_{CN} entering equation (1) can be estimated. P_{CN} is defined as the probability for CN formation from the configuration of two nuclei in contact. Being σ_{ER} typically neglectable with respect to σ_{CNF} (only a few picobarns for the reactions in Fig. 8) P_{CN} is the ratio between the number of events attributed to CN fission and the fissionlike events all together.

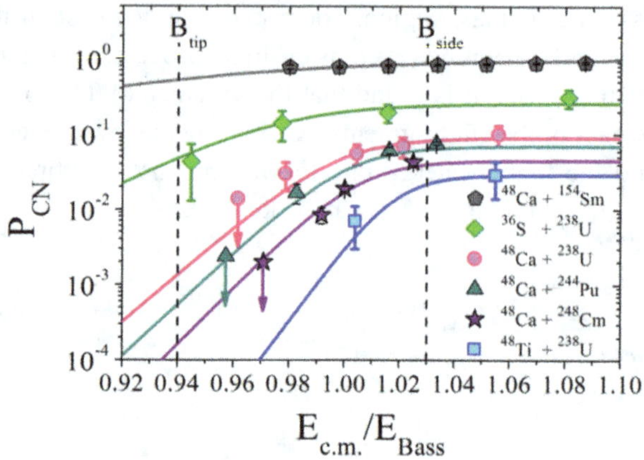

Fig. 9. The fusion probabilities below and above Bass barrier obtained from the analysis of mass-energy distributions of fissionlike fragments for the reactions with strongly deformed target nuclei: $^{48}Ca+^{154}Sm$, $^{36}S+^{238}U$,$^{48}Ca+^{238}U$,^{244}Pu, ^{248}Cm, and $^{48}Ti+^{238}U$. From [5].

Figure 9 shows the trend of the fusion probability versus the center of mass energy below and above the Bass barrier [5] for a number of reactions. There are few interesting features in this plot. The targets in Fig. 9 are all deformed in the ground state. This means that there are two Coulomb barriers, one B_{tip} when the collision occurs with the tip of the target nucleus, the other B_{side} when the collision occurs with the side of the target nucleus. By definition B_{tip} and B_{side} are the minimum and the maximum Coulomb barrier, respectively. For energies below B_{side}, side collisions are below barrier, and this explains the lower probability for $E_{c.m.} < B_{side}$. Above B_{side}, P_{CN} saturates as expected for all reactions. This means that the not only the choice of the system is important to maximize the fusion cross section, but also the choice of the bombarding energy is crucial. Another observation is that if the reactions on ^{238}U are considered, increasing the mass and charge of the projectile results in a drop of P_{CN} of about 2 orders of magnitude at energies above B_{side}. Last comment concerns with the fact that for all the reactions in Fig. 9 a trend is observed that may be brought back to some common empirical algebraic relationship. In the lack of reliable quantitative models for P_{CN}, to identify an empirical relation and some scaling parameter to predict P_{CN} for a given reaction would be a great progress. Zagrebaev and Greiner [11] indeed proposed the following energy dependence:

$$P_{CN}(E_{c.m.}) = P_{CN}^0/\{1 + exp[\alpha(\beta - E_{c.m.}/E_{Bass})]\} \quad (3)$$

which is overlapped to the data in Fig. 9. P^0_{CN}, α, and β are empirical constants. P^0_{CN} is the fusion probability above the barrier, β is the effective barrier for CN formation, and α is responsible for the sub-barrier suppression of fusion probability. Of course, each of the curves in Fig. 9 corresponds to different fitting parameters. Therefore equation (3) is still not the empirical law searched for.

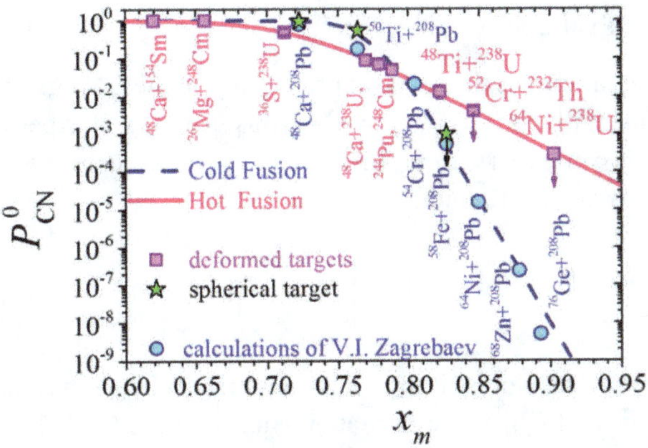

Fig. 10. Fusion probability in cold (spherical target nuclei) and hot fusion (strongly deformed target nuclei) reactions at energies above the Coulomb barrier versus the mean fissility parameter of the reaction. The circles are the fusion probabilities calculated for cold fusion reactions [11]. Adapted from [5].

However, if the data above the Bass barrier are plotted against the recently identified scaling parameter [12]:

$$x_m = 0.75x_{eff} + 0.25x_{CN}$$

as shown in Fig. 10, the following empirical law, also proposed in Ref [11]:

$$P^0_{CN}(E_{c.m.}) = 1/\left\{1 + exp\left(\frac{x_m-\xi}{\tau}\right)\right\} \tag{4}$$

reproduces surprising well the data for cold and hot reactions. x_{eff} is the effective fissility parameter and x_{CN} is the liquid drop model fissility. The values of the fitting parameters are the following [5]:

Reactions with Pb: $\tau = 0.0067 \pm 0.0007$; $\xi = 0.776 \pm 0.007$;

Reactions with actinide nuclei: $\tau = 0.0226 \pm 0.0006$; $\xi = 0.721 \pm 0.002$.

This new scaling parameter x_m, termed as "mean" fissility, seems to embed the main features of the entrance channels, such as the products Z_PZ_T or the mass asymmetry. However, the main difference in the trend between hot and cold reactions is to be searched in the orientation during the collision of the target with the projectile being important as the target nuclei are deformed in the case of hot fusion reactions (orientation effect) in Fig. 10. This feature cannot be included in a scaling parameter and hence the difference between the two types of reactions.

5. The Survival Probability

The last term in equation 1 is the survival probability of the CN against fission. It is defined experimentally as $W_{survival} = \sigma_{ER}/(\sigma_{ER} + \sigma_{CNF})$. If experimental data are not available it can be written in a form that explicitly uses the neutron and fission partial decay widths Γ_n and Γ_f:

$$W_{\text{survival}} = P_{\text{xn}}(E^*_{\text{CN}})\prod_{i=1}^{x}\left(\frac{\Gamma_n}{\Gamma_n+\Gamma_f}\right)_{i,E^*_{\text{CN}}},$$

where P_{xn} is the probability of emitting x neutrons. The index i runs on the number of emitted neutrons and E^*_{CN} is the excitation energy of the CN. The ratio Γ_n/Γ_f can be computed through the statistical model as:

$$\frac{\Gamma_n}{\Gamma_f} = \frac{4A_{CN}^{2/3}a_f\left(E^*_{CN}-B_n\right)}{ka_n\left[2a_f^{1/2}(E^*_{CN}-B^*_f)^{1/2}-1\right]}\exp\left[2a_n^{1/2}(E^*_{CN}-B_n)^{1/2}-2a_f^{1/2}(E^*_{CN}-B^*_f)^{1/2}\right]$$

where $B^*_f = B_f(E^*_{CN}) = B_f exp\left(-\frac{E^*_{CN}}{E_D}\right)$, $a_n = A_{CN}/10$, $a_f = 1.1a_n$, $E_D = 0.4\,A_{CN}^{4/3}/a_n$, k = 0.98 MeV.

This equation implies that to arrive to an estimate of σ_{ER} some knowledge of the fission barrier is needed as well. This is a critical point because the values of fission barriers rely on models. Differences between predictions vary from 0.5 up to 2 MeV. This translates in survival probabilities that can differ up to one order of magnitude from model to model. Conversely, if the ratio Γ_n/Γ_f or σ_{ER} is measured the fission barrier can be estimated. Indeed, this is still not enough because other ingredients of the model need to be fixed as the neutron binding energy B_n, a_f and a_n and the dumping of B_f as the excitation energy is raised. All of this means that the uncertainties in the estimates of $W_{survival}$ play an important role. Furthermore, the formula above for Γ_n/Γ_f is an approximation and a much more complex statistical model of particle evaporation and fission should be invoked along with dynamical effects such as nuclear viscosity [6, 13-16].

All of the above gives support to the necessity of measuring $W_{survival}$ over a wide range of masses and energies to establish a larger and larger set of data on which to test the different models [4, 13].

6. Perspectives

From the analysis shown in this presentation, it is clear that the synthesis of superheavy elements has to face many challenges. Since the predictive power of the currents models is quite limited, experimental campaigns should be designed to increase substantially the amount of data available on the fusion process, especially of the front of heavier target nuclei. Understandingly, this is not an easy task because heavier target nuclei such as Cm, Bk or Cf are difficult to make and to handle. From the trend of the capture and fusion cross sections, we can conclude that further progress in the field of synthesis of superheavy nuclei can be achieved using hot fusion reactions between actinide nuclei and Ti, Fe or Cr isotopes. Of course, for planning experiments on the synthesis of superheavy nuclei of up to Z=122, new research and more precise quantitative data obtained from the processes of FF and QF of these nuclei in reactions with Ti, Fe and Cr ions are needed.

Above all, the biggest challenge is to be able to detect evaporation residues with cross sections of the order of the pico or sub-pico barn. This task requires a major jump in the presently available intensity for stable beams and an update of the technology for targets and for detection. These are indeed the aims of the SHE factory being under construction at the FLNR at JINR in Dubna and soon to be ready to operate.

References

1. V. I. Zagrebaev and W. Greiner, *Nucl. Phys. A* **944**, 257 (2015).
2. M. G. Itkis, E. Vardaci, I. M. Itkis, G. N. Knyazheva and E. M. Kozulin, *Nucl. Phys. A* **944**, 204 (2015) and references therein.
3. Yu. Ts. Oganessian, *J. Phys. G, Nucl. Part. Phys.* **34** (2007) R165.
4. W. Loveland, *Eur. Phys. J. A* **51**, 120 (2015).
5. E. M. Kozulin *et al.*, *Phys. Rev. C* **94**, 054613 (2016).
6. V. I. Zagrebaev and W. Greiner, *"Giant Nuclear Systems of Molecular Type"*, Clusters in Nuclei, Lecture Notes in Physics 818, 267 (2010).
7. M. G. Itkis *et al.*, *Nucl. Phys. A* **734**, 136 (2004).
8. E. M. Kozulin *et al.*, *Phys. Rev. C* **96**, 064621 (2017).
9. M. G. Itkis, *"Impact of Quasifission on SHE Production"*, ECT* Workshop on Spontaneous and induced fission of very heavy and super-heavy nuclei, Trento Italy (2018).

10. I. M. Itkis *et al.*, *Phys. Rev. C* **83**, 064613 (2011).
11. V. I. Zagrebaev and W. Greiner, *Nucl. Phys. A* **944**, 257 (2015).
12. R. du Rietz *et al.*, *Phys. Rev. C* **88**, 054618 (2013).
13. Hongliang Lü *et al*, *Phys. Rev. C* **94**, 034616 (2016).
14. E. Vardaci *et al.*, *Phys. Rev. C* **92**, 034610 (2015).
15. E. Vardaci *et al.*, Eur. Phys. J. A **43**, 127 (2010).
16. A. Di Nitto *et al.*, *Eur. Phys. J.* **47**, 83 (2011).

Experimental Study of the $^{249-251}Cf + ^{48}Ca$ Reactions: Toward the Magic Neutron Number N=184

A. A. Voinov[1*], N. T. Brewer[2,3,4], V. K. Utyonkov[1], K. P. Rykaczewski[3], Yu. Ts. Oganessian[1], F. Sh. Abdullin[1], R. A. Boll[3], D. J. Dean[3], S. N. Dmitriev[1], J. G. Ezold[3], L. K. Felker[3], R. K. Grzywacz[3,4], M. G. Itkis[1], N. D. Kovrizhnykh[1], D. C. McInturff[3], K. Miernik[2,5], G. D. Owen[3], A. N. Polyakov[1], A. G. Popeko[1], J. B. Roberto[2], A. V. Sabelnikov[1], R. N. Sagaidak[1], I. V. Shirokovsky[1], M. V. Shumeiko[1], N. J. Sims[3], E. H. Smith[3], V. G. Subbotin[1], A. M. Sukhov[1], A. I. Svirikhin[1], Yu. S. Tsyganov[1], S. M. Van Cleve[2], G. K. Vostokin[1], C. S. White[3], J. H. Hamilton[6] and M. A. Stoyer[7]

[1]*FLNR, JINR, Dubna 141980, Russian Federation*
[2]*JINPA, ORNL, Oak Ridge, TN 37831, USA*
[3]*ORNL, Oak Ridge, TN 37831, USA*
[4]*Department of Physics and Astronomy, UT, Knoxville, TN 37996, USA*
[5]*Faculty of Physics, UW, Warsaw, Poland*
[6]*Department of Physics and Astronomy, VU, Nashville, TN 37235, USA*
[7]*Nuclear and Chemical Sciences Division, LLNL, Livermore, CA 94551, USA*
[*]*E-mail: voinov@jinr.ru*

Recently several complete-fusion reactions have been studied at the DGFRS setup (FLNR, JINR) with an aim of exploring the boundaries of the "Island of Stability" of the superheavy elements (nuclei around Z=114-126 and N=184). During the $^{249-251}Cf+^{48}Ca$ run with mixed Cf target (50.7% of ^{249}Cf, 12.9% of ^{250}Cf, and 36.4% of ^{251}Cf) just one decay chain was detected at the 252-MeV beam energy which was attributed to the previously observed ^{294}Og isotope, the product of the $^{249}Cf(^{48}Ca,3n)^{294}Og$ reaction; the corresponding reaction cross section was measured. For the $^{251}Cf(^{48}Ca, 3\text{-}4n)^{295,296}Og$ reactions no Og decays were detected during this run. Future possibilities for synthesis of new nuclides in the region of SHEs, namely more neutron-rich Lv, Ts and Og isotopes and new nuclides with higher Z (119 and 120), are discussed.

Keywords: Superheavy nuclei; alpha decay; compound nucleus; reaction cross section.

1. Introduction

The existence of an enhanced stability in the region of the superheavy nuclei, which has been developed in various theoretical approaches and hypothesized for about 50 years, has been validated by recent experiments. More than 50 new isotopes of the nuclei with Z = 104-118 from the predicted "Island of stability" of superheavy elements (SHE) were studied in pioneering experiments employing the Dubna Gas-Filled Recoil Separator (DGFRS), where performed complete-fusion reactions of ^{238}U, ^{237}Np, $^{242,244}Pu$, ^{243}Am, $^{245,248}Cm$, ^{249}Bk and ^{249}Cf targets with accelerated ^{48}Ca ions lead to the synthesis of the six new elements 113(Nh), 114(Fl), 115(Mc), 116(Lv), 117(Ts) and 118(Og) of the Periodic Table [1]. Also a few "hot fusion" reactions induced by ^{48}Ca ions and using actinide targets (^{238}U, ^{242,244}Pu, ^{243}Am, ^{248}Cm, and ^{249}Bk) were studied at IVO (FLNR, PSI), SHIP, TASCA (GSI), BGS (LBNL), and GARIS (RIKEN) setups then. Obtained there experimental data on isotopes ^{283}Cn, $^{285\text{-}289}$Fl, ^{287,288}Mc, $^{290\text{-}293}$Lv, and ^{294}Ts [2-7] are in full consistent with the results from DGFRS. Studied decay properties of these Z=112-118 nuclei point to a considerable stabilizing effect when approaching the predicted neutron magic number N=184.

There are few possible ways to expand our knowledge about more habitants from the SHE region. Since the isotopes of $^{249\text{-}251}$Cf are the heaviest material available in enough quantities for target production today, one can try to go further to higher Z of new elements in direct reactions with heavier than ^{48}Ca projectiles bombarding U – Cf targets. Five reactions ^{64}Ni + ^{238}U (SHIP [8]), ^{58}Fe + ^{244}Pu (DGFRS [9]), ^{54}Cr + ^{248}Cm (SHIP [8]), ^{50}Ti + ^{249}Cf and ^{50}Ti + ^{249}Bk (TASCA [10]) leading to $^{299,302}120^*$ and $^{299}119^*$compound nuclei were employed already, however, no events attributed to SHE were observed in these experiments. The upper cross section limits were set at 0.07 – 1.1 pb range depending on the reaction. Obviously, the sensitivities of present experiments are not enough and new modern facilities (like Superheavy Elements Factory – SHEF – in JINR) should start operating to satisfy these challenges soon (see [11] in this Proceeding for details).

Another way is to use ^{48}Ca-induced reactions and to study less probable reaction channels with evaporations of 2 and 5 neutrons from compound nuclei. Up to now ^{294}Ts (N=177) and ^{294}Og (N=176) are the two heaviest nuclei from the valley of SHE around N=184. Thus, performing the ^{248}Cm(^{48}Ca,2n)^{294}Lv and ^{249}Bk(^{48}Ca,2n)^{295}Ts reactions could lead to ^{294}Lv (N=178) and ^{295}Ts (N=178). Reactions with the heaviest available today target material, ^{251}Cf, could produce ^{295}Og, ^{296}Og, and ^{297}Og (N=177, 178, and 179). Their decay descendants will pass through nuclides with already known properties.

Reactions with lighter target materials (^{239,240}Pu, ^{241}Am) and ^{48}Ca beam could explore the left-hand edge of the "Island of stability" and observe decays of the lightest isotopes of $^{282\text{-}285}$Fl and $^{282\text{-}286}$Mc nuclei in corresponding 2-5n-evaporation reaction channels. Two new ^{284}Fl and ^{285}Fl isotopes have been already synthesized and studied in the reactions ^{239}Pu(^{48}Ca, 3n)^{284}Fl and ^{240}Pu(^{48}Ca, 4-3n)^{284,285}Fl [12].

In this work, we present the results of the experiment aimed at the synthesis of new Og isotopes in the $^{249\text{-}251}$Cf + ^{48}Ca fusion-evaporation reactions [13].

2. Experiment

The first identification of an isotope of ^{294}Og (Z=118) element was attempted and successful at JINR Dubna in 2002 using the ^{249}Cf+^{48}Ca reaction [14]. The ^{48}Ca beam energy was selected above the Coulomb barrier to create the ^{297}Og compound nucleus with an excitation energy E^* of about 29 MeV. With an accumulated beam dose of 2.5×10^{19} particles of ^{48}Ca at 245 MeV, a single decay chain of recoil tagged and correlated high-energy α-decays and spontaneous fission (recoil-α-α-SF) was detected. The estimated cross section for the $3n$-reaction channel based on one observed event was 0.3 pb.

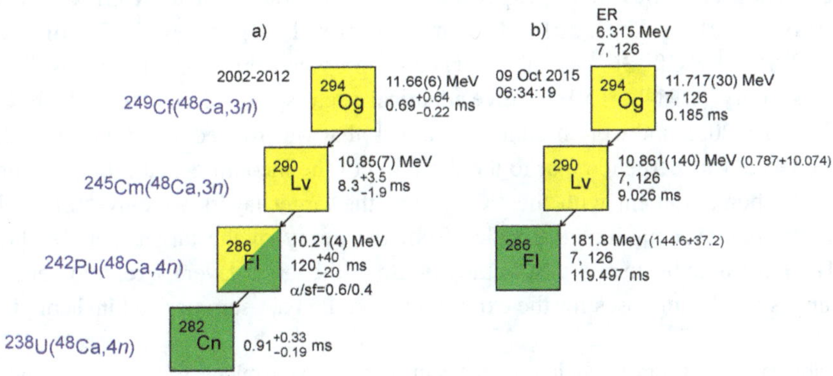

Fig. 1. (a) Average decay properties ($T_{1/2}$ and E_α) of ^{294}Og-^{282}Cn determined in the reactions ^{48}Ca + ^{249}Cf, ^{245}Cm, ^{242}Pu, ^{238}U [1]. (b) Properties of ^{294}Og-^{286}Fl decay measured in this work. The right rows provide energies, time intervals between events and their DSSD strip numbers for α-decay and SF. Energies of summed signals are given in parentheses.

The second experiment with a ^{249}Cf target and ^{48}Ca beam was performed in 2005 at a higher beam energy of 251 MeV corresponding to the compound nucleus excitation energy range between 32 MeV and 37 MeV. Two additional events consistent with the first decay chain were detected, and the properties of all three decay chains attributed to ^{294}Og decay were presented in Ref. [14]. The half-life of ^{294}Og based on three α decays at the average energy of 11.65 ± 0.06 MeV was determined to be 0.89 ms. Two decay chains consisted of recoil-α-α-SF events and one longer decay chain terminated with SF after three consecutive α-decays. These observations are consistent with independently studied decay properties of the granddaughter isotope of ^{294}Og, ^{286}Fl, with similar decay probabilities through α-emission and spontaneous fission. The cross section at the beam energies used to observe these two new ^{294}Og events was estimated to be 0.5 pb for the excitation energy range $E^* =$ 32.1–36.6 MeV [14].

One more event of ^{294}Og formation was observed in the experiment on Z=117 element synthesis as a result of the buildup of ^{249}Cf in the ^{249}Bk target material due to its β-activity [15]. The cross-section estimate for this single event of 0.3 pb for the excitation energy range of 26.6–37.5 MeV for ^{297}Og was also in good agreement with earlier deduced cross sections. The four-event average of ^{294}Og decays yielded $T_{1/2}$ of 0.69 ms and $E\alpha = 11.66 \pm 0.06$ MeV [15]. A summary of decay properties as they were experimentally known prior to mixed $^{249\text{-}251}$Cf target campaign is shown in Fig. 1a.

The new series of the experiments was undertaken at the DGFRS during October, 2015 – April, 2016. The target of mixed Cf isotopes (50.7% of ^{249}Cf, 12.9% of ^{250}Cf, 36.4% of ^{251}Cf) was provided by Oak Ridge National Laboratory (ORNL, TN). Twelve identical sectors were delivered to FLNR in February 2015 and were mounted on a disk that was rotated at 1700 rpm so that the target was perpendicular to the direction of the incoming beam. In the course of the bombardment with the ^{48}Ca beam, the target layers were systematically monitored by counting α-particles from the decay of the target isotopes [13]. The lab-frame beam energies in the middle of the target layers, excitation energy ranges and beam doses for the experiments studied are summarized in Table I.

Table I. The target material, lab-frame beam energies in the middle of the target, resulting excitation-energy intervals, total beam doses, and numbers of observed decay chains assigned to the parent nuclei ^{294}Og (3*n*) are listed.

Target nucleus	E_{lab} (MeV)	E_{exc} (MeV)	Beam dose × 10^{19}	Number of chains 2*n*/3*n*/4*n*/5*n*	Ref.
^{249}Cf	245	26.6–31.7	2.5	- / 1 / - / -	[14]
	251	32.1–36.6	1.6	- / 2 / - / -	[14]
^{249}Cf*	252	26.6–37.5	1.6	- / 1 / - / -	[15]
^{249}Cf	252	33.0–37.4	1.6	- / 1 / - / -	[13]
	258	37.8–42.4	1.1	- / - / - / -	[13]
^{251}Cf	252	34.2–38.6	1.6	- / - / - / -	[13]
	258	39.0–43.6	1.1	- / - / - / -	[13]

* ^{249}Cf as buildup result of ^{249}Bk β-decay ($T_{1/2}$ = 327 d).

The array of Silicon detectors at the DGFRS final focus has been modified to increase the position resolution of recorded signals and subsequently reduce the probability of observing sequences of random events that mimic decay chains of implanted nuclei. The new detection system includes a 0.3-mm thick Double-sided Silicon Strip Detector (DSSD) manufactured by Micron

Semiconductor Ltd. This large DSSD has 1-mm wide strips, 48 at the front side and 128 at the back side, creating over 6000 1-mm^2 pixels in one Silicon wafer. Such high pixilation helps to achieve superior position resolution for recoil-correlated decay sequences reducing potential random events. This new Si-detector array was designed, assembled, commissioned off-line and provided by ORNL. The signals from all detectors were processed using MESYTEC linear preamplifiers. Further, these analog signals were split into two independent measurement branches by special spectroscopic splitter-amplifier PA32-64 designed by the DGFRS group. Thus, all detectors' signals were processed simultaneously by analog electronics similar to those used in previous DGFRS experiments [1, 9, 12, 14, 15], and by digital electronics system based on XIA Pixie-16 modules provided by ORNL. This new DSSD assemblage and two independent registering systems were successfully applied recently in ^{239}Pu + ^{48}Ca and ^{240}Pu + ^{48}Ca experiments [12]. The FWHM energy resolution of the implantation detector was 34 to 78 keV depending on the strip, while the summed signals recorded by the side and implantation detectors had an energy resolution of 147 to 263 keV. Other experimental conditions, including the method of calibration of the detectors, were the same as in previous DGFRS experiments (see [1] and references therein).

3. Results

Present experiment with mixed $^{249\text{-}251}$Cf target was carried out at two ^{48}Ca-beam energies, see Table I. At the beam energy E = 252 MeV with accumulated beam dose of 1.6×10^{19} particles just one correlated decay chain (Fig. 1b) was observed which we assign to ^{294}Og, the product of 3*n*-evaporation channel of the ^{249}Cf(^{48}Ca, 3*n*)^{294}Og reaction which was studied at the DGFRS earlier. Its decay properties reproduce in full those observed in [1, 14, 15]: the new decay chain consists of implanted recoil nuclei, two following α-particles and terminates by SF of ^{286}Fl. The cross section of the ^{249}Cf(^{48}Ca,3*n*)^{294}Og reaction for 252-MeV projectiles was measured (Fig. 2) to be about 0.9 pb. For the ^{251}Cf(^{48}Ca,3*n*)^{296}Og reaction the upper cross section limit is set at the level of 3.4 pb.

At 258 MeV beam energy the bombardment was performed with the aim to produce ^{295}Og or ^{296}Og isotopes, the products of the ^{251}Cf(^{48}Ca,x*n*)$^{299\text{-x}}$Og reaction with evaporation of 4 and 3 neutrons. The beam dose of 1.1×10^{19} particles was accumulated during 56 days, no decay chains were observed at this energy which could be attributed to SHN. The upper bounds for the cross section are 4.1 pb for ^{294}Og and 5.7 pb for ^{296}Og [13].

Finally, this experiment was stopped due to significant pollution of the target material caused by melted glue from sector frames which covered all the

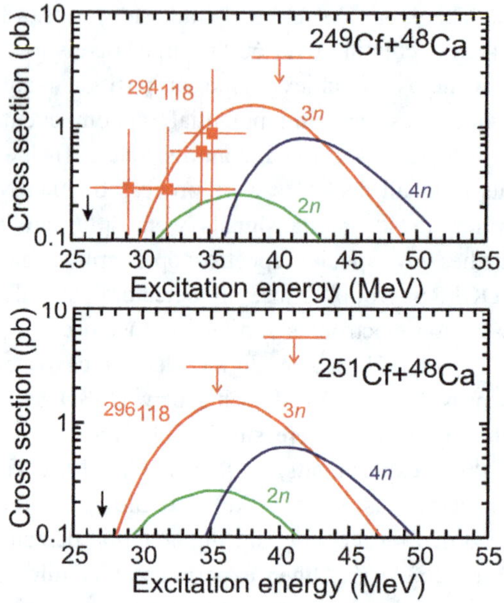

Fig. 2. The calculated excitation functions [16] of the $^{249}Cf(^{48}Ca,xn)^{294-x}Og$ and $^{251}Cf(^{48}Ca,xn)^{299-x}Og$ reactions and the experimental data [1, 13-15].

target sectors and prevented ERs escaping from the target (the target surface was monitored during the experiment by registering the α-particle count from the target). The target was sent back to ORNL for the regenerating. After recovering it looks optimistic to use mixed Cf target and ^{48}Ca beam for the producing the heaviest ^{295}Og and ^{296}Og isotopes.

Acknowledgments

We acknowledge the support of the RFBR Grant No. 16-52-55002. Research at ORNL was supported by the U.S. DOE Office of Nuclear Physics under DOE Contract No. DE-AC05-00OR22725 with UT-Battelle, LLC. Research at LLNL was supported by LDRD Program Project No. 08-ERD-030, under DOE Contract No. DEAC52-07NA27344 with Lawrence Livermore National Security, LLC. This work was also supported by the U.S. DOE through a Grant No. DE-FG-05-88ER40407 (Vanderbilt University).

References

1. Yu. Ts. Oganessian and V. K. Utyonkov, *Nucl. Phys. A* **944**, 62 (2015).
2. R. Eichler *et al.*, *Nature* **477**, 72 (2007); *Radiochim. Acta* **98**, 133 (2010).

3. S. Hofmann *et al.*, *Eur. Phys. J A* **32**, 251 (2007); *Eur. Phys. J A* **48**, 62 (2012).
4. J. M. Gates *et al.*, *Phys. Rev. C* **83**, 054618 (2011); D. Rudolph *et al.*, *Phys. Rev. Lett.* **111**, 112502 (2013).
5. P. A. Ellison *et al.*, *Phys. Rev. Lett.* **105**, 182701 (2010); J. M. Gates *et al.*, *Phys. Rev. C* **92**, 021301(R) (2015).
6. D. Kaji *et al.*, *J. Phys. Soc. Jpn.* **86**, 034201 (2017).
7. J. Khuyagbaatar *et al.*, *Phys. Rev. Lett.* **112**, 172501 (2014).
8. S. Hofmann *et al.*, *J. Phys. G: Nucl. Part. Phys.* **42**, 114001 (2015); *Eur. Phys. J A* **52**, 180 (2016).
9. Yu. Ts. Oganessian *et al.*, *Phys. Rev. C* **79**, 024603 (2009).
10. Ch. E. Düllmann, *EPJ Web of Conferences* **131**, 08004 (2016).
11. S. N. Dmitriev, this volume.
12. V. K. Utyonkov *et al.*, *Phys. Rev. C* **92**, 034609 (2015); ibid, **97**, 014320 (2018).
13. N. T. Brewer *et al.*, *Phys. Rev. C* **98**, 024317 (2018).
14. Yu. Ts. Oganessian *et al.*, *Phys. Rev. C* **74**, 044602 (2006).
15. Yu. Ts. Oganessian *et al.*, *Phys. Rev. Lett.* **109**, 162501 (2012).
16. V. I. Zagrebaev *et al.*, *Phys. Rev. C* **85**, 014608 (2012).

3

Nuclear reactions, fission and decays. Rare processes.

Toward Ternary Fission Accompanied by the ^{8}Be Nucleus

D. A. Artemenkov, V. Bradnova, M. V. Gustova, O. M. Ivanov, R. R. Kattabekov, K. Z. Mamatkulov, V. V. Rusakova, A. V. Sabelnikov, A. A. Zaitsev, P. I. Zarubin* and I. G. Zarubina[a]

Joint Institute for Nuclear Research, 141980, Dubna, Russia
E-mail: zarubin@lhe.jinr.ru

I. Ambrožova, M. Kakona, M. Lužova, O. Ploc and K. Turek

Nuclear Physics Institute, Czech Academy of Sciences, 250 68 Řež, Czech Republic

E. Firu, M. Haiduc and A. Neagu

Institute of Space Science, 077125 Magurele, Romania

R. Stanoeva

South-Western University, 2700 Blagoevgrad, Bulgaria

Experiments in preparation for search for uranium ternary fission by means of nuclear track emulsion are summarized. The study will be focused on the possible involvement of the unstable nucleus ^{8}Be in the suggested scenario of the collinear tri-partition in the fission.

Keywords: Ternary fission; nuclear track emulsion; uranium; thermal neutrons; unstable nucleus.

1. Introduction

Implantation of uranium compounds into nuclear track emulsion (NTE) allow one to expand experimental means of nuclear fission studies. It is worth recalling that for the first time ternary and quaternary fission induced by neutrons was observed in such a way [1]. For quite a while the ternary fission of the ^{252}Cf isotope introduced into NTE has been explored [2, 3]. Owing its origin to NTE the physics of the ternary fission can serve as one of the drivers of interest in this classical technique. Not providing timing and energy analysis NTE preserves its observational mission. Due to full sensitivity to charged fragments in the full solid angle at record resolution of 0.5 μm this technique provides spatial patterns of few-prong nuclear fission events not unavailable in any other approach. Being measured lengths and thicknesses of tracks allow one to classify last ones as produced by as α-particle, heavier or lighter fragment.

Search for the collinear cluster tri-partition (CCT) of fissionable nuclei is among current challenges (reviewed in [4]). Such a process is assumed to proceed through sequential binary fissions via formation of an intermediate state composed of two resulting fragments. Such a state is considered existing long enough with the respect to the fission time scale to be considered as a kind of a nuclear molecule. Decaying sequentially it could lead to alignment emission directions of the three fragment along the common axis. Therefore, there are obvious difficulties in separating pairs of fragments moving in the same direction. Orientation toward the ternary fission involving sufficiently long-lived isotope ^{8}Be allows one to overcome this difficulty. ^{8}Be emission in spontaneous decays ^{252}Cf established recently [5] supports this idea. It is worth noting that the ^{8}Be emission mimics 2α-particle radioactivity. In general, unbound configurations of lightest nuclei (α,*t*) originated from decays of light nuclei exited above relevant thresholds aren't excluded in the ternary fission.

The unstable ^{8}Be nucleus is considered as a loose bond of α-particles whose centers are a separated by a distance of about the α-particle size. So, it would be too little to refer it to exotic nuclei. Due to its size over deformation axis comparable to a heavy nucleus one ^{8}Be can be considered as an important participant among heavier fission fragments (Figure 1). Besides, ^{8}Be could serve as a temporary covalent bond in an emerging ensemble. A ^{8}Be accompanied fission can proceed like 4-body instantaneous decay or sequential one when ^{8}Be is kept by one of heavier fragments while the other one drifts away. The second option leads to the CCT pattern. Both scenarios seem intriguing and their interplay is possible. Thus, experimental examination of the pattern of the ternary fission involving ^{8}Be is an inspiring goal

Regarding the current status of the NTE technique the following can be noted. Being developed more than half a century ago it remains a status of a universal and cost-efficient detector. BR-2 type NTE with an unsurpassed

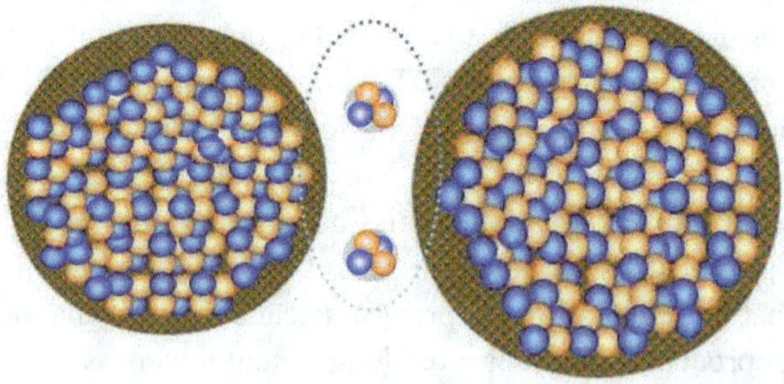

Fig. 1. Scenario of ternary fission of heavy nucleus involving ^{8}Be.

observation beginning from fission fragments up to relativistic particles. In the last decade, the NTE technique is actively applied in the BECQUEREL experiment [6] at the JINR Nuclotron allowing studying of light nuclei including radioactive ones nuclei at their relativistic dissociation [7]. Unstable nuclei ^{8}Be and ^{9}B were identified by invariant masses of their decay products in the cases of the isotopes ^{10}B and ^{10,11}C [8].

Meaning of the last fact is as follows. As is known, nucleosynthesis chains involving ^{8}Be and ^{9}B are suppressed due to an absence of the bound ground states. Nevertheless, this circumstance does not prevent the substantial structural contribution of ^{8}Be and ^{9}B. Therefore, it cannot be ruled out that they can serve as transient states persisting in the bound nuclei. The obtained experience was applied recently to identify for relativistic decays of the Hoyle-state [9].

At the time, production of NTE layers in Moscow that had lasted for forty years stopped more than decade ago. The interest in further application of NTE stimulated its reproduction at the workshop MIKRON, which is a part of the Slavich Company (Pereslavl-Zalessky) [10]. At present, NTE samples are produced by casting layers from 50 to 200 μm onto a glass substrate. The testing of the new NTE via irradiation with relativistic particles in above mentioned studies proved its similarity to NTE BR-2.

Next, recent applications the reproduced NTE are summarized as essential ingredients in the suggested study (more detailed in [11] and references herein). First of them is ability to reconstruct the ground ^{8}Be$_{0+}$ and first exited ^{8}Be$_{2+}$ states. The second one is sufficiency of angular resolution in direction determination of heavy ions relevant energy. Third one is preparation of NTE samples enriched with uranium. Last (and not least) is possibility of automatic counting of thermal neutron induced reactions in NTE.

2. Reconstruction of ^{8}Be$_{2+}$

NTE were exposed to ^{8}He nuclei with an energy of 60 MeV at the fragment separator ACCULINNA of the Flerov Laboratory of Nuclear Reactions (JINR). Figure 2 shows example of "hammer-like" decay of ^{8}He nucleus stopped in NTE as typical among about two thousand observed ones in this study. The search for β decays of ^{8}He nuclei was concentrated on the search for "hammers". Often, gaps were observed between stopping points of primary tracks and "hammer-like" decays in the so-called "broken" events. "Broken" events were assumed to take place owing to drift of the produced ^{8}He atoms.

The directions of ^{8}He arrival, the stopping points of their nuclei, the vertices of their decay, and the α particle stopping points were detected for 136 "whole" and 142 "broken" events. In "broken" events, the decay points were determined

Fig. 2. Mosaic macro photograph of hammer-like decay of the ^{8}He nucleus stopped in the nuclear track emulsion (horizontal track). The decay results in a pair of relativistic electrons (dotted tracks) and a pair of α particles (oppositely directed short tracks). The decay image is superposed on a macro photograph of a human hair 60 μm thick to illustrate the spatial resolution.

by extrapolation of electron tracks. The distribution of opening angles for α-particle pairs has an average value of (164.9 ± 0.7)°. Such a "hammer" kink connected with momentum carried away by an eν pair is illustrating angular resolution.

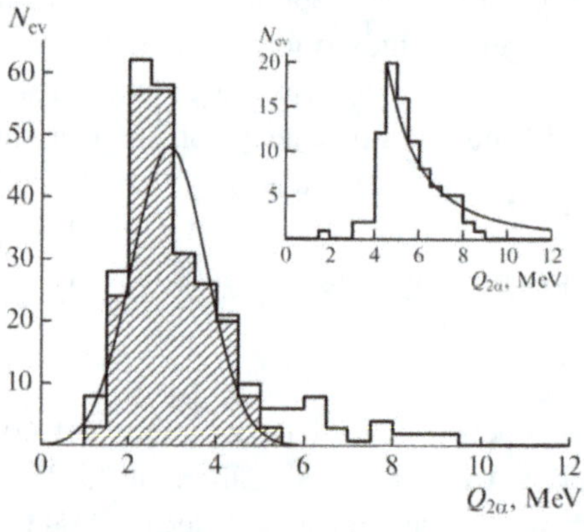

Fig. 3. Invariant mass $Q_{2\alpha}$ distribution for 278 pairs of α particles: shaded histogram corresponds α particles with ranges shorter than 12.5 μm; line corresponds to the Gaussian approximation; the inset shows the $Q_{2\alpha}$ distribution of 98 additional pairs with α-particle ranges above 12.5 μm.

The matching of an α-particle range to its energy is obtained via spline interpolation of calculations using the program SRIM. The energies and opening angles of α-particles yield the 2α invariant mass distribution $Q_{2\alpha}$. The variable Q

is determined as the difference between the invariant mass of the final system M^* and the mass of the primary nucleus M. M^* is determined as the sum of products of 4-momenta P_i, i.e., $M^{*2} = (\Sigma P_i)^2$. The $Q_{2\alpha}$ distribution (Figure 3) on the whole corresponds to the decay from the first excited state of $^8Be_{2+}$.

For events in which the ranges of both α particles are shorter than 12.5 μm and the opening angles are larger than 145°, the average value of $Q_{2\alpha}$ is equal to (2.9 ± 0.1) MeV with the RMS equal to 0.85 MeV, which corresponds to $^8Be_{2+}$. At the same time, this distribution has an extended "tail" which is not described by a Gaussian. The inset in Figure 3 shows the $Q_{2\alpha}$ distribution corresponding to both ranges longer than 12.5 μm. It is possible that its shape reflects the spatial structure of the $^8Be_{2+}$ state.

3. Reconstruction of $^8Be_{0+}$

NTE exposed to 14.1 MeV neutrons allows one to study the ensembles of triples of α-particles produced in disintegrations of ^{12}C nuclei of NTE composition. Exposure of NTE to 14.1 MeV neutrons was performed on one of devices DVIN of an applied destination. In 400 selected 3α disintegrations measurements of angles relative to plane of a NTE layer and its surface as well as their lengths were performed for all α-particle tracks.

Distribution over ranges of α-particles L_α has an average value $<L_\alpha>=$ (5.8 ± 0.2) μm at RMS (3.3 ± 0.1) μm. This distribution has an asymmetric shape described by the Landau distribution. Directly associated with it the distribution over energy of α-particles E_α defined by ranges L_α in the SRIM model has an average value $<E_\alpha> =$ (1.86 ± 0.05) MeV and RMS (0.85 ± 0.03) MeV.

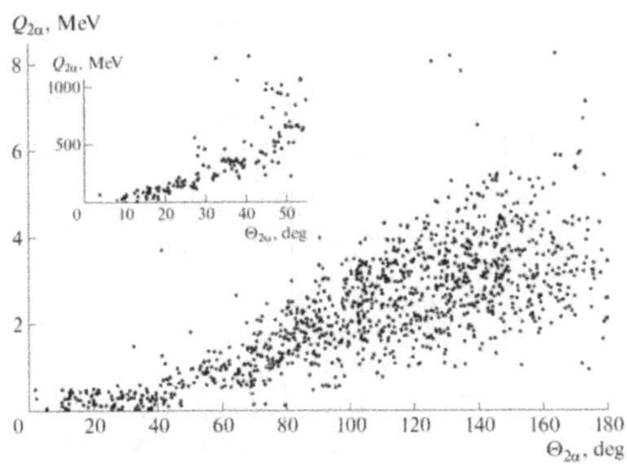

Fig. 4. Energy $Q_{2\alpha}$ and opening angle $\Theta_{2\alpha}$ correlations in α-particle pairs produced in ^{12}C splitting by 14.1 MeV neutrons.

Determination of angles and energy values versus ranges allows one to determine the invariant mass $Q_{2\alpha}$ of pairs. Correlation over $Q_{2\alpha}$ and opening angles $\Theta_{2\alpha}$ in α-particle pairs reveals features of the ^{8}Be nucleus. The region of large opening angles $\Theta_{2\alpha} > 90°$ is corresponding to $Q_{2\alpha}$ of $^8Be_{2+}$, while $\Theta_{2\alpha} <$ 30° – $^8Be_{0+}$. Distribution over $Q_{2\alpha}$ points to these states (Figure 4). Its right side meets the shape expected from the decay through $^8Be_{2+}$. Condition $Q_{2\alpha} < 200$ keV has allowed to allocate 56 decays $^8Be_{0+}$ (Figure 5). The $^8Be_{0+}$ selection reveals in the correlated peak relevant to the $^8Be_{2+}$ excited state.

Reconstruction of the decaying $^8Be_{0+}$ allows one to derive distribution over their values of kinetic energy E_{kin} (Figure 6). It has an average value

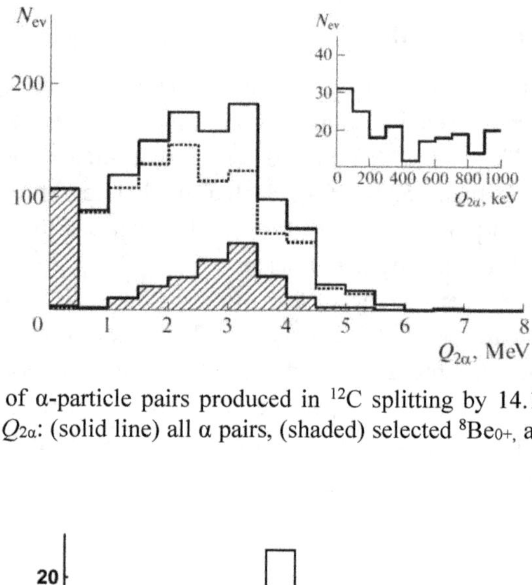

Fig. 5. Distribution of α-particle pairs produced in ^{12}C splitting by 14.1 MeV neutrons over invariant mass $Q_{2\alpha}$: (solid line) all α pairs, (shaded) selected $^8Be_{0+}$, and (dashed line) their difference.

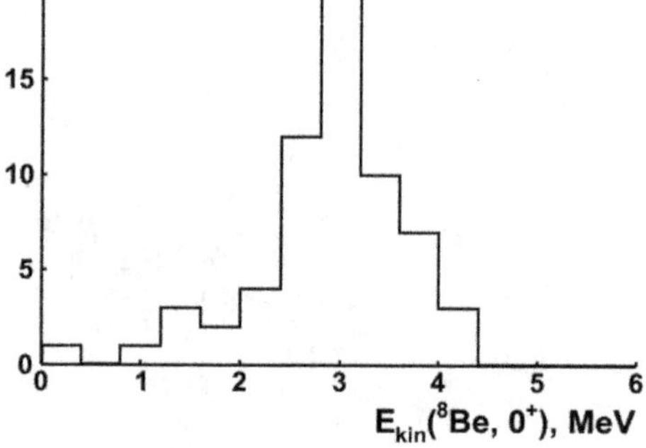

Fig. 6. Distribution $^8Be_{0+}$ of α-particle pairs produced in ^{12}C splitting by 14.1 MeV neutrons over kinetic energy E_{kin}.

$<E_{kin}>$ = (3.0 ± 0.1) MeV and RMS (0.75 ± 0.07) MeV. Since $^{8}Be_{0+}$ is reconstructed at lowest energy release it has to be reconstructed in the ternary fission when extra repulsion is provided by highly charged fragments.

4. Slow Heavy Ions in NTE

Surface irradiations of NTE samples at the Department of Radiation Dosimetry of the Nuclear Physics Institute were performed at first with manual movement of the ^{252}Cf source. Then a specially developed device was applied; the source was moved over the surface of this device automatically according to a convenient space and time pattern. When the NTE surface irradiated by the Cf source was examined, planar trios consisting of pairs of fragments and long-range α particles and trios of fragments were found (see Figure 7).

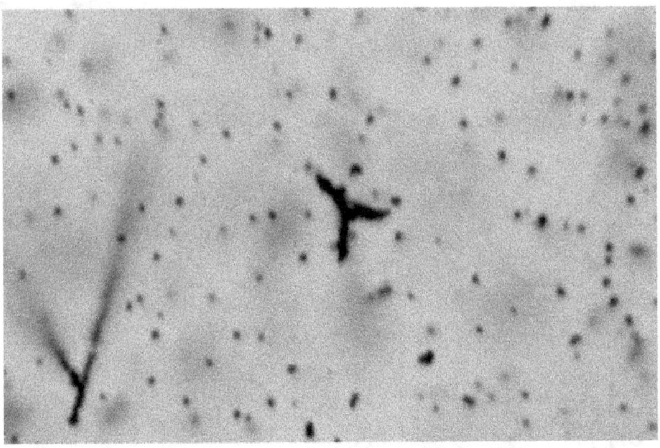

Fig. 7. Macro photograph of the first tri-partition event found in this study.

It should be underlined that the fact of observation of trios in NTE, rather than just pairs of fragments, is quite remarkable. For this to be possible, the vertices of these trios should be submerged to a depth not smaller than the typical layer thickness.

The fragment opening angles were also measured in these events (Figure 8). Their distribution is characterized by an average value of (111 ± 2)° and an RMS of 36°. It can be concluded that no candidates for collinear fission have been found yet, and their search should be continued. The ranges of all fragments were measured in 96 events of tri-partition without α particles. The average energy of fission fragments is about 400 A keV.

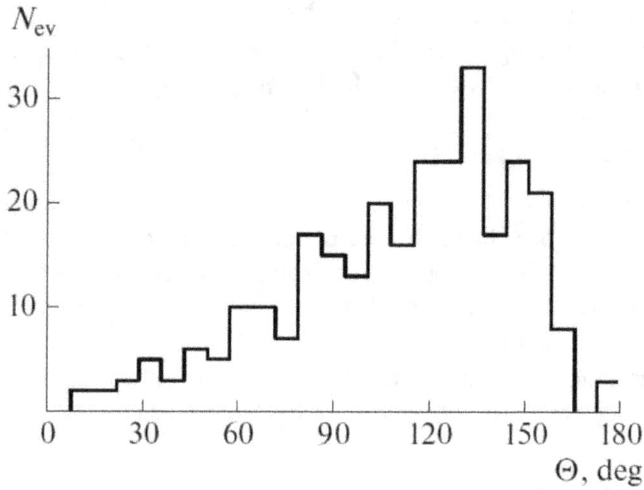

Fig. 8. Distributions of ^{252}Cf tri-partition events over opening angles of fragment pairs.

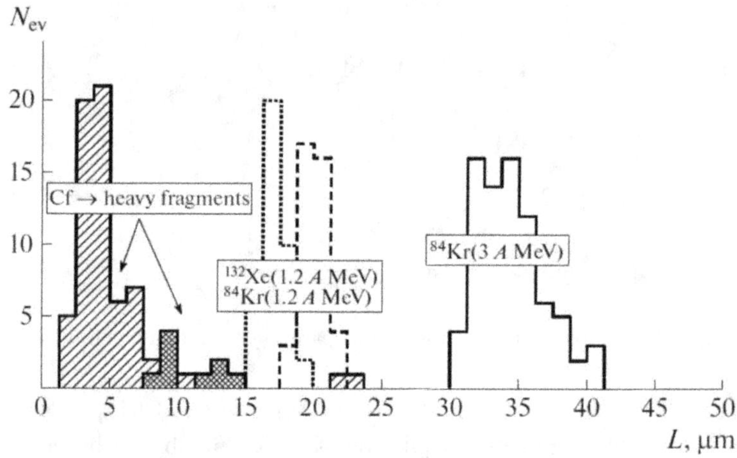

Fig. 9. ^{84}Kr, ^{132}Xe, and ^{86}Kr ion range distributions in decays Cf → 3 fragments and Cf → 2 fragments + long-range α.

Besides, NTE samples were exposed at the Flerov Laboratory of Nuclear Reactions (JINR) at the cyclotron IC-100 to 1.2 *A* MeV $^{86}Kr^{+17}$ and $^{124}Xe^{+26}$ ions and at the cyclotron U-400M to 3 *A* MeV ^{84}Kr ions. The irradiation was performed in vacuum irradiation chambers without a black paper using a red photo lamps as a light source. For better track observation, samples were installed at a large angle with respect to the beams. The sample irradiation density reached 10^6 tracks per cm^2 in several seconds. Figure 9 shows the range distribution for ions stopped in NTE without visible scattering. Then, the calibration of ion ranges in NTE needs to be extended much lower than 1 *A*.

Reducing energy of the IC-100 beam is possible in two ways. The first and most simple one is installation directly in front of a target aluminum film or Mylar desired thickness. Its advantage consists in immutability of the measured particle flux, and relatively fast replacement of degrader, but the drawback is that after degrader is producing complex flows over ion charges, angles and energy. The second method is carried out by installing a degrader immediately at the output of the accelerator at the beginning of a long transport channel. Its advantage lies in the fact that due to collimation in the channel ions fall on the target at normal, remaining mono-energy and mono-charge, and the lack of a long time of replacement of the degrader associated with the inlet of the atmosphere and its subsequent pumping, as well as a significant loss of the beam.

Recently, the idea of a smooth energy regulator was implemented. Both polymer and metal films (aluminum) can be used as a degrader in the regulator. When the frame degrader is rotated around the axis an effective path length of ions through the degrader varies according to the inverse cosine law. When the ion speed slows down, it is recharged. The recharge curve was calculated by the LISE program. This allowed to correctly predict the flux density of ions at the integrated current with the "quick flap of the bundle". This makes it possible to move the calibration downwards in energy.

5. Exposure to Thermal Neutrons

NTE samples enriched in boron (boric acid and borax) were irradiated for 30 min at the thermal neutron channel of the IBR-2 reactor (JINR). The subsequent sample analysis yielded an intensity estimate of 5×10^5 neutrons per second. The samples were produced by casting boron-enriched NTE to a thickness of about 60 μm onto a 2 mm glass substrate. The application of glass resulted in activation of sodium contained in it, which presented a problem, although inevitable.

The presence of boron in NTE allows one to observe charged products of the $n_{\mathrm{th}} + {}^{10}\mathrm{B} \rightarrow {}^{11}\mathrm{B}^* \rightarrow {}^{7}\mathrm{Li}^* + (\gamma) + \alpha$ reaction. The fragments take away energy of 2.8 MeV. A 478 keV photon is emitted with a probability of 93% by the ${}^{7}\mathrm{Li}$ nucleus from its only excited state. The coordinate measurements of tracks in 112 ${}^{7}\mathrm{Li} + {}^{4}\mathrm{He}$ events were performed. Their directions in pairs are no collinear and have an average angle $\Theta({}^{7}\mathrm{Li} + {}^{4}\mathrm{He})$ of $(148 \pm 14)°$ due to photon emission. The average energy $Q({}^{7}\mathrm{Li} + {}^{4}\mathrm{He})$ was (2.4 ± 0.2) MeV with RMS of 0.8 MeV in agreement with the energy carried away by the photon. The distribution of $\Theta(\gamma + {}^7\mathrm{Li})$ between the photon emission angles calculated from the momentum conservation and the ^{7}Li direction point to apparent anti-correlation (Figure 10). It would be useful to calibrate in the reaction of $n_{\mathrm{th}} + {}^6\mathrm{Li} \rightarrow t + \alpha$ in NTE enriched with the lithium borate.

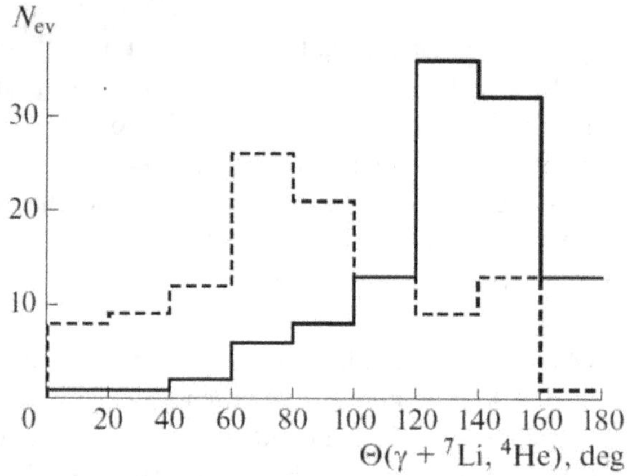

Fig. 10. Distribution over the angles $\Theta(\gamma + {}^7\mathrm{Li}, {}^4\mathrm{He})$ between the calculated photon directions and the directions of (solid line) ^{7}Li and (dashed line) ^{4}He nuclei.

6. Impregnation with Uranium

Recently, an impregnation with an aqueous solution of nitrate uranyl of NTE on a glass plate was mastered in the Flerov Laboratory. 20 samples of NTE applied to one side of glass plates of 6×4.5 cm^2 were used. The procedure was conducted in a dark room at room temperature. Drying of samples was carried out under the same conditions. A lamp with a red light filter were used for lighting. It was established in advance that saturation of the samples with water is achieved in no more than 15 minutes.

A freshly prepared aqueous solution of $UO_2(NO_3)_2$ was poured into a Plexiglas cuvette with grooves for placing samples in an upright position. A thickness of a solution layer in a cuvette was about 20 mm. An initial concentration of a uranium solution determined by x-ray fluorescence analysis (XRF) was 600 μg/ml.

A soaking was conducted in 2 stages. In the first stage, 12 samples were soaked, and in the second stage, the remaining 8. The first 12 samples were pre-unpacked and then sequentially one after another installed in a ditch with a solution. The installation procedure took no more than 10 minutes. In the established position, the samples did not touch each other, the distance between the two adjacent samples was about 3 mm. 60 minutes after the installation of the last sample, the samples were extracted from the cuvette with the solution in the same order. The extraction procedure took no more than 10 minutes. During the extraction process, each sample was placed in a vertical position at the bottom edge on a sheet of filter paper to remove the residual solution, and then moved to a dry cell of Plexiglas with grooves to install the samples in a vertical position. Further drying of the samples took place in an upright position. In the established position the samples did not touch each other, the distance between two adjacent samples was about 3 mm. After extraction of all 12 samples from the solution, an aliquot was selected from the cuvette with the solution to determine the possible change in the concentration of uranium in the solution at the first stage.

At the second stage, the remaining 8 samples were immersed in the same solution. The mode of placing the samples in the solution, impregnation and extraction of samples from the solution was similar to the mode for 12 samples. After the extraction of all 8 samples from the solution, an aliquot was selected from the solution cell to determine the possible change in the uranium concentration in the solution at the second stage. The analysis of the uranium solution aliquots selected at the first and second stages by XRF showed a change in the concentration of uranium in the solution from 600 μg/ml to 500 μg/ml at the first stage and from 500 μg/ml to 410 μg/ml at the second stage. This gives reason to assume the presence of sorption properties of the emulsion with respect to uranium.

The samples were developed in about two days. On macro photographs taken in one of the developed samples at immersion depths of 9, 15. 21 and 30 μm from the surface tracks of α-particles are visible (Figure 11). An observer can notice that their density decreases rapidly. Samples from both batches were scanned over an area of 16 points 1 cm apart. In each point were carried out the photographing depth of field with a step of 1 μm with a microscope objective ×60 and adapter for camera ×0.5. Calculation of traces lying in the plane of the sample was carried out visually on the obtained images with a step in depth of 3 microns. It was found that in the sample from the second batch the number of α tracks is less by (10 ± 3) % than in the samples from the first batch.

The disappearance of tracks of α-particles to a depth of no more than 40 μm was observed at all control points on the three studied samples. The total distribution of all found planar traces by depth (Figure 12) gives an idea of this effect. The average value of the penetration depth of uranium is 15 μm with 6 μm RMS. It can be concluded that the absorption of uranium had proceeded in the surface layer of gelatin (in fact, protein medium).

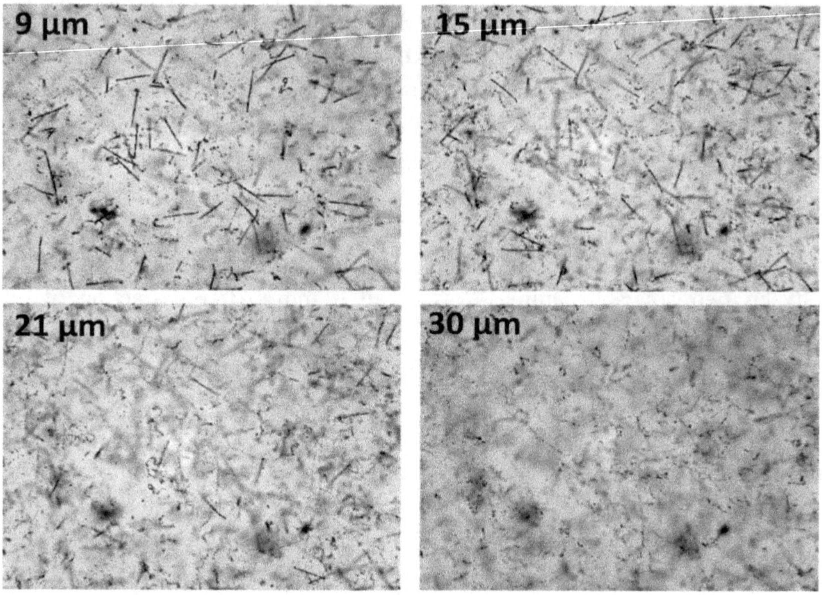

Fig. 11. Macro photographs of NTE soaked in solution of nitrate uranyl.

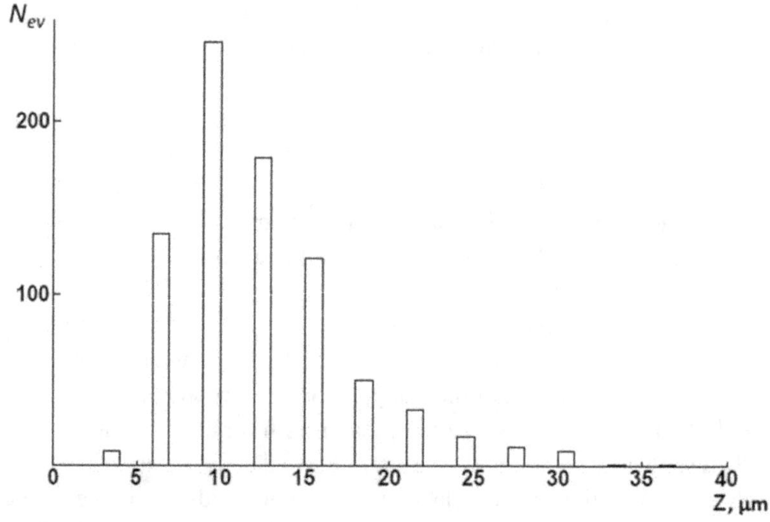

Fig. 12. Distribution of number of planar α-tracks per 1 μm over immersion depth in NTE sample soaked in solution of uranyl nitrate.

In the context of the proposed study, this effect indicates that the search for fission events should be almost planar, which is a fundamental simplification. In general terms, this observation is important for understanding the danger of uranium compounds to humans. At the same time, it allows us to offer methods of cleaning from contamination with gelatin.

7. Concluding Remarks

The competitive character of NTE was proved recently in a series of measurements of short tracks of α particles and slow heavy ions. For these purposes microscopes KSM-1 manufactured by Carl Zeiss (Jena) about half of century and still functioning well are applied in Laboratory of High energy Physics, JINR. The possibility of α spectrometry was verified and the ^{8}He atom drift effect was established in measurement of decays of ^{8}He nuclei implanted in NTE. Correlations of α particle trios in ^{12}C nucleus splitting by 14.1 MeV neutrons as well as ^{7}Li and ^{4}He nuclei produced in ^{10}B break up by thermal neutrons in boron-enriched NTE were studied. In this series of experiments, the NTE resolution proved to be perfect as expected physical effects in invariant mass distributions of reaction product ensembles were clearly observed. NTE samples are calibrated using 1.2 and 3 *A* MeV Kr and Xe ions. NTE surface exposures to a ^{252}Cf source allowed to find and measure events containing fragment pairs and long range α particles, as well as fragment triples.

Thus, the major prerequisites are available to argue application of NTE technique in the state-of-art problems of the nuclear fission physics. Due to perfect observation the ^{8}Be accompanied channel of the ternary fission can become the "Golden key" for experimental verification of the CCT hypothesis. It is possible to be identifiable ^{8}Be decays in NTE by measurements of decay α-particle ranges and emission angles. Such events could be found among 4-prong ones. Smaller thickness and larger ranges of α-particle ranges in them will allow to distinct them from quaternary fission cases. Observation of heavy fragments is a principal advantage NTE allowing studying angular correlations.

Use of microscopes is last but not least aspect The use of modern computerized and automatic microscopes for the analysis of nuclear emulsion, containing both fission events and associated traces of radioactivity, allows one to address to tasks of the past at the state-of-art level that were considered exhausted due to the limited possibilities of direct human analysis. Being provided with relevant image recognition programs they could be suggested to finding and measuring short nuclear tracks with the most precise spatial resolution at an unprecedented statistics level. However, recognition of few-prong events in NTE has not yet been achieved with an automated microscope, especially, when real study complications are taken into account. At the present stage, automatic analysis can very effectively complement the human search for rare triple and fourfold fission events in monitor counting of binary fission and α-radioactivity tracks. Computerization of measurements on the microscope and automatic changing lenses to higher magnification when a relevant event is found can radically enhance researcher capabilities. Suitable microscopes are available on the market.

Acknowledgments

Authors are grateful to Prof. S. N. Dmitriev for possibility to carry out described experiments in the Flerov Laboratory of JINR. Authors are especially indebted to Prof. Yuri Penionzhkevich (Flerov Laboratory, JINR) who attracted attention to possibility to use the NTE technique in the CCT problem. Saving of the NTE technique wouldn't be possible without support of Prof. A. I. Malakhov (Veksler & Baldin Laboratory, JINR).

References

1. Tsien San-Tsiang, Ho Zah-Wei, R. Chastel, and L. Vigneron, *Phys. Rev.* **71** (1947) 382.
2. E. W. Titterton and T. A. Brinkley, *Nature* **187** (1960) 228.
3. M. L. Muga, H. R. Bowman, and S. G. Thompson, *Phys. Rev.* **121** (1961) 271.
4. D. V. Kamanin and Y. V. Pyatkov, *Lect. Notes Phys.* **875 (3)** (2014) 184.
5. G. S. Ahmadov, Yu. N. Kopatch, S. A. Telezhnikov, F. I. Ahmadov, C. Granja, A. A. Garibov and S. Pospisil, Phys. Part. and Nuclei Lett. 12 (2015) 542.
6. The BECQUEREL Project. http://becquerel.jinr.ru/.
7. P. I. Zarubin, *Lect. Notes Phys.* **875 (3)** (2014) 51; arXiv:1309.4881.
8. D. A. Artemenkov, A. A. Zaitsev, P. I. Zarubin, *Phys. Part. Nucl.* **48** (2017) 147; arXiv: 1607.08020.
9. D. A. Artemenkov *et al.*, *Radiation Measurements* **119** (2018) 203; arXiv:1812.09096.
10. Slavich Company JSC. http://www.newslavich.com/.
11. P. I. Zarubin, *Phys. Atom. Nucl.* **79** (2016) 1525.

Isobaric Resonances, Beta Decay and Delayed Multineutron Emission in Very Neutron-Rich Nuclei

I. N. Borzov[1,2,*] and S. V. Tolokonnikov[1]

[1]*National Research Centre "Kurchatov Institute", 123182, Moscow, Russia*
[2]*Bogolubov Laboratory of Theoretical Physics, Joint Institute of Nuclear Research, 141980, Dubna, Russia*
[*]*E-mail: Borzov_IN@nrcki.ru*

The isobaric-analog resonances for nuclei around the neutron shell closures at N=20, 50, 82 are treated in the fully self-consistent Density Functional plus Continuum Quasiparticle Random Phase Approximation (DF+CQRPA). The aim is to check how the self-consistency is preserved in the calculations for long isotopic chains. The beta-decay half-lives and delayed multi-neutron emission branching are calculated for the reference Ni isotopic chain. The relative contributions of the GT and first-forbidden transitions are compared with that of the relativistic QRPA and Finite Amplitude Method. The accuracy of the global beta decay calculations performed within FRDM+RPA, DF+CQRPA and RHB+QRPA models are analysed.

Keywords: Exotic nuclei; isobaric-analog resonances; beta decay; self-consistent approach.

1. Introduction

The isospin and spin-isospin excitations in charge-exchange reactions and weak interaction processes are the hot topics of nuclear structure theory. The most simple is the isobaric-analog resonance (IAR), a non-spin excitation ($\Delta L=0$, $\Delta S=0$, $\Delta T=1$) related to the isospin operator τ. If the total spin transferred to the daughter nucleus is non-zero, the Gamow-Teller (GT) giant and pygmy resonances are induced via the spin-isospin operator $\sim\sigma\tau$. The low Ex modes are excited both in charge-exchange reactions and β-decay. This feature is favorable for studying the nuclear structure far from the β-stability line.

The IAR energy is important quantity for determining the nuclear symmetry energy, neutron skin and nuclear matter equation of state (EOS). It is also a sensitive indicator of poorly known Coulomb-nuclear correlations strength. A combined analysis of the IAR, β-decay half-lives ($T_{1/2}$) and β-delayed neutron emission probabilities (P_{xn}) helps one to elucidate the isospin and spin-isospin dependent components of nuclear density functional at high isospin-asymmetry regime. Also the β-decay properties are the input parameters for planning the RIB experiments and astrophysical r-process modeling.

The DF+CQRPA [1] calculations of the ground state properties, isobaric-analog resonances and beta-decay properties of nuclei in the vicinity of the neutron shell closures at N=20, 50, 82 are compared with the spherical relativistic Hartry-Bogolubov plus QRPA model [2] and deformed Finite Amplitude Method (FAM) [3]. All these frameworks include the allowed Gamow–Teller (GT) and first-forbidden (FF) beta decays. An emphasis is made on the quality of density functional in describing the beta decay properties. Special attention is paid to contribution of the first-forbidden transitions (%FF) in total beta-decay rates and delayed multi-neutron emission probabilities.

2. DF+CQRPA Description of the IAR and β-Decay Properties

Self-consistent models provide microscopically founded extrapolation of the β-decay properties to extreme N/Z ratios important for reliable predictions of beta decay strength function. The most realistic description of the β-decay strength distributions near the closed shells is given by the Interacting Shell-Model including the GT+FF decays [4,5]. These calculations can be considered as the standard benchmarks for the N=20, 50, 82 isotones. In a wider region of spherical nuclei including the super heavy ones, an efficient method is the Continuum Quasiparticle Random Phase Approximation (CQRPA) [1] based on the energy-density functional (EDF) approach.

In the DF+CQRPA model, the β-decay strength functions and integral properties are treated within the extended finite Fermi system theory [6] augmented with the self-consistency conditions [7]. For the ground state properties the Fayans EDF [8] is used in which explicit energy dependence arising from an effective account of the many-body correlations is excluded. This results in more sophisticated density dependence than the one of standard Skyrme functional

The DF+CQRPA model allows for fully self-consistent treatment of the isobaric-analog resonances. The effective interaction in the isospin charge-exchange sector (ΔL=0, ΔS=0, ΔT=1) is derived from the same EDF, as used for finding the single-particle hamiltonian and pairing potential. For simultaneous calculations of the spin-isospin resonances as well as GT and FF beta-decays, the DF+CQRPA model developed in [1] also employs the self-consistent ground states from the same Fayans DF3 functional [8]. As neglecting the spin-isospin dependent components of the DF induce a small error in the ground state properties, for the global beta decay calculations a well-founded spin–isospin effective NN-interaction can be used instead the one derived from the DF.

Thus, in the DF+CQRPA [1], the Landau-Migdal contact interaction augmented by the finite-range one-π and one-ρ exchange terms modified by nuclear medium [6] is chosen for the one-particle-one-hole (ph) channel. In the ground state calculations, the density-dependent zero-range isovector (T=1)

pairing force is used with the A-dependent strength. At the QRPA level, an induced isoscalar (T=0) proton-neutron effective NN-interaction in the particle-particle channel (dynamic pairing) is assumed to have a similar form. Importantly, the NN-interactions constants are kept the same for all A. The correlations beyond the QRPA are included by re-scaling the spin-dependent multipole operators by the unique energy-independent quenching factor $Q^{1/2} = (g_A/G_A)$. The same factor Q is used for the one-π effective interaction.

The allowed and first-forbidden transitions are treated in terms of the reduced multipole operators depending on the space and spin variables [1]. The full first-forbidden operators set is used. The relativistic operators α, Υ_5 are reduced to their space-dependent counterparts via CVC and PCAC relations. This gives a possibility to develop the pnCQRPA on non-restricted ph-base. A provision is added which allows one to fix (before variation) the proper ground state spin-parity in the odd-A isobaric partners.

3. DF+CQRPA Description of IAR in Ca, Sn and Pb Isotopes

For fully self-consistent description of the IAR several important ingredients have to be included. First, an accurate treatment of the two-body spin-orbit interaction is needed for providing self-consistency. In the present calculations its parameters are taken the same, as in the FaNDF^0 functional [9].

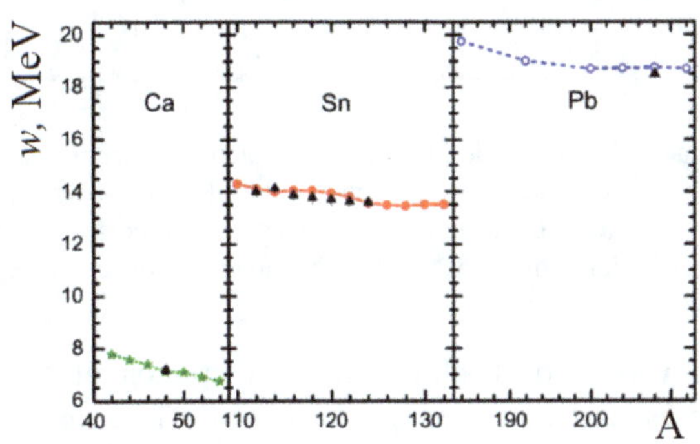

Fig. 1. The calculated and experimental IAR energies (relative to the parent ground state) for the Ca, Sn and Pb isotopic chains [10].

Second, an important ingredient is the Coulomb interaction. In addition to the direct and Slatter exchange terms, the Coulomb-nuclear correlation (CNC) term [9] is included. If the parameters of CNC are set to provide full screening of the Slatter term, the so-called Nolen-Shiffer anomaly is practically eliminated. Finally, the pairing part of the EDF is chosen to depend on the normal density and its gradient. The DF3f set [10] obtained in such a way from the DF3 functional reduces an underestimate of IAR energy which typically exists in the microscopic models. The deviation of the calculated IAR energies from the reference data in the Ca, Sn and Pb chains (Fig. 1) is $\Delta E \leq 200$ KeV [10].

Figure 2 shows that only in the fully self-consistent RPA or QRPA, the spread single-particle strengths are collected in a single (degenerate) peak. With no Coulomb exchange included (Fig. 2), the IAR energies are too high. Notice that with no dynamic pairing, the Fermi strength function still would have contained spurious satellites with large contribution to the sum rule which is not supported by the (p,n)-type experiments.

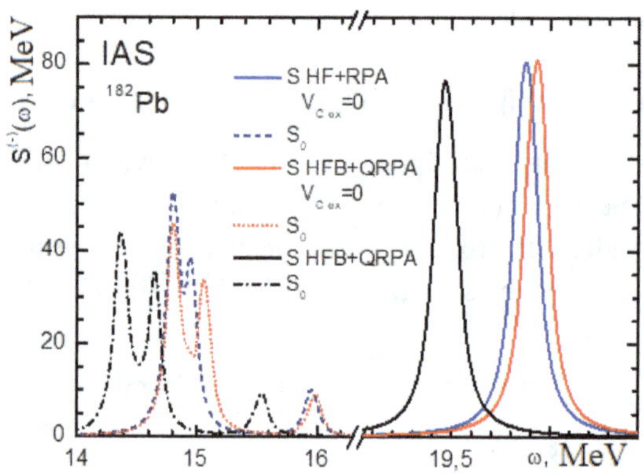

Fig. 2. The peaks shown in solid lines are calculated from the HF+RPA (blue), HFB+QRPA(red) without Coulomb exchange and HFB+QRPA with Coulomb exchange (black). The dotted, dash-dotted and dashed lines are the corresponding single-particle responses. (The artificial width of 100KeV is included for convenience.)

Our framework also enables a good description of the neutron matter characteristics, as well as the masses and radii. Thus, no contradiction exist between the estimates for the IAR energies, neutron skin and symmetry energy, as often happened with the Skyrme parametrizations of the EDF [11].

4. Beta Decay Half-Lives and Multi-Neutron Emission Rates

The total half-lives for the reference Ni chain calculated from the DF3+CQRPA [1], RHB+CQRPA [2] and Skyrme+FAM [3] are compared with the experimental data in Fig. 3. The latter two calculations significantly overestimate the

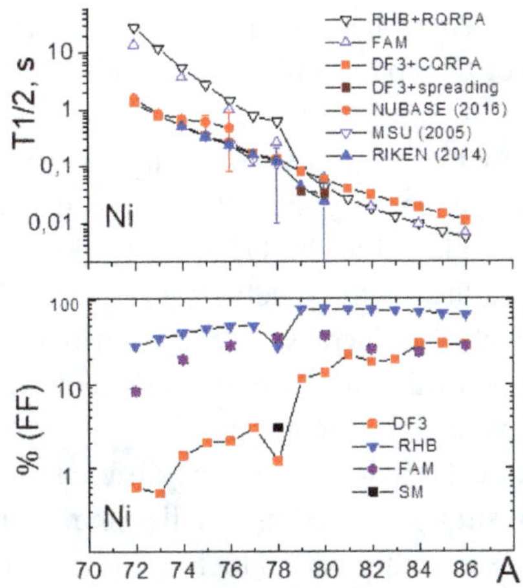

Fig. 3. (Top) The experimental half-lives of Ni isotopes from NUBASE-2016, MSU (2005), RIKEN (2014) compared to the theoretical calculations: DF+CQRPA (red squares), RHB+QRPA [2] (down triangles). (Bottom) The % FF values: DF+CQRPA (red squares), RHB+QRPA [2] (down triangles), FAM [3] (brown squares), Shell Model [4] (black square).

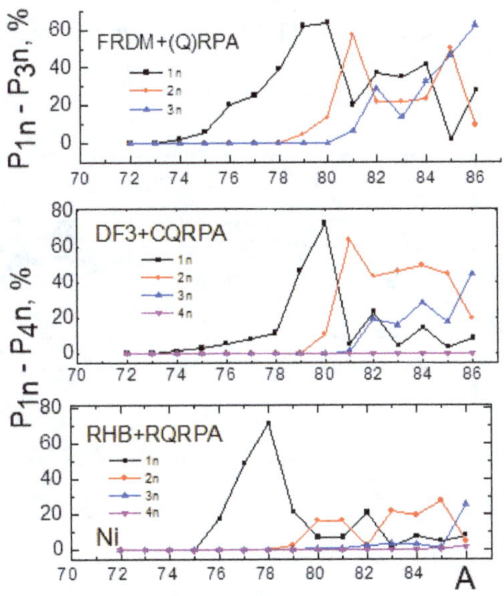

Fig. 4. Comparison of beta-delayed neutron emission probabilities of Ni isotopes obtained with three different models: the FRDM [9], DF3+CQRPA [1] and RHB+RQRPA [2].

half-lives for isotopes below the closed neutron shell at N = 50. In the lower panel of Fig. 3 we plot the contribution of the first-forbidden transitions to the total decay rates.

It shows that below the neutron shell closure N=50, the contribution of the FF transitions are enhanced in the two models, while the DF3a+CQRPA predicts lower FF contributions relative to the GT ones (this is supported by available decay schemes). Above the neutron shell closure N=50, the RHB+RQRPA and the DF3a+CQRPA display an increase of the contribution of the FF transitions due to rising occupation of the orbits in the next neutron shell, while the %FF-values in Skyrme+FAM are nearly constant.

Above the closed neutron shell N=50, the differences in the contributions of the first-forbidden transitions are getting smaller and the half-lives predicted by all models are close to each other within their accuracy.

From Fig. 4 it can be seen that the P_{xn} values obtained in the FRDM [9] and DF3a+CQRPA [1] are basically similar but differs from the RHB+RQRPA [2]. Namely, the maxima of the one-neutron emission probabilities occur at different masses and P_{2n}-values in the first two models are much higher.

5. Global Calculations of Beta Decay Properties

In Fig. 5, the global calculations [1,12] are compared with the experimental data.

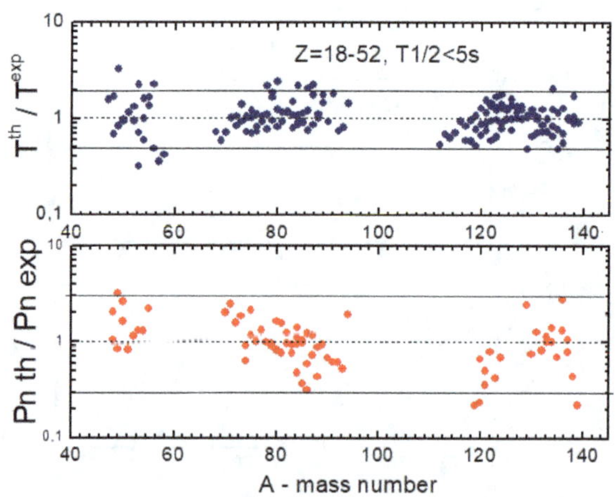

Fig. 5. Ratio of theoretical and experimental values of half-lives and Pn-values of (quasi-) spherical nuclei in Z=18-21, 27-35 and 44-51 regions with $T_{1/2}$<5s calculated using the DF3+CQRPA formalism, plotted versus the atomic number.

The ratio of the half-lives calculated within the spherical CQRPA based on non-relativistic Fayans functional to the experimental data is plotted versus the mass numbers. Here (near-) spherical nuclei in Z = 18-21, 27-35 and 44-51 regions are selected with $T_{1/2}$ ranging from 10 ms to 5 s. The constraint is imposed, as the formalism employs spherical ground state description. Importantly, it can be seen from Fig. 5 that the most of the half-lives are in agreement within the factor of two with the available data.

A similar ratio for the total neutron emission probabilities (P_n) of (near-) spherical nuclei is shown in Fig. 4. It can be concluded that experimental data are reproduced basically within the factor of two-three with the exceptions of weakly deformed nuclei in Z = 31-35, A ~85 and Z = 44-45, A ~120 regions with small total P_{tot}-values. In these isotopes non-account for spreading of the beta decay strength due to the interplay of the deformation and complex configurations [13] causes a systematic underestimate of the total P_n-values.

Notice that the FRDM [9] describes the half-lives within the factor of 10. The RHB+QRPA [2] describes them mostly within the same accuracy but systematic deviations up to a factor of 100 occur in the regions of nuclear chart between closed shells where the ground state deformations is not weak. For the total Pn-values, the deviations from the data in [9,2] basically exceed the factor of 10.

6. Conclusions

The self-consistent approach to nuclear beta-decay allows to reliably describe both ground state properties, as well as isospin and spin-isospin excitations of (quasi)spherical [1,2,10,12] and deformed [3] nuclei in a wide region of the nuclear chart. These self-consistent models based on extended energy density functional are more founded for predicting the β-decay properties than the semi-microscopic approach [9] used as a standard for the r-process modelling.

Reliable description of the Q_β-values, multi-neutron emission thresholds and transition energies are the key to a successful prediction of unknown beta decay properties. The errors in the phase-spaces defined by these quantities impact the beta decay strength distributions propagating further into the half-lives and delayed multi-neutron emission probabilities.

A significant amount of data on very neutron-rich nuclei expected from the new generation radioactive beam facilities can provide a base for verification and further improving of the theoretical predictions.The measurements of the IAR in non-magic stable and unstable nuclei are needed for better constraining the isospin components of the EDF. A simultaneous analysis of the β-decay integral characteristics ($T_{1/2}$, P_{xn}) helps to reconstruct the incomplete β-strength functions.

Acknowledgments

Discussions with T. Marketin and F. Minato are appreciated. Participation in the Symposium EXON-2018 was supported by the grant of Russian Scientific Foundation (RSF 16-12-10161). The full IAEA support of participation in the CRP «Development of a Reference Database for Beta-Delayed Neutron Emission» during the RCM-2 and RCM-3 Meetings is gratefully acknowledged.

References

1. I. N. Borzov, *Phys. Rev.* **C67**, 025802 (2003).
2. T. Marketin, L. Huther, G. Martinez-Pinedo, *Phys. Rev.* **C93**, 025805 (2016).
3. M. T. Mustonen, J. Engel, *Phys. Rev.* **C93**, 014304 (2014).
4. Q. Zhi, E. Caurier, G. Martinez-Pinedo, *Phys. Rev.* **C87**, 025803 (2013).
5. S. Yoshida, Y. Utsuno, N. Shimizu, T. Otsuka, *Phys. Rev.* **97**, 054321 (2018).
6. A. B. Migdal, *Finite Fermi-system theory*, 2nd. Ed., Nauka, Moscow, 1981.
7. S. A. Fayans *et al.*, *Nucl. Phys.* **A676**, 49 (2000).
8. S. A. Fayans, *JETP Letters* **68**, 169 (1998).
9. P. Moeller *et al.*, *Phys. Rev.* **C67**, 055802 (2003).
10. I. N. Borzov, S. V. Tolokonnikov, *Phys. At. Nucl.*, (2019), in print.
11. Li-Gang Cao, X. Roca-Mazo, G. Colo, H. Sagawa, *Phys. Rev.* **C92**, 034308 (2015).
12. I. N. Borzov, *Phys. At. Nucl.*, **81(6)**, 680 (2018).
13. A. P. Severyukhin, V. V. Voronov, I. N. Borzov, N. Van Giai, *Phys. Rev.* **C90**, 044320 (2014).

Additional Arguments in Favor of True Quaternary Fission of Low Excited Actinides

D. V. Kamanin*, A. A. Alexandrov, I. A. Alexandrova, Z. I. Goryainova, E. A. Kuznetsova, A. O. Strekalovsky, O. V. Strekalovsky and V. E. Zhuchko

Joint Institute for Nuclear Research, 141980 Dubna, Russia
**E-mail: kamanin@jinr.ru*

Yu. V. Pyatkov and A. V. Tomas

National Nuclear Research University MEPhI (Moscow Engineering Physics Institute), 115 409 Moscow, Russia, Joint Institute for Nuclear Research, 141980 Dubna, Russia

V. Malaza

University of Stellenbosch, Faculty of Military Science, Military Academy, Saldanha 7395, South Africa

Specific linear structures in the region of a big missing mass in the fission fragments mass correlation distributions were revealed due to effective cleaning of this region from the background linked with scattered fragments. One of the most pronounced structures looks like a rectangle bounded by the magic nuclei. The fission events aggregated in the rectangle show a very low total kinetic energy. We propose possible scenario of forming and decay of the multi-cluster pre-scission configuration decisive for the experimental findings.

Keywords: Multibody decays; clustering; magic nuclei.

1. Introduction

In our experiments dedicated to the search for manifestations of multibody decays of low excited nuclei [1–3] two fragments were actually detected while deficit of their total mass, comparing to the mass of the mother system ("missing" mass), served as a sign of at least ternary decay. The detected fragments flied apart almost collinearly and at least one of them showed magic nucleon composition. We dubbed this decay channel "collinear cluster tri-partition (CCT)" in order to underline the likeness with the known cluster decay or "lead radioactivity". One of the most pronounced of the observed decay modes was already discussed in our previous publications [4, 5]. The revealed features of the process suggest that true quaternary fission likely takes place in this case. The term "true" is used to underline that all the decay products have comparable masses (by analogy with the known term "true ternary fission").

Such kind of decay of low excited heavy nuclei of ^{252}Cf and ^{235}U was observed for the first time, hence nothing was known about the mechanism of the process. Presented below more thorough analysis of the experimental data provides additional arguments in favor of the quaternary nature of the decay.

2. Experiments and Results

The spectrometer used in the experiment was based on the modules of the FOBOS setup [1]. The TOF-E (time-of-flight vs. energy) method for the measurements of two fission fragments (FFs) masses in coincidence with two detectors placed at 180 degrees was applied. The TOF of the fragment was measured over a flight path of 50 cm between the "start" detector, placed next to the ^{252}Cf source, and the "stop" detectors formed by position-sensitive avalanche counters (PSAC). PSACs also provided the fragment emission angle with a precision of 1^0. The energies of those coincident fragments which passed through the PSACs were measured in the Bragg ionization chambers (BIC). For detecting the isotropic component of the neutrons emitted in fission the "neutron belt" [2] consisting of 140 ^3He filled neutron counters was assembled in a plane perpendicular to the symmetry axis of the spectrometer which at the same time served as the mean fission axis.

The FFs mass correlation distribution obtained in the experiment is shown in Figure 1a. The events involved were selected by applying the selection conditions illustrated in the Figures 1c and 1d. The rectangular structure marked by the arrow in the center of the graph 1a attracts attention. The low vertex of the structure lies in the vicinity of the point (68, 68) amu associated with the magic Ni isotope. The opposite corner of the rectangle presumably corresponds to the ^{84}Se nucleus (Figure 1b). The yield of the events in the structure does not exceed 10^{-5} per binary fission.

Application of the different selection gates allowed to obtain the mass correlation distribution shown in Figure 2a. Only fission events lying inside the box w1 (Figure 2b) and satisfying the condition that at least one neutron was detected (n = 1) are presented. Specific rectangular structure pointed out in Figure 1a is reproduced, but the selection rule used in this case turns out less rigorous and the points inside the rectangle become also selected. A total yield of the filled rectangle has approximately doubled. The total kinetic energy (TKE) of the events in the structure lies in the range (90–140) MeV.

It should be noted that in both cases (Figures 1 and 2) the selection gates were chosen heuristically within criterion of observing any graphic cluster [2, 6]. Special attention was paid to the robustness of the revealed structures against a variation of the selection gates.

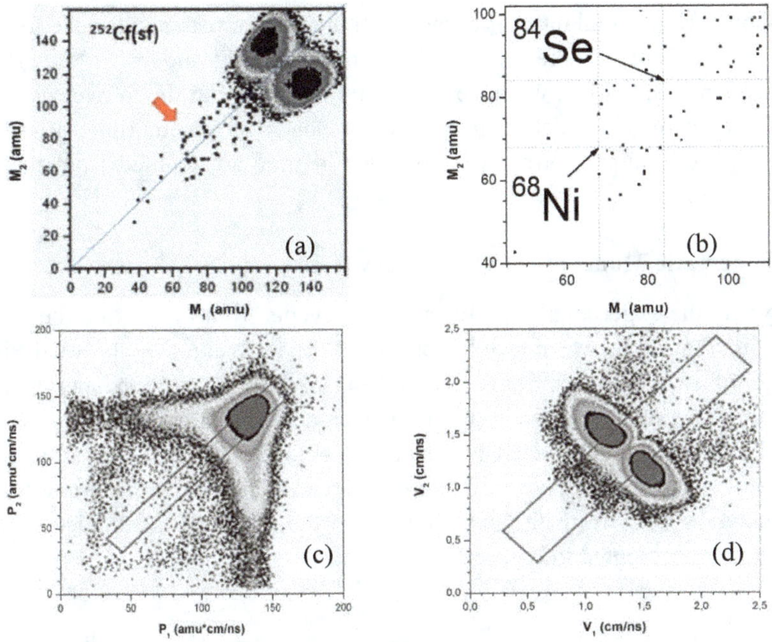

Fig. 1. Mass correlation distribution for the FFs from ^{252}Cf(sf) – (a). The rectangular structure in the center of the graph (marked by the arrow) is under analysis. It is shown in the larger scale in (b). The boxes in (c) and (d) illustrate the selection rule by the momenta and velocities correspondingly. Appling of such selection to the initial FFs mass-mass distribution results in the plot (a).

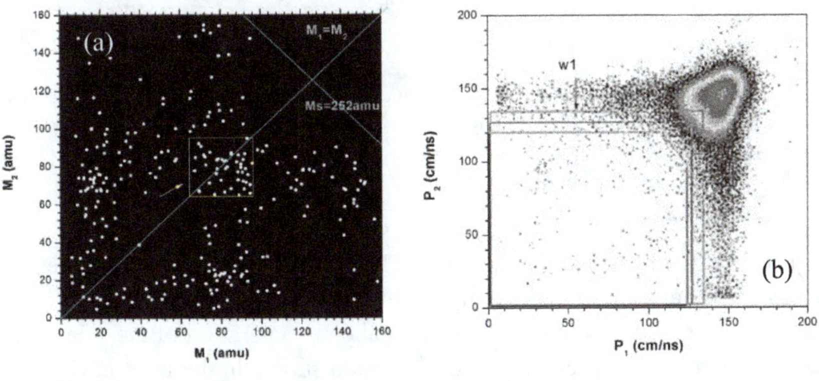

Fig. 2. FFs mass correlation distribution from ^{252}Cf(sf) – (a). The events to be analyzed are marked by the arrow. For all shown points the condition is met that the FFs momenta lie in the box w1 (b) and n = 1 (more than one neutron was detected in fission event).

Comparing a total number of events in the plot 2a with similar one for n = 2 (not shown here) in the frame of the mathematical model of the neutron registration channel [2], we have come to the conclusion that a neutron source which could provide such difference corresponds to the multiplicity n = 2, if neutrons are emitted isotropically, or n = 7 if they are emitted from the fully accelerated FFs.

3. Kinematic Model of the Quaternary Decay

Analyzing the experimental findings for the events from the box marked by the arrow in Figure 2a, we noticed a proximity of the values of the total kinetic energy of two detected fragments TKE_2 and their interaction energy at the Coulomb barrier E_b. As an example, the parameters for the point where two nuclei of ^{85}As were detected are analyzed below. The nuclear charge of the fragment was calculated according to unchanged charge density hypothesis. The parameters known from the experiment and the graph of the nucleus-nucleus potential are presented in Figure 3a.

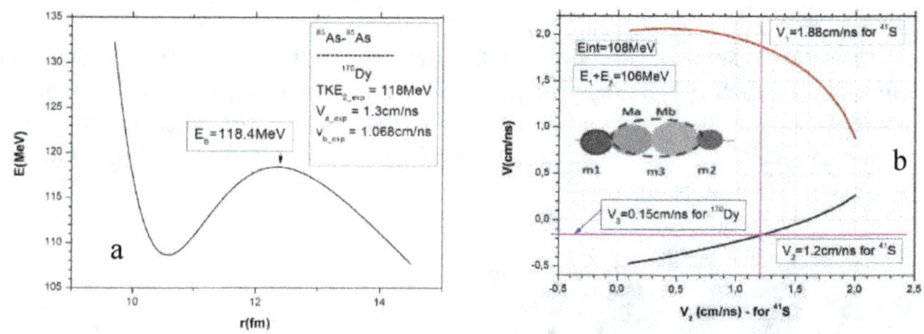

Fig. 3. Interaction potential between the nuclei ^{85}As-^{85}As – (a). (b) – velocities of the side fragments V_1, V_3 satisfying the system of equations (1) at the interaction energy E_{int} =108 MeV between the partners of the ternary prescission configuration ^{41}S-^{170}Dy-^{41}S. See text for details.

The observed correlation between TKE_2 and E_b could exist if the ^{170}Dy nucleus, being at rest, undergoes the break-up into two ^{85}As nuclei. In its turn ^{170}Dy nucleus could stay at rest after simultaneous separation of the side clusters of equal masses in the prescission configuration shown in the inset in Figure 3b. Certainly, this is an ideal scenario which could play differently in reality.

There is a very simple criterion for testing the proposed scenario. If the central fragment m3 moves as a whole before the brake-up, the velocities of the resultant fragments V_a and V_b should meet the conditions:

$$V_a=V^{(a)}+V_0;\ V_b=V^{(b)}-V_0,$$

where $V^{(a)}$ and $V^{(b)}$ are the velocities of the fragments after scission of the central fragment m3 being at rest. Corresponding energies of these fragments are supposed to be equal to their interaction energy E_b at the Coulomb barrier: $E^{(a)} + E^{(b)} = E_b$. V_0 is the drive velocity of the fragment m3 before its brake-up. In the case under analysis $V^{(a)} = V^{(b)} = 1.15$cm/ns, $V_0 = 0.15$cm/ns, $E_0 \approx 2$MeV. The results of the similar calculations for all 45 points from the box marked by the arrow in Figure 2a are presented in Figure 4a.

Clear and expected correlation seen in Figure 4a shows that really at the moment of the scission of the central fragment m3, it flies as a single body with the drive velocity not exceeding 0.3cm/ns.

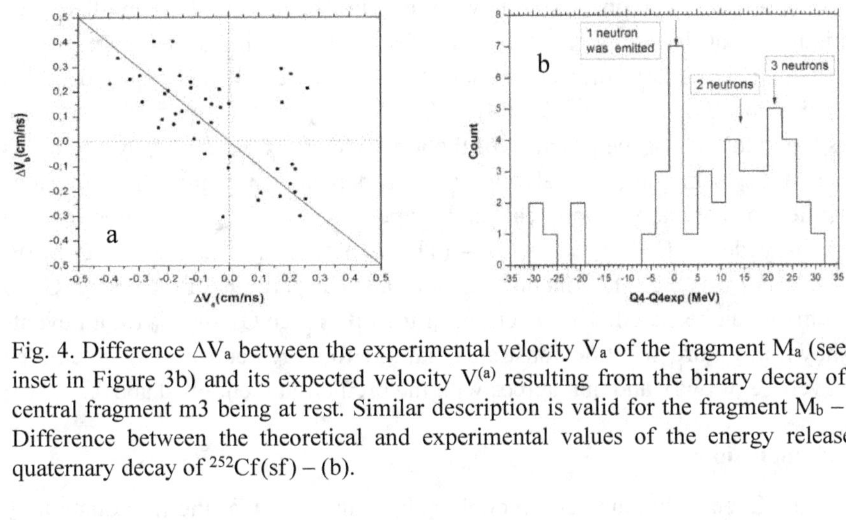

Fig. 4. Difference ΔV_a between the experimental velocity V_a of the fragment M_a (see the inset in Figure 3b) and its expected velocity $V^{(a)}$ resulting from the binary decay of the central fragment m3 being at rest. Similar description is valid for the fragment M_b – (a). Difference between the theoretical and experimental values of the energy release in quaternary decay of ^{252}Cf(sf) – (b).

Thus, we deal with a sequential process. At the first stage, a ternary decay takes place whereby the central and two side fragments become free. At the second stage, the central fragment undergoes binary brake-up. In order to test energy conservation in the proposed scenario of the quaternary decay we used the following approach.

In case of ternary decay, after full acceleration (at infinity) of all the fragments, both energy and momentum conservation laws should be met:

$$\begin{aligned} E_1 + E_2 + E_3 &= E_{int} \\ \vec{p}_1 + \vec{p}_2 + \vec{p}_3 &= 0, \end{aligned} \tag{1}$$

where E_{int} is the interaction energy between the fragments at the beginning of the acceleration; E_i and p_i are their energies and momenta, respectively. Thus, there are three unknown velocities, and one has only two equations for their determination. However, changing step-by-step one of the velocities or energies, we can solve the set of Equations (1) for each fixed value, for instance, V_2 (Figure 3b). According to the algorithm, any vertical line intersecting both the V_2 axis and the curves above, provides a trio of parameters, namely $\{V_1, V_2, V_3\}$, satisfying the system (1).

Directly from the experiment only missing mass is known. We suppose it to be the total mass of two side fragments m1 and m2, equal by masses. The energy released for different stages of the considered decay are as follows: $Q_2(Dy \rightarrow 2\ ^{85}As)$ = 71 MeV, $Q_3(Cf \rightarrow\ ^{41}S/^{170}Dy/^{41}S)$ = 165 MeV, $Q_4(Cf \rightarrow\ ^{41}S/^{85}As^{85}/As/^{41}S)$ = 235 MeV. Expected excitation energy of the ^{170}Dy nucleus after separation of the side fragments $E^*(Dy) = TKE_{2exp} - Q_2 + B_n$ = 55MeV under condition that one neutron with binding energy B_n was emitted. We remind that at least one neutron was detected in coincidence with fission fragments (Figure 2a). Interaction energy of the partners of the ternary decay in the scission point $E_{int}(^{41}S/^{170}Dy/^{41}S) = Q_3 - E^*(Dy) - E_0(Dy)$ = 108MeV. All possible velocities of the products of the ternary decay $Cf \rightarrow\ ^{41}S/^{170}Dy/^{41}S$ at this value of E_{int} satisfying the relations (1) are depicted in Figure 3b. Now it is possible to compare theoretical and experimental energy releases of the quaternary decay $Q_4 - Q_{4exp} = Q_4 - (TKE_{4exp})$ = (235–232) MeV = 3 MeV, where TKE_{4exp} is the total kinetic energy of all four partners of the decay. Good agreement can be stated. The spectrum of the difference $Q_4 - Q_{4exp}$ for all events is shown in Figure 4b. It demonstrates that one to three neutrons can be emitted (an average of two), and that agrees with the prediction mentioned above.

4. Conclusions

The experimental findings and model analysis indicate that the true quaternary decay of the $^{252}Cf(sf)$ goes via very elongated prescission configurations of the mother nucleus. Similar conclusion was drawn in [7] concerning the mechanism of the collinear cluster tri-partition (CCT) of the same nucleus. Presumably the quaternary decay is due to preformation in the valley of the symmetric nuclear shapes of the potential energy surface of ^{252}Cf specific nuclear shape which looks like a deformed magic core (strong shell minima at $N \sim 88$, 100 and $Z \sim 60$) and two light clusters on the sides. Further elongation of such system gives rise to the true quaternary decay observed for the first time.

Acknowledgments

This work was supported by the research project № 18-32-00538 of the Russian Foundation for Basic Research (RFBR), by the MEPhI Academic Excellence Project (Contract No. 02.a03.21.0005, 27.08.2013) of the Russian Science Foundation, and by the Department of Science and Technology of the Republic of South Africa (RSA).

References

1. Yu. V. Pyatkov et al., *Eur. Phys. J. A* **45**, 29 (2010).
2. Yu. V. Pyatkov et al., *Eur. Phys. J. A* **48**, 94 (2012).
3. D. V. Kamanin, Yu. V. Pyatkov, "Clusters in Nuclei – Vol. 3" ed. by C. Beck, *Lecture Notes in Physics* **875**, 183 (2013).
4. D. V. Kamanin et al., Int. Symposium on Exotic Nuclei "EXON-2016", Kazan, Russia, 04-10 September 2016. Conference proceedings, Editors: Yu. E. Penionzhkevich, and Yu. G. Sobolev. Published by World Scientific Publishing Co. Pte. Ltd., 2017. p. 243-248.
5. Yu. V. Pyatkov et al., *Journal of Physics: Conference Series, V.* **863**, 012046 (2017).
6. Yu. V. Pyatkov et al., *NIM A* **488**, 381 (2002).
7. Yu. V. Pyatkov et al., *Phys. Rev. C* **96**, 064606 (2017).

Confirmation of High Neutron Yields for Ba-Mo from SF of ^{252}Cf

B. M. Musangu,[1] A. H. Thibeault,[1] T. H. Richards,[1,2] E. H. Wang,[1] J. H. Hamilton,[1] C. J. Zachary[1] and A. V. Ramayya[1]

[1]Department of Physics and Astronomy, Vanderbilt University, Nashville, TN 37235, USA
[2]Department of Physics and Astronomy, University of Alabama, Tuscaloosa, AL 35487, USA

J. O. Rasmussen[3] and Y. X. Luo[1, 3]

[3]Lawrence Berkeley National Laboratory, Berkeley, CA 94720, USA

G. M. Ter-Akopian and Yu. Ts. Oganessian

Joint Institute for Nuclear Research, RU-141980 Dubna, Russian Federation

S. J. Zhu

Department of Physics, Tsinghua University, Beijing 100084, China

The neutron multiplicity yields of Ba-Mo, Ce-Zr and Nd-Sr have been studied with improved precision by using γ–γ–γ–γ as well as γ–γ–γ data and the latest level structures of these nuclei. The results clearly confirm that the Ba-Mo yield data have a hot fission mode where 8, 9, and 10 neutron evaporation channels are observed which are not in other channels. This mode can indicate that ^{144}Ba is likely hyper deformation at scission to give rise to such high neutron multiplications. A new experiment is being planned to do fission fragment–γ–γ coincidence studies to investigate details of the fission process and to study new more neutron-rich nuclei.

Keywords: Yield matrix; Ba-Mo; Ce-Zr; Nd-Sr.

1. Introduction

About 3.1 percent of Californium-252 (^{252}Cf) nuclei undergo spontaneous nuclear fission, in which the nucleus splits into two daughter nuclei, followed by a number of neutrons evaporated. The remaining 96.9 percent of nuclei α-decay to ^{248}Cm, whose analysis is not part of this experiment. This number of neutrons is often referred to as the "neutron channel" number, N_c, which may be computed from the nuclear masses of ^{252}Cf and the daughter isotopes produced:

$$N_c = 252-(A_1+A_2). \quad (1)$$

where A_1 and A_2 are the mass numbers of the two daughter isotopes. Empirically, the majority of ^{252}Cf spontaneous fission events evaporate 3 or 4 neutrons, averaging at 3.7. [1] However, neutron channel numbers as low as zero to as high as 10 or greater can be observed in some cases. In previous research, the case of barium and molybdenum was shown to have strong 8, 9, and 10 neutron channels. [1] The primary motivation of this research is to test the uniqueness of the Ba-Mo case by studying its neutron channel yields along with those of two other fragment pairs.

Under the nuclear shell model, nucleons occupy certain "shells" with defined energies, much like electrons do in an atom. If all the nucleons in a nucleus occupy the shells of lowest possible energy, the nucleus is said to be in its ground state. Otherwise, the nucleus is in an excited state and will emit γ-rays, or high energy photons, as it decays to lower states. The daughter nuclei of spontaneous fission often exist in excited energy states following their production. These γ-rays can be measured by a detector (see Section 2.1). Much like how an atom may emit photons of specific energy as its electrons descend to lower energy orbitals, a nucleus will emit γ-rays of specific energy depending on its shell structure. Thus, the presence of specific isotopes may be identified by the γ-rays detected in a sample. This is analogous to identifying elements by their spectral lines.

Of particular interest in this experiment are the ground state transition γ-rays. Some isotopes have a single ground state γ-ray, but others have multiple ones. The largest ground state γ-ray is generally the highest intensity γ-ray emitted by an isotope, because all daughter nuclei will emit this γ-ray, excluding the very unlikely case they were produced in the ground state during the fission process. By measuring the intensity of ground state γ-rays, we can deduce how likely specific isotopes of fission partner elements be produced in the spontaneous fission of ^{252}Cf. If the ground state γ-ray is inconvenient to measure, it is also possible to use a higher transition to make this calculation, as long as its intensity relative to the ground state transition is known.

It was shown and confirmed in earlier work [2] that the total intensities of the lowest $2^+ \rightarrow 0^+$ ground-state transitions observed in the de-excitation of even-even fission products reflect, to a high degree of accuracy (<5%), the total independent yields of these isotopes. Additionally, it was also shown that the values of related yields of pairs of fission fragments, as they appear after neutron evaporation, can be extracted from the analysis of γ–coincidences detected from prompt fission γ-ray with new large detector arrays.

2. Experimental Setup

Triple-coincidence and quadruple-coincidence data used in the present work were taken by Gammasphere, at Lawrence Berkley National Laboratory in November 2000. The Gammasphere consisted of 101 Hyper-Pure-Germanium (HPGe) detectors arranged in a spherical configuration. Each HPGe crystal also has seven Sodium Iodide (NaI) Compton shields. These Compton shields detect gamma rays that escape the HPGe crystal after undergoing Compton scattering, thus reducing many of the background detections. About $5.7*10^{11}$ γ–γ–γ and higher fold, and $1.9*10^{11}$ γ–γ–γ–γ and higher fold coincidence events were collected by Gammasphere. A ^{252}Cf sample was placed in the center of the Gammasphere and allowed to decay naturally, emitting γ-rays measured and counted by the detectors. Critical to the experiment is recording the relative time at which a γ-ray was detected; as this allows us to set coincidence windows (see Section 3).

3. Experimental Procedure

A powerful tool in nuclear analysis is generating a spectrum of γ-rays that fall within a coincidence time window (in our experiment it is 1 μs) of one or more other γ-rays of specified energy. The analysis of coincidence counting method is referred to as "gating" on an energy. In order to find peaks for the yield computations, we generally set double gates on the two most intense transition energies of an isotope of a fission product element. In the generated coincidence spectrum, we then identify and measured the areas and energies of the ground state transition peaks for the isotopes of the fission partner element. We then record the parameters for further calculations. The area under the peak was estimated using the method described in the previous work [1]. These fission pairs involved even-even fragments therefore, the yields were determined from the area of the peaks corresponding to the coincidences between $2^+ \rightarrow 0^+$ transitions in both fragments. The relative yields of these fragment pairs were obtained after corrections for the known detection efficiency and internal conversion probabilities. The internal conversion probabilities were obtained from [3]. By gating on one isotope and measuring the transition energies of other possible partner isotopes, we find how often the different partner isotopes were produced in coincidence with one another. In some cases, the ground state transition peaks are difficult to identify. For these exceptions, we measure other transitions within the same isotope and scale their yields according to published measurements of their relative intensities from our data. See Section 4 for detailed description of our steps.

We repeat the above steps for each isotope of a fission product so that a yield for each measurable pair of isotopes is calculated and finally recorded in a yields matrix. Some isotopes have multiple ground state transitions. For these cases, we measured the relative intensities of each ground state transition. In many cases, a single transition dominated over the others, allowing us to treat it as a single ground state transition. For the other cases, we took the sum of the relative yields from each transition.

We use the Radware software for setting coincidence gates and measuring the areas and energies of γ-ray peaks. We measured the areas of most peaks by fitting them to Gaussian distributions. For more poorly-shaped peaks, we simply took a sum of the counts found within its bounds.

4. Results and Discussion

We found some peaks unusable because of contamination or similar transition energies found in other isotopes. In order to measure yields in these cases, we used peaks found in higher transitions and scaled them appropriately. In order to calculate the scaling factor, we set a clean gate with no contamination on energy transitions of other isotopes and measure the intensities of the ground state transition and of the higher transition. By taking the ratio of these intensities, we compute the scaling factor needed. In the Sr-Nd yields matrix (see Table I), we left most of the items in the ^{148}Nd and ^{149}Nd columns blank because those isotopes are weakly populated in the spontaneous fission of ^{252}Cf, so it is difficult to measure Sr transition energies. Some similar energies, such as the ground state transitions of ^{98}Sr and ^{100}Sr, are therefore not distinguishable with these gates. Moreover, ^{150}Nd and ^{100}Sr have the same first E2 energies. We also made a plot of the extracted yields against the fission's neutron channel number (see Fig. 1). There is no evidence for the 9 and 10 neutron emission yield.

Table I. New yield matrix for Neodymium and Strontium from the spontaneous fission of ^{252}Cf.

Yield	^{148}Nd	^{149}Nd	^{150}Nd	^{151}Nd	^{152}Nd	^{153}Nd	^{154}Nd	^{155}Nd	^{156}Nd
^{91}Sr							0.004(1)		0.005(1)
^{92}Sr							0.020(3)	0.019(4)	0.030(5)
^{93}Sr					0.005(1)		0.027(5)	0.012(2)	0.022(4)
^{94}Sr			0.010(2)	0.026(4)	0.088(15)	0.180(31)	0.132(22)	0.047(8)	0.043(7)
^{95}Sr			0.017(3)	0.033(6)	0.117(20)	0.305(51)	0.109(18)	0.023(4)	0.012(2)
^{96}Sr			0.040(7)	0.060(10)	0.159(27)	0.234(40)	0.062(10)	0.010(2)	
^{97}Sr			0.030(5)	0.043(7)	0.061(10)	0.071(14)	0.021(4)		
^{98}Sr	0.010(2)	0.031(6)	0.080(15)	0.044(7)	0.063(11)	0.025(5)			
^{99}Sr			0.020(4)	0.017(3)	0.015(3)				
^{100}Sr			0.026(4)	0.006(1)					

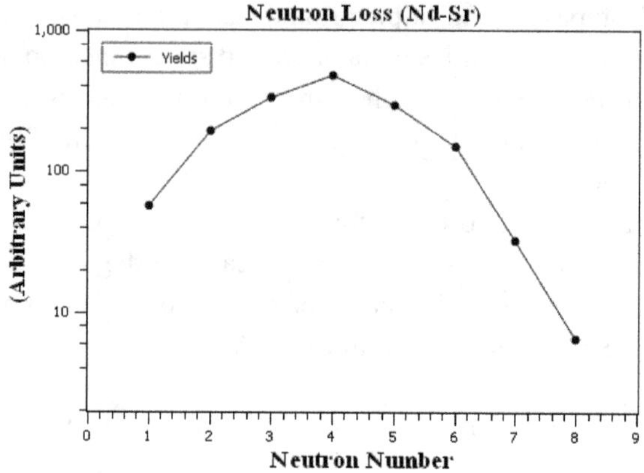

Fig. 1. New relative yield curve of Nd-Sr.

In the case of determining the Ce-Zr yield matrix, measurements of multiple γ-rays emitted by the Ce-Zr fission fragment pair formed in spontaneous fission of ^{252}Cf were used to extract the yields. A two-dimensional matrix was created from the initial data by selecting the γ-ray coincidences occurring within 50ns. The peaks observed in this two-dimensional spectrum arise from the coincidences between γ-ray emitted promptly by both complementary fission fragments of different fragment pairs. Table II below displays the relative yields data that were collected. These yields of the correlated fragment pair of Ce-Zr in the spontaneous fission of ^{252}Cf were next scaled according to Ter-Apkopian's independent yield [1] and summed for each isotope of Ce. This summation and Ter-Akopian's calculated data for Ce-Zr, were both normalized such that ^{148}Ce had a value of 100. Then these two data sets were compared to see if Ter-Akopian's calculations could be verified. As can be seen below in Fig. 2, the present relative yield data are in agreement with the previous ones, and thus are experimentally confirmed.

Table II. New yield matrix for Cerium and Zirconium from the spontaneous fission of ^{252}Cf.

Yield	^{144}Ce	^{145}Ce	^{146}Ce	^{147}Ce	^{148}Ce	^{149}Ce	^{150}Ce	^{151}Ce	^{152}Ce
^{97}Zr							0.015(3)	0.025(5)	0.020(4)
^{97}Zr					0.028(4)	0.026(5)	0.074(12)	0.082(15)	0.050(9)
^{98}Zr					0.065(10)	0.142(26)	0.212(35)	0.145(27)	0.058(10)
^{99}Zr			0.018(3)	0.056(9)	0.202(32)	0.251(45)	0.297(49)	0.122(24)	0.033(6)
^{100}Zr	0.011(2)	0.032(6)	0.125(20)	0.283(45)	0.552(87)	0.393(71)	0.278(46)	0.113(22)	0.017(3)
^{101}Zr		0.107(19)	0.255(41)	0.329(52)	0.540(85)	0.225(41)	0.168(30)		
^{102}Zr	0.024(4)	0.161(29)	0.430(68)	0.401(63)	0.395(62)	0.099(18)	0.029(5)		
^{103}Zr	0.033(6)	0.144(26)	0.220(40)	0.142(26)	0.115(19)				
^{104}Zr	0.005(1)	0.069(13)	0.008(1)	0.001(0)					

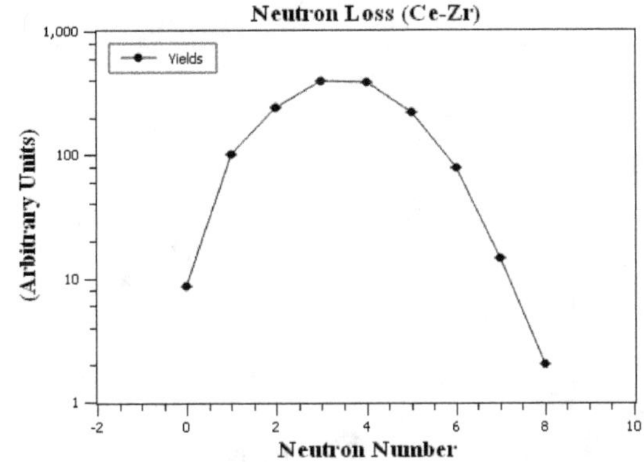

Fig. 2. New relative yield curve of Ce-Zr.

The results from the Ce-Zr study were similar to those from the previous study [1], with smaller error limits. Again, there is no evidence for the 9 and 10 neutron emission yield.

The relative intensities of all transitions feeding the ground states of the barium isotopes for each molybdenum isotope were determined based on new levels schemes [4-6] to produce a new set of relative yields. The new yields of Mo-Ba are given in Table III and Fig. 3. The errors are significantly reduced because of the improved statistics, the use of quadruple coincidence data to further remove accidental coincidences and improved knowledge of level schemes. As seen in Fig. 3, a similar deviation from a Gaussian fit to the data for the 0 to 6 neutron emission channels is seen at neutron numbers 7, 8, 9, 10. Also shown in Fig. 3 is a Gaussian fit of the 8, 9, 10 neutron data with the Gaussian fit to the 0-6 neutron emission subtracted for the data. This new analysis, coupled with the new analysis of the Zr-Ce and Nd-Sr yields which do not exhibit 8, 9, 10 neutron emission confirms the existence of this extra hot mode only in Mo-Ba yield.

Table III. New yield matrix for Cerium and Zirconium from the spontaneous fission of ^{252}Cf.

Yield	^{138}Ba	^{139}Ba	^{140}Ba	^{141}Ba	^{142}Ba	^{143}Ba	^{144}Ba	^{145}Ba	^{146}Ba	^{147}Ba	^{148}Ba
^{102}Mo			0.004(0)		0.013(1)	0.036(6)	0.147(13)	0.279(25)	0.321(27)	0.222(29)	0.051(5)
^{103}Mo			0.010(1)	0.012(1)	0.056(7)	0.151(10)	0.440(25)	0.576(35)	0.370(20)	0.052(6)	0.005(1)
^{104}Mo	0.008(1)	0.002(0)	0.012(4)	0.051(42)	0.241(13)	0.529(29)	1.059(55)	0.763(41)	0.375(20)	0.100(12)	
^{105}Mo	0.010(1)	0.006(2)	0.061(6)	0.161(11)	0.535(33)	0.895(55)	1.059(66)	0.661(41)	0.171(11)		
^{106}Mo	0.013(5)	0.011(2)	0.145(15)	0.267(24)	0.836(53)	0.915(56)	0.695(42)	0.280(19)	0.096(16)		
^{107}Mo	0.041(5)	0.069(7)	0.346(1)	0.454(40)	0.705(60)	0.472(42)	0.237(21)				
^{108}Mo	0.010(1)	0.024(2)	0.266(22)	0.162(13)	0.150(12)	0.077(6)	0.027(2)				

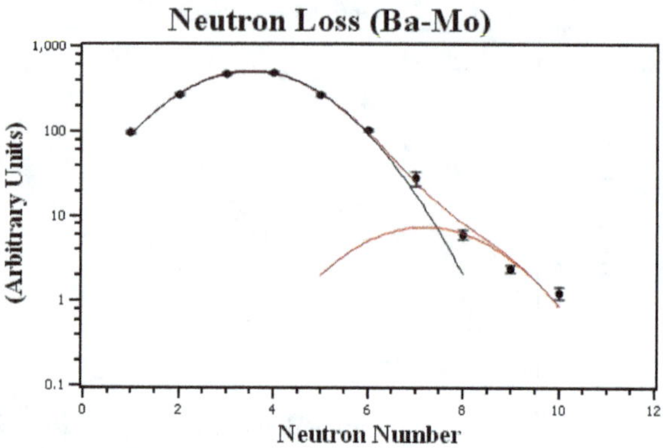

Fig. 3. Relative intensities of the 1 to 10 neutron emissions in the Mo-Ba matrices from analysis Gammasphere data. The smooth lines are Gaussian fits to the 1-7 and 6-10 neutron emission data.

5. Conclusion

New yield matrices were determined for Ba-Mo, Ce-Zr and Nd-Sr fission partners from the spontaneous fission of ^{252}Cf. A similar deviation from the Gaussian fit to the normal fission mode was found in Ba-Mo for the 8, 9, and 10 neutron channels as found in previous analyses, to confirm the existence of the proposed "extra-hot-fission" mode. Errors are reduced in this newest analysis compared to previous studies because of the greater number of statistics of the latest Gammasphere experiment, the use of quadruple coincidences in the analysis and improved level schemes. A new experiment is being planned to do fission fragment–γ–γ coincidence studies to investigate details of the fission process and to study new more neutron-rich nuclei.

Acknowledgment

From the Department of Physics and Astronomy at Vanderbilt University I would like to thank Joseph H. Hamilton, Akunuri V. Ramayya, Enhong Wang, and Christopher J. Zachary for their guidance on this project. I would also like to thank Thomas Richards from University of Alabama for his contribution. He was also supported by the NFS REU program here at Vanderbilt University. I would also like to thank my family and friends for their support. The work at Vanderbilt University and Lawrence Berkeley National Laboratory are supported by the US Department of Energy under Grant No. DE-FG05-88ER40407 and Contract No. DE-AC03-76SF00098. The work at Tsinghua University was supported by the National Natural Science Foundation of China under Grant No. 11175095. The work at JINR was supported by the Russian Foundation for Basic Research Grant No. 08-02-00089 and by the INTAS Grant No. 03-51-4496.

References

1. G. M. Ter-Akopian *et al.*, *Phys. Rev. C* **55**, 1146 (1997).
2. C. Goodin *et al.*, *Phys. Rev. C* **74**, 017309 (2006)
3. Bricc.anu.edu.au
4. M. R. Bhat, Evaluated Nuclear Structure Data File (ENSDF).
5. Experimental Unevaluated Nuclear Data List, www.nndc.bnl.gov/xundl.
6. Private communication from Vanderbilt University.

Fission Fragment Distribution and the r-Process*

I. V. Panov
National Research Center "Kurchatov Institute", 123182, Moscow, Russia
NRC "Kurchatov Institute – ITEP", 117218, Moscow
E-mail: panov.iv@gmail.com

I. Yu. Korneev
National Research Center "Kurchatov Institute", 123182, Moscow, Russia
E-mail: Ivan_K@abbyy.com

In the present report the influence of fission mass distribution model on the final abundance of heavy elements in the matter, ejected during neutron star mergers is discussed. Different models of fission fragment mass distribution for exotic nuclei were applied to the r-process, calculated for the scenario of neutron star mergers, and the significant influence of fission fragment mass distribution model on the abundance of heavy nuclei was shown. We have found that fission fragment mass distribution model is most important for fissioning nuclei with A > 260 from the region near to neutron drip-line and for superheavy nuclei. It was shown that tri-partition or ternary fission, the role of which can strongly rise in neutron rich hot dense matter, strongly affected on the abundance of heavy nuclei.

Keywords: Kilonova; nucleosynthesis; r-process; fission.

1. Introduction

The first registration of neutron star mergers process and simultaneous registration of the elements produced in the r-process [1] have confirmed theoretical predictions that the main scenario of the r-process nucleosynthesis should be rather jets appeared during compact binary mergers in close binaries than the supernova explosion [2].

As it was shown for the first time in numerical calculations [3] of the r-process during neutron star mergers [4], fission in such a scenario of the main r-process leads to formation of heavy r-elements from the second peak at abundance curve till uranium and thorium and the r-process pattern is close to the abundances of heavy elements for all different scenario of neutron stars mergers, where the quasi-equilibrium between second and third peaks exists. And agreement with observations strongly depend, apart of scenario, on nuclear physics input – decay characteristics, delayed processes probabilities and fission rates.

* This work is supported by Russian Science Foundation, Grant No. 16-12-10519.

In the present report we discuss the influence of fission fragment mass distribution (FFMD) model on the final abundance of heavy elements in the matter, ejected during neutron star merger [3]. Recently different models of FFMD for exotic nuclei [5] were applied to the r-process, calculated for the scenario of neutron star mergers [6], where the significant influence of fission fragment mass distributions on the abundance of heavy nuclei was shown. But the physics details of such a difference is still unclear as well as the shift of second and third peaks. We considered different FFMD models and their application to the r-process with the aim to understand of the main reason leading to different abundances when the different FFMD models are used. We have considered either the influence of possible strengthening of triple or ternary fission on the final abundance of heavy nuclei formed in the r-process [7] as additional possible channel of fission of heavy ultra-neutron-rich nuclei with $Z > 92$ and $A > 260$.

2. Fission Fragment Mass Distribution and the r-Process

Direct inclusion of fission processes in the r-process calculations also requires, except the fission rates, the knowledge of the resulting mass distribution of fission fragments, that is important because newly formed fission fragments have come back to the r-process as a new source of seeds for the r-process.

In the previous work [5] we considered three well known mass distributions of fission products. The first FFMD model is based on micro-macro approach (ABLA model) [8] and two other phenomenological models [9,10] are based on nuclear systematics. Phenomenological model [9] was based on experimental data and used linear dependence of number of fission neutrons on A and (A-Z) and consider both asymmetric fission for nuclei with $A < 260$, and asymmetric one for heavier nuclei. We have considered a number of extreme cases of the FFMD model [9] with different widths in distribution of light and heavy fission fragments, different number of fission neutrons ν_f and transformation of asymmetric fission into symmetric one when $A > 260 + \nu_f$. The other utilized phenomenological model was based on simple normal distribution of A and Z for every fission fragment, and neglect fission neutrons [10].

When these models were applyed to the r-process calculation, the results (Fig. 1) confirm the strong dependence of abundance Y_A on the fission fragments mass distribution, emphasized earlier [6]. In used scenario this dependence even more strong, and depend rather on FFMD model, than on variations in model parameterization. Smaller widths of FFMD leads mainly to changes of Y_A for nuclei with $A < 130$. Suppression of fission neutrons number ν_f leads to more worse description of Y_A for 130 and 196 peaks and FFMD, described as in [10], leads to more worse agreement in Y_A mainly for region of A~130, and heavier nuclei are strongly underabundant in comparison with basic distribution.

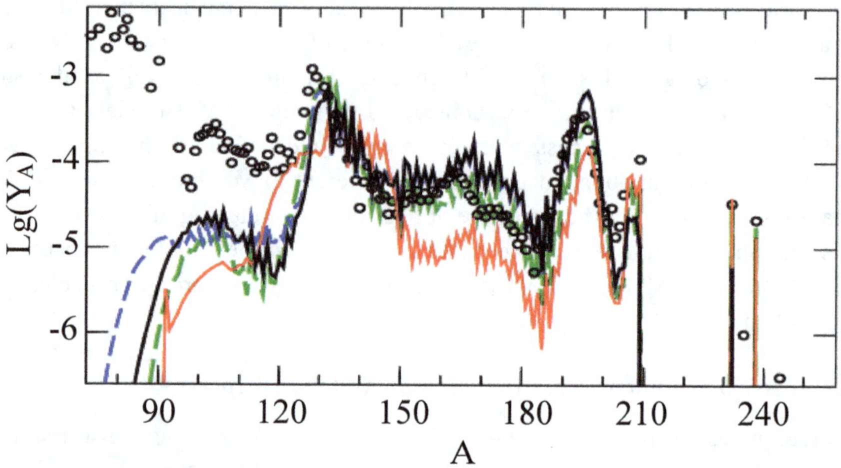

Fig. 1. Influence of FFMD on the final r-process abundances Y_A in NSM scenario. Abundance of heavy nuclei Y_A calculated with basic FFMD [9] – green dashed line; on the basis of the same model [9] but with increased widths – dashed blue line; FFMD [9], but with reduced fission neutron number – black line.

3. Discussion and Conclusions

Influence of different FFMD models utilized in the r-process nucleosynthesis is strong and models with symmetric fission for nuclei with $A > 260$ [5,9], describe r-process nucleosynthesis better than the other phenomenological model, describing FFMD close to mass distributions predicted in [10]. Different character of Y_A dependence on FFMD model between present and previous [6] results probably emerge in part due to different scenario considered, but additional analysis should be done.

Earlier [4] we have discussed the influence of ternary fission on the r-process. Triple or ternary fission [5,7,11] create only nuclei with $A < 130$ that is very important for nuclear abundances Y_A, produced in the r-process and may be considered additionally to binary one along r-process path in transuranium region. Ternary and quaternary fission were observed experimentally since 1947 [12]. Theoretical evaluations of the possibility of fission into n equal fragments (ternary fission when $n = 3$) were done [14]. For example, when the fissility parameter exceeds 0.61 (that is usual for neutron rich transuranium nuclei), ternary fission energetically becomes more preferable to binary fission. Though the experimental evaluation of the ternary fission probability in neutron induced reactions at low energy shows that this process is at least by 3 orders smaller than binary fission and the formation probability of a heavy third fragment ($A > 40$) decreases below 10^{-6} [11,14], in fission of very neutron-rich nuclei this probability can rise [7,15]. The experiments with heavy ions [7] shows that the yields of ternary fission fragments of stable or long-lived isotopes of Th, U and Bi can be of the same order as in binary fission. And the masses of the true ternary fission fragments can be approximately equal. Existed experimental facts [16,17] and theoretical evaluations [18] let us consider that in fission of extremely neutron rich nuclei (when the energy of beta-decay is about 15 MeV or more and is of the order of shell corrections), the probability of ternary fission of nuclei can be rather high. We have compared the r-process abundances computed with taking into account triple fission with observations.

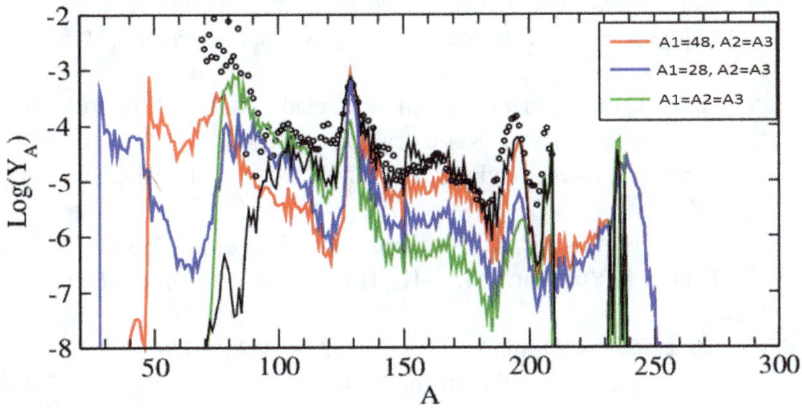

Fig. 2. r-process calculations in NSM scenario with different models of FFMD for triple fission: 1) for true ternary fission when A1=A2=A3 (green line); 2) triple fission, when A1=28 (blue line) or A1=48 (red line). Black line - only binary fission is taken into account.

In the r-process calculations we considered 3 extreme cases for the ternary fission when fragment masses are close to experimentally observed [16,17] ones and theoretically considered [15,18]: (1) true ternary fission (green line at Fig. 2) and cases 2,3) when 3d fragment have smaller mass: (2) A1=48, A2=A3 (red line); (3) A1=28, A2=A3 (blue line), have shown at the Figure 2. The probability of triple (or ternary) fission [7,11] is highly uncertain but for nuclei with $A > 260$ near the neutron drip-line, fission into 3 fragments is energetically possible and could occur. Present [16,17] and forthcoming experiments as well as theory predictions [15] should clarify the probabilities of such a processes, but superposition of present abundances can describe the observed abundance curve since 1st till 3d peaks.

Acknowledgement

The work was done with financial support of Russian Scientific Foundation, Grant No. 16-12-10519.

References

1. N. R. Tanvir, A. J. Levan, C. González-Fernández et al., *Astrophys. J. Letters*, **848**, L27 (2017).
2. L. Hudepohl, B. Mueller, H.-T. Janka, et al., *Phys. Rev. Lett.*, **104**, 251101 (2010).
3. I. V. Panov, F.-K. Thielemann, *Astron. Lett.* **29**, 510 (2003).
4. C. Freiburghause, S. Rosswog, F-K. Thielemann, *Astrophys. J.* **525**, L121 (1999).
5. I. V. Panov, I. Yu. Korneev, T. Rauscher, et al., *A&A*. **513**, A61 (2010).
6. M. Eichler, A. Arcones, A. Kelic et al., *Astrophys. J.* **808**, 30 (2015).
7. V. P. Perelygin, N. H. Shadieva, S. P. Tretiakova et al., *Nucl. Phys. A*, **127**, 577 (1969).
8. K.-H. Schmidt, A. Kelić, M. Ricciardi, *Europhys. Lett.*, **83**, 32001 (2008).
9. I. V. Panov, I. Yu. Korneev, F-K. Thielemann, *Astronomy Lett.*, **34**, 189 (2008).
10. T. Kodama, K. Takahashi, *Nucl. Phys. A*, **239**, 489 (1975).
11. I. Tsekhanovich, Z. Büyükmumcu, M. Dav et al., *Phys. Rev. C*, **67** 034610 (2003).
12. H. Diehl, W. Greiner, *Nucl. Phys. A*, **229**, 29 (1974).
13. Tsien San-Tsiang, Ho Zan-Wei, R. Chastel et al., *Phys. Rev.* **71**, 382 (1947).
14. W. J. Swjatecky, *Phys. Rev.*, **101**, 651; **104**, 993 (1956).
15. A. V. Karpov, *Phys. Rev. C*, **94**, 064615 (2016)
16. A. V. Karpov, V. I. Zagrebaev, W. Greiner, *EPJ Web of Conferences*, **17**, 10002 (2011).
17. Yu. V. Pyatkov, D. V. Kamanin, A. A. Alexandrov et al., *Phys. Rev. C*, **96**, id.064606 (2017).
18. Yu. V. Pyatkov, D. V. Kamanin, J. E. Lavrova et al., *Journal of Physics: Conf. Series* **863**, 012046 (2017).

Light Clusters in Supernova Matter*

I. V. Panov
National Research Center "Kurchatov Institute", 123182, Moscow, Russia
NRC "Kurchatov Institute – ITEP", 117218, Moscow
E-mail: panov.iv@gmail.com

A. V. Yudin
NRC "Kurchatov Institute – ITEP", 117218, Moscow, Russia
E-mail: yudin@itep.ru

In present report we explore the appearance of light very neutron-rich clusters at high densities of collapsing stellar cores. Special attention is devoted to the superheavy hydrogen isotopes and dineutron and tetraneutron as well, which were not considered in previous studies. We have found that for light nuclei it is important to use exact information about its properties (values of spins and energies of known excited states) to obtain a reliable EoS. In our analysis we consider the influence of nuclear parameters, binding energy in first turn, on the abundances of nuclei under consideration. We discuss the domains in supernova where the light nuclei influence on the isotope abundances can have most important consequences.

Keywords: Supernova; equation of state; dineutron; nucleosynthesis.

1. Introduction

The history of equation of state (EoS) development and modification is closely connected with supernovae modelling. In first calculations of stellar gravitational collapse very simplified EoSs were used: the matter was considered as an ideal gas of free neutrons, protons, alpha-particles and ^{56}Fe, surrounded by electron-positron pairs and blackbody radiation [1]. Later this approach was modified (see, for example Langer et al. 1969 [2]): instead of using one «heavy» nucleus with fixed parameters, an «average» nucleus was considered, representing the full chemical composition. Its parameters (A,Z, etc.) were determined by minimizing the total free energy of the system. This approach is very popular, and also applied in the most widely used current EoSs [3,4].

In parallel to this, another approach based on the concept of nuclear statistical equilibrium (NSE) was developed. It explicitly includes an ensemble of different nuclei (hundreds or even thousands). A general finding, present in all these models, is that deuterons, tritons and helions appear abundantly for typical

* This work is supported by Russian Science Foundation, Grant No. 16-12-10161.

conditions of supernova matter in addition to α-particles, see, e.g., Pais et al. [5]. The presence of light clusters can modify weak interaction rates and therefore the dynamics of astrophysical processes [6]. This report is devoted to «forgotten» superheavy hydrogen isotopes, such as $^{4\text{-}6}$H, and especially to the role of ^{4}H and light neutron clusters as well.

2. Light Clusters and NSE

It is generally accepted that unstable states must be included in the NSE approach. Here we want to present some simple arguments to ensure the possibility for inclusion of some additional superheavy hydrogen isotopes (with A=4-6) and such neutral clusters as ^{2}n and ^{4}n [7, 8] into the NSE approach. One of most abundant among them – ^{4}H isotope, has negative neutron separation energy (although positive binding energy) and very short lifetime. It ordinary breaks by neutron decay with the characteristic time $\tau_{H4} \approx 1.4 \times 10^{-22}$ sec [9]. Three known exited states of ^{4}H have the same order of magnitude lifetime (see Table 1 in [10]). Simplest way to create ^{4}H is the reaction, inverse to neutron decay, via neutron-capture. But it is not the only way for generating ^{4}H [11]. Besides this, there would be much more production and destruction channels in the hot and dense medium, e.g. fission or multifragmentation of heavy nuclei. Isotopes ^{5}H and ^{6}H also were included into consideration, in spite of uncertainties in their parameters [12].

Besides superheavy hydrogen isotopes, some neutral clusters like dineutron, tetraneutron etc [12-14] also can be formed in nuclear reactions, in part during decay reactions: ^{7}H(^{4}n), ^{16}Be(^{2}n), ^{26}O(^{2}n), ^{28}O(^{4}n). There were a lot of confirmations of ^{2}n and ^{4}n existence during decay of exotic nuclei or in the field of nuclei [15]. More than that, such clusters can exist under sub-nuclear densities, as in the neutron stars [16]. In present work ^{2}n and ^{4}n also were included into consideration.

As it was shown in extended comparison of the results for different EoSs [10], all of them have shown very close results, especially for light nuclei in broad range of temperature and density values. As a base case for present research we use extended NSE EoS model [17] (EoS YN hereinafter). All nuclei are assumed to form ideal gas except free neutrons and protons for which degeneracy effects are also included. Nuclei parameters (binding energies, known excited states etc) are taken from online database of the NNDC. The original set of nuclides in EoS YN [17] are extended mostly by inclusion of neutron-rich isotopes from hydrogen to iron peak nuclei. All in all the total number of nuclei (mainly light and iron-peak elements) under consideration is 398. For nuclei the first excited states were taken into account – from experiment, when exists, or due to Fermi-gas model. Besides nuclei, this model accounts for electron-positron pairs of arbitrary degeneracy and relativism and blackbody radiation according to [18].

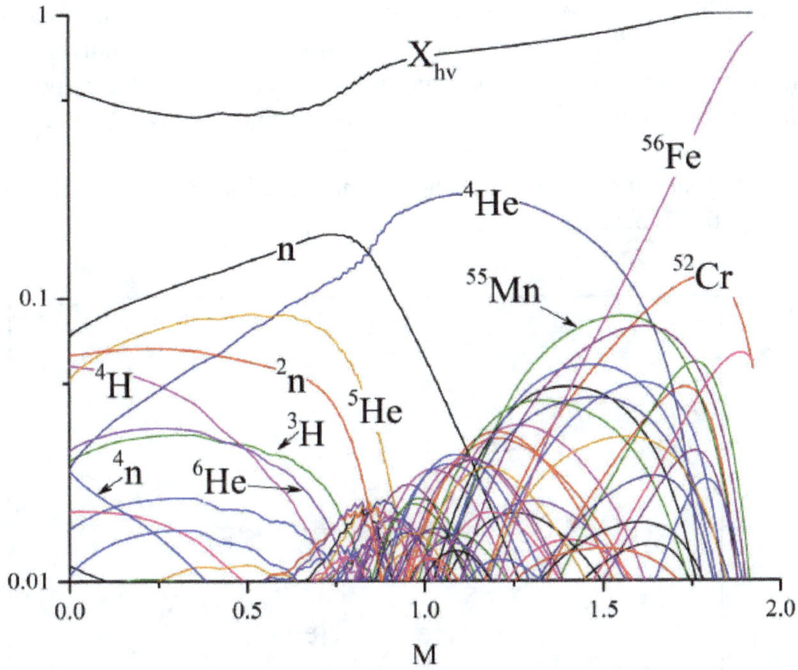

Fig. 1. Mass fractions of nuclei X_i as a function of mass coordinate M (in solar masses) for the moment of 1 ms before bounce. Black line marked X_{hv} shows the total mass fraction of elements with $Z > 2$. The most abundant isotopes of Fe, Mn and Cr are shown as well as light neutral clusters, hydrogen and helium isotopes. Calculations with NY EoS [10].

In this report we present the results, based on different EoSs, produced in the framework of EoS NY [17], Hempel et al., later EoS HS [19], and by Blinnikov et al., later BRPS [20]. The last experimental data [21] are used for the binding energies and for H and He isotopes the experimentally known excited states are explicitly included, also with updated data on spin degeneracy.

3. Infall and Post-Bounce Stages of Collapse

As an example we calculated the chemical composition of matter with YN EoS for the moment of time 1 ms before bounce approximately, when the central density of a collapsing stellar core reaches roughly 3×10^{13} g/cm^3. The profiles of temperature, density, and electron fraction Y_e inside the central part of the collapsing stellar core were taken from [19]. Figure 1 shows the chemical composition (mass fractions X_i) for the same moment of time. Black line marked X_{hv} also shows the total mass fraction of all elements with $Z > 2$. In the outer regions with mass coordinate (m>1.5 $M_\odot$) the core consists mainly from heavy iron-peak nuclei, while in the central, neutronized part, the individual mass fractions of light nuclei are dominant. Besides free neutrons there are a lot of helium and hydrogen isotopes. The most peculiar thing is a high abundance of ^{4}H (shown by purple line) and ^{2}n (red) in the central region. X_i of ^{4}H and ^{2}n reaches almost 10% and it can be more abundant than α-particles and free protons whose mass fraction is below 10^{-2}.

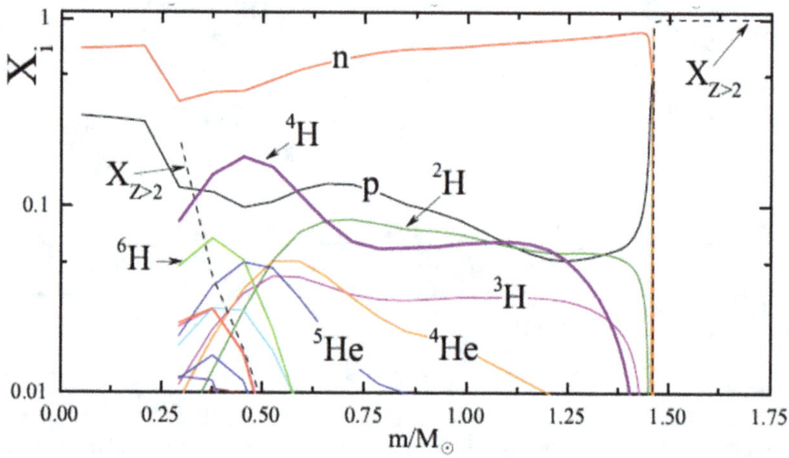

Fig. 2. Mass fractions X_i of hydrogen and helium isotopes as a function of m. The black dashed line marked $X_{Z>2}$ shows the total mass fraction of all rest nuclei. Stellar profile corresponds to 200 ms after bounce. Calculations according to modified HS EoS [19].

To complete our discussion about the role of light clusters and especially hydrogen neutron-rich clusters for the supernova EoS, we calculated the chemical composition of matter for the star profile corresponding to approximately 200 ms after bounce (Fig. 2 [10]). We show mass fractions of H and He isotopes together with total mass fraction X_Z>2 of all nuclei with Z>2. The used equation of state is modified EoS HS, which, contrary to other EoS's [17,20], can be applicable for supernuclear densities also. Multineutrons are not considered in this EoS. Central part of stellar core with 0<m/$M_\odot$<0.25 is occupied with pure nuclear matter from neutrons and protons. Lepton fraction there is $Y_e \leq 0.3$. Above this area a region passed and heated by expanding shock wave is placed. Behind this shock, at m<1.5$M_\odot$ the matter is neutronized up to Y_e~0.1. There are a lot of superheavy hydrogen isotopes, including ^{4}H. The region above shock consists mainly of nuclei with Z>2. We can conclude that light clusters remain to be an important ingredient of matter on the post-bounce stage of collapse also.

4. Importance of Light Clusters

After the previous discussions the main question is: what is the most important effect of light clusters for the supernova matter problem? Do we have to describe them accurately, or are more crude approximations sufficient? An answer to this question can be obtained. For the same before-bounce star profile like in our Fig. 1 we calculated the relative changes of a few quantities due to ^{4}H. The relative changes are calculated with our base NSE EoS [17] with and without ^{4}H included. All other nuclei and its parameters were the same. The change in pressure is negligible, or at most relatively small, well below 1%. The effect for entropy is more pronounced than for the pressure, but still small, only few percents in the central region of the core. In other works, (see [22]), it was also found that the entropy is more sensitive to the chemical composition.

An important part of the neutrino-matter interaction rates in supernovae is the coherent scattering of neutrinos on nuclei and free nucleons. Despite the fact that this process is approximately elastic (especially for heavy nuclei), it gives an important contribution to the opacity of matter for neutrinos. Its cross-section is roughly proportional to the square of the nucleus mass number and its change, caused by the presence/absence of only one nucleus ^{4}H, can reach 7 percents, i.e. almost twice the effect for entropy and 20 times that of for pressure. This is because neutrino-matter interaction rates are highly sensitive to the nuclear composition.

The main conclusions from above are the following: thermodynamic quantities are not so sensitive to the appearance of superheavy hydrogen isotopes, including ^{4}H, and neutral clusters. The change in pressure and hence the direct change of the collapse dynamics will be rather minor. But the change of the neutrino transport (e.g., the onset of trapping and the neutronization) and the evolution of the neutrino quantities can be significant.

5. Conclusions

We performed a global study of light asymmetric clusters effect for the supernova matter conditions. One of our most peculiar finding is a high abundance of superheavy hydrogen isotopes and dineutron and tetraneutron as well, which were not taken into account previously. We explore a wide domain of thermodynamic parameters, representative for the matter of collapsing stellar core on infall stage of collapse as well as during post-bounce phase. It appears that for light nuclei it is important to use exact information about its properties (values of spins and energies of known excited states) to obtain a reliable EoS. For the heavy nuclei the effect of the whole partition functions occurs to be of the same importance.

By comparison of three EoSs [17, 20, 22] with different underlying physics we ensured the stability of light clusters effect for various conditions. As can be seen from our Fig. 2, superheavy hydrogen isotopes with A > 4 [14] also can be abundant at the post-bounce part of collapse.

Acknowledgment

The work was done with financial support of Russian Scientific Foundation, Grant No. 16-12-10161.

References

1. V. S. Imshennik and D. K. Nadezhin, *Astron. Zh.*, **42**, 1154 (1965).
2. W. D. Langer, L. C. Rosen, J. M. Cohen, A. G. W. Cameron, *Ap&SS*, **5**, 259 (1969).
3. J. M. Lattimer and F. D. Swesty, *Nucl. Phys. A*, **535**, 331 (1991).
4. H. Shen, H. Toki, K. Oyamatsu, K. Sumiyoshi, *Nucl. Phys. A*, **637**, 435 (1998).
5. H. Pais, S. Chiacchiera, C. Providencia, *Phys. Rev. C*, **91** 055801 (2015).
6. T. Fischer, G. Martnez-Pinedo, M. Hempel et al., *EPJ Web Conf.*, **109**, 06002 (2016).
7. F. M. Marqués, N. A. Orr, N. L. Achouri et al., *Phys. Rev. Lett.*, **109**, 239201 (2012).
8. A. M. Shirokov, G. Papadimitriou, A. I. Mazur et al., *Phys. Rev. Lett.*, **117** id.182502 (2016).
9. M. Wang, G. Audi, F. G. Kondev et al., *Nuclear Data Sheets*, **120**, 6 (2014).
10. A. V. Yudin, M. Hempel, S. I. Blinnikov et al., *MNRAS*, **483**, 5426 (2018).
11. S. I. Sidorchuk, D. D. Bogdanov, A. S. Fomichev et al., *Phys. Lett. B*, **594**, 54 (2004).
12. Yu. E. Penionzhkevich, *Physics of Atomic Nuclei*, **79**, 549 (2016).
13. A. I. Baz, V. I. Gol'danskiĭ, Ya. B. Zel'dovich, *Soviet Physics Uspekhi*, **8**, 177 (1965).
14. L. V. Grigorenko, M. S. Golovkov, S. A. Krupko et al., *Physics-Uspekhi*, **59**, 321 (2016).
15. G. M. Ter-Akopian, A. M. Rodin, A. S. Fomichev et al., *Phys. Lett.*, **426**, 251 (1998).
16. J. P. Kneller and G. C. McLaughlin, *Phys. Rev. D*, **70**, 043512 (2004)
17. A. V. Yudin and D. K. Nadyozhin, *Astronomy Letters*, **30**, 634 (2004).
18. S. I. Blinnikov, N. V. Dunina-Barkovskaya, D. K. Nadyozhin, *Astrophys. J. Suppl.*, **106**, 171 (1996).

19. M. Hempel, T. Fischer, J. Schaffner-Bielich, M. Liebendorfer, *Astrophys. J.*, **748**, 70 (2012).
20. S. I. Blinnikov, I. V. Panov, M. A. Rudzsky, K. Sumiyoshi, *Astronomy & Astrophysics*, **535**, id.A37, 13 (2011).
21. G. Audi, M. Wang, A. H. Wapstra et al., *Nucl. Data Sheets*, **120**, 1 (2014).
22. M. Hempel and J. Schaffner-Bielich, *Nucl. Phys. A*, **837**, 210 (2010).

Fission and Quasi-Fission Dynamics Near the Coulomb Barrier: γ Rays as Probe for their Timescale*

A. Pulcini[1,2,†], E. Vardaci[1,2], E. M. Kozulin[3], M. Ashaduzzaman[1,2], C. Borcea[4], A. Bracco[6], S. Brambilla[6], S. Calinescu[4], F. Camera[6], M. Ciemala[7], F. Davide[1,2], B. de Canditiis[1,2], A. Di Nitto[1,2], S. Dmitriev[3], O. Dorvaux[8], I. M. Harca[3,5], I. M. Itkis[3], V. V. Kirakosyan[3], G. Knyazheva[3], N. Kozulina[3], I. V. Kolesov[3], G. La Rana[1,2], A. Maj[7], I. Matea[9], K. Novikov[3], C. Petrone[4], D. Quero[1,2], P. K. Rath[1,2], E. Saveleva[3], C. Schmitt[10], G. Sposito[1,2], O. Stezowski[11], W. H. Trzaska[12] and J. Wilson[9]

[1]Dipartimento di Fisica "Ettore Pancini", Università degli Studi di Napoli Federico II, Napoli, 80126, Italy
[2]Istituto Nazionale Fisica Nucleare, Sezione di Napoli, Napoli, Italy
[3]Flerov Laboratory of Nuclear Reactions, Dubna, Russia
[4]Horia Hulubei National Institute of Physics and Nuclear Engineering, Bucharest - Magurele, Romania
[5]Faculty of Physics, University of Bucharest, Romania
[6]INFN and Università di Milano, Italy
[7]Institute of Nuclear Physics PAN, Krakow, Poland
[8]IPHC, CNRS/IN2P3, Univ. de Strasbourg, France
[9]IPN, CNRS/IN2P3, Orsay, France
[10]GANIL, CEA/DSMCNRS/IN2P3, Caen, France
[11]Univ. de Lyon, CNRS/IN2P3, IPNL, France
[12]Accelerator Laboratory of University of Jyvaskyla, Finland
†E-mail: pulcini@na.infn.it

The overlap in the mass symmetric region of the reaction products from fusion-fission and quasi-fission complicates the assignment of symmetric events to complete fusion on the basis of the mass distribution alone. Additional observables, besides mass distribution, should be used. The method proposed here relies on the fact that fusion-fission and quasifission are characterized by a different timescale. Within this framework we performed a detailed study to find out if timescales can be probed via angular momentum as measured via γ rays multiplicity. The proof of principle was carried out by measuring the γ rays in coincidence with two fragments in the reaction ^{32}S + ^{197}Au at beam energy near the Coulomb barrier. The experiment has been performed at the Tandem ALTO facility at IPN Orsay using a detection setup consisting of ORGAM and PARIS γ detector arrays coupled with the CORSET time-of-flight spectrometer. Preliminary results on the sensitivity of this method to distinguish reaction channels with different dynamics are discusses.

Keywords: Nuclear reactions; fusion-fission; quasi-fission; prompt gamma-rays.

* This work was supported by FLNR, JINR, INFN, the IN2P3-JINR Agreement No 14-90, the ENSAR2 project "N-SI-88", by the financial contribution of Romania to the Joint Institute of Nuclear Research, Dubna, Russia to the project no. 03-5-1094-2010/2016, by the Polish National Science Centre under Contract No. 2013/08/M/ST2/00257 and the French LEA COPIGAL project.

1. Introduction

Heavy-ion induced fusion reactions have been very successfully used for the production of superheavy elements (SHE) [1]. In this regard, the knowledge of fusion cross sections is essential to define the most promising reaction for the production of new elements and/or new isotopes. In fusion events a compound nucleus (CN) is produced that evolves toward fission or evaporation of light particles which brings to an evaporation residue. The fusion cross section can be measured by summing the cross sections of these two decay channels. In reactions involving medium mass nuclei the fission and evaporation cross sections can be of about the same magnitude because other reaction mechanisms are negligible [2,3]. This particular condition constitutes an advantage in the study of fission dynamics of medium mass nuclei [4].

In the case of the massive nuclei necessary for the search of superheavy elements, the evaporation residues cross section is negligible with respect to fission one. Therefore, it is sufficient to select and count the fission events to estimate the fusion cross section [5]. At energies around the Coulomb barrier, the quasifission (QF) reaction mechanism becomes dominant and counteracts the complete fusion [5]. Even if the compound nucleus is produced, the observed mass distribution of binary products in QF reactions can have a dominant component in the symmetric mass region where fusion-fission (FF) events are expected. Therefore, the estimate of fusion cross sections based on counting symmetric or nearly symmetric mass events is biased because QF and FF events are overlapped. Usually, in the asymmetric region of the mass distribution it is possible to disentangle, to some extent, the components of QF and asymmetric fission modes, whereas this is not possible in the symmetric region where the two components are overlapped. This overlap constitutes an inescapable problem when FF cross sections have to be estimated [6].

Within this work, we consider additional observables possibly capable to unambiguously separate the FF and QF products. These observables should reflect the differences between the slower dynamical evolution of the system passing through an equilibrium stage giving rise to the FF products, and the faster QF reactions where the large mass transfer and energy dissipation are strongly driven by shell effects [6]. A possible difference can be found in the angular momentum available during the decay stage, when the composite system separates in two parts. In FF, the orbital angular momentum is transferred into internal degrees of freedom of the compound nucleus, whereas only part of it is available for the QF products, being the system not thermalized at the separation stage. Hence, by measuring the γ rays in coincidence with the fission-fragments an indication on the angular momentum transfer can be obtained. Light charged particles can provide insights on dynamical evolution of the FF [4], however the emission probability is negligible in the reactions for SHE production. Therefore, the observables we considered here are the discrete γ rays and their multiplicity M_γ in the binary channel. It is worth to note that using discrete γ rays it is also possible to identify the fragments' charge.

In order to test this method, the reaction $^{32}S + ^{197}Au$, at the energy near the Coulomb barrier, E_{lab} = 166 MeV (E_{CM}/V_{Coul} = 1.01), was carried out at the Tandem ALTO accelerator at IPN Orsay, France. This reaction is characterized by a large FF cross section and a negligible contribution from QF [7]. At the same time this reaction has a significant contribution from quasi-elastic (QE) reactions. The FF and QE channels can be well separated by exploiting the large differences in the mass-TKE distributions. Therefore, this experiment represents an optimal benchmark because it is possible to measure within the same experiment the differences existing among the γ rays in coincidence with the binary fragments assigned to the QE and FF, i.e. from the fast and slow processes respectively. After a brief description of the experimental setup and validation of the methods adopted to extract the experimental observables the promising results are shown.

2. Experimental Setup

Binary reaction products were detected in coincidence by the two-arms TOF spectrometer CORSET [8]. Each arm consists of a micro channel plate (MCP) acting as start detector and a position-sensitive MCP acting as stop detector. The time resolution of each MCP is ~150 ps and the mass resolution achievable is ~3 u. The positions of CORSET arms were chosen to maximize symmetric fragments detection.

The prompt gamma rays were detected with ORGAM [9], an array of high-resolution Ge detectors, individually surrounded by BGO anti-Compton shields, and PARIS [10], an array of high-efficiency $LaBr_3$(Ce)-NaI(Tl) phoswichs. For this experiment 10 ORGAM detector units were used, all located at backward angles, and 10 PARIS units, 9 packed together plus a single unit, all located at forward angles. The photopeak efficiencies in the low energy range were ~1.5% and ~0.6% for ORGAM and PARIS, respectively. ORGAM has a better energy resolution (~4 keV at 1408 keV) than PARIS (~60 keV at 1332 keV), however PARIS provides additional information by spanning a larger dynamical range (up to 20 MeV), compared to ORGAM (up to 2.4 MeV).

3. Results

The ultimate goal of the experimental program is to find a signature of the reaction dynamics in order to separate FF and QF products using the gamma ray observables. In this study we widely profit of the measurements of mass and TKE of the coincident fragments.

The M-TKE matrix for binary events is reconstructed from CORSET coincidence data. The measurements of the times-of-flight and the flight paths of the fragments are used to reconstruct the event-by-event velocity vectors $\overrightarrow{v_1}$ and $\overrightarrow{v_2}$ of the two emerging fragments; then, from the two velocity vectors, by considering mass and momentum conservation laws, masses and energies of each couple of fragments are reconstructed. The M-TKE matrix for the ^{32}S + ^{197}Au reaction at E_{lab} = 166 MeV is shown in Fig. 1 (left).

The QF component is negligible in this reaction. By considering only the full momentum transfer (FMT) and the V_2 *vs.* V_1 events it is possible to isolate the FF reaction products. The FMT selection is obtained by requiring that the sum of velocity vector projections on the reaction plane is equal to the center of mass velocity (see ref. [11] for more details). Thus, the M-TKE distribution in Fig. 1 (right), obtained requiring the FMT and selection of FF events in the velocity plot, can be considered as originated in FF reactions. The two loci on the left and right of the FF region Fig. 1 (left) can be ascribed to the quasi-elastic (QE) reaction.

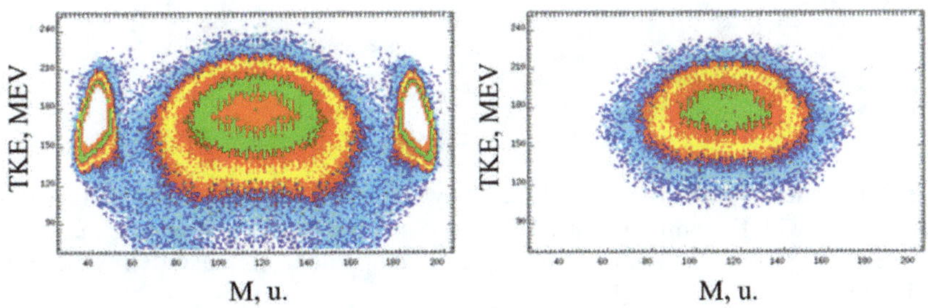

Fig. 1. Mass-TKE matrix of the binary events within the ^{32}S + ^{197}Au reaction at E_{lab} = 166 MeV without conditions (left) and applying the gate to select those with full momentum transfer (right).

After making this separation between events it is possible to extract the γ fold — the average number of γ rays detected per event — associated to each reaction mechanism, namely QF or FF. To extract folds, however, γ rays events have to be properly processed. For instance, for both PARIS and ORGAM, gates have been considered on time signals to isolate the prompt γ rays component of the energy spectra. The anti-coincidence condition between Ge and BGO detectors suppresses the Compton component of the ORGAM energy spectrum. Moreover, the fold extracted from PARIS detectors has been corrected for cross-talk among the crystals of the 3×3 cluster. The correction has been done using a simple algorithm: if two or more neighbouring detectors produce signals in a single event, their total contribution to the multiplicity is considered to be 1.

Figure 2 shows the comparison between PARIS and ORGAM fold distributions in coincidence with FF. To check the consistency between the results of the two setups, the fold distribution measured with PARIS was multiplied by the ratio between the measured efficiencies of the two detection arrays. The distribution values are in agreement, as shown in Fig. 2. This comparison not only provides the fold values but represents also a stringent test for the corrections applied to the different arrays in the current analysis.

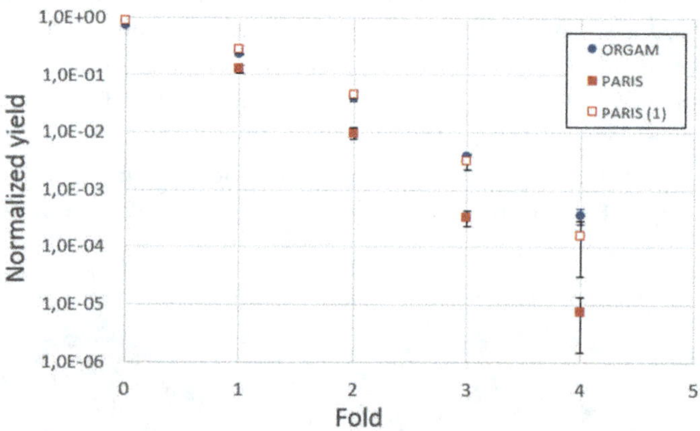

Fig. 2. Fold distributions for γ rays detected by ORGAM (circles) and PARIS (squares) in coincidence with fission fragments. The PARIS (1) distribution (empty squares) is equal to the PARIS fold distribution (full squares) multiplied for the ORGAM to PARIS efficiency ratio.

The fold distributions of γ rays detected with PARIS array in coincidence with FF and QE products were extracted. The QE and FF yields, normalized to the yields without γ rays in coincidence (fold = 0), are shown in Fig. 3 in logarithmic scale. The γ-fold distribution of QE events drops faster than the FF one. Furthermore, in FF events, up to 4 γ rays can be detected in coincidence, whereas in QE events the maximum is 1. Hence, these data show that the γ fold distribution represents a probe sensitive to the time scale of the processes, being less the orbital angular momenta transfer to the fragments in the faster reactions.

In order to have an observable independent from the detection setup, we reconstructed the multiplicity distributions from the measured fold by simulating the fission process and the related γ decay multiplicity distribution with the GEF

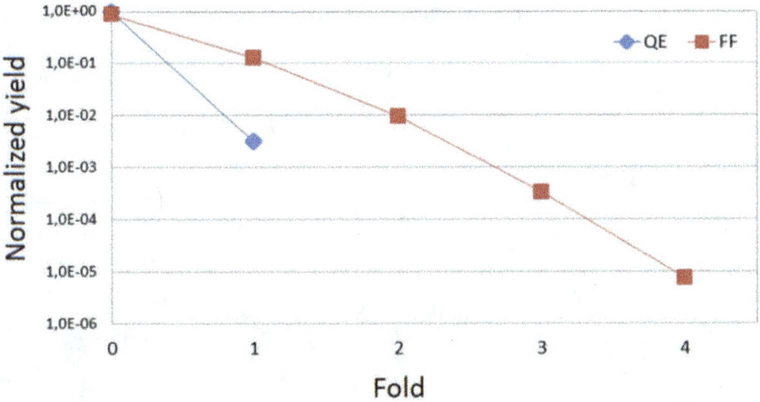

Fig. 3. Gamma fold distributions detected with PARIS in coincidence with QE products (full diamonds) and FF (full squares). Counts are normalized for comparison and are shown in logarithmic scale.

code [12]. The multiplicity distribution obtained by GEF was converted to the fold distribution. The input parameters in GEF have been fixed by considering the interaction potential. In the simulations the angular momentum of the compound nucleus (l_{rms}) is a free parameter. The conversion of the simulated multiplicity distribution to the fold distribution was made by writing a new code that takes into account the efficiency of the detection setup and its change with the γ ray energy. The results obtained with the new conversion code are in good agreement with the fold conversion as proposed in ref. [13] by Maj et al. Fig. 4 shows the multiplicity distributions obtained with GEF for $l_{rms} = 12.4, 17, 25$, and 40 ħ. The average multiplicity ⟨Mγ⟩ ranges from 12.8 to 21 by changing l_{rms} from 12.4 to 40 ħ. These simulated multiplicities have been converted to fold and directly compared to the data, see Fig. 5

The GEF simulation with l_{rms} 40 ħ provides the best agreement with the measured fold distribution, therefore we can assume that the Mγ distribution in the FF channel resembles the one shown in Fig. 4 with the green line.

Independently from the l_{rms} used and the conversion method used, we can conclude that the fold distribution assigned to the FF events reaches higher values than the one for QE events and that events with γ-fold >2 can be reasonably assigned to FF events. Even if further investigation is needed to confirm the validity of this method for the separation of FF and QF products, the results obtained in the present analysis support the conclusion that the γ fold observable seems to be adequate to probe the time scales of reaction processes.

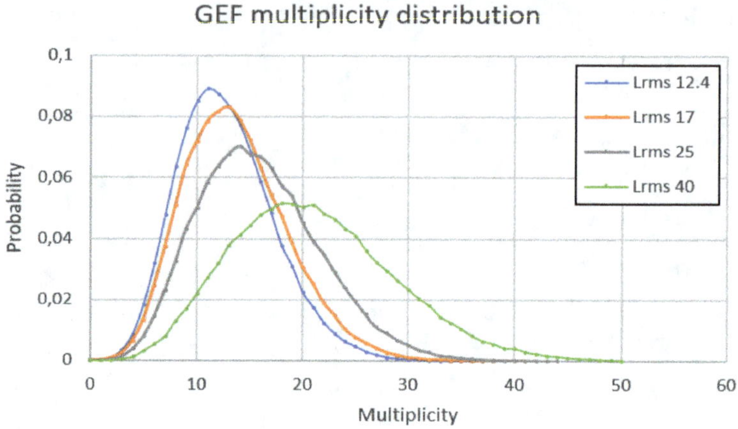

Fig. 4. γ rays multiplicity distributions in FF channel simulated by GEF for 4 different values of l_{rms}.

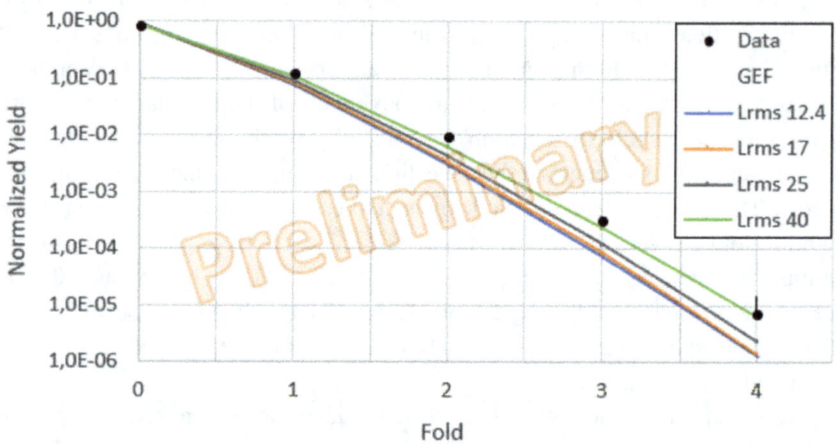

Fig. 5. γ fold distribution in coincidence with FF with PARIS detectors (black dots) compared to GEF simulations for 4 different values of l_{rms}. (lines).

References

1. Yu. Ts. Oganessian, A. Sobiczewski and G.M. Ter-Akopian, *Phys. Scr.* **92**, 023003 (2017) and references therein.
2. A. Di Nitto *et al.*, *Eur. Phys. J. A* **47**, 83 (2011).
3. E. Vardaci, *Eur. Phys. J. A* **43**, 127 (2010).
4. E. Vardaci, *Phys. Rev. C* **92**, 034610 (2015).

5. M. G. Itkis, E. Vardaci, I. M. Itkis, G. N. Knyazheva and E. M. Kozulin, *Nucl. Phys. A* **944**, 204 (2015) and references therein.
6. E. M. Kozulin *et al.*, *Phys. Rev. C* **89**, 014614 (2014).
7. B. B. Back *et al.*, *Phys. Rev. C* **32**, 195 (1985).
8. E. M. Kozulin *et al.*, *Instr. and Exp. Tec.* **51** (2008).
9. P. J. Nolan, F. A. Beck, D. B. Fossan, *Annu. Rev. Nucl. Part. Sci.* **45**, 561 (1994).
10. A. Maj *et al.*, *Acta Phys. Pol. B* **40**, 565 (2009).
11. E. M. Kozulin *et al.*, *Phys. Rev. C* **86**, 044611 (2012).
12. K.-H. Schmidt, B. Jurado, C. Amouroux, and C. Schmitt, *Nucl. Data Sheets* **131**, 107 (2016).
13. A. Maj *et al.*, *Nucl. Phys. A* **571**, X35-220 (1994).

Manifestations of Pear-Shaped Clusters in Collinear Cluster Tri-Partition of ^{252}Cf

Yu. V. Pyatkov
National Nuclear Research University MEPhI (Moscow Engineering Physics Institute), 115409 Moscow, Russia, Joint Institute for Nuclear Research, 141980 Dubna, Russia
E-mail: yvp_nov@mail.ru

D. V. Kamanin, A. A. Alexandrov, I. A. Alexandrova, Z. I. Goryainova, E. A. Kuznetsova, A. O. Strekalovsky, O. V. Strekalovsky and V. E. Zhuchko
Joint Institute for Nuclear Research, 141980 Dubna, Russia

V. Malaza
University of Stellenbosch, Faculty of Military Science, Military Academy, Saldanha 7395, South Africa

Specific mode of the collinear cluster tri-partition of the ^{252}Cf nucleus is discussed. The structure manifests itself as a rhombic-spiral structure in the fission fragments correlation mass distribution.

Keywords: Ternary fission; shell corrections; nuclear clusters.

1. Introduction

Most of our experiments in recent years were dedicated to the study of new multibody at least ternary decay channel of low excited heavy nuclei [1, 2]. Due to specific features of the effect, it was called by us collinear cluster tri-partition (CCT). Now we have an entire collection of different CCT manifestations, observed through the linear structures in the mass correlation distributions of the decay products. Some of the structures look like the straight lines satisfying the condition $M_1 + M_2 =$ const, where M_1 and M_2 are the fission fragments (FFs) detected in coincidence while their total mass is less than the mass of the fissioning nucleus. The mass of the missed fragment is obviously constant for the fission events forming each such line. As far as we know, similar structures had been never discussed in the past.

2. Experiments and Results

The experiment Ex1 was carried out at the COMETA-F spectrometer in the FLNR (JINR, Dubna, Russia). The layout of the spectrometer is shown in Figure 1.

It is double-armed time-of-flight FFs spectrometer consisting of four mosaics of PIN diodes, and two timing detectors. Signals from all the detectors were digitized in 0.2 ns increments by the fast flash-ADC (Amplitude to Digital Convertor) CAEN DT5742. For the proper reconstruction of the FF mass, if a PIN diode is used for the measurement of both FF energy and velocity one should take into account so called "pulse height defect" and "plasma delay" [3, 4]. This is exactly what is done in the data processing approach proposed by us in [5, 6]. In the latter publication new "parabola time-pickoff" method used in Ex1 is described.

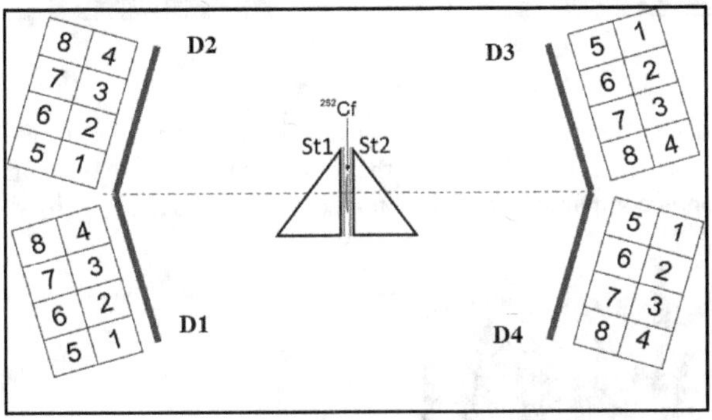

Fig. 1. Layout of the COMETA-F setup. A spectrometric source of ^{252}Cf (sf) is located in between two microchannel plates based timing detectors St1 and St2 giving "start" signals. Four mosaics (*D*1–*D*4) of eight PIN diodes each provide the signals for measurement of both the FF time-of-flight and energy. The FF mean flight pass does not exceed fifteen centimeters.

FFs mass correlation distributions in the region of the "Ni-bump" obtained in Ex1 and Ex2 [1] are presented in Figure 2. Due to the background conditions of the experiment Ex1, the events with the energy of the light fragment in the range $E_2 = (6 \div 30)$ MeV were selected (Figure 2a).

The data from Ex1 along with the presence of the lines at the mass numbers M_2= (128, 68, 72) u marked in the Figure by numbers 1, 2, 3 correspondingly, show some additional structures. A rhombic-spiral structure in the lower right corner of Figure 2(a) that resembles a rose depiction we called "nuclear rose". It consists of the family of lines $M_1 + M_2$ = const and several lines almost perpendicular to them. The lines of the first sort (marked by the numbers 1–6 in Figure 3(a)) will be analyzed in the next section while a discussion of the latter is below the scope of this paper.

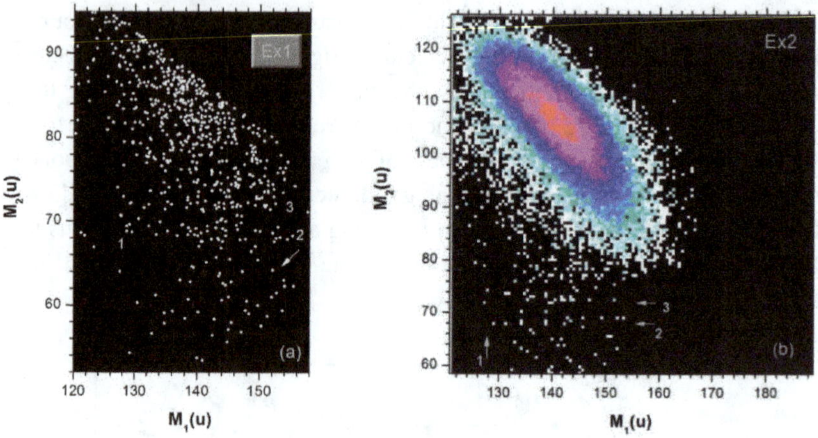

Fig. 2. FFs mass correlation distributions from $^{252}Cf(sf)$ obtained in Ex1 – (a). Specific rhombic-spiral structure in the bottom of the figure (“nuclear rose”) is marked by an arrow. The distribution is compared with that obtained earlier in Ex2 [1] – (b).

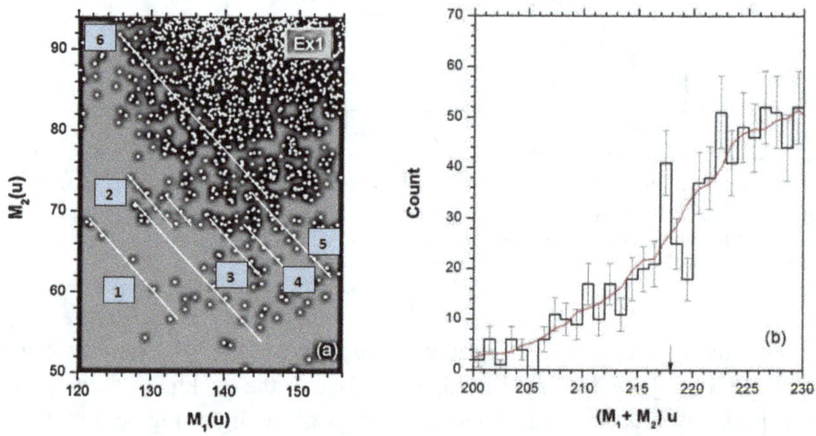

Fig. 3. Mass-correlation distribution from Ex1 – (a). The tilted numbered lines in white meet the condition $M_1 + M_2 = \text{const}$. Projection of the distribution along such direction in the vicinity of line 6 – (b). The peak centered at 218u marked by the arrow exceeds by more than two sigma the level obtained by smoothing the original spectrum.

Two structures from Figure 3(a) are presented in Figure 4 in the larger scale. Expected prescission cluster composition of the decaying system for the events marked by arrows are shown in insets. Each list of the clusters begins with the “missed” nucleus. Its mass corresponds to the missing mass in the expression $M_1 + M_2 = \text{const}$.

Different projections of the mass correlation distribution from Ex1 are shown in Figure 5. As can be inferred from the Figure 5a, the statistics in Ex1 is approximately three times more than that in Ex2. A total yield of two Ni peaks in Ex1 does not exceed 10^{-4} per binary fission which agrees with our previously obtained value. The data of Ex2 indicated that the heavy clusters in the ternary prescission configurations are predominantly magic nuclei (Table 1 in [7]). Noticeably large statistics in Ex1 allowed us to confirm previous observation. The projection of the distribution shown in Figure 5(b) onto M_1 axis for the range of M_2 = (65–76) u clearly demonstrates increased yield of the heavy fragments corresponding to the magic isotopes of ^{128}Sn, ^{134}Te, ^{140}Xe, ^{144}Ba, ^{150}Ce, ^{154}Nd (their masses are marked in Figure 5(b) by arrows).

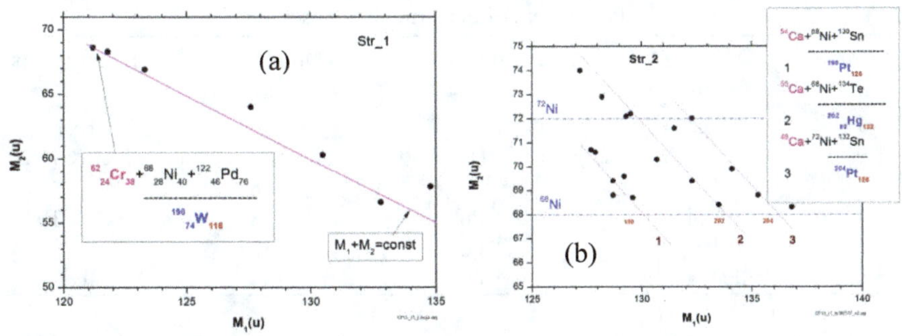

Fig. 4. Two structures from the Figure 3(a) in the larger scale. Cluster compositions for the events marked by arrows are shown in insets.

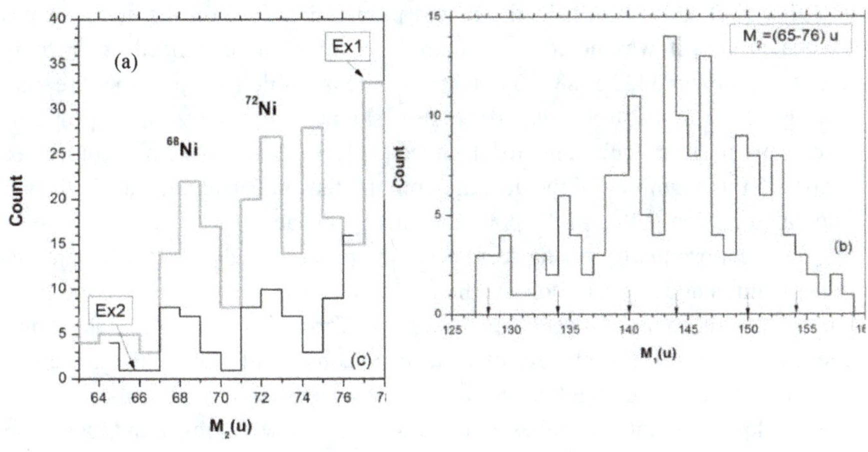

Fig. 5. Projections of the mass-correlation distributions obtained in Ex1 and Ex2 onto M_2 axis — (a). Projection of the mass-correlation distribution from Ex1 onto M_1 axis under condition M_2 = (65–76) u — (b).

3. Discussion

For the first time we had observed linear structures meeting the condition $M_1 + M_2 = \text{const}$ in our experiment (Ex3) at the FOBOS setup based on the gas-filled detectors [8]. The linear ridges corresponding to the constant missing masses were revealed as a fine structure in the mass correlation plot collected with good statistics. Later, using COMETA spectrometer based on the mosaics of PIN diodes, we also observed the structures under discussion (Ex2), but the data suffered from small statistics [1]. Parameters of all linear structures observed so far in the region of so called "Ni-bump" are presented in Table 1.

Table 1. Parameters of the linear structures shown in Figure 3a.

Parameters of the structures observed				
Str №	Missing fragment	Heavy magic core	Number of neutrons	Experiments
1	^{62}Cr	^{190}W	116	Ex1
2	$^{48, 50, 54}$Ca	^{198}Pt, ^{202}Hg, ^{204}Pt	120, 122, 124	Ex1 & Ex2
3	^{44}S	^{208}Pb	126	
4	^{40}Si	^{212}Po	128	Ex1 & Ex3
5	^{36}S	^{216}Po	132	
6	^{34}Si (N = 20)	^{218}Po	134	Ex1

As can be inferred from Figure 6, showing the shell corrections map [9], the numbers of neutrons in the heavy magic cores (the forth column in Table 1) manifest themselves through the discussed structures correspond to the valley of negative shell corrections. In the recent publication [10] on the study of pear-shaped nuclei, it was noted that octupole correlations enhanced at the magic neutron numbers 34, 56, 88, 134. Thus, it is reasonable to expect that the nuclei having 116–134 neutrons would be pear-shaped and show magic properties. According to the calculations, for instance [11], fissioning system becomes pear-shaped at the beginning of the descent from the fission barrier, at least in the most populated fission valley of the potential energy surface.

We propose the following scenario of the process leading to the formation of the structures under discussion. At the initial stage of the descent from the fission barrier, the fissioning nucleus undergoes transformation from a prolate to more complicated shape. The nucleus of such shape consists of a heavy magic octupole deformed core and the light cluster in contact with the thin edge of the "pear". At further elongation the core takes dumbbell-like shape with longer and longer neck between the heavy and light parts of the dumbbell. The mass of the light cluster stays unchanged. After the first rupture in the neck, a heavy fragment and a

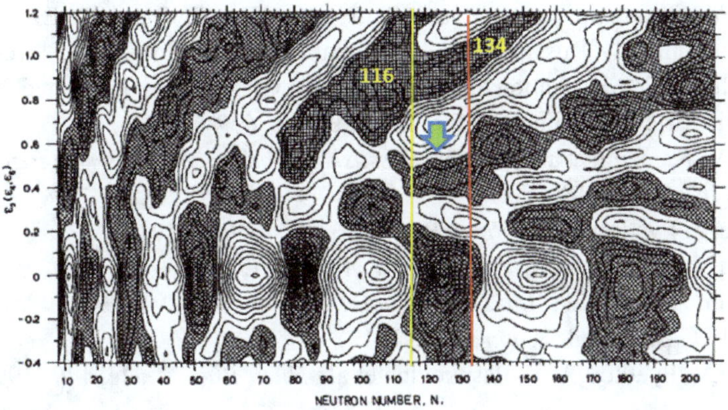

Fig. 6. Shell energy diagram [10] depending on neutron number and deformation. Areas corresponding to negative shell energy are shaded, and the contour separation is 1 MeV.

di-nuclear system consisting of the light cluster and some part of the neck fly apart. Later, due to the second rupture, the light cluster and the light FF become free. In general, the kinematics of the process seems to be similar to that discussed in [7].

4. Conclusions

Based on the new data we clarify a mechanism of the CCT mode similar to heavy ion radioactivity [8]: octupole deformed magic core plays the same role as magic Pb cluster in the "Lead radioactivity".

Acknowledgments

This work was supported by the research project No. 18-32-0053 of the Russian Foundation for Basic Research (RFBR), and by the MEPhI Academic Excellence Project (Contract No. 02.a03.21.0005, 27.08.2013) of the Russian Science Foundation.

References

1. Yu.V. Pyatkov et al., *Eur. Phys. J. A* **48**, 94 (2012).
2. D.V. Kamanin, Yu.V. Pyatkov, "Clusters in Nuclei – Vol. 3" ed. by C. Beck, *Lecture Notes in Physics* **875**, 183 (2013).
3. S.I. Mulgin et al., *Nucl. Instrum. Meth.* **388**, 254 (1997).
4. H.O. Neidel and H. Henschel, *Nucl . Instr. and Meth.* **178**, 137 (1980).

5. D.V. Kamanin et al., Int. Symp. on Exotic Nuclei "EXON-2014", Kalaningrad, Russia, 08-13 September 2014. Conference proceedings, Editors: Yu.E. Penionzhkevich, and Yu.G. Sobolev. Published by World Scientific Publishing Co. Pte. Ltd., 2015. p. 677-684
6. D.V. Kamanin et al., *Izvestiya Rossiiskoi Akademii Nauk. Seriya Fizicheskaya*, **82** (**6**), 804 (2018).
7. Yu.V. Pyatkov et al., *Phys. Rev. C* **96** (**6**), 064606 (2017).
8. Yu.V. Pyatkov et al., *Eur. Phys. J. A* **45**, 29 (2010).
9. Sven Aberg et al., *Annu. Rev. Nucl. Part. Sci.* **40**, 439 (1990).
10. L.P. Gaffne et al., *Nature* **497** (2013).
11. Yu.V. Pyatkov, V.V. Pashkevich et al., *Nucl. Phys. A* **624**, 140 (1997).

Pygmy Dipole Resonance of ^{134}Sn

A. P. Severyukhin and N. N. Arsenyev
Bogoliubov Laboratory of Theoretical Physics,
Joint Institute for Nuclear Research,
141980 Dubna, Moscow Region, Russia

Yu. E. Penionzhkevich and D. Testov
Flerov Laboratory of Nuclear Reactions,
Joint Institute for Nuclear Research,
141980 Dubna, Moscow Region, Russia

F. Ibrahim and D. Verney
Institut de Physique Nucleaire, CNRS-IN2P3
and Universite Paris-Sud, 91405 Orsay, France

Starting from a Skyrme interaction with tensor terms, the low-energy 1^- spectrum populated in ^{134}Sn from the ^{134}In β-decay has been studied within a microscopic model including the two-phonon configuration effects. We assume an amplification of the intensity distribution by the integral Fermi factor and the most important contributions come from the allowed Gamow-Teller transitions.

Keywords: QRPA; Pygmy dipole resonance; Tin.

1. Introduction

Recently, with the advent of advanced radioactive beam facilities and novel experimental techniques, unexplored regions of exotic nuclei and new phenomena became accessible for detailed spectroscopic studies. The concentration of the $E1$ strength around the particle separation energy is commonly called the pygmy dipole resonance (PDR) because of its weak strength in comparison with the giant dipole resonance, which dominates the $E1$ strength distribution in nuclei [1]. There is a special structure of the PDR which appears as a new collective motion in neutron-rich nuclei [2]. The occurrence of non-negligible low-lying $E1$-strength can influence the radiative neutron capture cross section by orders of magnitude and, consequently, also the rate of the astrophysical r-process nucleosynthesis [3]. It was also shown that in the neutron-rich nuclei, the $E1$ strength could be significantly more important than in nuclei close to the valley of β-stability. The ability to correctly reproduce the A = 130 solar-system isotopic r-process abundance peak, which acts as the main bottleneck in the

r-process matter ow, has served as a key test of nucleosynthesis models designed to calculate the total r-abundance pattern. The fact is that for nuclei near $A = 130$, the $E1$ strength was measured up to date only in stable nuclei with small or medium neutron excess. In particular, the $E1$ strength distribution of ^{130,132}Sn have been studied extensively [4], but little is known about low-energy 1^- spectrum of ^{134}Sn. The probabilities of β-delayed neutron emission help to identify $E1$ strength distribution of ^{134}Sn and it can be extracted using ^{3}He long counter TETRA [5] for β-decays studies at ISOL facility ALTO (JINR-IN2P3 collaboration). For a more firm interpretation of the experimental results, we have studied the $E1$ strength distribution of ^{134}Sn and the β-feeding intensities of ^{134}In calculated within a microscopic model including the two-phonon configuration effects.

2. The Method

One of the successful tools for studying the $E1$ strength distribution is the quasiparticle random phase approximation (QRPA) with the self-consistent mean-field derived from a Skyrme energy-density functional (EDF) [2, 6]. Also, many charge-exchange versions of it were developed during the last decade, see e.g., Refs. [7, 8]. Their common feature is that they allow to relate the properties of the ground states and excited states through the same EDF. Taking into account the basic ideas of the quasiparticle-phonon model [9], the Hamiltonian is then diagonalized in a space spanned by states composed of one and two QRPA phonons. The complexity of calculations beyond standard QRPA increases rapidly with the size of the configurational space, so one has to work within limited spaces. It is well known that, if the QRPA matrix elements take a separable form the QRPA energies can be obtained as the roots of a relatively simple secular equation. Making use of the finite rank separable approximation (FRSA) [10, 11] for the residual interaction, the approach has been generalized for the effect of phonon- phonon coupling (PPC) [12]. Recently, the FRSA in the cases of the charge- exchange excitations and the β-decay was introduced in Refs. [13, 14] and in Refs. [15, 16], respectively. In the case of the β-decay of ^{134}In, we use the EDF T43 which takes into account the tensor force added with refitting the parameters of the central interaction [17]. The set T43 gives a reasonable description of properties of the Gamow-Teller (GT) and charge-exchange spin-dipole resonances [18] and the β-decay half-life of the neutron-rich doubly-magic nucleus ^{132}Sn [16]. The pairing correlations are generated by a surface-peaked zero-range force with a strength of -870 MeV·fm3 and a smooth cut-off at 10 MeV above the Fermi energies [11]. This value of the pairing strength is fitted to reproduce the experimental neutron pairing gaps of ^{126,128}Cd and ^{130,134}Sn obtained by the three-point formula [16].

3. Results of Calculations

In an appropriate β-decay model, the correct amount of the integral β-strength should be placed within the properly calculated Q_β window provided that the spectral distribution is also close to the "true" β-strength function.

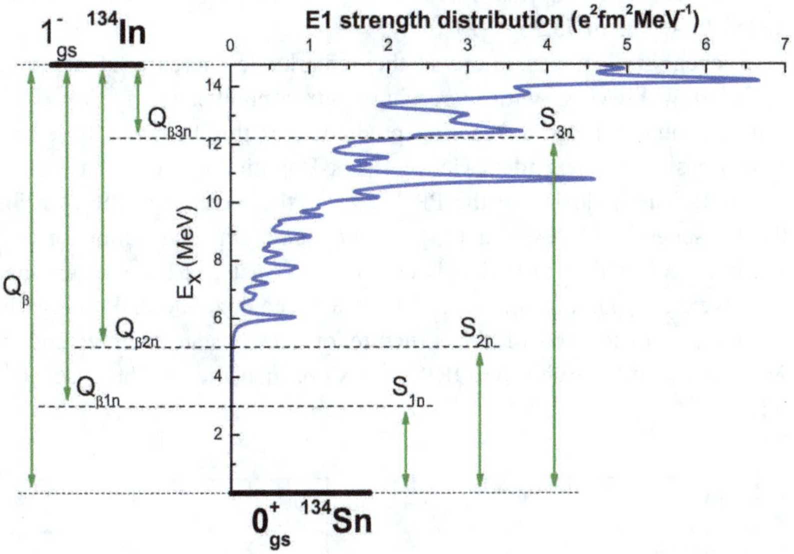

Fig. 1. The strength distribution of $E1$ transitions from the ground state to 1⁻ states of ^{134}Sn obtained with the T43 EDF. $Q_{\beta xn}$ and S_{xn} (with $x = 1, 2, 3...$) energies are calculated with the same EDF.

The correct description of the Q_β value for the parent nucleus ^{134}In and the neutron separation energies (S_{xn}) for the daughter nucleus ^{134}Sn is the important ingredient for the reliable prediction of the probabilities of multi-neutron emissions (P_{xn}). The calculated Q_β, S_n, S_{2n} and S_{3n} values are shown in Fig. 1. Our results are in reasonable agreement with available evaluated energies [19].

The multi-neutron emission is basically a multistep process consisting of (a) the β-decay of the parent nucleus ^{134}In which results in feeding the excited states of the daughter nucleus ^{134}Sn followed by the (b) γ-deexcitation to the ground state or (c) multi-neutron emissions to the ground state of the final nucleus $^{134-x}$Sn. The difference in the characteristic time scales of the β-decay and subsequent neutron emission processes justifies an assumption of their statistical independence. For P_{1n} we have $Q_{\beta 2n} \leq EGT \leq Q_{\beta n}$, while for P_{xn} this becomes $Q_{\beta x+1n} \leq EGT \leq Q_{\beta xn}$, where EGT is the GT transition energy, $Q_{\beta xn} \equiv Q\beta - S_{xn}$. We assume an amplification of the intensity distribution by the integral Fermi factor and the most important contributions come from the allowed GT transitions. We find that the main contribution to the total probability of the multi-neutron emission comes from the $\beta 2n$-decay. The inclusion of the PPC effects plays the key role in our calculation. P_{2n}-value of 46.8% was predicted within the relativistic QRPA calculation [20]. The DF3+cQRPA calculation predicted P_{2n}-value of 53.35% [21].

For even Sn isotopes, there is the remarkable increase of summed $E1$ strength below 11 MeV, with increasing neutron number from ^{132}Sn till ^{134}Sn. It is worth mentioning that the first prediction of this behavior based on the canonical basis time-dependent Hartree-Fock-Bogoliubov theory has been done in Ref. [22]. The inclusion of the PPC leads to the increase of the contribution of the low-energy $E1$ strength [23]. The $E1$ strength distribution of ^{134}Sn is shown in Fig. 1. The distribution is computed by using a Lorentzian smearing with an averaging parameter Δ = 0:2 MeV. We recognize the PDR energy below 11 MeV. The difference of the structure of the 1^- states below and above 11 MeV is illustrated with the QRPA transition densities in the case of ^{134}Sn, see Fig. 2.

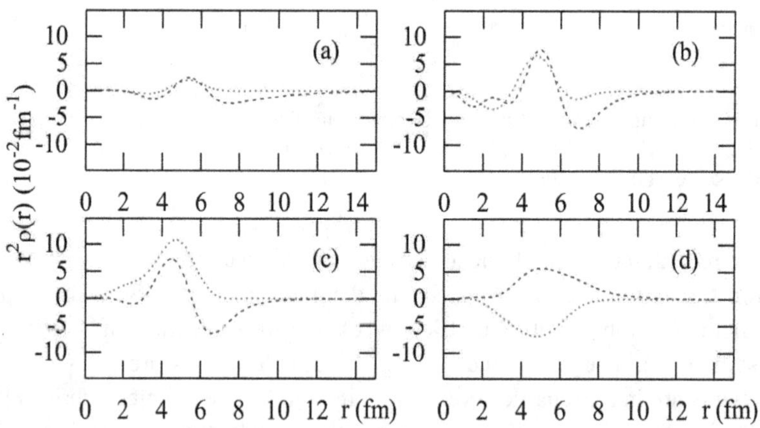

Fig. 2. Transition proton (dotted line) and neutron (dashed line) densities to selected QRPA 1^- states of ^{134}Sn. Panels (a), (b), (c) and (d) correspond to the transition densities for the states at Ex =7.8 MeV, 9.2 MeV, 11.3 MeV and 14.4 MeV, respectively. All transition densities are multiplied by $r2$.

For the 1^- states below 11 MeV, the proton and neutron densities in the nuclear interior region are in phase. However one can see the neutron dominance when $r > 7$ fm (see Fig. 2(a) and (b)). For the 1^- states about 11 MeV we found that states above this energy have an intermediate behavior. In particular, the next fairly collective 1^- state at 11.3 MeV has neither the isoscalar character nor the isovector character. This is not the PDR where the neutron skin oscillates against the inner core. Increasing further the excitation energy we observe the low-energy tail and directly of the giant dipole resonance (see Fig. 2(d)). Thus, the PDR of ^{134}Sn is located within the $\beta 2n$-decay window.

4. Conclusions

By means of the QRPA calculations and considering the coupling between the phonons, we have studied the GT strength distribution of the ^{134}In β-decay and the $E1$ strength distribution of ^{134}Sn. Using the Skyrme interaction T43 in conjunction with the surface-peaked pairing interaction, the β-feeding intensities and the low-energy 1^- spectrum are calculated. The main contribution of the PDR is concentrated in the $\beta 2n$-decay window. We obtain also the strong two-neutron emission that was found within the QRPA with Fayans functional and relativistic lagrangians before.

The low production yields of such species in present and future radioactive ion-beam facilities makes it difficult to investigate the PDR in neutron-rich nuclei via the standard charge-exchange or Coulomb-excitation reactions. For ^{134}In, it would be desirable to experimentally study the β- feeding intensities. The probabilities of β-delayed neutron emission can be measured using ^{3}He long counter TETRA for β-decays studies at ISOL facility ALTO (JINR-IN2P3 collaboration).

Acknowledgments

We thank I. N. Borzov and Nguyen Van Giai for useful discussions. This work is partly supported by the CNRS-RFBR agreement No. 16-52-150003, the IN2P3-JINR agreement, and the Russian Science Foundation (Grant No. RSF-17-12-01170).

References

1. D. Savran, T. Aumann, A. Zilges, *Prog. Part. Nucl. Phys.* **70**, 210 (2013).
2. N. Paar, D. Vretenar, E. Khan, G. Colo, *Rep. Prog. Phys.* **70**, 691 (2007).
3. M. Arnould, S. Goriely, K. Takahashi, *Phys. Rep.* **450**, 97 (2007).
4. P. Adrich et al. (*LAND-FRS Collab.*), *Phys. Rev. Lett.* **95**, 132501 (2005).

5. D. Testov et al., *Nucl. Instrum. Methods A* **815**, 96 (2016).
6. J. Terasaki et al., *Phys. Rev. C* **71**, 034310 (2005).
7. M. Bender, J. Dobaczewski, J. Engel, W. Nazarewicz, *Phys. Rev. C* **65**, 054322 (2002).
8. S. Fracasso, G. Colo, *Phys. Rev. C* **76**, 044307 (2007).
9. V. G. Soloviev, *Theory of Atomic Nuclei: Quasiparticles and Phonons* (Institute of Physics, Bristol and Philadelphia, 1992).
10. Nguyen Van Giai, Ch. Stoyanov, V. V. Voronov, *Phys. Rev. C* **57**, 1204 (1998).
11. A. P. Severyukhin, V. V. Voronov, Nguyen Van Giai, *Phys. Rev. C* **77**, 024322 (2008).
12. A. P. Severyukhin, V. V. Voronov, Nguyen Van Giai, *Eur. Phys. J. A* **22**, 397 (2004).
13. A. P. Severyukhin, V. V. Voronov, Nguyen Van Giai, *Prog. Theor. Phys.* **128**, 489 (2012).
14. A. P. Severyukhin, H. Sagawa, *Prog. Theor. Exp. Phys.* **2013**, 103D03 (2013).
15. A. P. Severyukhin et al., *Phys. Rev. C* **90**, 044320 (2014).
16. A. P. Severyukhin et al., *Phys. Rev. C* **95**, 034314 (2017).
17. T. Lesinski et al., *Phys. Rev. C* **76**, 014312 (2007).
18. C. L. Bai et al., *Phys. Rev. C* **83**, 054316 (2011).
19. M. Wang et al., *Chin. Phys. C* **41**, 030003 (2017).
20. T. Marketin, L. Huther, G. Martínez-Pinedo, *Phys. Rev. C* **93**, 025805 (2016).
21. I. N. Borzov, *Phys. of Atom. Nucl.* **79**, 910 (2016).
22. S. Ebata, T. Nakatsukasa, T. Inakura, *Phys. Rev. C.* **90**, 024303 (2014).
23. N. N. Arsenyev et al., *Phys. Part. Nucl.* **48**, 876 (2017).

Clustering and Emergent Collectivity in Multidimensional Potential Energy Surface Calculations*

A. V. Unzhakova[1†] and O. N. Granichin[1,2]
[1]Saint Petersburg State University,
[2]IPME RAS
Saint Petersburg, Russia
†E-mail: a.unzhakova@spbu.ru

Over the last three decades both the quantity and quality of the experimental data on the properties of nuclear-decay modes have increased tremendously. Theoretical models must be able to treat the multiple nuclear shapes seen in new decay modes with changing intrinsic structures of a primordial clusters formed from the nucleons present in a fissile nucleus. Clustering and nuclear collective states play important role in determining nuclear potential energy surfaces (PES's), fission barriers and landscape. In this work PES's are calculated by the macroscopic-microscopic method using realistic nuclear potential with constant thickness of the surface layer for strongly deformed nuclear shapes in high-dimensional deformation space. Complex multimodal structure of the heavy nuclei fission potential energy surfaces represents effect of a large-amplitude collective motion of the nucleons and their clustering.

Keywords: Fission; potential energy surfaces; clustering.

1. Introduction

Recent studies of nuclear fission are revealing an unexpected richness and diversity of nuclear shapes, structures, and decay modes, which are challenging our understanding of the fission process [1,2].

Theoretical models must be able to trace over long sequences of fission process stages with nuclear system changing its structure and to treat the multiple nuclear shapes seen in the new decay modes [3,4]. Generally, dynamics of the fission process is described as the evolution of collective degrees of freedom of the system such as the main fission elongation value and the mass-asymmetry value. PES depends on the collective degrees of freedom, so the choice of the relevant coordinates and necessary constraints is a fundamental challenge.

2. Clustering in Fission

A large number of coordinates involved into a theoretical description of nuclear collective states and clustering effect leads to complexity of calculations and to a

* This work was supported by RSF (project 16-19-00057).

unique opportunities to reproduce fine structures of the PES landscape at the same time. Multidimensional PES topology around the saddle points leading to the possible ternary and quaternary fission modes remain an open question.

For the first time multidimensional calculations of PES's were presented by Vitaly Pashkevich in [5]. New features of the PES structure emerging in seven-dimensional deformation space have been compared in this work to the three-dimensional PES. The distinction between few-dimensional and high-dimensional PES topologies have been explained using the catastrophe theory. Possibility of clustering and emergent collectivity influenced by nuclear deformation was shown in [6,7].

3. Results of Calculations

In our work, multi-dimensional potential energy surfaces for heavy nuclei have been calculated by the Strutinsky shell correction method. The Woods-Saxon type potential with the constant thickness of the surface layer was used and the liquid-drop model was employed for calculating the macroscopic energy [7]. In our theoretical approach the theoretical description of the transition from the one nuclear shell structure into two or even several cluster shell structures presenting

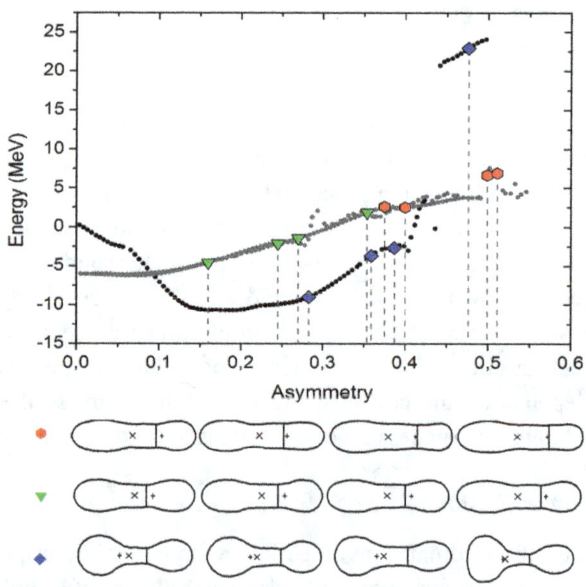

Fig. 1. Potential energy surface minima for ^{236}U with clustered configurations calculated using asymmetry as a constraint. The energy is minimized with respect to the ten deformation parameters.

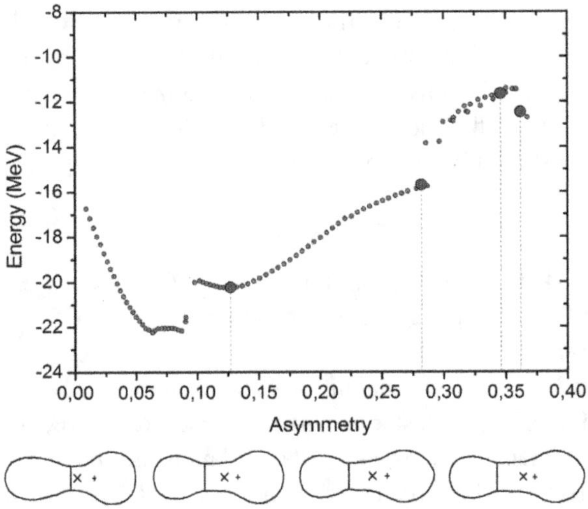

Fig. 2. Potential energy surface minima for ^{252}Cf with different asymmetrical configurations calculated using asymmetry as a constrain. The energy is minimized with respect to the ten deformation parameters.

in the fissile system at large deformations is calculated as a chain of the discrete equilibrium states with unlimited flexibility of the nuclear surface.

Arbitrary nature of the nuclear shape variation and stochastic generation of the abundance of the high-dimensional system shapes, both are essential for the eventual cluster structures emergence. Changing curvature of the nuclear system surface plays a role of a load common for nonequilibrium cluster formation theories [8]. Exogenous nuclear shape distortions applied in the energy minima searching procedure with asymmetry constraint allow us to probe cluster configurations lying higher in energy than conventional fission valley bottoms.

The equilibrium points of the high-dimensional minimization procedure obtained in asymmetry-constrained calculations are presented in Fig. 1 for ^{236}U and in Fig. 2 for ^{252}Cf. The nuclear shapes calculated with ten deformation parameters for these points of PES demonstrate various possible clustering configurations at high nuclear system elongation in close proximity to the scission point.

4. Conclusions

In fission process nuclear system transforms into two, in rare cases, into three or four nuclear subsystems with near magic shell structures. It could be considered as a process of recursive emerging collectivity in terms of the complex system consisting of many identical particles. Strong influence of clustering on the multimodal PES structure formation has been demonstrated by the use of the state space of high-dimensionality. Fundamental importance of the constant thickness of the surface layer of the nuclear mean field for the cluster formation calculations was also demonstrated in the time-dependent framework [9].

References

1. G.M. Ter-Akopian, J.H. Hamilton, Yu.Ts. Oganessian *et al.*, Yields of correlated fragment pairs in spontaneous fission of ^{252}Cf , *Phys. Rev. C* **55**, 1146 (1997).
2. Yu.V. Pyatkov, D.V. Kamanin, A.A. Alexandrov *et al.*, Examination of evidence for collinear cluster tri-partition, *Phys. Rev. C* **96**, 064606 (2017).
3. T.M. Shneidman, G.G. Adamian, N.V. Antonenko *et al.*, Manifestation of cluster effects in the structure of actinides, *EPJ Web of Conferences* **107**, 03009 (2016).
4. H. Paşca, A.V. Andreev, G.G. Adamian *et al.*, Possible origin of transition from symmetric to asymmetric fission, *Phys. Lett. B* **760**, 800 (2016).
5. V.V. Pashkevich, Shell effects in collective degrees of freedom, *Low Energy Nuclear Dynamics, EPS XV Nuclear Physics Divisional Conference*, (St. Petersburg, Russia, 1995) p. 177.
6. R.V. Jolos,. Yu.V. Palchikov, V.V. Pashkevich, A.V. Unzhakova, Reflection asymmetric deformation and clustering in heavy nuclei, *Nuovo Cimento A* **110**, 941 (1997).
7. V.V. Pashkevich, Yu.V. Pyatkov, A.V. Unzhakova, Structure of Fission Potential Energy Surfaces in Ten-dimensional Spaces, in *Proc. of Fission and Properties of Neutron-Rich Nuclei (ICFN4) 2007*, eds. J. Hamilton, A. Ramayya, H. Carter (World Scientific, 2008) p. 218.
8. A.V. Unzhakova, T.A. Khantuleva, O.N. Granichin, Cluster degrees of freedom in fission of actinides, in *Proc. of Fission and Properties of Neutron-Rich Nuclei (ICFN6) 2016*, eds. J. Hamilton, A. Ramayya, P. Talou (World Scientific, 2017) p. 582.
9. B. Schuetrumpf and W. Nazarewicz, Cluster formation in precompound nuclei in the time-dependent framework, *Phys. Rev. C* **96**, 064608 (2017).

Study of the Beyond-Neutron-Threshold Structure in the N=50 Region from β-Delayed Neutron and γ Spectroscopy at ALTO

D. Verney and the BEDO/TETRA* Collaboration

Institut de Physique Nucléaire, IN2P3-CNRS/Université Paris Saclay Orsay, FR-91406, France

We have recently launched an experimental program at ALTO to investigate the structure of the neutron-threshold region. This endeavor was further encouraged by the unexpected observation of "ultra"-high-energy γ-rays (8-9 MeV) in the β-delayed emission products of ^{83}Ga (Z=31; N=52; $T_{1/2}$=312 ms; Q_β=11.7 MeV) sources collected at the BEDO station. These results clearly point towards the existence of structures in the threshold region of the daughter-nucleus excitation spectrum, governing the decay properties in the ^{78}Ni region.

Keywords: β-strength; Pygmy modes; nuclear Pandemonium.

1. Physics Context: β-Decay, Collective Structures and Pandemonium

The description of β decay as a weak interaction process has reached such a high level of precision that it has become a powerful tool in the search for evidence of physics beyond the Standard Model [1]. But we are still far from a satisfactory quantitative description of this process at the global nuclear level. This is essentially due to the fact that in the highly-correlated, many-body nuclear system the transformation of a neutron into a proton (or vice versa) described by the Fermi and Gamow-Teller (GT) operators gives rise to collective states involving large nucleon sub-systems: GT resonances (GTR) accompanied by low-energy satellite components possibly forming "Pygmy" GTR. These low-lying components play a crucial role because they are located at low energy in the daughter nuclei excitation spectra and their effects in the β-strength distribution are magnified by the Fermi phase function. The problem is that, precisely due to the fact that they represent only a small fraction of the GTR strength, their microscopic description may strongly depend on the details of the correlation treatment and density functional used and remain a challenging theoretical task.

The possible existence of these low-lying structures in the β-strength function, above the neutron-separation energy has actually been questioned, giving rise to a long-standing debate opposing those who believed in a

* Collaboration between FLNR/JINR Dubna and IPN/IN2P3 Orsay.

completely statistical nature for the decay of states located above the threshold (the "nuclear Pandemonium") and those who, on the contrary, believed that structure effects should drive the observed particle and γ emissions. The theoretical background of the Pandemonium concept [16] seems actually to ultimately take root in Bohr's compound nucleus model: the idea would be that, in complete analogy to fusion reaction, a hot, unstructured compound system is formed just after the spontaneous β decay: this analogy is shown in a pictorial way with Fig. 1.

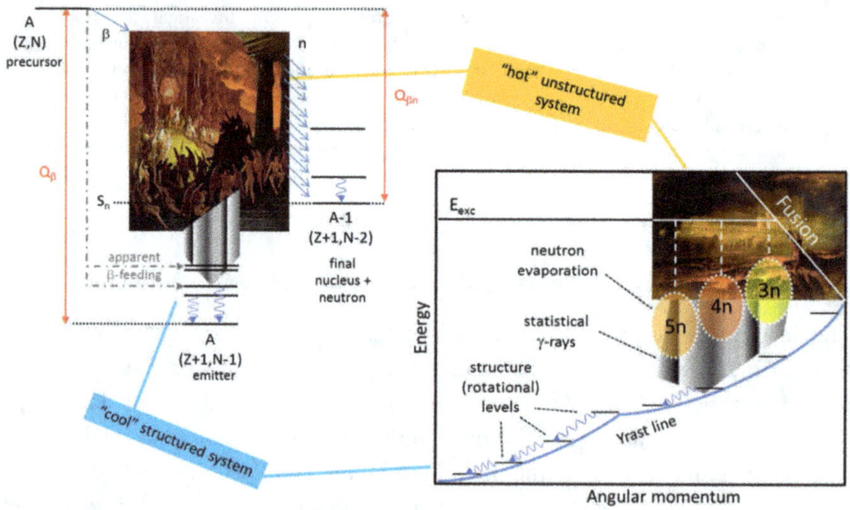

Fig. 1. Pictorial illustration of the analogy between the decay mode of the nuclear Pandemonium formed after β decay spontaneous reaction, and the decay mode of a compound nucleus after fusion reaction.

As illustrated in Fig. 1, one of the major consequences of the Pandemonium is the near experimental impossibility to collect all the statistical γ-strength, resulting in an quasi inescapable over-estimation of the true direct β-feeding of low-lying structure levels.

However, there has been recently growing experimental evidence, from β-delayed γ and neutron spectroscopy measurements, seemingly pointing towards the existence of specific regions of the daughter nucleus excitation spectrum that concentrate larger fractions of the β-strength [2-4]. These energy regions significantly overlap with regions where highly collective excitation modes (typically E1) of the nucleus may lie and the question of the nature of unbound states populated by high-Q_β radioactivity was raised again. This interplay may result in unusually strong γ emission from levels located well above the neutron threshold. Should this strong γ/n competition turn out to be more general than expected e.g. throughout the *r*-process mass region (and particularly close to magic numbers) part of commonly admitted nucleosynthesis scenarios would be affected [4]. In particular, the relative importance of first-forbidden transitions in the N=50 region [5] may have long been overestimated.

In the ^{78}Ni region these questions were introduced by two recent studies of neutron emissions following the β-decay of very neutron rich Ga (Z=31) isotopes: (*i*) the time-of-flight neutron spectroscopy of the ^{83,84}Ga precursors [3] allowing a reconstruction of the β-strength function, and (*ii*) the integrated P_n measurements of the same precursors obtained with the TETRA long-counter at ALTO [6]. But the γ-emission of high-lying states in the ^{83,84}Ge daughters remained so far uninvestigated.

In this paper we report on the recent observations of unexpected “ultra”-high-energy γ-rays (8-9 MeV) in the β-delayed emission products of ^{83}Ga sources. Part of these results were already communicated in [7] and we believe it is worthwhile providing additional material in the present paper.

This unexpected observation further suggests that structure effects, at and well above the daughter-nuclei neutron threshold, governs the decay properties in the ^{78}Ni region. It has triggered a wider experimental effort at ALTO, involving the ^{3}He neutron-counter TETRA [8], the time-of-flight neutron spectrometer MONSTER [9] and high energy γ spectroscopy array PARIS [10].

2. β-Delayed High-Energy γ Spectroscopy of ^{83}Ge

The measurement was performed using low-energy radioactive ^{80,83}Ga ion beams produced at the photofission Isotope Separation On Line (ISOL) facility ALTO, operated at the IPN in Orsay [11]. The beams were collected on a movable tape collector at the center of the BEDO setup [12] (see Fig. 1 in Ref. [7]). A cylindrical plastic scintillator with an average efficiency of 56% surrounding the collection spot was used for β electron detection. Four Ge detectors and a 2×2×4 inch $LaBr_3$ scintillator were used for γ spectroscopy. While the Ge detectors were set with an energy range at around 2.5 MeV, the $LaBr_3$ scintillator had an 11 MeV range, matching roughly the 11.7 MeV Q-value of the ^{83}Ga β decay. Ge detectors and the $LaBr_3$ scintillator in the BEDO setup were calibrated in energy and efficiency using standard γ-ray sources up to 1.5 MeV and for the scintillator with known γ rays from ^{80}Ga radioactivity up to 5.5 MeV as well. The response function and energy linearity of the $LaBr_3$ detector were investigated up to 11 MeV using the ^{27}Al(p,γ)^{28}Si reaction at the ARAMIS accelerator operated at the CSNSM in Orsay [13]. The resolution of the scintillator is 30 keV at 1 MeV and 120 keV at 6–7 MeV.

At high energy, the γ-spectrum recorded by the 2×2×4 inch $LaBr_3$ scintillator detector is obviously dominated by Compton scattering events, and escape and double-escape events originating from pair creations and the resulting incomplete energy collection due to the limited size of the crystal. A simulation of the detector response function was done with GEANT4 code and the model was validated up to 11 MeV using the aforementioned ARAMIS γ source.

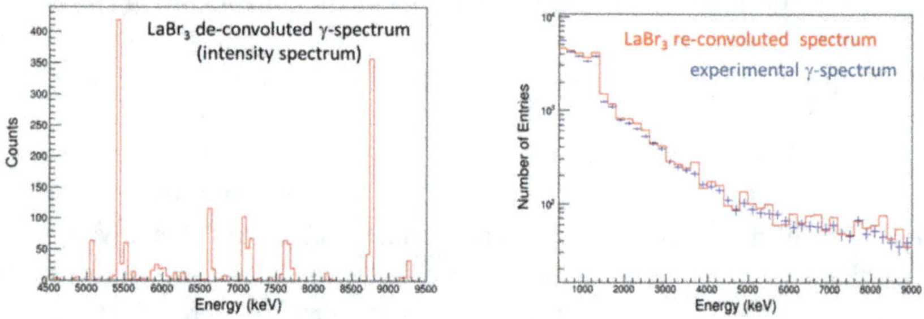

Fig. 2. Left panel: number of occurrence of photon events at a given energy after correction for the detector's response function, forming the de-convoluted γ spectrum. Right panel: experimental γ spectrum in blue, re-convoluted spectrum in red (see text for the details) [14].

The experimental data corresponding to the high-energy part of the spectrum were then de-convoluted from this response function. To achieve this, the Gold algorithm was employed (as implemented in the Root code from CERN) with different numbers of cycles. The result is shown in the left-panel of Fig. 2. An accumulation of intensities is observed at 5-6 MeV and then at 7-8 MeV and beyond. When this distribution is re-convoluted with the scintillator response function (see Fig. 2, right panel) it gives origin to a smooth distribution (in red), compatible with the experimental spectrum (error bars in blue). This deconvolution analysis was also used to have a correct evaluation of the γ efficiency for the 5-9 MeV range, where the summation of photo peaks, single and double escape peaks and Compton shoulders must be taken into account. While the exact number of γ transitions varies with the number of cycles, the general features (i.e. energy regions of accumulation) remain similar and the γ intensity between 5-8 MeV varies only with 1% when obtained with different cycle repetitions. This has been taken as a systematic error summed quadratically to the statistical one to obtain an estimate of the total β feeding proceeding through beyond-threshold γ-ray emission relative to those of known intensity for low-energy γ transitions. The 1348 keV line in the β-delayed neutron daughter ^{82}Ge, with an absolute γ intensity of 62.8(3)%, was taken as a reference. With the hypothesis of direct β feeding for all transitions observed, the total $I_{\beta\gamma}$ value for states in the 5 to 9 MeV interval works out at 16(4)%. This value, is to be compared to the integrated probability of neutron emission of 85(6)% [6]. This means that the vast majority of the ^{83}Ga β strength lies above the neutron sepration threshold of ^{83}Ge. Due to the phase factor it is likely that most of it is associated to (strongly) allowed GT transitions. Said in other words, only a very small fraction of the β decay of the ^{83}Ga ground state (presumably $J^{\pi}=5/2^{-}$) proceeds through first-forbidden transitions towards the low-lying positive parity states of ^{83}Ge. This fraction cannot exceed the uncertainties obtained on the two previous numbers i.e. 4-6%.

3. Conclusion and Perspectives

Our result clearly points toward a very strong n/γ emission competition from levels located nearby to far above the neutron threshold of ^{83}Ge. Noteworthy is the fact that this very neutron rich nucleus is located just next to the N=50 shell closure. The question is now to try and understand the microscopic origin of such a feature, in particular the structure of these high-lying emitting states. In Ref. [7] we have shown, from beyond mean-field QRPA calculations, that a scenario where the β-decay actually populates some components of the Pygmy electric dipole resonance (PDR) of ^{83}Ge could explain these experimental findings. This scenario was investigated because it seemed (at least to the authors of Ref. [7]), that only high-energy E1 (collective in a sense) transitions to the ground and very low-energy states could be fast enough to compete with the neutron emission. However, other (yet unknown) scenarios may be as relevant. The second question opened by this study is to understand if this situation is unique to ^{83}Ge (because of the valence topology for instance) or if this could be a more general feature, common to e.g. other shell closures' vicinities. The very recent discovery of similar high-energy transitions from levels located well above the neutron-separation threshold of the N=83 isotone ^{133}Sn seems to show that such intriguing features extend to the N=82 region. This further justifies an additional experimental effort at ALTO, jointly exploiting in the future the performances of the TETRA, PARIS and MONSTER spectrometers, as mentioned in section 1. This very strong n/γ emission competition from levels located nearby to far above the neutron threshold may indeed constitute a new facet of radioactivity of exotic nuclei.

References

1. N. Severijns, M. Beck and O. Naviliat-Cuncic, *Rev. Mod. Phys.* **78** (2006) 991.
2. Y. Fujita, B. Rubio, W. Gelletly, *Prog. Part. Nucl. Phys.* **66** (2011) 549; Y. Fujita et al., *Phys. Rev. Lett.* **112** (2014) 112502; J.L. Tain et al., *Phys. Rev. Lett.* **115** (2015) 062502.
3. M. Madurga et al., *Phys. Rev. Lett.* **117** (2016) 092502.
4. A. Spyrou et al., *Phys. Rev. Lett.* **117** (2016) 142701.
5. T. Marketin, L. Huther, and G. Martínez-Pinedo et al., *PRC* **93** (2016) 025805.
6. D. Verney, D. Testov et al., *Phys. Rev. C* **95** (2017) 054320.
7. A. Gottardo, D. Verney et al., *Phys. Lett. B* **772** (2017) 359.
8. D. Testov, D. Verney et al., *Nucl. Instrum. Methods Phys. Res A* **815** (2016) 96.
9. T. Martinez, D. Cano-Ott et al., NDS 120 (2014) 78.
10. A. Maj et al., The PARIS project, *Acta Phys. Pol. B* **40** (2009) 565.
11. F. Ibrahim et al., *Nucl. Phys. A* **787** (2007) 110.
12. A. Etilé et al., *Phys. Rev. C* **91** (2015) 064317.
13. E. Cottereau et al., *Nucl. Instrum. Methods Phys. Res. B* **45** (1990) 293.
14. Analysis performed by A. Gottardo, figures reproduced here with his kind permission.
15. M. Piersa et al., *Phys. Rev. C* **99** (2019) 024304.
16. J. Hardy, *Phys. Lett. B* **71** (1977) 307.

Clusterization and Crystallization of Complex Nuclei*

Y. A. Zaripova†, V. V. Dyachkov and A. V. Yushkov
Scientific Research Institute of Experimental and Theoretical Physics, Almaty, Republic of Kazakhstan
†E-mail: ZJ_KazNU@mail.ru

Cluster and crystallization forms of nuclei are considered in their manifestation at half-lives and on the parameters of quadrupole nuclear deformation, between which an anticorrelation for elongated spheroids and a correlation for oblate spheroids was found in isotopic chains.

Keywords: Quadrupole nuclear deformation parameter; exotic nuclei; half-lives.

1. Correlation of Half-lives of Nuclei with their Deformation

A number of experiments are direct evidence of the crystal structure of nuclei [1, 2]. Therefore, in the series of isotopic chains, which personify the "growth" of nuclear crystals with the addition of more and more new neutrons to stable isotopes, the quadrupole nuclear deformation parameter β_2 develops dramatically.

1.1. *Experiment*

Figure 1 clearly shows the anti-correlation of the non-sphericity parameter β_2 with the half-life $T_{1/2}$ for elongated spheroid nuclei with $\beta_2 > 0$ for cerium isotopes. For exotic nuclei, this leads to the limiting tendency of the curvature of the Riemannian nuclear space to zero:

$$\lim_{T_{1/2} \to 0} \varkappa \to 0. \tag{1}$$

With the recent discovery of Riemannian nuclear physics [3-6], the radioactivity micromechanism has been revealed. This is the establishment of the Lobachev hyperbolic open-loop geodesic with the negative curvature of space in the upper Lyapunov spatial layers from the center of the nucleus. That happens with the distance from the axis of the "Valley of stability" and the distance from the last stable isotope in the Mendeleev table – $^{208}Pb_{82}$. This is the actual implementation of the ideas of Einstein and Friedman on the geometrization of physics [7, 8].

* Work partially supported by grant IRN AP05131884 of the Ministry of Education and Science of the Republic of Kazakhstan.

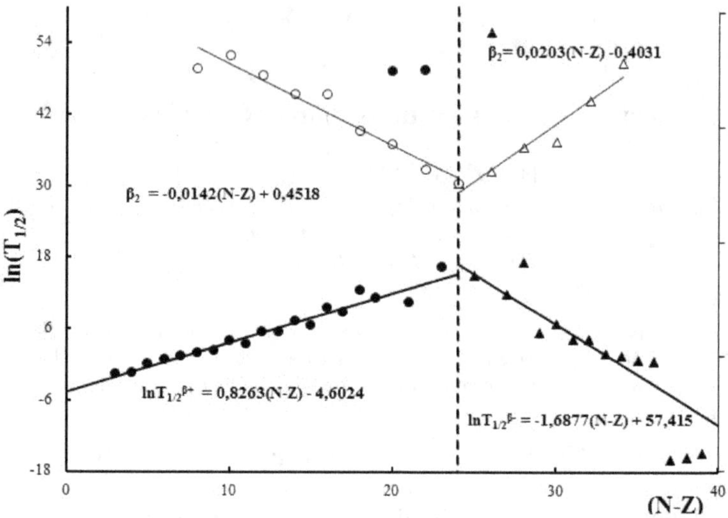

Fig. 1. The phenomenon of β_2 anti-correlation with $T_{1/2}$ for elongated nuclei.

In [9], for the first time, two regularities were found for the logarithmic dependence of the half-lives of $T_{1/2}$ for β^+- and β^--radioactivity on the argument (N–Z). The corresponding formulas for finding still unmeasured $T_{1/2}$ for remote exotic nuclei have been derived [2,9]

$$T_{1/2}(\beta^+,\varepsilon) = C_1 exp[2{,}30(5{,}81 - 1{,}84Z + 1{,}65N - 1{,}50 \cdot 10^{-2}NZ + 1{,}50 \cdot 10^{-2}10^2)];$$

$$T_{1/2}(\beta^-) = C_2 exp[2{,}30(-2{,}14 \cdot 10^{-1}Z - 0{,}67N - 2{,}30 * 10^{-3}NZ + 230 \cdot 10^{-3}Z^2)],$$

where C_1 and C_2 – normalizing dimensional multipliers.

Recently, the accelerating technique of heavy ions [10], detecting systems for detecting exotic heavy ions [11], and methods for measuring the ultrashort half-lives of exotic nuclei [12] have been booming. Therefore, for exotic nuclei in the framework indicated above, the paradigms of curvilinear non-Euclidean nuclear spaces, the actual problems are the measurements of their radii R and the parameters of quadrupole nuclear deformation β_2.

1.2. *Theory*

The parameter β_2 in the simplest approximation of the rotational spheroid is represented as follows

$$\beta_2 = 2\frac{b-a}{a+b}, \tag{2}$$

where a and b – small and large axis of rotation spheroid. In Figure 1, for cerium isotopic series, the regularities for β^+- and β^--decays in correlation with the new function $\beta_2 = \beta_2(N\text{-}Z)$ are shown. Indeed, a clear anti-correlation is visible for both decay branches in these isotopic chains.

For the first time, from the found anti-correlation, we obtained the dependence of quadrupole nuclear deformation on half-lives $\beta_2 = \beta_2(T_{1/2})$, which is described with high accuracy by formulas.

$$\beta_2\left(T_{1/2}^{\beta+}\right) = -0{,}0109 \cdot lnT_{\frac{1}{2}}{}^{\beta+} + 0{,}3444 \tag{3}$$

$$\beta_2\left(T_{1/2}^{\beta-}\right) = -0{,}008 \cdot lnT_{\frac{1}{2}}{}^{\beta-} + 0{,}2823 \tag{4}$$

From Figure 2, it can be seen that, indeed, the catastrophic development of deformation, both elongated and oblate nuclei, leads them to instability and radioactivity.

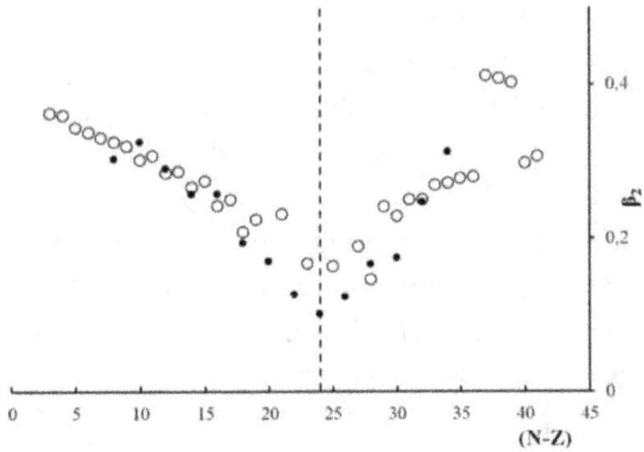

• – experimental data; ○ – β_2 values calculated from half-life $T_{1/2}$ values

Fig. 2. The phenomenon of the catastrophic development of the non-sphericity of exotic nuclei as they approach the boundaries of new "lines of unbound" states of neutron-deficient and neutron-rich nuclei cerium nuclei.

References

1. A.V. Yushkov, S.J. Aisina, K.A. Kuterbekov, N.N. Pavlova, Experimental-evidence in favor of the existence of nuclear-matter, *IZVESTIYA AKADEMII NAUK SSSR SERIYA FIZICHESKAYA* **53(1)**, 37 (1989).
2. M. Zhusupov, A. Yushkov, *Physics of Atomic Nuclei* (Parus, Almaty, 2006).

3. A. Yushkov, V. Dyachkov, Y. Zaripova, New laws in nuclear physics and microworld structures, *Monthly Scientific Journal* **9(20)**, 35 (2015).
4. V. Dyachkov, Y. Zaripova, A. Yushkov, Cluster structure of nuclei and new spatial regularities, *KazNU Bulletin. Physics Series* **2(57)**, 88 (2016).
5. V.V. Dyachkov, Y.A. Zaripova, A.V. Yushkov, Experimental foundations of new nuclear physics in non-Euclidean spaces, *in Materials of the International Scientific Conference Dedicated to the 80th Anniversary of Academician of the NAS RK Abdildin M.M,* (Almaty, Kazakhstan, 2018).
6. V.V. Dyachkov, Y.A. Zaripova, A.V. Yushkov, Experimental Foundations of Nuclear Physics in Non-Euclidean Spaces, in *Book of Abstracts IX International Symposium on Exotic Nuclei* (*EXON 2018*), (Petrozavodsk, Russia, 2018).
7. G.I. Shipov, *The Theory of Physical Vacuum, Theory of Experiments and Technology* (Science, Moscow, 1997).
8. A.A. Friedman, Selected works "*On the Curvature of Space*", "*On the Possibility of a World with a Constant Negative Curvature of Space*", "*Peace as Space and Time*" (Moscow, 1966).
9. A.M. Ivanov, A.V. Yushkov, Isotopic law of change of half-lives of atomic nuclei, *Izv. Academy of Sciences of the Kazakh SSR. Physics and Mathematics Series* **4**, 39 (1979).
10. Yu.E. Penionzhkevich, V.M. Vakhtel, *Modern Accelerators of Charged Particles and their Application* (Publishing house of VSU, Voronezh, 2018).
11. A.G. Artukh, et al., Radioactive nuclear beams of COMBAS facility, *Physics of Elementary Particles and Atomic Nuclei* **46(1)** 96 (2016).
12. Yu.E. Penionzhkevich, R.G. Kalpakchieva, *Light Nuclei at the Neutron Stability Boundary* (JINR, Dubna, 2016).

4

Scientific projects and experimental facilities

Decay Spectroscopy of the Heaviest Nuclear Species — Perspectives at S^3

D. Ackermann

Grand Accélérateur National d'Ions Lourds - GANIL, CEA/DRF-CNR, S/IN2P3,
Bd. Becquerel, BP55027, F-14076 Caen, France
E-mail: dieter.ackermann@ganil.fr
http://pro.ganil-spiral2.eu

After more than half a century of research addressing the synthesis and nuclear structure of superheavy nuclei (SHN) a boost for its progress is expected from the advent of new instrumentation. An order of magnitude in beam intensity increase is envisaged to be provided by new powerful accelerators like the new DC280 cyclotron at the SHE factory of FLNR/JINR or the superconducting LINAC at SPIRAL2 of GANIL. In addition new ion-optical installations like the separator-spectrometer set-up S3 with two complementary detection systems, SIRIUS and LEB, will provide a substantial sensitivity increase for decay spectroscopy after separation (DSAS), as well as for alternative and complementary methods like high precision mass measurements and laser spectroscopy. DSAS has proven in the past to be a powerful tool to study the low lying nuclear structure of heavy and superheavy nuclei. Single particle levels and other structure features like *K* isomerism, being important in the fermium-nobelium region as well as for the route towards spherical shell stabilized SHN, have been investigated almost up to the limit posed by the sensitivity of present-day instrumentation. Precision mass measurements and laser spectroscopy will offer the possibility to study alternative features like binding energies, charge radii and quadrupole moments. At the magnetic spectrometer VAMOS of GANIL with the recently improved mass resolution and the development of *Z* identification, deep-inelastic reactions like multi-nucleon transfer can be used to reach more neutron-rich nuclei in the region of light actinides, possibly being extended towards higher *Z*.

Keywords: Superheavy nuclei; nuclear structure; α-, γ-, conversion electron spectroscopy; nuclear fission; heavy ion separators/spectrometers.

1. Introduction

Nuclei beyond fermium-rutherfordium owe their existence solely to quantum mechanics, where the fission barrier in a liquid drop approach would vanish, which gives rise to the nuclear physicists' definition of a superheavy nucleus (SHN). These features, taken into account as shell corrections in so-called macroscopic-microscopic models (see e.g. [1]), cause the extension of the chart of Segré towards high Z and A. These nuclei are, in turn, an ideal laboratory to study the strong nuclear interaction by in-beam methods as well as decay spectroscopy after separation [2]. This paper will focus on decay spectroscopy after separation (DSAS) of these deformed nuclei at Z=100-112 and N=152-162.

These studies have the potential to link to the next heavier spherical closed shell nuclei, by investigating single particle levels [3] close to the fermi energy which according to some models play a major role in defining the spherical shell gaps in the region of the so-called island of stability. For in-beam spectroscopy which gives access to nuclear structure at higher spins like e.g. rotational bands, I refer to the recent review by Theisen et al. and references therein [4].

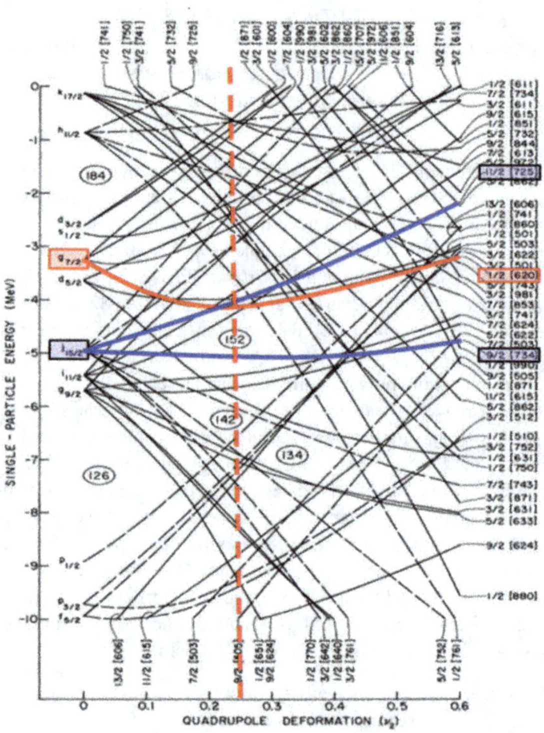

Fig. 1. Neutron single particle energies as a function of quadrupole deformation from a momentum-dependent Woods-Saxon model by R.R. Chasman et al. [9]; figure was taken from there and adopted as in reference [2].

Particularly interesting features like *K* isomers, can be used to trace deformation towards the island of stability [5]. A number of cases have been found in the region of fermium isotopes and beyond mainly for isotopes of all even-*Z* elements up to darmstadtium (Z=110) apart from seaborgium (Z=106), but also for some even-odd and odd nuclei (see figure 4). High intensity accelerators, targets withstanding those high beam intensities and in particular actinide targets, efficient in-flight separators and spectrometers, and highly efficient detectors with fast electronics are the essential ingredients for the success of the field. The new SPIRAL2 facility and, in particular, the superconducting separator-spectrometer setup S³ [6] presently under construction at GANIL will offer great perspectives for the field [5]. A number of ground-breaking experiments are envisaged, including DSAS examining features like the single particle structure, deformation and the search for K-isomers for those exotic nuclear species.

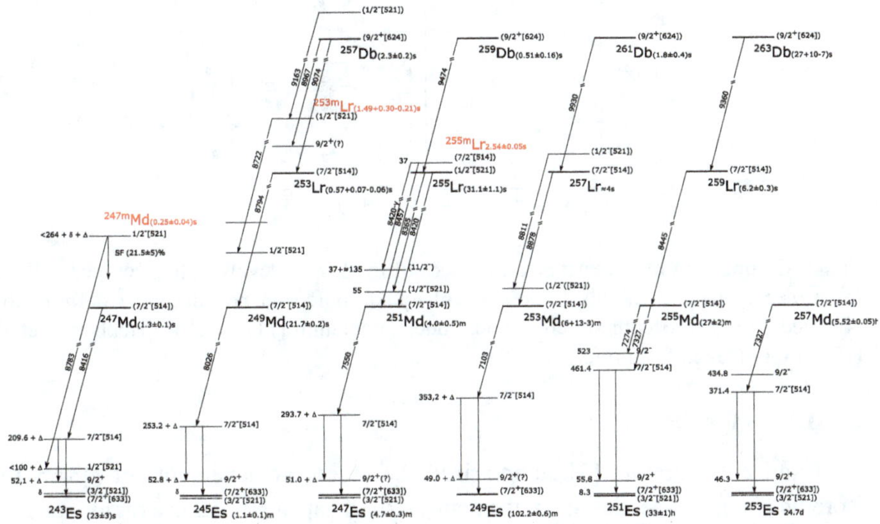

Fig. 2. Decay scheme for odd-Z isotopes from einsteinium to dubnium in the vicinity of the N=152 closed shell. For the origin of the data see references in [2]. (Figure taken from reference [2])

The experimental tools classically employed in the field are recently complemented by alternative approaches with precision mass measurements, using devices like Penning traps [7] and multi-reflection time of flight spectrometers (MRToF), and laser spectroscopy [8], which give access to basic properties of those heavy nuclei like binding energies, charge radii and quadrupole moments. S³ will profit from high beam intensities provided by the superconducting LINAC, similar to the ones expected from the new DC280 cyclotron of the new SHE-factory at FLNR/JINR. The highly sensitive DSAS set-up SIRIUS and the low energy branch (LEB) will provide a powerful set of instrumentation for studies at the extremes of nuclear stability like the region SHN or N+Z nuclei.

Some of the recent achievements obtained by DSAS for SHN in section 2. *K* isomerism will be the subject of section 3. The planned investigation of binary reactions at VAMOS+AGATA will be outlined in section 4. An outlook showing the opportunities offered by S3 will be given in section 5.

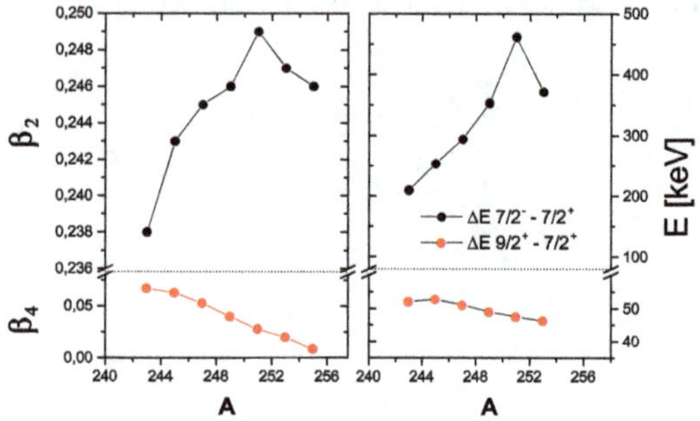

Fig. 3. Comparison of the energy differences of the lowest levels in the isotopes ^{243}Es to ^{253}Es (see main part of this figure 2) with quadrupole and hexadecapole deformation obtained by a macroscopic-microscopic model calculation [11]. See also Heßberger et al. [12]. (Figure taken from reference [2])

2. DSAS for SHN

Asai and collaborators [3] summarized the recent achievements of DSAS for isotonic chains in the vicinity of the neutron shell gap at N=152 in a region ranging from fermium to meitnerium. A general problem to investigate single particle states is the admixture of other components to the initial and final states for which transitions are observed. In regions of low-level density, however, the single-particle character is enhanced. Moreover, a spin and parity assignment is supported by the selectivity provided by transition selection rules in stepwise construction of level schemes, built on states for which quantum numbers are known. A more realistic picture can be achieved by employing models which take into account correlations beyond the single particle picture by coupling to the relevant configurations which involve e.g. vibrational states. In particular, the N=151 and 153 isotones with one hole in and one particle more than the sub-shell closure, respectively, are suited to trace single particle levels which are important for the predicted spherical shell gaps. The Nilsson diagram in figure 1 shows the situation for calculated neutron single particle trends in the vicinity of N=152.

To illustrate a possible single particle scenario by an instructive example, three single particle levels (SPL) are highlighted in figure 1. One the two SPL, originating from the *j*15/2 shell, which define the deformed shell gap at *N*=152, is crossing the 1/2$^-$[620] at a quadrupole deformation of $\approx$0.25. To confirm this configuration a direct observation of this 11/2$^-$[725] SPL in ^{253}No is essential as pointed out ref. [10], where attempt was made to populate this state by α decay of ^{257}Rf. Apart from α decay, ^{257}Rf has an EC decay branch into ^{257}Lr with branching ratios b_{EC} of 0.094±0.014 and 0.115±0.015 for the g.s. and the isomer decay, respectively [10]. The subsequent α decay then populates the g.s. and excited states in ^{253}Md, providing access to the low-lying structure of this odd-*Z* even-*A* isotope. This data is shown together with the at present known experimental nuclear structure information of odd-*Z* odd-*A* isotopes in the region from einsteinium to Dubnium in figure 2 [2]. An interesting feature here was pointed out by Heßberger et al. [12] with the consistent behavior of the energy difference for the two SPLs 7/2$^-$[514] and 7/2$^+$[633] in einsteinium isotopes and the g.s. deformation of these nuclei, predicted by the macroscopic-microscopic model calculations by Parkhomenko and Sobiczewski [11] (figure 3).

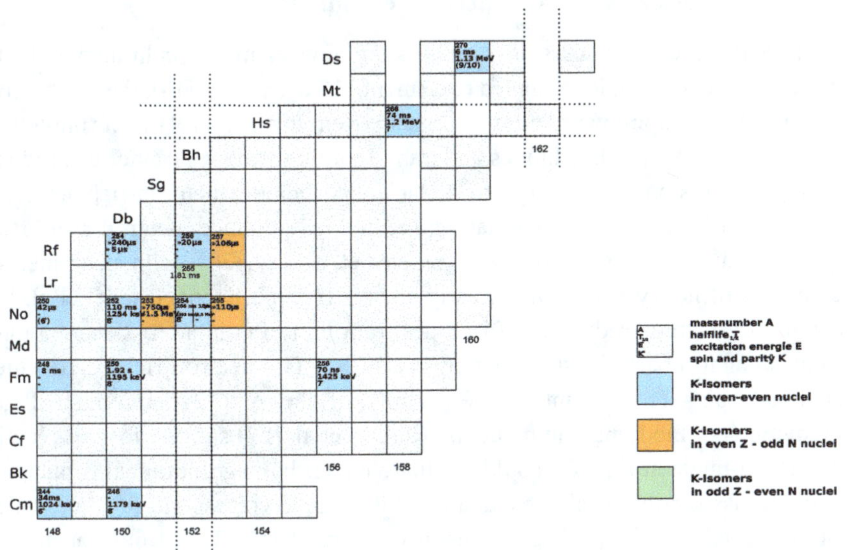

Fig. 4. Excerpt of the chart of nuclides indicating the *K*-isomers observed for heavy nuclei in the region Z ≥ 96. Half-life, decay energy, spin and parity values are given for *K*-isomers only. (Figure and caption are taken from reference [2])

3. *K* isomerism for SHN

K isomers can be caused by a large difference in angular momentum and possibly opposite parity of the initial and final state. They have been predicted as quasi-particle excitations for the region Z=100 to 116 [13].

The heaviest nucleus for which such an isomeric state has been observed is ^{270}Ds [14]. Recently such a state has been reported also for its α-decay daughter ^{266}Hs [15]. For both the isomeric state is longer-lived than the g.s., a rare relatively feature. As deformation is the necessary precondition *K*-isomer formation, they can serve to as a tracer towards high Z and the predicted spherical shell stabilized SHN. Together with the nuclear structure information on the involved quasi-particle states, this can be used to improve model predictions to eventually localize the long sought-for island of stability of SHN. Table 2 in ref. [5] gives an update of the table of known *K*-isomers in even-even nuclei in heavy and SHN from [16] in the region from curium to darmstadtium. A summary of observed *K*-isomers in that region is shown in figure 4 [2]. An outlook into the expectations for *Z*=104 to 110 and the next deformed neutron shell gap at N=162 is given by V. Prassa et al. [17], employing energy density functional (EDF) calculations.

4. Binary Reactions — An Alternative Approach

Despite the recent success in synthesis of new elements culminating in the detection of decay chains assigned to element 118 oganesson [18], the perspectives to push the fusion-evaporation approach even further, are rather limited. In particular, neutron-rich actinides and trans-fermium isotopes cannot be produced in complete fusion reactions due to the lack of sufficiently neutron-rich projectile-target combinations. An alternative method to produce neutron-rich heavy systems has been proposed by Zagrebaev et al. [19, 20], with deep-inelastic collisions of heavy nuclei. In an early chemistry experiment Schädel et al. [21] could observe the production of isotopes from Pu to Fm in a cross-section range from mbar to nbarn for the reaction ^{238}U+^{238}U. The neutron-rich part of these isotope chains reaches into close vicinity of the *N*=152 neutron shell gap. Predictions of the Langevin model by Karpov et al. [22] suggest for this reaction that multi-nucleon transfer could populate nuclei having additional 5 nucleons, with a cross section of about 1 μbarn, which would correspond for the uranium isotopic chain to ^{243}U, being one neutron short of 152. This would open those isotopes for in-beam nuclear structure studies at the large solid angle magnetic spectrometer VAMOS in combination with the multi-detector γ-tracking array AGATA. The mass resolution of ΔA=1/500, recently achieved for VAMOS [23], together with a *Z* identification by X rays should provide the necessary identification in *A* and *Z* to extend isotope-tagged γ-ray spectroscopy further towards *N*=152. In a recent investigation of the reaction ^{136}Xe+^{238}U at the combination of the magnetic spectrometer PRISMA at LNL and AGATA, Birkenbach et al. could establish a rotational band up to 20-24 $\hbar$ for ^{240}U [24]. The heaviest uranium isotope, ^{242}U, for which the first 2^+ state is known, is two neutrons away from N=152 [25]. An experiment to study the reaction $^{238}U+^{238}U$ has been accepted by the GANIL Program Advisory Committee (PAC) and is planned to be carried out in the nearest future at VAMOS+AGATA. To extend the reach of this type of measurements to higher Z heavier actinide targets like e.g. ^{248}Cm [26] are needed.

5. Outlook: Opportunities at S3

At SPIRAL2/GANIL a high intensity super conducting linear accelerator (SC LINAC) is presently under construction with specifications which respond to the need of highest intensities [27]. The S^3 set-up [6] combined with the DSAS array SIRIUS (Spectroscopy and Identification of Rare Isotopes Using S^3) and the high intensity beams from SC LINAC will be one of the most powerful facilities for SHN/SHE research worldwide together with the SHE Factory presently under construction at JINR/FLNR. A summary of the scientific program envisaged for the field of SNH is given e.g. in refs. [2, 5]. The possible features to be studied for the various heavy and superheavy nuclei range from cross section measurements to detailed decay spectroscopy. X-ray spectroscopy has the potential of settling the still open question of Z identification for the heaviest nuclei produced for the first time at the FLNR (see e.g. refs. [18, 28]. Apart from allowing for the investigation of K-isomers, isotopic and isotonic trends of low lying nuclear excitations by exploiting γ-electron- α/fission and X-ray coincidences, SIRIUS is also an ideal tool to study delayed processes like isomeric states and β-delayed fission. In addition, a low energy set-up including a gas stopping cell, laser spectroscopy instrumentation and a multi-reflection time-of-flight spectrometer (MR ToF) will be used to study nuclei in the N=Z region as well as the heaviest nuclear species. In a farer future the synthesis and investigation of, also so far unknown, highest-Z systems is envisaged, for which the earlier experiments will establish the basis.

Acknowledgments

The research and development reported here was performed and/or planned with the SHE research group at SHIP and TASCA of GSI, and the French scientific community interested in SHN/SHE research. The author would like to express his gratitude, in particular to the SHIP and TASCA groups with F.P. Heßberger, A. Mistry, S. Heinz, S. Hofmann, J. Khuyagbaatar, M. Block, M. Laatiaoui and A. Yakushev; the S^3 team at GANIL with D. Boilley, L. Caceres, F. Déchery, P. Delahaye, N. Lecesne, M. Lewitowicz, J. Piot, H. Savajols, Ch. Stodel, and the colleagues from the French SHN/SHE community O. Dorvaux, A. Drouart, B. Gall, K. Hauschild, A. Korichi, W. Korten, A. Lopez-Martens, B. Sulignano,

Ch. Theisen, M. Vandebrouck, M. Zielinska, and my colleagues from theory E. Litvinova, B. Lu, V. Prassa and D. Vretenar.

References

1. A. Sobiczewski and K. Pomorski, *Prog. Part. Nucl. Phys.* **58**, 292 (2007).
2. D. Ackermann and C. Theisen, *Phys. Scr.* **92**, 083002 (2017).
3. M. Asai, F.P. Heßberger, and A. Lopez-Martens, *Nucl. Phys. A* **944**, 308 (2015).
4. C. Theisen, P. Greenlees, T.-L. Khoo, P. Chowdhury, and T. Ishii, *Nucl. Phys. A* **944**, 333 (2015).
5. D. Ackermann, *Nucl. Phys. A* **944**, 376 (2015).
6. F. Déchery et al., *Eur. Phys. J. A* **51**, 1 (2015).
7. M. Block et al., *Nature* **463**, 785 (2010).
8. M. Laatiaoui et al., *Nature* **538**, 495 EP (2016).
9. R.R. Chasman, I. Ahmad, A.M. Friedman, and J.R. Erskine, *Rev. Mod. Phys.* **49**, 833 (1977).
10. F.P. Heßberger et al., *Eur. Phys. J. A* **52**, 1 (2016).
11. A. Parkhomenko and A. Sobiczewski, *Acta Phys. Pol. B* **35**, 2447 (2004).
12. F.P. Heßberger et al., *Eur. Phys. J. A* **48**, 75 (2012).
13. F. Xu, E. Zhao, and P.W.R. Wyss, *Phys. Rev. Lett.* **92**, 252501 (2004).
14. S. Hofmann et al., *Eur. Phys. J. A* **10**, 5 (2001).
15. D. Ackermann et al., *GSI Sci. Rep.* **2011**, 208 (2012).
16. R.-D. Herzberg and P. Greenlees, *Prog. Part. Nucl. Phys.* **61**, 674 (2008).
17. V. Prassa, B.-N. Lu, T. Nikšić, D. Ackermann, and D. Vretenar, *Phys. Rev. C* **91**, 034324 (2015).
18. Y.T. Oganessian and V.K. Utyonkov, *Nucl. Phys. A* **944**, 62 (2015).
19. V.I. Zagrebaev and W. Greiner, *J. Phys. G: Nucl. Part. Phys.* **31**, 825 (2005).
20. V.I. Zagrebaev, Y.T. Oganessian, M.G. Itkis, and W. Greiner, *Phys. Rev. C* **73**, 031602 (2006).
21. M. Schädel et al., *Phys. Rev. Lett.* **41**, 469 (1978).
22. A.V. Karpov and V.V. Saiko, *Phys. Rev. C* **96**, 024618 (2017).
23. D. Ralet et al., *Phys. Scr.* **92**, 054004 (2017).
24. B. Birkenbach et al., *Phys. Rev. C* **92**, 044319 (2015).
25. T. Ishii et al., *Phys. At. Nucl.* **70**, 1457 (2007).
26. M. Schädel et al., *Phys. Rev. Lett.* **48**, 852 (1982).
27. E. Petit on behalf of the SPIRAL2 collaboration, in Proceedings of NA-PAC2016, Chicago, IL, USA, number TUA1IO02, 2016.
28. Y.T. Oganessian, A. Sobiczewski, and G.M. Ter-Akopian, *Phys. Scr.* **92**, 023003 (2017).
29. D. Rudolph et al., *J. Rad. Nucl. Chem.* **303**, 1185 (2015).

Highlights of the Day-One Experimental Program at the Gamma-Beam System of ELI-NP

Dimiter L. Balabanski
Extreme Light Infrastructure – Nuclear Physics,
Horia Hulubei National Institute for R&D in Physics and Nuclear Engineering,
Reactorului Str. 30, 077125 Bucharest – Magurele, Romania
E-mail: dimiter.balabanski@eli-np.ro

The emerging experimental program with brilliant gamma beams at the Extreme Light Infrastructure – Nuclear Physics facility (ELI-NP) is presented with emphasis on the prepared day-one experiments. Experiments at ELI-NP will cover nuclear resonance fluorescence measurements, studies of large-amplitude motions in nuclei, photo-fission and photonuclear reactions of astrophysics interest, and measurements of photonuclear reaction cross sections. The physics cases of the flagship experiments at ELI-NP and the performance of the related instruments, which are under construction for their realization, are discussed.

Keywords: Photonuclear reactions; photo-activation; photo-fission; nuclear resonance fluorescence; giant and pigmy dipole resonances; nuclear spectroscopy.

1. Introduction

A versatile research program related to physics and applications with brilliant gamma beams at the Extreme Light Infrastructure – Nuclear Physics facility (ELI-NP) has been defined in several Technical Design Reports (TDR) [1] and reviewed periodically [2-5]. Nuclear resonance fluorescence (NRF) experiments, photo-neutron reaction measurements, studies of (γ,α) or (γ,p) reaction cross sections and photo-fission experiments are considered. Gamma-beam applications include the production of intense beams of slow positrons, NRF analytical techniques, γ-ray based radiography and tomography and research related to new medical radioisotopes. Here some of the flagship day-one experiments, which are under consideration, and the expected performance of the instruments, which are constructed for their realization, are discussed. In addition, a family of beam diagnostic detectors is developed, which will monitor the spatial, temporal, spectral and power parameters of the gamma beam, and will provide this information to the experiments upon request.

2. Ideas for Day-One Experiments with High-Brilliance Gamma Beams

The experimental program will target NRF measurements which provide a specific research niche for the ELI-NP facility. In particular, the pencil-size gamma beams at ELI-NP will provide access to targets that are available in small quantities and will open the actinide region for studies of low-lying E1 and M1 excitations. Detailed high-resolution studies of the dipole strength distribution in the region of the pygmy dipole resonance (PDR) will be done. Related to this experimental program, the ELIADE detector array of segmented Ge Clover detectors, large-volume $LaBr_3$:Ce detectors and digital data acquisition system is under construction. The array will have eight CLOVER Ge detectors which will be able to detect with high efficiency gamma rays with energies up to several MeV in the presence of the high radiation background produced by the gamma beams. Gamma-ray energies, angular distributions and γ-ray polarization will be measured with high accuracy. The γ-ray photo-peak efficiency of ELIADE is about 6%. A 3D drawing of ELIADE is presented in Fig. 1. The Ge Clover detectors are mounted at backward angles in two rings of four detectors each.

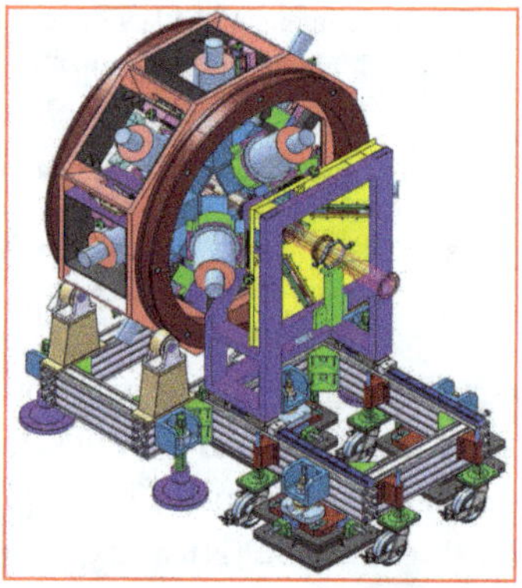

Fig. 1. A 3D model of the ELIADE Ge Clover detector array (see details in the text).

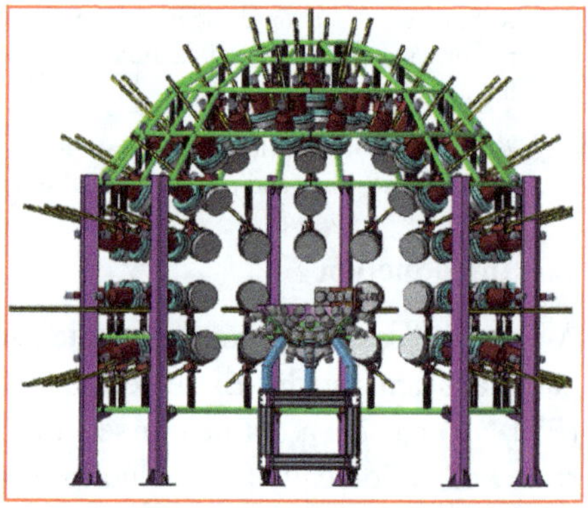

Fig. 2. A 3D model of the ELIGANT-GN detector array (see details in the text).

A photo-activation measurement of ^{180}Ta with the ELIADE array, *i.e.* a measurement of the subsequent γ decay as a function of the gamma-beam energy, is discussed. This experiment is closely related to nuclear astrophysics research, since the detailed knowledge of the doorway states and the flux which passes through them, provides a sensitive thermometer for studying the star conditions during the nucleosynthesis. Indications from a bremsstrahlung experiment about the doorway states, through which the $I^{\pi} = 9^{-}$, $T_{1/2} > 1.2 \cdot 10^{15}$ year isomer is de-excited, have been reported [6,7]. At ELI-NP this measurement can be done with the needed sensitivity and precision. It can be performed already during the commissioning of the GBS.

Extensive photonuclear reaction cross-section measurements were performed during the 1960-1980 at the Lawrence Livermore National Laboratory in USA and the Centre d'Etudes Nucleaires de Saclay in France. The two data bases have inconsistencies, which occur in a non-systematic way. This asks for new more precise measurements, which ideally can be done with a high-resolution gamma beam. Photo-neutron (γ,xn) cross sections with x = 1,2 will be measured using an in-beam neutron multiplicity technique utilizing the high-and-flat efficiency neutron detector ELIGANT-TN. An example for the existing discrepancies is the photodisintegration cross section in ^{9}Be, which has been measured in several experiments. The last three measurements were performed at laser Compton backscattering (LCS) facilities and provide discrepant results within the energy ranges where the cross section varies significantly within the energy spread of the incident γ-ray beam, while for the slow energy variation regions all three data sets are in relatively good agreement [8-10]. The current understanding is that the discrepancies between the three measurements are due to the energy unfolding of the measured cross sections, a procedure which relies on the precise determination of the incident LCS γ-ray beam energy spectra. A day-one experiment at ELI-NP can focus on the ^{9}Be(γ,n) reaction cross section within the 1.6 MeV – 3.0 MeV energy range. The expected narrow bandwidth will allow a complete mapping of the complex ^{9}Be resonant structure with little need for energy unfolding.

Studies above the particle evaporation threshold address open questions related to nuclear structure and astrophysical abundances of nuclear species, *e.g.* photodisintegration cross-sections for low-abundance nuclei relevant to the *p*-process nuclear synthesis and nuclear structure of the giant (GDR) and pigmy (PDR) dipole resonances, as well as the magnetic dipole resonances (MDR), from studies of gamma and neutron decays. The ELIGANT-GN detector array will be used for these measurements (see Fig. 2). ELIGANT-GN is a 4π spectrometer which consists of 30 $LaBr_3$:Ce and $CeBr_3$ large volume 3″ x 3″ scintillation detectors mounted at backward angles for detection of γ rays, and of up to 60 BC501A liquid and GS20 ^{6}Li glass scintillators for detection of neutrons [1,11]. The liquid scintillators will be used for measurements of the neutrons of energies above 1 MeV, while the ^{6}Li glass detectors are the most efficient at lower energies. The energy range below 1 MeV is important because large fraction of the strength of PDR and MDR is observed from neutron threshold up to the first excited states in residual nuclei in the measurement with the neutron threshold technique. The experimental set-up is arranged in such way, as to allow simultaneous γ-ray and neutron measurements, as well as coincidence gamma measurements. The large

γ-ray efficiency of the array will enable detailed studies of the fragmentation of the E1 and M1 strength of collective nuclear excitations. The neutron detectors will allow measurements of neutron spectra, which will complement the studies with the ELIGANT-TN array.

The GDR γ decay in ^{208}Pb is a case for a day-one experiment [1,2] because very detailed measurement on the absorption cross sections with fine structures is available for this case, while only few data with large uncertainties exist for the γ-decay branch. Details related to such measurement with the ELIGANT-GN array were discussed recently and detailed simulations of the detector response were presented [11]. The combination of ELIADE and ELIGANT-GN arrays is ideal for studies of the nuclear dipole polarizability, which is related to the slope of the symmetry term of the nuclear equation of state (EoS) through the neutron skin thickness. Thus, ELI-NP will provide accurate and unambiguous data for the E1 strength below and above the neutron threshold.

Studies of photonuclear reactions with emission of charged particles are related to nuclear astrophysics research. Day-one experiments, which are considered at ELI-NP, are the photodisintegration of ^{7}Li and the ^{16}O(γ,α)^{12}C reaction at energies close to the Gamow window. The motivation for studying the ^{7}Li(γ,t)^{4}He reaction is not only a final check on one of the reactions related to the "cosmological Li problem", but also an approach to the recently pointed disagreement between theoretical models and the experimental data. The mirror alpha capture reactions ^{3}H(α,γ)^{7}Li and ^{3}He(α,γ)^{7}Be have been at odds in the last five years as the different theoretical models agree very well with the ^{3}He(α,γ)^{7}Be experimental cross section but could not reproduce the ^{3}H(α,γ)^{7}Li experimental results. A study of the ^{16}O(γ,α)^{12}C will help understanding He burning in massive stars. The cross section of the time-reversal ^{12}C(α,γ)^{16}O reaction has been measured down to energies around 1 MeV but must be extrapolated to helium-burning energies around 300 keV. There are two energy ranges, which are important for the study of the ^{12}C(α,γ)^{16}O cross section, at energies below 1 MeV to approach the Gamow peak or at higher energies to constrain the R-matrix extrapolation. Modelling the evolution and explosion of massive stars requires a 10% uncertainty on the ^{12}C(α,γ)^{16}O reaction cross section but recent analysis of the world data estimated this uncertainty at around 20% [12]. The approach for such studies is the measurement of the time-reversal ^{16}O(γ,α)^{12}C reaction.

Two instruments will be used for the realization of these experiments, a 4π array of Si strip detectors [1,13,14] and a time-projection chamber (TPC) with an electronic read-out [1,15,16].

Experiments in the field of photo-fission address mapping of the fission barriers, as well as studies of rare fission modes. A flagship experiment is the investigation of fission transmission resonances in the second and third potential minima in the actinides. Studies of the photo-fission cross-section and fragment identification, will result in mapping the fission barrier landscape as a function of the photon-beam energy in the range from 4 MeV to 8 MeV [17].

Pilot experiments on ^{238}U and ^{232}Th [18,19] have been carried out at the HIγS LBS facility, but the available beams there lack resolution to resolve the resonance states. Further experiments at ELI-NP will profit from the narrow bandwidth of the γ beam.

The photofission the research programme includes also measurement of kinetic energy, angular, mass and charge distribution of fission fragments, measurements of absolute photofission cross-sections, studies of rare photofission events, such as triple fission, highly asymmetric fission, clusterization phenomena, the predicted cold valleys of fission potential, etc [1].

3. Conclusion

A diverse research program is been prepared at ELI-NP, which covers all possible directions of photonuclear studies. State-of-the-art instrumentation is constructed for the realization of these experiments. All detector arrays will be ready in time to take first available beams and perform the defined day-one experiments.

Acknowledgements

This work is supported by the Extreme Light Infrastructure Nuclear Physics (ELI-NP) Phase II, a project co-financed by the Romanian Government and the European Union through the European Regional Development Fund – the Competitiveness Operational Programme (1/07.07.2016, COP, ID 1334)

References

1. ELI-NP Technical Design Reports, *Rom. Rep. Phys.* **68**, Supplement (2016).
2. D. Filipescu et al., Perspectives for photonuclear research at the Extreme Light Infrastructure – Nuclear Physics (ELI-NP) facility, *Eur. Phys. J. A* **51**, 185 (2015).
3. S. Gales et al., New frontiers in nuclear physics with high-power lasers and brilliant monochromatic gamma beams, *Phys. Scr.* **91**, 093004 (2016).
4. D. L. Balabanski et al., New light in nuclear physics: The extreme light infrastructure, *Europhys. Lett.* **117**, 28001 (2017).
5. S. Gales et al., The extreme light infrastructure – nuclear physics (ELI-NP) facility: New horizons in physics with 10-PW ultra-intense lasers and 20-MeV brilliant gamma beams, *Rep. Prog. Phys.* **81**, 094301 (2018).

6. D. Belic et al., Photo-induced depopulation of the ^{180}Ta^{m} isomer via low-lying intermediate states: Structure and astrophysics implications, *Phys. Rev. C* **65**, 035801 (2002).
7. D. Belic et al., Photoactivation of ^{180}Ta^{m} and its implications for the nucleosynthesis of the Nature's rarest naturally occurring isotope, *Phys. Rev. Lett.* **83**, 5242 (1999).
8. H. Utsunomiya et al., Photodisintegration of ^{9}Be with laser-induced Compton backscattering γ rays, *Phys. Rev. C* **63**, 018801 (2000).
9. C. W. Arnold et al., Cross-section measurement of ^{9}Be(γ,n)^{8}Be and implications for the $\alpha + \alpha + n \rightarrow {}^{9}$Be in the *r* process, *Phys. Rev. C* **85**, 044605 (2012).
10. H. Utsunomiya et al., Photodisintegration of ^{9}Be through the $1/2^{+}$ state and cluster dipole resonance, *Phys. Rev. C* **92**, 064323 (2015).
11. M. Krzysiek et al., Simulation of the ELIGANT-GN array performances at ELI-NP for gamma beam energies larger than neutron threshold, *Nucl. Instr. Meth. Phys. Res. A* (2018), https://doi.org/10.1016/j.nima.2018.11.58.
12. R. J. deBoer et al, The ^{12}C(a,γ)^{16}O reaction and its implications for stellar helium burning, *Rev. Mod. Phys.* **89**, 035007 (2017).
13. C. Matei, D. Balabanski, D. Filipescu, O. Tesileanu, Photodisintegration reactions for nuclear astrophysics studies at ELI-NP, *J. Phys. Conf. Series* **940**, 012025 (2018).
14. S. Chesnevskaya et al., Performance studies of X3 silicon detectors for the future ELISSA array at ELI-NP, *JINST* **13**, T05006 (2018).
15. M. Cwiok et al., A TPC detector for studying photonuclear reactions at astrophysical energies with gamma-beams at ELI-NP, *Acta Phys. Pol. B* **49**, 509 (2018).
16. M. Gai et al., Time-projection chamber (TPC) detectors for nuclear astrophysics studies with gamma beams, *Nucl. Instr. Meth. Phys. Res. A* (2019), https://doi.org/10.1016/j.nima.2019.01.006.
17. A. Krasznahorkay, Tunneling through triple-humped fission barriers, in *Handbook of Nuclear Chemistry, vol. 1*, eds. A Vertes et al. (Dordrecht: Springer, 2011) pp. 281-318.
18. L. Scige et al., Exploring the multihumped fission barrier in ^{238}U via sub-barrier photofission, *Phys. Rev. C* **87**, 044321 (2013).
19. J. A. Silano and H. A. Karwowski, Near barrier photofission in ^{238}U and ^{232}Th, *Phys. Rev. C* **98**, 054609 (2018).

Electrospun Metal Oxide Nanofibers for Applications in Electronics and Photonics

O. Ya. Berezina*, N. P. Markova, A. L. Pergament,
E. N. Kolobova, D. A. Kirienko and D. I. Balashov

Petrozavodsk State University, Petrozavodsk, 185910, Russia
**E-mail: berezina@petrsu.ru*

Metal oxide nanofiber matrices are synthesized on SiO_2 covered silicon and glass-ceramic substrates by the electrospinning method. The materials studied are indium-zinc oxide (IZO), vanadium dioxide, and Er-doped sodium-potassium niobate. The results of studying the electrical properties of IZO nanofibers, as well as their sensitivity to UV radiation depending on the In to Zn concentration ratio, are presented. It is shown that the highest sensitivity to UV is observed at the indium content of about 50 atomic %. The photocurrent increment with respect to the dark current is more than 4 orders of magnitude. The response and recovery times are 60 and 500 sec., respectively. The results obtained suggest that IZO nanofibers can find application as UV sensors with improved characteristics. Vanadium and niobium oxide nanofibers possess applied potentialities for electronic and photonic devices and as biocompatible materials.

Keyword: Metal oxide nanofiber matrices.

1. Introduction

Sol-gel electrospinning routes have been actively exploited for the synthesis of a variety of polymeric, metallic, and metal oxide nanofibers. The surface functionality of electrospun nanofibers could be greatly manipulated by the immobilization of metallic and metal oxide catalytic nanoparticles, graphenes, and functional polymers to provide greatly enhanced electrochemical, structural, electrical, and optical properties [1].

In this work we report on the study of the physical properties of metal oxide nanofibers, such as indium-zinc oxide (IZO), a material which is able to change its electrical resistance when exposed to UV or absorbed gases, vanadium dioxide (VO2), a material with metal-insulator transition, and sodium-potassium niobate NaxK1-xNbO3 (NKN), a material exhibiting ferroelectric and piezoelectric properties. The choice of these materials is conditioned by the fact that all of them exhibit phenomena interesting from the viewpoint of physics, viz. phase transitions, photo- and gas-sensory properties, and on the other side, they are all possess applied potentialities in oxide electronics and photonics. Meanwhile, the implementation of these materials in the form of nanostructures can significantly modify the manifestation of these effects and improve the corresponding functional properties, which is important for technical applications.

2. Experimental Details

The samples under study were obtained by means of electrospinning of corresponding gel precursor solutions containing metal compounds, namely, hydrated indium nitrate and zinc acetate dehydrate, vanadium acetyl acetonate, and sodium and potassium acetates with niobium ethoxide and erbium nitrate pentahydrate, respectively, in case of IZO, VO_2, and NKN:Er nanofibers. Polyvinylpyrrolidone (PVP) was used as a polymer binder, and after deposition, the nanofiber samples were annealed to remove PVP. Sample preparation procedures have been described in more detail in our previous works [2-4].

The nanofibers' dimensions and morphology were characterized by scanning electron microscopy (Hitachi SU 1510) and optical microscopy. Current-voltage (*I-V*) characteristics were measured with a Keithley 2410 SourceMeter. For the *I-V* characteristic measurements, gold contacts (1 mm in diameter with a 1.5 mm distance between contacts) were deposited onto the nanofiber matrices on Si-SiO_2 substrates using a Cressington 108 AUTO sputtering system.

To assess the photosensitivity of the IZO nanofibers, a source of weak UV radiation in the wavelength region 230-290 nm, based on a mercury lamp, was used. It was placed directly over the sample at a distance of 5 cm. The sample surface irradiance, measured with a TKA-PCM 12 radiometer, was approximately 1 mW/cm^2 in the UVC (λ = 100-280 nm), 0.055 mW/cm^2 in UVA (320-400 nm), and 0.026 mW/cm^2 in UVB (280-320 nm) regions.

3. Results and Discussion

The diameter of the obtained IZO nanofibers ranges from 250 to 320 nm (Fig. 1a), and after annealing it reduces down to 50-100 nm (Fig. 1b). The same is true for niobium and vanadium oxides; for instance, the average diameter of NKN fibers after annealing is 100-200 nm (Fig. 1c).

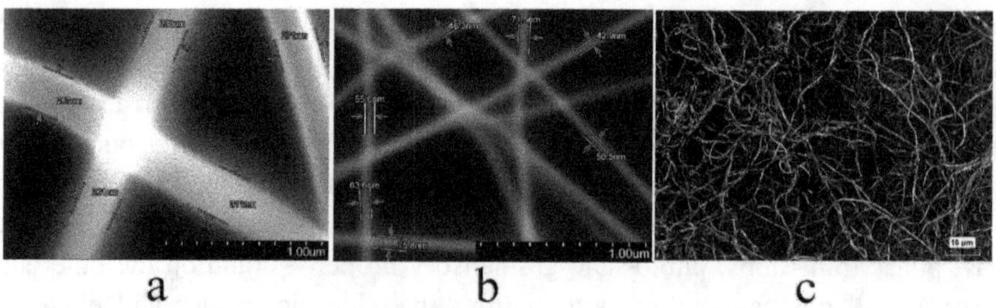

Fig. 1. SEM images of IZO nanofibers before (a) and after (b) annealing. (c) Large depth-of-field optical image of Er:(Na,K)NbO_3 fiber mat captured by 3D laser scanning microscope Keyence VK-9710 [4].

In the case of IZO, a series of samples with different In to Zn concentration ratio have been prepared. It is shown that the highest sensitivity to UV is observed in the IZO nanofiber samples at the indium content of about 50 at.% (Fig. 2).

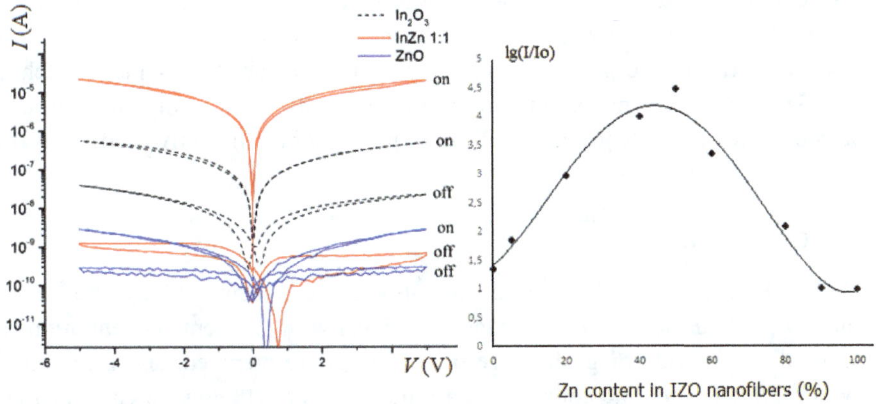

Fig. 2. (left) *I-V* characteristics of samples with In content of 100% (In_2O_3), 50% (InZn 1:1), and 0% (ZnO) when they are irradiated with UV ("on") and under natural illumination ("off"). (right) UV sensitivity (with I_0 being the dark current) of IZO nanofibers as a function of zinc percentage.

The photocurrent increment with respect to the dark current is more than 4 orders of magnitude. The response and recovery times are 60 and 500 sec., respectively.

A slow rise and decline of the conductivity might be associated with the photochemical processes of oxygen desorption and adsorption on the fibers' surface. Oxygen molecules from the environment are easily absorbed on the surface of nanofibers by capturing free electrons from the conduction band, and the molecules become negatively charged ions in this case. When exposed to UV radiation, electron-hole pairs are generated in the IZO nanofibers. Then, the photogenerated holes recombine with adsorbed O_2^- ions to form oxygen molecules that desorb from the nanofibers' surface [5]. Simultaneously, an increase in the number of electrons in the conduction band occurs. Under the influence of the applied bias voltage, free electrons move to the anode, and the photocurrent thus appears. The larger the surface area of IZO nanofibers, the higher the rate of adsorption (desorption) of oxygen molecules from their surface at UV irradiation and therefore the higher the photocurrent is. At the switching off the source of UV radiation, oxygen from air is adsorbed to the surface of the fibers, captures excess electrons from the conduction band, and the resistance of the fibers gradually increases to the initial value.

The obtained vanadium oxide nanofiber samples demonstrate the metal–insulator transition (MIT) at T_t ~60 °C and electrical switching with an S-shaped *I-V* characteristic [3]. In the NKN nanofiber samples, Er doping with the concentration of 2 at. % provides readily detectable room-temperature broad-band photoluminescence. Impedance spectroscopy and static electrical tests revealed ferroelectric properties and electric field induced resistance switching [4]. Note that NKN nanofibers are promising lead-free biocompatible strongly ferroelectric materials [6], and VO_2, in turn, might be efficiently used in diverse devices of oxide electronics [7].

4. Conclusion

IZO, NKN, and vanadium oxide nanofibers are synthesized by the sol-gel electrospinning method. Zn-In oxide nanofibers with different concentrations of In oxide (in a wide range from zero to 100 %) are prepared, and the samples with the indium to zinc concentration ratio of about 1:1 show an enhanced UV sensitivity. In such IZO samples, the ratio of the photocurrent to the dark current reaches 4.5 orders of magnitude.

Intense room-temperature photoluminescence is achieved in Er-doped $(Na,K)NbO_3$ nanofibers. Ferroelectric properties, bipolar resistance switching, and strong rectification effect in a sandwich Au/Er:NKN/Pt capacitive cell are demonstrated. Vanadium oxide nanofibers are shown to exhibit metal-insulator transition and electrical switching with an S-type *I-V* characteristic.

The obtained metal-oxide nanofibers possess applied potentialities for electronic and photonic devices and as biocompatible materials. Particularly, IZO nanofibers can find application as UV sensors with improved characteristics. Photoexcited luminescence adds to NKN fibers great potential to serve as multimodal bioimaging tensile and torsion sensors, energy harvesting biocompatible nanogenerators, material for bioelectronic implants, electrically polarizable scaffolds for bioengineering, repair, and regeneration of damaged tissue. MIT and switching in VO_2 might be used in Mott-FETs, memory cells, oscillatory neural networks, and other various devices of oxide electronics.

Acknowledgements

This work was supported by the RF Ministry of Education and Science, project no. 16.5857.2017/8.9 (state program) and by the Flagship University Development Program of Petrozavodsk State University (2017–2021).

References

1. Il-Doo Kim, Seon-Jin Choi. Nanofibers and Nanotubes. In: The Sol-Gel Handbook. Editors: David Levy, Marcos Zayat. 2015 Wiley-VCH Verlag GmbH & Co. KGaA.
2. N. P. Markova, O. Ya. Berezina, A. L. Pergament. Dependence of the photoconductivity of nanofibers based on indium-zinc oxide on the concentration of components. *Journal of Physics: Conf. Series.* **929**, 012046 (2017).
3. O. Berezina, D. Kirienko, A. Pergament, G. Stefanovich, A. Velichko, V. Zlomanov. Vanadium oxide thin films and fibers obtained by acetylacetonate sol–gel method. *Thin Solid Films.* **574**, 15 (2015).
4. A. M. Grishin, N. P. Markova. Broadband luminescent ferroelectric biocompatible Er:(Na,K)NbO_3 nanofibers. *J. Am. Ceram. Soc.* **100**, 1051 (2017).
5. C. Soci, A. Zhang, B. Xiang, S. A. Dayeh, D. P. Aplin, J. Park, X. Y. Bao, Y. H. Lo, and D. Wang. ZnO nanowire UV photodetectors with high internal gain. *Nano Letters.* **74**, 1003 (2007).
6. A. Jalalian, A. M. Grishin. Biocompatible ferroelectric (Na,K)NbO_3 nanofibers. *Appl. Phys. Lett.* **100**, 012904 (2012).
7. V. A. Pergament, A. Velichko, M. Belyaev, Putrolaynen. Electrical switching and oscillations in vanadium dioxide. *Physica B.* **536**, 239 (2018).

Determination of Separation Efficiency of the Mass-Spectrometer MASHA by Means of Measurement of Absolute Cross-Sections of Evaporation Residues

E. V. Chernysheva, A. M. Rodin, S. N. Dmitriev, A. V. Gulyaev, A. B. Komarov, A. S. Novoselov, Yu. Ts. Oganessian, A. V. Podshibyakin, V. S. Salamatin, S. V. Stepantsov, V. Yu. Vedeneev and S. A. Yukhimchuk

Flerov Laboratory of Nuclear Reactions, Joint Institute for Nuclear Research, Dubna, 141980, Russia

L. Krupa, M. Holik, S. Pospisil and I. Stekl

Institute of Experimental and Applied Physics, Czech Technical University in Prague, Horska 3a/22, Prague, 128 00 Praha 2, Czech Republic

J. Kliman and D. Kamas

Institute of Physics SASc Dubravska Cesta 9, 84228 Bratislava, Slovakia

A. Opíchal and J. Pechousek

Faculty of Science of Palacký University in Olomouc, 17, Listopadu 1192/12, 779 00 Olomouc, Czech Republic

A. Maher

Centre for Theoritical Physics (CTP), Faculty of Engineering, The British University in Egypt (BUE) El Sherouk City, Misr-Ismalia Road, Egypt

The separation efficiencies of the mass-spectrometer MASHA were determined for short-lived mercury isotopes produced in *xn*-evaporation channels of complete fusion reaction $^{40}Ar+^{144}Sm \rightarrow ^{184}Hg$. For the determination of the efficiencies two independent techniques were used: the modified catcher foil technique [1] for the measurement of the absolute cross-sections and the solid state ISOL technique using the mass-spectrometer MASHA [2]. The absolute cross-sections of evaporation residua in the reactions $^{40}Ar+^{144}Sm \rightarrow ^{184}Hg$ and $^{48}Ca+^{142}Nd \rightarrow ^{190}Hg$ were obtained with use of moving absorber array. The beam interruption method was applied. During the accumulation time (beam is on) five thin aluminum foils of absorber array were placed one under another so the reaction products were stopped there. After that the absorber array was moved and extended to its full length so that each absorber foil was placed opposite to the corresponding silicon detector during the measurement time (beam is off). Then the α-decays of synthesized isotopes were detected and the energy and the exact time of α-particle registration were recorded. As a result, the method allowed reliable identification of reaction products from their α-decay energies and decay times.

Keywords: ISOL technique; mass-spectrometer; separation efficiency; absolute cross-sections.

1. Introduction

Mass Analyzer of Super Heavy Atoms (MASHA) is the mass-spectrometer designed for identification of superheavy nuclei by their masses with simultaneous detection of their α-decays or spontaneous fission [2, 3].The spectrometer combines ISOL method [4] of nuclide extraction from a hot catcher and classical mass-spectrometry methods for registration of the reaction products. The separation efficiency is a very important parameter in the experiments with superheavy nuclei which have extremely low production cross-sections.

Two independent experiments were carried out aimed at measuring the separation efficiencies of mercury isotopes. In the first one the absolute cross-sections of xn-evaporation channels of the reactions $^{40}Ar + ^{144}Sm$ and $^{48}Ca + ^{142}Nd$ were measured using array of moving aluminum absorbers (the modified well-known catcher-foil experiment [1]). In the second experiment the yields of isotopes in the same reactions were obtained with the mass-spectrometer MASHA using the solid state ISOL method. The ratio of the cross-sections measured in these experiments allowed us to extract the separation efficiencies of the selected isotopes.

2. Experiment

The experiments were carried out at the Flerov Laboratory of Nuclear Reactions (JINR, Dubna). The beams of ^{40}Ar and ^{48}Ca ions from U400-M cyclotron were used. As a target the layers of ^{144}Sm and ^{142}Nd with thicknesses of 320-330 μg/cm^2 were placed on the 1.5 μm titanium backings. The scheme of the experiment with the moving absorbers is shown in Fig. 1. On the left-hand side (a) the foil array is in the accumulation regime when all foils are placed one

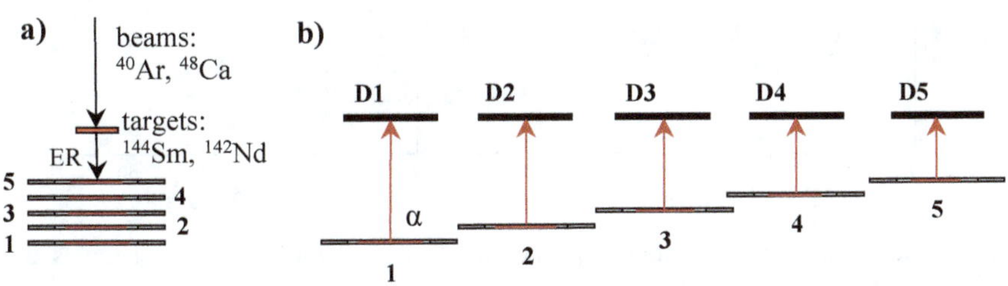

Fig. 1. Scheme of the experiment with the moving aluminum absorbers: a) the positions of absorber foils are one under another in the accumulation regime when beam is on; b) absorbers array is extended to its full length, each foil is placed opposite to the corresponding detector in the measurement regime when beam is off.

under another. The evaporation residua from the complete fusion reactions $^{40}Ar + ^{144}Sm$ and $^{48}Ca + ^{142}Nd$ were stopped according to their energies and ranges in different foils of the array. Then the beam was switched off and the array of five foils was extended to its full range of 15 cm, so each foil was placed opposite to the corresponding silicon detectors D1-D5 (shown in Fig. 1 on the right-hand side) during the accumulation time. The time of moving of aluminum absorbers between their extreme positions was 0.3 s. The α-particles emitted by short-lived mercury isotopes were registered by silicon detectors. The measurement and accumulation total cycles were 20 s for $^{179\text{-}183}Hg$ isotopes produced as the evaporation residua in the complete fusion reaction $^{40}Ar+^{144}Sm$ and 120 s for the $^{183\text{-}185}Hg$ isotopes produced in the reaction $^{48}Ca+^{142}Nd$.

The nuclide identifications were performed using both the energies of α-particles emitted at the decay of reaction products and their times of life. The energy and time characteristics of the reaction products are presented in Fig. 2. To calculate the times of life of nuclides we used the time difference measured between the start of measurement cycle and the time when α-particle was detected. In Figs. 2c,d these times are shown for different energy intervals. The lines show the fits of experimental data by the decay curve. Fig. 2b presents the decomposition of the energy spectrum of registered α-particles into the

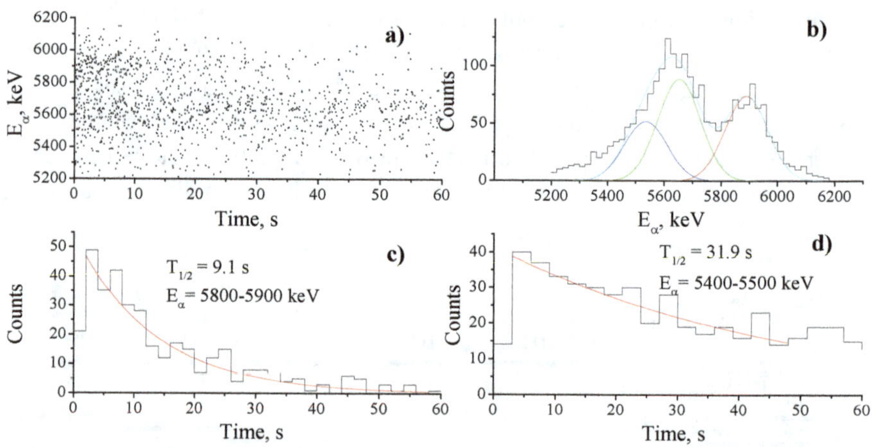

Fig. 2. Identification of mercury isotopes produced in *xn*-evaporation channels of the reaction $^{48}Ca+^{142}Nd \rightarrow ^{190}Hg$ at the energy E_{lab}=240 MeV, using energies and times of registration of emitted α-particles: a) two dimensional matrix (T_α,E_α); b) energy spectrum of α-particles; c) and d) times of registration of α-decays of ^{183}Hg and ^{185}Hg isotopes with their calculated times of lives.

components corresponding to *5n, 6n* and *7n* channels for the reaction $^{48}Ca + ^{142}Nd$ at a beam energy of $E_{lab} = 240$ MeV.

To calculate the absolute cross sections the number of events of each isotope was corrected on its α-decay probability and the coefficients dealt with the detector geometries. Besides the ratio of the nuclides half-lives to the time of measurement cycle was taken into account.

In the second experiment the mercury isotopes, produced in the same complete fusion reactions, passed through the hot catcher, the ECR ion source, the magneto-optical system of the spectrometer, and were detected by 192-strip detector in the focal plane of the spectrometer. The mass resolution of the spectrometer is $M/\Delta M = 1300$. As a hot catcher the nano-graphite paper was used. The effective cross-sections obtained in this experiment were used further for determination of the separation efficiencies of the mass-spectrometer.

3. Results of Experiment

The absolute cross-sections for *2n-4n* channels of the reaction $^{40}Ar+^{144}Sm$ at the energies $E_{lab} = 150$-210 MeV are shown in Fig. 3 together with Channel Coupling calculations performed with the NRV code [5]. The influence of the beam energy spread on the excitation functions was taking into account. The deconvolution of the data was performed with the procedure [6].

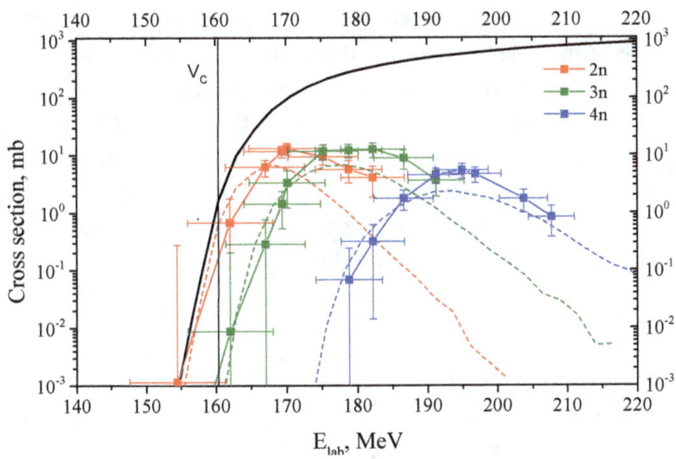

Fig. 3. Squares — the absolute cross-sections of 2*n*-4*n* evaporation channels in the reaction $^{40}Ar + ^{144}Sm \rightarrow ^{184-xn}Hg + xn$ obtained by the method of moving absorber foils. Lines — theoretical calculations of 2*n*-4*n* channels from the Channel Coupling model [5].

It is seen from Fig. 3, that shapes of excitation functions are similar to predicted curves, but the absolute values are approximately 2 times higher. The positions of maxima of *xn*-channels of the reaction $^{40}Ar + ^{144}Sm$ are presented in Table 1. The separation efficiency of nuclides depends mainly on the extraction efficiency from the hot catcher and the times of life of the nuclides. The total separation efficiencies of the mass-spectrometer MASHA were found as the ratio of the effective cross-sections of the residua measured with the mass-spectrometer MASHA to the absolute cross-sections obtained in the independent experiment with the moving absorbers. The dependence of these separation efficiencies on the nuclide half-lives were observed. The efficiency was found to be 0,81% for ^{182}Hg, 1,11% for ^{181}Hg and 1,04% for ^{180}Hg.

Table 1. Absolute cross sections for $2n$ - $4n$ evaporation channels of the complete fusion reactions $^{40}Ar+^{144}Sm$.

	$^{182}Hg(2n)$	$^{181}Hg(3n)$	$^{180}Hg(4n)$
σ, mb	13.0±1.5	12.0±2.1	5.4±0.9
E, MeV	169	179	195

4. Conclusion

The mass-spectrometer MASHA separation efficiencies of the of mercury isotopes were determined in the complete fusion reaction $^{40}Ar + ^{144}Sm$. The absolute cross-sections were found in the experiment with moving aluminum absorbers. The beam interruption method allows to define the times of life of nuclides and hence provides a reliable identification of reaction products. The absolute cross-sections were compared with the Couple Channels calculations [5] taking into account the influence of beam energy spread on excitation functions. The separation efficiencies depend on the nuclide times of lives. It was found to be 0,81% for ^{182}Hg, 1,11% for ^{181}Hg and 1,04% for ^{180}Hg isotopes.

References

1. D. Vermeulen *et al.*, Z. Phys. A, 1984, V. 318, p. 157-169.
2. A. M. Rodin *et al.*, Instr. Exp. Tech., 2014, V. 57, No. 4, p. 386-393.
3. A. M. Rodin *et al.*, Hyperfine Interaction, 2014, V. 227, p. 209-221.
4. S. Lukic *et al.*, Nucl. Instr. Meth. in Phys. Res. A, 2006, V. 565, p. 784-800.
5. http://nrv.jinr.ru/nrv A.V. Karpov *et al.*, NRV web knowledge base on low-energy nuclear physics // Phys. of Atomic Nuclei, 2016, V. 79, p. 749-761.
6. M. Morhac *et al.*, Nucl. Instr. Meth. in Phys. Res. A, 1997, V. 401, p. 385-408.

Experimental Foundations of Nuclear Physics in Non-Euclidean Spaces*

V. V. Dyachkov[†], Y. A. Zaripova and A. V. Yushkov
Scientific Research Institute of Experimental and Theoretical Physics, Almaty, Republic of Kazakhstan
[†]*E-mail: slava_kpss@mail.ru*

An analysis of a number of modern experiments with nuclei having a deformation in the framework of the Riemannian geometry with positive and negative geodesic curvature is given.

Keywords: Non-euclidean space; curvature; geodesic; Euler characteristic.

Introduction

In a number of previous works by the authors [1–8], the necessity was substantiated of a transition in the recent history of the development of nuclear physics to the compilation of the equations of motion in a true adequate physical space. Which, the authors proclaimed non-Euclidean Riemannian elliptic closed space. This corresponds to the program outlined by B. Riemann as early as 1854 in his trial lecture (On the hypotheses underlying the geometry): "We are on the threshold of a field belonging to another science — physics". Einstein A. had a similar problem statement when he spoke about the geometrization of physics. Both of these programs came together when solving problems of motion in the microcosm. In fact, the asymptotically giant density of nuclear matter ρ=0,147 fm-3 unambiguously establishes the Riemannian character of intra- and near-nuclear space. The second experimental fact: the non-radiative motion of electrons around the nucleus along closed trajectories and the non-radiative motion of valence nucleons in closed orbits around the even-even foundation of the nucleus clearly indicate the elliptical metric of the geodesic in both cases. Finally, the third experimental fact is the finiteness of the charge number Z of the Mendeleev's table and the finiteness of the isotopic chains of exotic nuclei, both toward the neutron deficit and toward the neutron excess, clearly indicates a change in the signs of the curvature of space $\varkappa$ in all three of these cases. Namely, at a radial distance from the center of gravity of the nucleus, the

* Work partially supported by grant IRN AP05131884 of the Ministry of Education and Science of the Republic of Kazakhstan.

positive sign of the Riemannian space $\varkappa>0$ is replaced to the negative sign of the Lobachevsky space $\varkappa<0$.

1. Latest Experimental Evidence

Among the newest experimental facts prompting a change of the Schrödinger motion paradigm in the microworld to the Riemannian paradigm in Einstein space-time, we note, first of all, not the infinite removal of the Fermi level in the excited states of the nuclei, but its final position according to one estimate at ~150 MeV, others – 25-50 MeV.

Another experimental evidence is the existence of halo-nuclei and a clearly defined dependence of the size of the nucleus on the energy of the excited levels [9]. Finally, we note the sum of the experimental facts that reinforce conclusions about the existence in nuclei, in addition to the gas and liquid-droplet phases of nuclear matter, also of the solid-state quasicrystalline phase [10].

The ordered symmetry arrangement in the volume of nuclei of nucleons, alpha clusters and multiclusters contributes to the phenomenon of dense spherical packages, which is widely known in classical crystallography. The indicated circumstance throws a bridge from the macro-world of classical crystals into the micro-world of quasicrystal nuclei. In fact, out of 230 known Fedorov crystal symmetries, according to the Belov theorem, only 8 meet the conditions of the densest globular packages. This remarkably coincides with the eight magic numbers of the nuclei 2, 8, 20, 28, 50, 82, 126, 184, which undoubtedly indicates that the nature of the fundamental cause of the magic of the nuclei are the densest ball packages of nuclides with multiclusters.

2. Action Program

The program for constructing a new theory of Riemannian nuclear physics naturally follows from the program for constructing Riemannian geometry and tensor analysis. Such a program is firmly based on the enormous experimental legacy accumulated since the discovery of the atomic nucleus in 1911 and on the extremely high elaboration of the Riemannian-Lobachev geometry itself. These two points require a radical revision of the system of physical concepts in the microcosm and, above all, those that are amenable to nuclear experiment and accurate measurements.

Let us consider such a general new system of concepts on the example of the fundamental consequences of the new concept of “curvature” of nuclear space.

The first consequence: the curvature of the space $\varkappa$. This quantity, new to nuclear physics, is measurable in its well-established definition in Riemannian geometry for spherical and spheroidal bodies:

$$\varkappa = \frac{1}{\rho^2}, \tag{1}$$

where ρ-the radius of curvature of space, which we, in the first approximation, identify with the radius of the nucleus $\rho \equiv R_N$.

The second consequence: the limit radius of curvature of space ρ_{max}, at which the Riemannian space with positive curvature $\varkappa_R>0$ transitions to the Lobachev space with negative curvature $\varkappa_L<0$. The value of ρ_{max} today can be, if not precisely measured, then sufficiently accurate (within 10%) estimate for the asymptotic decrease of half-lives in isotopic chains.

The third consequence: the limits of the existence of bound states of nuclei by Z-N–matrix - $\rho_{max}(Z,N)$, as the boundaries of the elements of the construction of the substance of the Universe.

This concept naturally extends to the limits of the existence of planetary systems near stars; the limits of the existence of bound states of galaxies, depending on the density of the nucleus of galaxies and, ultimately, on the limits of the existence of universes.

3. Conclusion

Thus, we can safely assume that in our day the Schrodinger path of development of nuclear physics is completed and a new Riemannian path begins. Such a movement of exotic nuclei and exotic spaces is not a manifestation of crisis phenomena in nuclear physics. This is a normal natural evolutionary return to the Newtonian deterministic paradigm of movement of microparticles, but not in the Newtonian, but in the Einstein paradigm of space-time.

The program of actions outlined in this short note will undoubtedly define a new big leap in knowledge about the microworld: in the physics of atoms, in the physics of atomic nuclei and in particle physics.

References

1. A.V. Yushkov, V.V. Dyachkov, Y.A. Zaripova, New laws in nuclear physics and microworld structures, *Monthly Scientific Journal* **9(20)**, 35 (2015).
2. V.V. Dyachkov, Y.A. Zaripova, A.V. Yushkov, Cluster structure of nuclei and new spatial regularities, *KazNU Bulletin. Physics Series* **2(57)**, 88 (2016).

3. V.V. Dyachkov, Y.A. Zaripova, A.V. Yushkov, Experimental foundations of new nuclear physics in non-Euclidean spaces, *in Materials of the International Scientific Conference Dedicated to the 80th Anniversary of Academician of the NAS RK Abdildin M.M.* (Almaty, Kazakhstan, 2018).
4. A.V. Yushkov, V.V. Dyachkov, Y.A. Zaripova, Ya.V. Sidorov, Topological typology of nuclei, *in Materials of the International Scientific Conference Dedicated to the 80th Anniversary of Academician of the NAS RK Abdildin M.M.* (Almaty, Kazakhstan, 2018).
5. Yu.E. Penionzhkevich, A.V. Yushkov, Clusterization and crystallization of nuclei, *in Proc. LXYIII International Conference NUCLEUS 2018*, (Voronezh, Russia, 2018).
6. A.V. Yushkov, V.V. Dyachkov, Y.A. Zaripova, Ya.V. Sidorov, Topological typology of nucle, in *Book of Abstracts IX International Symposium on Exotic Nuclei* (*EXON 2018*), (Petrozavodsk, Russia, 2018).
7. Y.A. Zaripova, V.V. Dyachkov, A.V. Yushkov, Clusterization and Crystallization of Complex Nuclei, in *Book of Abstracts IX International Symposium on Exotic Nuclei* (*EXON 2018*), (Petrozavodsk, Russia, 2018).
8. V.V. Dyachkov, Y.A. Zaripova, A.V. Yushkov, Experimental Foundations of Nuclear Physics in Non-Euclidean Spaces, in *Book of Abstracts IX International Symposium on Exotic Nuclei* (*EXON 2018*), (Petrozavodsk, Russia, 2018).
9. A.S. Demyanova, Identification of dilute cluster states in ^{11}B, ^{12}C, ^{13}C, in *Book of Abstracts IX International Symposium on Exotic Nuclei* (*EXON 2018*), (Petrozavodsk, Russia, 2018).
10. N.N. Pavlova, A.M. Ivanov, A.V. Yushkov, K.A. Toktarov, Some rules in isotopic changes of the structural form of light, average and heavy-nuclei, *IZVESTIYA AKADEMII NAUK SSSR SERIYA FIZICHESKAYA* **43(11)**, 2317 (1979).

Current Status of New Fragment Separator ACCULINNA-2 and the First Day Experiments

A. V. Gorshkov[1,2,*], S. G. Belogurov[1], A. A. Bezbakh[1,2,3], D. Biare[1], W. Beekman[4], V. Chudoba[1,3], A. S. Fomichev[1,5], M. S. Golovkov[1,5], E. M. Gazeeva[1], L. V. Grigorenko[1,2,6], G. Kamiński[1,7], S. A. Krupko[1,2], B. Mauyey[1,2,8], I. A. Muzalevsky[1,2], E. Yu. Nikolskii[1,2], Yu. L. Parfenova[1,9], A. Serikov[1], S. I. Sidorchuk[1], R. S. Slepnev[1], P. G. Sharov[1,2], G. M. Ter-Akopian[1,5], R. Wolski[1] and B. Zalewski[1,7]

[1]*Flerov Laboratory of Nuclear Reactions, JINR, 141980 Dubna, Russia*
[2]*National Research Center "Kurchatov Institute", 123182 Moscow, Russia*
[3]*Institute of Physics, Silesian University in Opava, 74601 Opava, Czech Republic*
[4]*SIGMAPHI, 56000 Vannes, France*
[5]*Dubna State University, 141982 Dubna, Russia*
[6]*National Research Nuclear University "MEPhI", 115409 Moscow, Russia*
[7]*Heavy Ion Laboratory, University of Warsaw, 02-093 Warsaw, Poland*
[8]*L.N. Gumilyov Eurasian National University, 010000 Astana, Kazakhstan*
[9]*Skobeltsyn Institute of Nuclear Physics, MSU, 119991 Moscow, Russia*
[*]*E-mail: agorshkov@jinr.ru*

Commissioning of the new fragment separator ACCULINNA-2 at FLNR JINR is accomplished. The separator is destined to expand the possibilities in studies of dripline nuclei performed with the exotic secondary radioactive ion beams (RIBs) at energies of 5 – 50 *A*MeV. The projected high transmission and purification level were confirmed experimentally for a number of RIBs in last two years. The ACCULINNA-2 setup will become a backbone facility at FLNR for the research in the field of light exotic nuclei. This report shows the current status of the separator, describes the obtained RIB's parameters and first experiments as well, provides the overview of the developing detection, monitoring and control subsystems.

Keywords: Fragment separator; secondary beams; exotic nuclei.

1. Present Status of ACCULINNA-2 Experimental Complex

1.1. *Introduction*

The new fragment separator ACCULINNA-2 in the Flerov Laboratory of Nuclear reactions, JINR Dubna is in the commissioning phase since 2016 [1, 2]. The separator and a part of the primary beam line were designed and built in collaboration with SIGMAPHI company, France [3–5]. It is 36 meters long achromatic separator consisting of two 45-degree dipole magnets, 14 quadrupoles, 8 multipoles (3 octupoles and 5 sextupoles) and 4 steering magnets.

The high intensity primary beams are delivered by the U-400M cyclotron to the rotating production target module installed in the first intermediate focal plane F1 (see Fig. 1) to produce radioactive ion beams in fragmentation reactions via in-flight method.

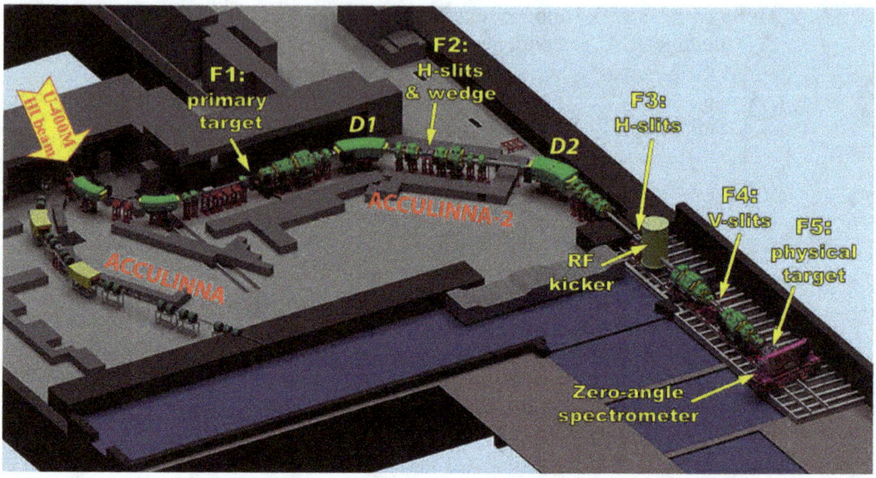

Fig. 1. The ACCULINNA-2 layout at U-400M cyclotron cave. The separator focal planes are indicated by F1-F5. The concrete radioactive protection, the dipole magnet of zero-angle spectrometer and planned RF-kicker are also shown.

The target module integrates a vacuum chamber with water cooled beryllium target mounted on rotating magnetic liquid feedthrough and a set of water-cooled diaphragms. The target module is designed to work with heating power up to 2 kW. Using the combination of magnetic analysis and energy losses in achromatic wedge degrader located at the intermediate dispersive focal plane F2, the secondary ions are separated in flight. Then, the secondary beam is delivered into the low-background experimental area with full particle-by-particle identification.

1.2. *Separator equipment and scientific tools*

In order to become fully-featured measuring complex for the studies of exotic nuclei, ACCULINNA-2 should be equipped with a numerous devices. The separator equipment and scientific tools for ACCULINNA-2 facility are under active development. Some parts were already mounted and took part in first scientific experiments, some of them are still in R&D or production stages.

For full-scale operation of ACCULINNA-2 complex, it is required to control not only the magnetic elements of the fragment separator, but also all gas-vacuum equipment and moving mechanisms. The control system made by ACCULINNA-2 group allows remotely control all movable devices (pneumatically actuated and motor driven motion feedthroughs), all vacuum and gas equipment and provides beam current measurement from all Faraday cups. Interlocks are performed at hardware level with programmable relays and in case of incident guarantee 30 ms reaction time. The user-friendly LabVIEW graphical interface, with the additional logging, protections and alerts allow to operate setup with minimal risks of equipment and detectors damages.

For primary beam control in addition to Faraday cups two primary beam profile monitors, mounted on pneumatically actuated motion feedthroughs, are provided for precise measurements of primary beam shape.

To adjust the momentum acceptance of fragment separator, the motor driven horizontal slits in F2 and F3 are installed. Additional vertical slits will be placed into focal plane F4 during RF-kicker installation (see section 1.3.1). Stepper motors provide 0.1 mm slits gap accuracy. The slits have changeable cranks optimized for usage with proton- or neutron-reach beams.

1.2.1. RIB diagnostic instruments

RIB diagnostic system of ACCULINNA-2 complex consists of time-of-flight (ToF) and tracking detectors. For precise particle identification, two ToF scintillation detectors are installed in F3 and F5 focal planes of ACCULINNA-2 at a distance of 12.3 meters. Depending on the experimental program these detectors can be equipped by organic scintillator (*e.g.* BC404) with a thickness down to 25 μm with an active area of 60 mm in diameter. Each scintillator is coupled with four Hamamatsu R7600-200 photomultiplier tubes (PMT). In the case of scintillator thickness of 125 μm, the time resolution of 100 ps (sigma) can be obtained.

Two multi-wire proportional chambers (MWPC) serve as a tracking detectors. In current experiments old type of MWPCs with 32x32 wires and 1.25 mm wire distance are used [6]. These detectors operate at atmospheric pressure of CF_4 and CH_4 gas mixture with a ratio 9 to 1, respectively. The signals registered by wires, passing through the low level discriminators (LLD) are recorded by homemade CAMAC registers. A new fragment separator ACCULINNA-2 has bigger image size in final focal plane compere to the previous generation one. Production of new multi-wire proportional chambers with 48x48 wires is nearly finished and they will be tested soon with Mesytec MQDC32 read out.

1.2.2. Physical targets and particle detectors

The studies of exotic nuclei at both Dubna fragment separators are mainly carried out with cryogenic targets (all stable helium and hydrogen isotopes including even tritium). The available cryogenic target cells 25 or 30 mm in diameter with nominal thicknesses from 0.4 to 6 mm have modular construction and can have a thickness up to 10^{22} at/cm^2. The cells have stainless steel windows with thickness from 6 to 9 μm.

Cryogenic target equipped by automated gas system with safety interlocks and cryogenic temperature stabilization. The system will be implemented to the common ACCULINNA-2 control system and will prevent possibility to blow up foils of cryogenic cells, ensure safe usage of cryogenic target in any emergency situations and decrease losses of rare and expensive gases.

In collaboration with Federal State Unitary Enterprise Russian Federal Nuclear Center (FSUE RFNC - VNIIEF) in Sarov the ACCULINNA-2 team started development of new tritium target complex. The maximum volume of liquid or solid tritium will be about 1 cm^3, which corresponds to activity about 2.7 kCi. Whole tritium gas system will be mounted on the top of the cold head in specially designed chamber with radiation safety monitoring. The filling and maintenance of tritium gas system will be performed in specially designed room at second floor of F5 area of ACCULINNA-2 with special ventilation, radiation monitoring and access control. First test of new tritium target complex is planned in 2021.

The ACCULINNA-2 team has a full set of charged particle detector telescopes. We have annular and square shaped single sided and double sided silicon strip detectors with thickness from 20 μm to 1.5 mm. The CsI(Tl) segmented arrays with square and annular shape are complementing well with corresponding Si telescopes. In addition, 40 neutron detection modules, based on stilbene crystals with corresponding PMTs assembled into the “neutron wall” [7] can by effectively used for broad variety of experiments.

1.2.3. DAQ

All particle detectors and RIB diagnostic system of ACCULINNA-2 with appropriate electronics are implemented to one distributed multi-processor data acquisition (DAQ) system based on Multi Branch System (MBS) [8]. Main DAQ combines VME crates with installed CES RIO4 or RIO3 VME controllers [9], distributed along the beam line. One VME crate have CVC-GTBV4 VME-CAMAC brunch [10] to read out old school electronics. Each controller operates under LynxOS real-time operation system. Via TRIVA modules [10] all crates connected to the common trigger bus for event synchronization. An additional

Linux based MBS event builder server allows to merge all data flows from each VME controller into single data stream of the reconstructed events. Flexible user readout interface of MBS allows to implement almost any configuration of the detectors spaced from each other by more than the length of existing fragment separator.

The processing of the beam profilometer signals is carried out by two modules of 32 channel readout system for low-current Profile Acquisition Digitizer (POLAND) based on ASIC current-to-frequency converters [11]. MBS is responsible for data collection and management. The beam profile monitoring is constructed as an independent subsystem but can be easily integrated into the main data acquisition system of ACCULINNA-2.

The online monitoring and visualization of experimental data proceed by the Go4 analysis Framework [12]. For further offline data analysis ROOT or any ROOT-based frameworks can be easily applied.

1.3. *Future subsystems development*

1.3.1. RF kicker

The purification level of ACCULINNA-2 separator is typically enough for preparation of quite pure beams in final focal plane F5 in the case of neutron-reach fragmentation products. The neutron-deficit radioactive beams suffer from large number of contaminations. To deal with it, in 2019 the vertically deflecting radio-frequency (RF) kicker [13] will be installed in a distance of 25 meters from production target just after the achromatic focal plane F3 (see Fig. 1).

It has 70 mm gap between electrodes and operates at the maximum field of 15 kV/cm with a frequency from 15 to 22 MHz. RF kicker phase correction has synchronization with cyclotron RF or with a signal from pickup device installed in a primary beam line. The most important projected parameters of RF kicker together with the example of its expected operation can be found in Ref. [1] for the case of ^{28}S isotope.

1.3.2. Zero-angle spectrometer

The ACCULINNA-2 experimental complex devoted to studies of exotic nuclear systems and its decay products. To distinguish between different charged particles, to measure their energies and angles, in 2017 at ACCULINNA-2 facility dipole magnet for zero-angle spectrometer was installed (see Fig. 1). It is 20 tones open-frame magnet with 180 mm gap. The reaction products enter the front aperture and bent by the well mapped magnetic field providing that their magnetic rigidities (and hence the momentum values) can be defined if their individual trajectories are measured accurately. The tracking detectors of ACCULINNA-2 provide 1.25-mm accuracy in the hit positions of RIB nuclei on the physics target. It means that, to use this dipole as zero-angle spectrometer, it should be equipped by specially designed position-sensitive detectors, which will provide similar precision in measuring coordinates of particles at the entrance and exit of the magnetic field. It is also important to use a ToF detector installed at the exit border of the magnet. This will give access to the velocity measurement and, taking into account the measured momentum values, to the determination of particle mass. Development of these detectors and appropriate electronics are in progress.

2. Basic Parameters of RIBs Obtained at ACCULINNA-2

The projected high transmission and purification level of ACCULINNA-2 were confirmed experimentally in set of tests with ^{15}N [1, 2] and ^{11}B primary beams in last two years. In the last one, the fragment separator was tuned for the ^{8}He secondary beam produced in fragmentation reaction of ^{11}B with energy of 33.4 *A*MeV on 1-mm-thick beryllium production target placed in F1 focal plane. The secondary beam was cleaned by natural beryllium 1-mm-thick achromatic wedge degrader (angle 2 mrad), located at F2.

Time of flight, energy losses and spatial distribution of the reaction products were registered by ACCULINNA-2 RIB diagnostics detectors. ToF plastic scintillators had 250 μm thickness. Typical two-dimensional plots of beam spot measured in the final focal plane F5 and the ΔE-ToF identification plot obtained when the separator was tuned for the ^{8}He (26.1 *A*MeV) secondary beam are shown in Fig. 1. Slits at F2 was fully opened (±68 mm) and in F3 slits gap was ±15 mm.

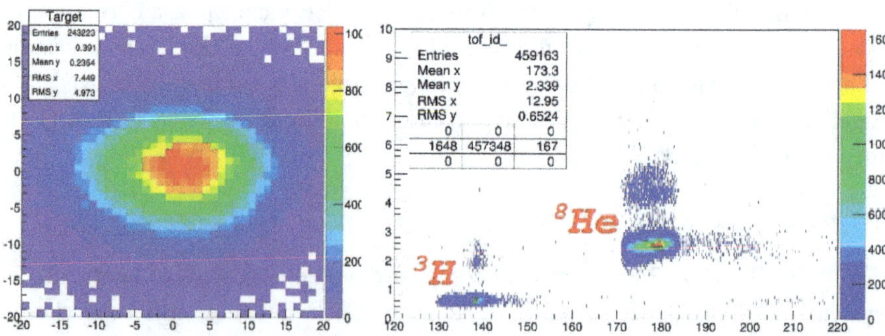

Fig. 2. The profile of ^{8}He secondary beam at F5 plane obtained via MWPC (left panel). The identification plot resulting the measurement of energy loss in thin plastic *versus* time of flight (right panel). The results were obtained with 1 mm beryllium target and $\Delta p/p = 7$ % at F2.

At F5 final focal plane ACCULINNA-2 provided the RIB's spot size of 16x11 mm, as shown in Fig. 2, while the beam size on the production target was 7 mm in diameter. In these conditions the beam purity of 87% was obtained. The measured ^{8}He intensity at F5 was $5{,}8 \cdot 10^4$ 1/s with 1 pμA ^{11}B primary beam on the production target, which is agree well with simulation given by LISE++ [14]. The discrepancy around factor of two between experiment and prediction could be explained by definition in a production mechanism at intermediate energy range for the reaction ^{11}B (33.4 *A*MeV) on Be target.

3. First Day Experiments

The first full scale experiment at ACCULINNA-2 fragment separator to study elastic and inelastic scattering of ^{6}He nuclei at the deuterium target was done in March 2018 [15]. The averaged intensity of ^{6}He beam (E = 26 *A*MeV) was about 10^5 1/s. The angular distributions were measured in a wide range from 25° to 130° in center-of-mass system. The data were taken in three runs by two telescopes, namely, d-telescope and He-telescope which worked in coincidence mode. The obtained angular correlations of particles in d-telescope with ^{6}He in He-telescope allowed to separate elastic channel on deuteron and proton (as a background). The analysis to obtain the final cross-section data is ongoing.

In October 2018 first test run aimed at a study of superheavy ^{7}H produced in the transfer reaction $d(^{8}He,^{3}He)^{7}H$ with 26 *A*MeV ^{8}He secondary beam on deuterium gas cryogenic target was performed. To detect ^{3}He recoil nuclei two silicon telescopes consisting of 22 μm and 1-mm-thick position-sensitive detectors were used. The energies and positions of tritons from the decay $^{7}H \rightarrow t+4n$ were measured at forward angles by ΔE-E telescope consisted of double-sided 1.5-mm-thick silicon strip detector and CsI(Tl)/PMT array (4 by 4 units). The data analysis is in progress.

In December 2018 a study of low-energy states of ^{10}Li in the reaction $d(^{9}Li,p)^{10}Li \rightarrow {}^{9}Li+n$ with 30 *A*MeV ^{9}Li beam on deuterium cryogenic target is planned. The key point of this measurement is the use of "combined mass" method based on detection of protons in backward direction in coincidence with fast neutrons emitted at forward angles. The neutrons will be measured by multi-detector array (5-cm-thick stilbene crystals with diameter of 8 cm coupled with 3 inch PMTs) which was recently commissioned and tested [7].

4. Conclusion

The first radioactive beams produced at the ACCULINNA-2 fragment separator at U-400M cyclotron complex were obtained. The basic ion-optical characteristics of ACCULINNA-2 described in the Letter of Intent [13] were confirmed experimentally for a number of RIBs. In 2018 the first day experiments devoted to study ^{6}He+d scattering and to search for the ^{7}H nucleus were carried out. Several experiments are planned to be performed at this facility in the near future including study of ^{7}H, ^{10}Li, ^{17}Ne and ^{26}S exotic systems. New separator equipment and scientific tools for ACCULINNA-2 is under active development.

Acknowledgments

The work is carried out with partial financial support of the Russian Science Foundation Grant No. 17-12-01367 and FAIR-Russia Research Center.

References

1. A.S. Fomichev *et al.*, Eur. Phys. J. A **54**, 97 (2018).
2. A.A. Bezbakh *et al.*, EPJ Web of Conferences **177**, 03001 (2018).
3. L.V. Grigorenko *et al.*, Nucl. Phys. News **24**, 22 (2014).
4. S.A. Krupko *et al.*, EPJ Web of Conferences **66**, 11021 (2014).
5. W. Beeckman *et al.*, Proc. 13th Intern. Conf. on Heavy Ion Accelerator Technology, HIAT2015, Japan, Sept. 7-11, 2015 (JACoW, 2016).
6. A.M. Rodin *et al.*, Nucl. Instrum. Methods B **204**, 114 (2003).
7. A.A. Bezbakh *et al.*, Instrum Exp Tech **61**, 5 (2018).
8. Multi Branch System (MBS), https://www.gsi.de/mbs.
9. Creative Electronic Systems, http://www.ces.ch.
10. GSI department of Experiment Electronics, https://www.gsi.de/en/work/research/experiment_electronics.htm.
11. S. Lochner *et al.*, GSI Scientific Report **2014-1**, p. 385 (2013).
12. Object Oriented On-Line Off-Line system (Go4), http://go4.gsi.de.
13. A.S. Fomichev *et al.*, JINR Communication E13-2008-168 (2008), http://aculina.jinr.ru/acc-2.php.
14. O.B. Tarasov and D. Basin, Nucl. Instrum. Methods B **266**, 4657 (2008), http://lise.nscl.msu.edu/lise.html.
15. B. Zalewski *et al.*, the present proceedings.

New Front and Back-End Electronics for the Upgraded GABRIELA Detection System

K. Hauschild,* R. Chakma, A. Lopez-Martenz, K. Rezynkina, V. Alaphillipe, L. Gibelin, N. Karkour and D. Linget

CSNSM, IN2P3-CNRS, UMR 8609, F-91405, Orsay, France
**E-mail: karl.hauschild@csnsm.in2p3.fr*
https://www.csnsm.in2p3.fr/GABRIELA

A. V. Yeremin, A. G. Popeko, O. N. Malyshev, V. I. Chepigin, A. I. Svirikhin, A. V. Isaev, E. A. Sokol, M. L. Chelnokov, Yu. A. Popov, D. E. Katrasev, A. N. Kuznetsov, A. A. Kuznetsova and M. S. Tezekbayeva

FLNR, JINR, 141980 Dubna, Russia

O. Dorvaux, B. Gall, P. Brionnet, K. Kessaci and C. Mathieu

IPHC,UMR 7178, Université de Strasbourg, CNRS-IN2P3, F-67037, Strasbourg

The GABRIELA [1] set-up is used at the FLNR to perform detailed nuclear structure studies of transfermium nuclei. Following the modernization of the VASSILISSA separator (SHELS) [2] the GABRIELA detection system has also been upgraded. The characteristics of the upgraded detection system will be presented along with results from some recent electronics tests.

Keywords: Digital signal processing; digital electronics; particle discrimination; focal plane decay spectroscopy.

1. Introduction

At the Flerov Laboratory for Nuclear Reactions (FLNR) experiments to perform detailed nuclear structure studies of transfermium nuclei are carried out using the GABRIELA setup (**G**amma **A**lpha **B**eta **R**ecoil **I**nvestigations with the **EL**ectromagnetic **A**nalyser) [1]. In 2006 the French Agence Nationale de la Recherche (ANR) awarded a grant to modernize the VASSILISSA separator which then became SHELS: the **S**eparator for **H**eavy **EL**ement **S**pectroscopy [2]. This crystallized additional funds from the JINR. Since the recoil implantation detector at the focal plane of SHELS is larger than that used for VASSILISSA (100×100 mm^2 cf 60×60 mm^2) the whole detection system of GABRIELA had to be rescaled. A second ANR grant and further funds from the FLNR were used for this purpose. Not only has the detection hardware been upgraded, but, we are also in the process of choosing new front-end electronics for the silicon detectors and implementing a fully digital DAQ using state of the art digitizers from National Instruments (NI) [3]. These developments will improve both the timing (allowing shorter lived isomeric states to be studied) and energy resolutions (improving the signal-to-background ratio allowing weaker channels to be observed). The characteristics of the upgraded detection system will be presented along with results from some of the electronics tests. In particular, the ability to discriminate 'escape' alpha particles from Internal Conversion Electrons (ICE). The new electronics will also allow us to detect fast pile-up events in the implantation Double-sided Silicon Strip Detector (DSSD) and perform time of flight measurements between the Micro-Channel Plate (MCP) and the DSSD when only one emissive foil is used (which is crucial to a successful experimental program using asymmetric reactions such as $^{18}O+^{242}Pu$).

2. GABRIELA and SHELS at the FNLR

The GABRIELA upgrade was completed in the first half of 2016. It consists of a $100\times100\times0.5$ mm^3 DSSD implantation detector at the focal plane of SHELS. This DSSD is surrounded by eight $50\times60\times0.7$ mm^3 silicon detectors in the backward hemisphere forming a square sectioned tunnel with two detectors on each side. One hyper-pure germanium (Ge) "clover" detector (CLODETTE) is placed in a collinear geometry just behind the recoil implantation DSSD and four single crystal coaxial Ge detectors are placed opposite each face of the tunnel [2]. New BGO Compton-suppression shields were developed for both Ge detector types. In front of the germanium detectors the vacuum chamber has 1 mm thin Dural windows to maximize L X-ray transmission. The single crystal detectors also have C-fiber entrance windows ~0.7 mm thick.

3. Electronics Tests

In moving from analogue to digital back-end electronics we expect to achieve the following goals: 1) reduced dead time, 2) improved event timing (from the current micro second level to sub-nano second), 3) improved energy resolution and 4) to be able to distinguish ICE from degraded escape alpha particles detected in the tunnel detectors. An additional requirement was to not require VHDL experts in order to facilitate on site maintenance and development. Currently there are only two options available: one from Nutaq, formally Lyrtech, which provided the VHS-ADC16 digitizers used to instrument SAGE [4] and the other from NI. A summary of the characteristics of the digitizers tested is given in Table 1. While Nutaq [5] can provide up to 768 channels in a single chassis at a competitive price it was unclear that the FPGA was large enough to deal with our (future) signal processing needs. Included in the table are values for NUMEXO2 [6], chosen for SIRIUS@S3 [7], for which we know that the size of the FPGA is problematic. The solution provided by NI should have a large enough FPGA for future developments. Additionally, we found the NI graphical interface to program the FPGA more user friendly. A number of tests were carried out at the CSNSM, IPHC and the FLNR to ensure that the future digital system would perform at least as well as our current analogue system. We report on some of these in the following sections.

Table 1. Pertinent characteristics of the digitizers considered.

Digitizer	NUTAQ Perseus 611x + M125	National Instruments PXIe-5170R	NUMEXO2
Channels/module	64	8	16
Modules/chassis	10	18	12
FPGA	Virtex-6 550T[a]	Kintex-7 325T	Virtex-6 130T
Logic cells per channel	8.6 k	40.6 k	8.0 k
Input voltage range(s)	±1 V	±0.1, ±0.2, ±1, ±2.5V	±1, ±4V
Sampling frequency	125 MHz	250 MHz	200 MHz
ENOB[b] (at 125 MHz)	11.3	11.5	-
ENOB (at 200 MHz)	-	-	10.6
ENOB (at 250 MHz)	-	11.1	-

[a]The largest compatible FPGA available.

[b]ENOB: Effective Number Of Bits can be defined using the IEEE standard 1057 as

$$ENOB = log_2\left\{(input\ V\ range)/(V\ RMS\ noise) * \sqrt{12}\right\}$$

3.1. *Ge energy resolution measurements*

Of utmost importance is the ability of the digital electronics and associated signal processing to provide energy resolutions at least equal to those of the current analogue chain. Excellent energy resolutions, FWHM < 1.9 keV at 1332 keV for CLODETTE have been achieved under experimental conditions with the PXIe-5170R digitizers from National Instruments using the Moving Window Deconvolution algorithm performed on the digitized signal recordrd to disc [8,9].

3.2. *MCP-MCP time-of-flight*

Our time-of-flight (ToF) system typically uses two thin emissive foils (EF). Electrons emitted from the foils, following the passage of a heavy ion, are focused on to four MCP detectors by a magnetic field. The set-up is shown schematically in Figure 1(a). The preamplifier (PA) output was passed through an SBLP-39 low-pass filter to increase the rise-time of the signals to be compatible with the 4 ns sampling period of the NI-PXIe-5170R. A passive RF splitter and a 2 ns delay was used to obtain two identical signals and hence an effective digitizer rate of 500 MHz. The LabVIEW data acquisition was triggered by the detection of an alpha particle in the DSSD. Analysis was performed off-line on the data written to disc. In Figure 1(b) a representative signal derived for MCP4 is shown from which the zero-crossing point is extracted. The ToF is given by difference between (MCP1.or.MCP2) and (MCP3.or.MCP4). The difference between the zero-crossing points for MCP4 and MCP2 as a function of alpha particle energy and the projection onto the time axis are presented in Figure 1(c) and (d) respectively. Averaged over the ^{233}U, ^{239}Pu, ^{238}Pu alpha lines the timing resolutions obtained were σ=156(8) ps and 139(11) ps at 250 and 500 MHz sampling rates respectively. With the implementation of a new lower noise and larger bandwidth preamplifier and a 1 GHz digitizer we hope to improve on this.

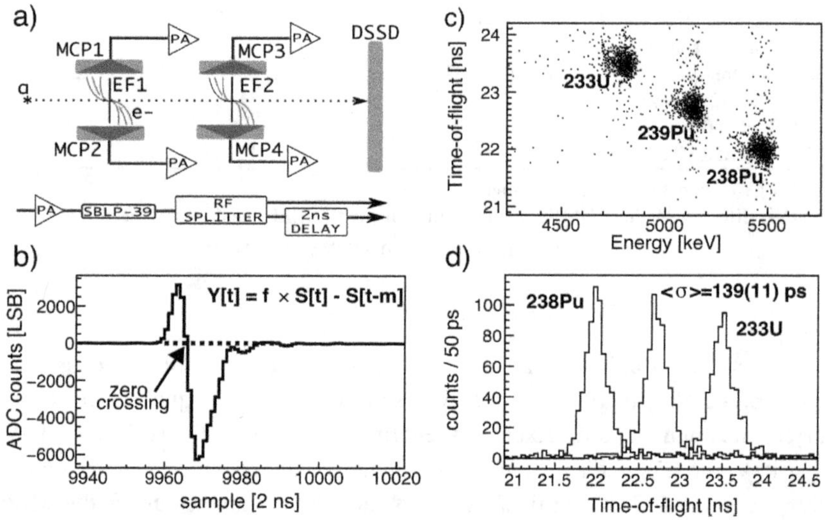

Fig. 1. (a) Schematic of the time-of-flight measurement system (acronyms are explained in the text). (b) A typical digital CFD signal for MCP4. The linearly interpolated zero-crossing point defines the event time. (c) A time-of-flight vs alpha particle energy matrix and (d) the projected time-of-flight.

3.3. *MCP-DSSD time-of-flight*

For very asymmetric reactions the foil furthest from the DSSD must be removed and the ToF is determined between either MCP3 or MCP4 and the DSSD. While our current preamplifiers from TechInvest provide excellent energy resolution (15-18 keV FWHM at 5499 keV and a full-scale of ~250 MeV) [10] their rise-time are not fast enough to achieve sub-ns timing. Two solutions are under investigation: modifications to the preamps developed for the tunnel detectors of SIRIUS and custom CREMAT preamps developed for the SIRIUS test bench at the IPHC [11]. Both can provide impressive energy resolutions as shown in Figure 2(a). Despite having ~75 cm cables from the vacuum chamber to the preamplifiers we have been able to obtain MCP-DSSD timing resolution, again averaged over the three alpha sources, of $\sigma = 800(45)$ ps in fully digital system. Future tests will be performed with shorter cables (and hence lower parasitic capacitance) in order to improve on these encouraging results.

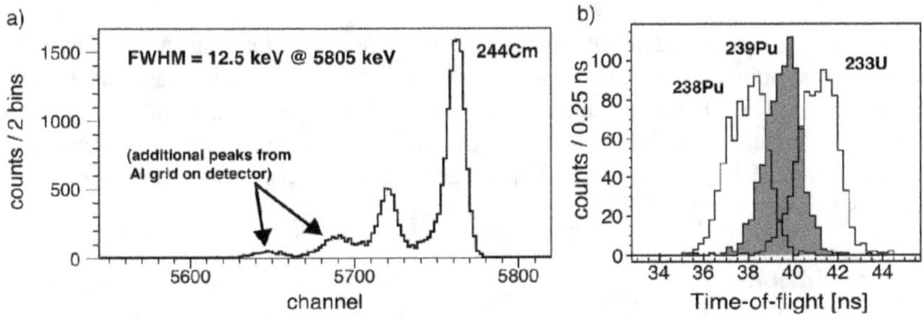

Fig. 2. (a) An example ^{244}Cm spectrum measured with the GABRIELA DSSD in situ using a custom built CREMAT preamplifier. (b) MCP4-DSSD time-of-flight for a triple alpha source.

3.4. *Particle discrimination*

Due to the detection geometry, α-particles from other reaction channels that escape the DSSD and hit the tunnel (leaving energy in both detectors) can mask the real fine structure α–ICE coincidences under study. An example is given in Figure 3(a) of the ground state alpha decay of ^{254}No masking α(^{255}No)-ICE(^{251}Fm).

Nevertheless, α-particles and electrons of the same energy have different ranges in Si and hence the charge collection times and shape of the preamp signals should be different. Using SHELS ^{233}U α-particles were transported from the target position to the GABRIELA focal plane. This ensured an e-free source of energy degraded alphas with energies overlapping the ICE from a standard ^{133}Ba source. In Figure 3(b) it can be clearly seen that ICE and low energy α particles can be distinguished from one another purely on the basis of the preamplifier rise-time. To our knowledge this is the first demonstration of ICE/α discrimination at an energy range appropriate to decay spectroscopy. With faster preamplifiers it should be possible to lower the discrimination threshold, but already, it will be possible to remove a large fraction of the escape α particles and thus improve the signal-to-noise for the ICE.

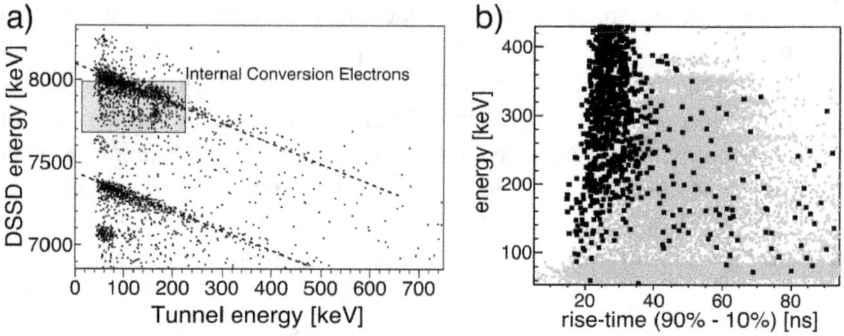

Fig. 3. (a) Prompt coincident DSSD-tunnel energy matrix following the implantation of evaporation residues of the reaction ^{208}Pb(^{48}Ca, xn). Dashed lines indicate alpha particles that escape the DSSD and are detected in the tunnel. ICE in ^{251}Fm from the alpha decay of ^{255}No are indicated by the shaded box. (b) Particle energy as a function of preamplifier signal rise-time. Black squares: ^{233}U alphas degraded in energy. Grey circles: conversion electrons from the decay of ^{133}Ba.

4. Conclusions

Excellent energy resolutions for Ge and Si detectors have been achieved with the NI-PXIe-5170R. The time-of-flight measurements, either between two MCPs or MCP-DSSD, can be performed in fully digital system with the required resolution. We have also demonstrated that ICE can be discriminated from escape alphas at energies relevant to decay spectroscopy.

Acknowledgments

The modernization of the VASSILISSA separator, now SHELS, was financed by the French Agence Nationale de la Recerche (Contract No. ANR-06-BLAN-0034-01 and ANR-12-BS05-0013) and by JINR. Work at FLNR was performed partially with the financial support of the Russian Foundation for Basic Research, Contract No. 17-02-00867 and 18-52-15004. M. Afif, M. Kropachev and H. Soghoyan of National Instruments are thanked for the technical support provided.

References

1. K. Hauschild *et al.*, *Nucl. Instr. Methods* **A 560**, 388 (2006).
2. A.G. Popeko *et al.*, *Nucl. Instr. Methods* **B 376**, 140 (2016).
3. http://www.ni.com/white-paper/53390/en/#toc4
4. J. Pakarinen *et al.*, *European Physics Journal* **A 50**, 53 (2014).
5. https://www.nutaq.com/nuclear-imaging-digitizers/

6. F.J. Egea *et al.*, *IEEE Transactions on Nuclear Science* **60**, 3526 (2013).
7. F. Dechery *et al.*, *European Physics Journal* **A 51**, 66 (2015).
8. A. Georgiev and W. Gast, *IEEE Trans. Nucl. Sci.* **40**, 770 (1993).
9. J. Stein *et al.*, *Nucl. Instr. Methods* **B 113**, 141 (1996).
10. A.V. Isaev *et al.*, *Instruments and Experimental Techniques* **54**, 37 (2011).
11. H. Faure, PhD Thesis, Strasbourg Unive

Charged Particle Trajectories in the MAVR High-Resolution Magnetic Spectrometer

V. A. Maslov, V. A. Zernyshkin*, Yu. E. Penionzhkevich,
I. V. Kolesov and O. B. Tarasov

Joint Institute for Nuclear Research,
Dubna, Moscow oblast, 141980 Russia
**E-mail: 94zernyshkin@gmail.com*

Charged-particle trajectories in the MAVR high resolution magnetic spectrometer [1] are simulated. The latter has been constructed at the U400 accelerator for purifying the incident beam and identifying the products of nuclear reactions induced by the beams of stable and radioactive nuclei. In this paper we present the MAVRPC++ software package, aimed at reconstructing the spectrometer focal plane, optimizing the fields of quadrupole and dipole magnets, and estimating the solid-angle acceptance for secondary particles.

Introduction

The mechanisms of nuclear reactions induced by beams of stable and radioactive nuclei and the properties of nuclei close to the boundaries of the stability region are among the research priorities of the JINR Laboratory of Nuclear Reactions (LNR). Corresponding experiments rely on the U400 and U400M cyclotrons at JINR and on the machines operating at other research centers: GANIL (France), RIKEN (Japan), and Q3D (Germany) [4–6]. All these experiments employ magnetic spectrometers and separators where by the primary beam is purified to a high degree and the reaction products are discriminated with high resolutions in momentum, energy, and mass.

At the JINR Laboratory of Nuclear Reactions, magnetic spectrometry was pioneered by V.V. Volkov in 1970 for investigating the multinucleon-transfer reactions and for synthesizing new neutron-excessive isotopes of light elements [7–8].

The method of magnetic spectrometry (with the MSP-144 analyzer [2]) was also employed at LNP for measuring the binding energies and probing the stability and structure of super-neutron-excessive isotopes of the lightest elements ($^{7-10}$He, $^{9-11}$Li, and $^{12-14}$Be). These measurements relied on the missing-mass technique. Energy levels of the ^{7,8,9}He, ^{10,11}Li, and ^{13}Be nuclides were measured for the first time [5, 6].

The commissioning of the DRIBs radioactive beam system [9] opened up new possibilities for such investigations. Relying on relatively high intensities of radioactive beams and on the good momentum resolution of the MSP-144 spectrometer, studies of ^{6}He induced reactions at energies in the barrier region were initiated. The excitation functions of the ^{6}He-induced fusion and neutron-transfer reactions were measured with a 300-keV precision [10]. Interesting data on deep-subbarrier fusion of the ^{6}He neutron-halo nucleus with heavy nuclei were obtained [11–13].

These measurements are of fundamental importance for nuclear physics and astrophysics and have been followed by similar investigations performed elsewhere. We should emphasize that the aforementioned experiments were carried out in close collaboration with the JINR member laboratories: NPI ASCR (Rez, Czech Republic), IFIN-HH (Bucharest, Romania), and NINP PAS (Krakow, Poland).

The MAVR high-resolution magnetic spectrometer will enable us to do the following:

(1) estimate the degree of nucleon stability for exotic nuclei;
(2) measure the masses of these nuclei with the missing-mass technique;
(3) investigate the excited and resonant states of exotic nuclei;
(4) search for nucleon clusters in light nuclei.

1. Experimental Setup

The MAVR high-resolution magnetic analyzer is comprised of an upgraded system of magnetic optics based on the MSP-144 magnet, a doublet of quadrupole lenses Q1 and Q2 upstream of the magnetic spectrometer, and a detector block located in the analyzer focal plane. MAVR is deployed in the experimental hall of the U400 accelerator.

The target placed at a distance of d_1 = 2576 mm from the front face of the spectrometer's first bending magnet is irradiated by an extracted ion beam. The distances between the target and the upstream quadrupole lens and between the two quadrupoles are r_1 = 307 mm and r_2 = 180 mm, respectively. The quadrupole pair has an effective length of r_{eff} = 350 mm. The distance between the downstream quadrupole lens and the front shutter of the magnet amounts to r_3 = 873 mm. The analyzing magnet consists of two parts, of which the second one generates a magnetic field that is 1.55 times stronger than the first one. The first and second magnets provide bending angles of φ_1 = 60° and φ_2 = 51°, respectively, so that for a detected particle the total bending angle amounts to 111°. The interpole distance equals 47 and 30 mm for the first and second magnets, respectively.

The conceptual scheme of the MAVR spectrometer is in Fig. 1.

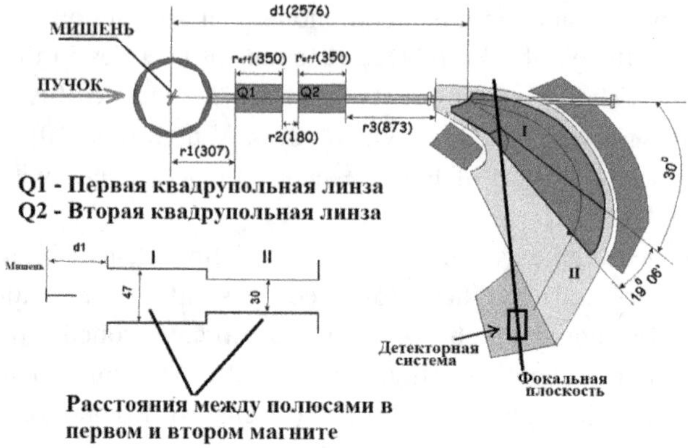

Fig. 1. Conceptual scheme of the MAVR high-resolution spectrometer.

With a system of detectors deployed in the spectrometer focal plane, reactions products are detected and precisely discriminated by charge Q, atomic number Z, and mass A by measuring their energy losses (ΔE), times of flight (T), and total energies (E). For this, detectors of different types including position-sensitive chambers and semiconductorstrip sensors will be used. In addition, an important part of data analysis is the reconstruction of particle trajectories in the analyzer, without which accurate identification of products of nuclear reactions is not possible.

The upgraded magnetic spectrometer features a high energy resolution of $\Delta E/E \sim 10^{-3}$, a large bending angle of 111°, and a dispersion along focal plane 2 cm/%. Thereby, products of nuclear reactions are detected with a high charge resolution of $\Delta Z/Z \sim 1/60$ essential for efficiently discriminating massive secondary nuclei. The basic parameters of the MAVR spectrometer are listed in Table 1.

Table 1. Basic parameters of the MAVR spectrometer.

Interpole distance for the first magnet, mm	47
Interpole distance for the second magnet, mm	30
Maximum magnetic rigidity $B\rho$, T m	1.5
Dispersion along focal plane, cm/%	2
Energy resolution $\Delta E/E$	1×10^{-3}
Mean radius of particle trajectory R, m	1.875
Solid-angle acceptance, msrad Up to	5
Particle's total bending angle in the spectrometer, deg	111

2. The MAVRPC++ Software Package

Since the spectrometer features a quadrupole-lens doublet, the position of the magnet focal plane depends on the quadrupole field strengths. For this reason, the MAVRPC software package [16] was developed. The MAVRPC software package is aimed at reconstructing the focal-plane position and estimating the spectrometer solid-angle acceptance as functions of the parameters of investigated secondary particles. This software package has been developed on the COSY INFINITY platform [3] aimed at designing optical systems, spectrometers and accelerators in particular. The latter relies on the methods of differential algebra: particle propagation through each element of an optical system is described in terms of a corresponding transition matrix. Since the quadrupole transition matrices depend on their induction gradients, that of the spectrometer as a whole and its focal-plane position also depend on the quadrupole fields.

MAVRPC++ is an enhanced version of MAVRPC. The main innovation of MAVRPC++ is that the calculations of the position of the focal plane were made with an accuracy of the 7th order (in MAVRPC, the accuracy of calculations did not exceed the 2nd order, the focal plane was approximated by a straight line). In addition, a function of calculation the dependence of the effective solid angle of the MAVR installation on the relative stiffness of the BP/BP0 particles was added.

Table 2 presents the results of calculations of the focal planes positions of the MAVR analyzer in MAVRPC++ for a central particle with BP0 = 0.91Tm for different fields of quadrupoles, and also provides an estimate of the solid angle of the analyzer for each of the options considered. The radius of the central trajectory is Rref = 1.875 m, Figure 2.

Thus, MAVRPC++ allows you to compare various settings of the MAVR analyzer and select the best options for a specific experiment. In the considered example, for a central particle with BP0 = 0.91Tm, the optimal choice of quadrupole fields is: Q1 = -0.5Tm, Q2 = 0.4Tm, which corresponds to the solid

Table 2. Calculations of the focal planes positions of the analyzer MAVR in MAVRPC++ for the central particle with BP0 = 0.91Tm.

Rref, m	Q1, T	Q2, T	Ω, msr
1.875	-0.5	0.4	5
1.875	-0.2	0.3	1.9
1.875	-0.1	0.3	4
1.875	-0.1	0.1	2.9
1.875	0.0	0.2	4

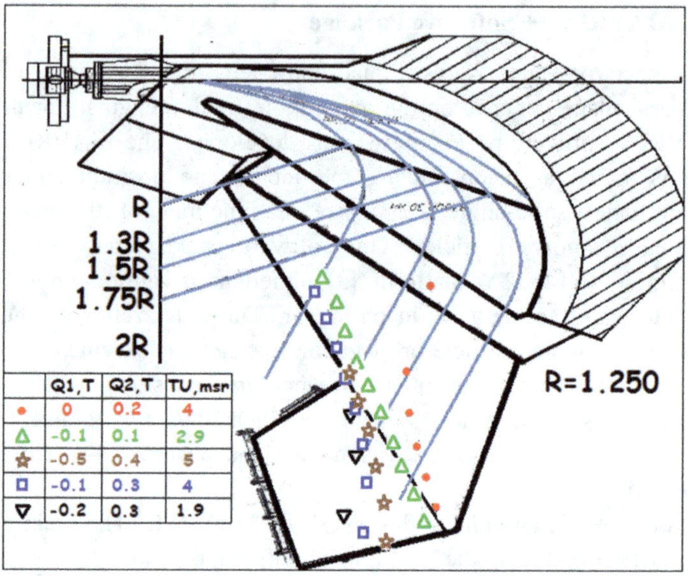

Fig. 2. Graphic representation of the reconstructed focal planes of the MAVR analyzer in MAVRPC++ for the central particle with BP0 = 0.91Tm with different fields of quadrupoles.

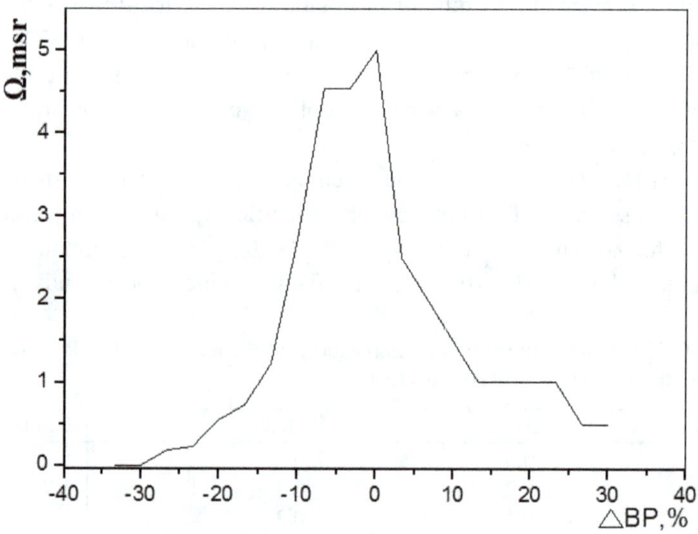

Fig. 3. Dependence of the effective solid angle of the MAVR facility on the relative stiffness of the BP/BP0 particles for a 60x60 mm detector system.

angle Ω = 5 msr and the central particle hits the detector system. An ordinary standard-size silicon detector (60x60 mm) was considered as a detector system. These calculations show that an increase in solid angle compared with the MSP-144 analyzer (Ω MMP-144 = 0.49 msrd [17]) is expected to be 10 times.

Figure 3 shows the dependence of the effective solid angle of the MAVR facility on the relative rigidity of the particles BP/BP0 (where BP0 is the rigidity of the central particle) for the chosen setting of the fields of installation of MAVR (central particle with BP0 = 0.91Tm, fields of quadrupoles: Q1 = -0.5Tm, Q2 = 0.4Tm).

Figure 3 clearly demonstrates that the selected setting of the spectrometer provides the maximum transmission of particles with BP0 = 0.91Tm through the spectrometer to a fixed point of the focal plane.

3. Conclusions

In this paper, charged-particle trajectories in the MAVR high-resolution magnetic spectrometer are simulated using the MAVRPC++ software package. Thereby, the spectrometer energy resolutions are estimated as $\Delta E/E \sim 1 \times 10^{-3}$, respectively, and, in the charge of nuclear-reaction products, $\Delta Z/Z \sim 1/60$. The solid angle acceptance of the MAVR spectrometer is found to exceed that of MSP-144 by a factor of up to ten.

Acknowledgments

This work was supported by the Russian Science Foundation under grant no. 17-12-01170 and by authorized representatives of the Czech Republic, Poland, and Kazakhstan under corresponding grants.

References

1. V. A. Maslov, V. I. Kazacha, I. V. Kolesov, S. M. Lukyanov, V. N. Melnikov, N. F. Osipov, Yu. E. Penionzhkevich, N. K. Skobelev, Yu. G. Sobolev, and E. I. Voskoboinik, *J. Phys.: Conf. Ser.* **724**, 012033 (2016).
2. N. K. Skobelev et al., *Nucl. Instrum. Methods Phys. Res.* **227**, 471 (2005).
3. www.cosyinfinity.org/.
4. S. M. Lukyanov et al., *J. Phys. G* **28**, L41 (2002).
5. R. G. Kalpakchieva and Yu. E. Penionzhkevich, *Phys. Part. Nucl.* **33**, 629 (2002).
6. Yu. E. Penionzhkevich, R. G. Kalpakchieva, and S. M. Lukyanov, *Nucl. Phys. A* **722**, 170c–175c (2003).

7. A. G. Artukh, G. F. Gridnev, V. L. Mikheev, V. V. Volkov, and J. Wilczynski, *Nucl. Phys. A* **192**, 170 (1972).
8. G. Artukh, V. V. Avdeichikov, J. Ero, G. F. Gridnev, V. L. Mikheev, V. V. Volkov, and J. Wilczynski, *Nucl. Phys. A* **160**, 511 (1971).
9. Yu. Ts. Oganessian, W. von Oertzen, and R. G. Kalpakchieva, in *Proceedings of the Conference on Nuclear Shells—50 Years, Dubna, Russia, 1999* (World Scientific, Singapore, 2000), p. 61.
10. Yu. E. Penionzhkevich, R. A. Astabatyan, N. A. Demekhina, G. G. Gulbekian, R. G. Kalpakchieva, A. A. Kulko, S. M. Lukyanov, E. R. Markaryan, V. A. Maslov, Yu. A. Muzychka, Yu. Ts. Oganessian, R. V. Revenko, N. K. Skobelev, Yu. G. Sobolev, D. A. Testov, and T. Zholdybaev, *Eur. Phys. J. A* **31**, 185–194 (2007).
11. Yu. E. Penionzhkevich, Yu. A. Muzychka, S. M. Lukyanov, R. G. Kalpakchieva, N. K. Skobelev, V. P. Perelygin, Yu. G. Sobolev, L. V. Mikhailov, V. Yu. Ugryumov, V. Vincour, Z. Dlouhy, L. Kostov, Ya. Mrazek, and N. O. Poroshin, *Phys. At. Nucl.* **65**, 1563 (2002).
12. Yu. E. Penionzhkevich, R. A. Astabatyan, N. A. Demekhina, Z. Dlouhy, R. G. Kalpakchieva, A. A. Kulko, S. P. Lobastov, S. M. Lukyanov, E. R. Markaryan, V. A. Maslov, Yu. A. Muzychka, Yu. Ts. Oganessian, D. N. Rassadov, N. K. Skobelev, Yu. G. Sobolev, V. Yu. Ugryumov, J. Vincour, and T. Zholdybaev, *Phys. Part. Nucl. Lett.* **3**, 362 (2006).
13. S. M. Lukyanov, Yu. E. Penionzhkevich, R. A. Astabatian, N. A. Demekhina, Z. Dlouhy, M. P. Ivanov, R. Kalpakchieva, A. A. Kulko, E. R. Markarian, V. A. Maslov, R. V. Revenko, N. K. Skobelev, V. I. Smirnov, Yu. G. Sobolev, W. Trazska, and S. V. Khlebnikov, *Phys. Lett. B* **670**, 321–324 (2009).
14. Yu. E. Penionzhkevich, R. G. Kalpakchieva, A. A. Kulko, S. M. Lukyanov, V. A. Maslov, and N. K. Skobelev, *Int. J. Mod. Phys. E* **17**, 2349–2353 (2008).
15. http://lise.nscl.msu.edu/lise.html.
16. V. A. Zernyshkin, V. A. Maslov, Yu. E. Penionzhkevich, V. I. Kazacha, I. V. Kolesov, O. B. Tarasov, *Physics of Particles and Nuclei Letters* **15**(5), 531–536 (2018).
17. A. V. Belozyorov et al., *Nuclear Instruments and Methods in Physics Research A* **411**, 343–350 (1998).

Exotic Nuclides at the Reactor PIK: PITRAP Project

Yu. N. Novikov[1,2*], Yu. I. Gusev[1], S. V. Chenmarev[1], S. A. Eliseev[1], P. E. Filyanin[1], T. V. Koneva[1], D. A. Nesterenko[1], A. W. Popov[1], D. Simonovski[2], O. I. Bezrodnova[1,2], N. S. Martynova[1,2] and Yu. V. Nechiporenko[1,2]

[1]Petersburg Nuclear Physics Institute, Research National Center "Kurchatov Institute", Gatchina, 188300, Russia
[2]St. Petersburg State University, St. Petersburg, 199034, Universitetskaya nab. 7/9, Russia
[*]E-mail: novikov_yn@pnpi.nrcki.ru

The PITRAP-project for mass measurements of exotic nuclides is based on the use of high flux-reactor PIK and a high precision Penning trap mass spectrometry. This combination gives a synergy in exploration of different problems of fundamental physics: from astrophysics to neutrino physics

Keywords: Exotic nuclides; astrophysical r-process; mass landscape in neutron-rich nuclides; high flux reactor; Penning trap mass spectrometry.

1. Introduction

A nuclear reactor is the unique installation for production of neutron-rich nuclides and among them of very exotic ones in the fission region of the Nuclear Chart. Nearly all of these nuclides that can exist as such participate in different astrophysical processes, unfolded as a result of stellar explosions. The exact number of these nuclides is unknown and masses (total binding energies) covering the energetics of these processes are mostly also unknown. The information on them is provided presently only by the theoretical predictions. However, they are not exact and considerably scattered for very neutron-rich nuclides. Fig. 1 shows how different are the predictions of the neutron drip-line that frames the area of nucleon-stable nuclides participating in the stellar processes. It is obvious that any experimental information on the mass values towards the drip-line is of paramount importance.

Success in experiments with nuclides of an exotic proton/neutron composition depends on the intensity of the producing primary beam as well as the sensitivity of the measuring setup. The new PIK reactor which is under commissioning at the PNPI NRC "Kurchatov institute" is expected to be a high-flux facility [1, 2], with a powerful production of such exotic nuclides. A Penning ion trap is an ultrasensitive apparatus capable of operating with a single ion. This combination provides synergy and a unique possibility for studying exotic nuclei. Penning ion traps are widely used to solve different physical problems [3].

The Penning ion-trap systems ISOLTRAP (CERN), SHIPTRAP (Germany), JYFLTRAP (Finland), LEBIT (USA), and TITAN (Canada) operate in combination with particle accelerators and are intended for the on-line direct mass measurements.

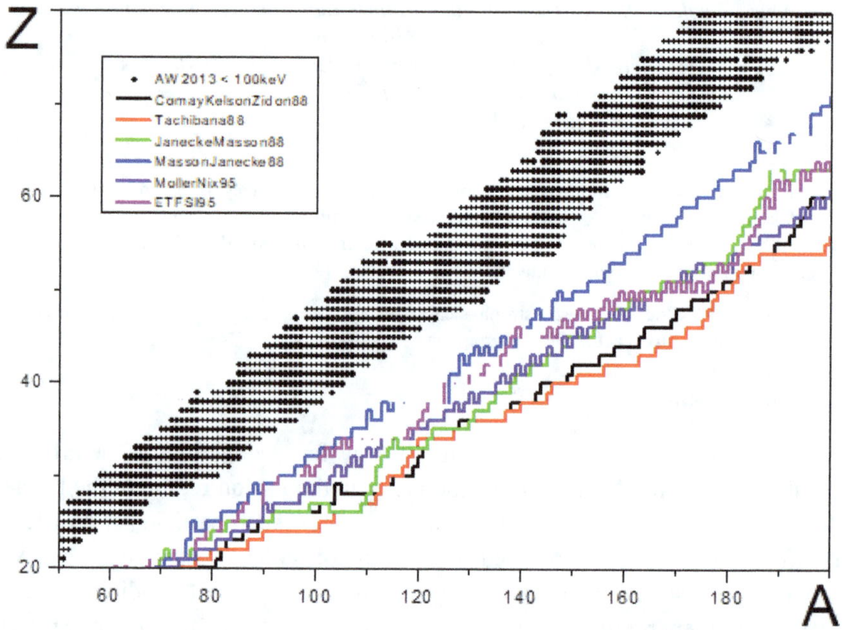

Fig. 1. Nuclear chart with the indicated neutron-drip-lines (in color) obtained by different mass formulae listed in the inset.

The PITRAP-project proposed at the St. Petersburg Nuclear Physics Institute incorporates many years of experience gained by the authors on some of the facilities listed above.

2. Physical Problems

High productivity with high sensitivity of PITRAP will make it possible to cover wide regions of the Nuclear Chart and obtain unique information with nuclear physical and astrophysical content.

We will focus here on the astrophysical aspect of the problem. For fast neutron capture (r-process) caused by explosion or neutron stars merge, the density distribution of isotopes relates [4] the astrophysical parameters (neutron density and temperature of the stellar medium) with the nuclear parameter – the separation energy of a neutron in nuclei, which depends on the mass difference of the nuclei. The latter quantity determines the path of the process under prescribed astrophysical conditions. For this reason nuclide mass becomes the key quantity that sets the path of the r-process. It is important to know the true 'nuclide path' of this process in order to determine the parameters characterizing the explosion of a star. The mass of nuclides in the initial and middle regions of the r-process at N = 50 and N = 82, which are important for determining and studying the starting path of this astrophysical process, can be just measured at the PIK reactor with the Penning Trap Mass Spectrometry (PTMS).

The mass of a nucleus, which is directly related with the nuclear total binding energy, is one of the fundamental characteristics reflecting the sum of all nucleonic interactions in nucleus. The measurement of mass in long isotopic and isotonic chains makes it possible to study the fine structure of the mass landscape. An important question in direct measurements is the possibility to resolve the isomeric and ground states. In [5] the mass doublet ^{164}Er–^{164}Dy with the small mass difference Q = 25.07(12) keV was resolved by the method of octupolar excitation in a Penning trap; the resolution attained was about 2×10^{-7}. A new method of phase determination of the cyclotron frequency making high resolution possible was also adopted in the SHIPTRAP setup [6]. Both these methods can be used at PITRAP to study the energy of isomeric states of nuclides. A large volume of new information on the mass of nuclides that can be obtained with PITRAP will improve the parameterization of the empirical mass formulae with the further possibility of using them directly to predict even more exotic regions of nuclides.

Another block of problems is associated with the high-precision off-line measurements of long-lived radionuclides. They can be produced by irradiating targets in the PIK reactor. The measurements can be performed after this irradiation in a trap independently of the reactor operation. High-precision measurements of the masses of long-lived nuclides can open up an entire strata of problems in fundamental physics ranging from neutrino physics to cosmochronology [7,8].

3. Production and Transportation of Exotic Nuclides

Nuclides formed during the irradiation of external targets in the PIK reactor [1, 2] (Fig. 2) can be extracted by several methods:

1. transporting fission products in a gas stream outside of the reactor core by means of aerosols (so called He-Jet transportation) followed by ionization and delivery of ions into a Penning trap,
2. or by use of a "cold target" in the area of the reactor core in combination with the IGISOL-type mass separation [9];
3. or employing a 'hot target' together with the IRINA electromagnetic mass separator [10].

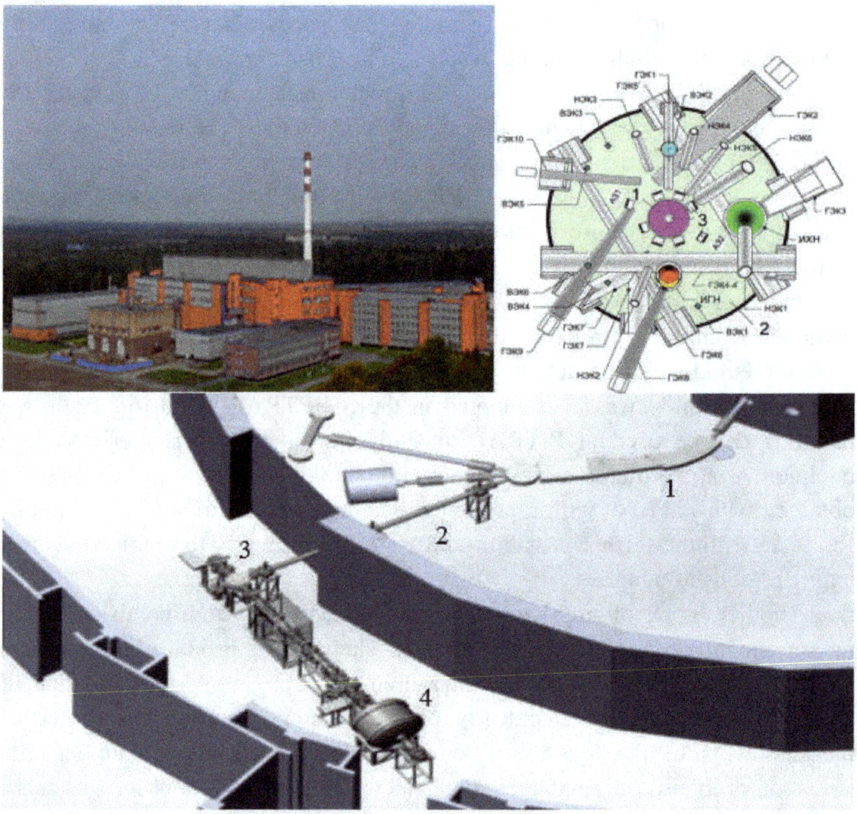

Fig. 2. Collage of the reactor PIK properties [1, 2] with basic characteristics: power: 100 MW; thermal neutron flux: 5×10^{15} n/cm^{2}sec; horizontal channels: 13; inclined and vertical channels: 14; neutron guide channels: 8; 50 positions for neutron instruments. In the bottom of figure - arrangement of the setup PITRAP (4) is shown. It will use the horizontal channel ГЭК 5 and the electromagnetic mass separator (1) followed by ion channel (2) and bending dipole magnet (3).

Two variants (1) and (3) are examined in the present work. The planned arrangement of a PITRAP at the PIK reactor complex is shown in Fig. 2. In the first method a capillary with the carrier gas and aerosols delivering the reaction products is connected to a chamber, where a skimmer, an ion source, and extracting electrodes are located. A magnet (3) that turns the ion beam towards the PITRAP channel (4) and performs selection by mass numbers is located farther.

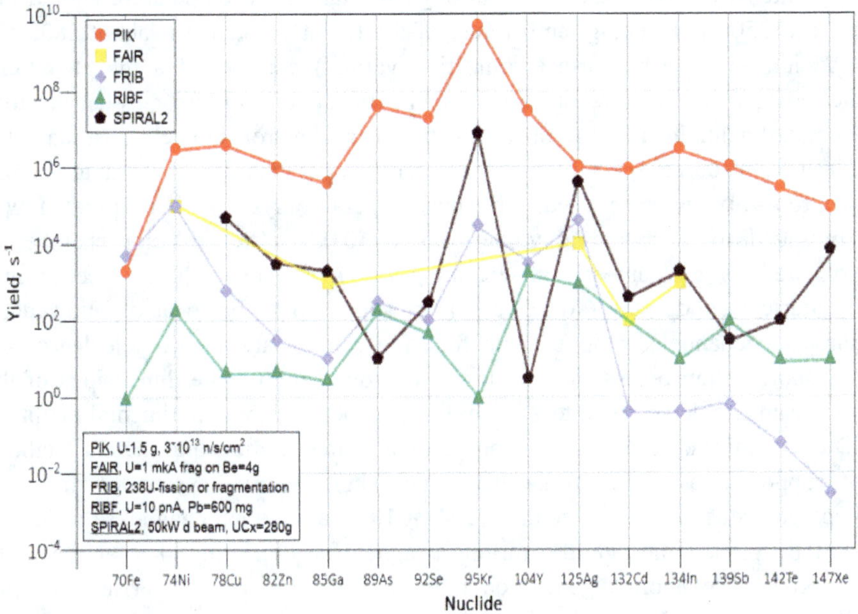

Fig. 3. Comparative production yields expected from different on-line installations in the world.

The fission products in the "cold" chamber containing a target in the form of an approximately 10 μm thick ^{235}U metal foil with an area of 150 cm^2 coated with nickel will be thermalized in a buffer gas, e.g., in helium at pressure of 0.2÷0.3 MPa. After this they are secured to aerosol particles and delivered by means of a capillary with inner diameter equal to about 1 mm and length of several meters towards the ion source. Then the ions are extracted and accelerated by the electric field created with a set of electrodes. The expected efficiency of the removal of nuclides through the capillary up to the ion source equals 50÷70%. The ionization efficiency depends on the type of ion source as well as the particular isotope and can vary from 0.1 to 10%. The power release in the target is approximately 1.5 kW. The target is cooled by water circulating in a closed loop. The details of this method can be found in [11].

The third proposed variant uses a massive hot target [10]. It is proposed that the target material in the form of ^{235}U carbide with a mass of 4 g in a graphite capsule be placed in a hermetic rhenium or tungsten container in a horizontal channel of the reactor (presumably ГЭК 5). The entire assembly is located in an evacuated volume cooled by distilled water. In the working regime with neutron flux intensity 3×10^{13} sec$^{-1}\cdot$cm^{-2} through the target its temperature should not exceed 2500 K (melting temperature of the target substance is about 2800 K). The ion source (the surface ionization variant) consists of a tungsten tube, heated to a temperature of 2600 K either by direct current or by an electron bombardment. In the case of an ionization by electron impact, a variant of a plasma ion source, specially developed for the operating conditions in an intense neutron flux, will be used. The mass-separator system (1 in Fig. 2) was envisaged for accelerating voltage up to 50 kV. The collector chamber is followed by a chamber for separating the ion beams where three turning cylindrical condensers (switchyards) are supposed to be located followed by three ion channels equipped with focusing electrostatic quadrupole lenses for transporting ion beams into the low-background room. The dimensions of the ion beam should be equal to 2 × 2 mm at the locations of experimental facilities. The transmission of the beam from the ion source to the experimental facilities is supposed to be 60÷90%. The vacuum in the entire system of the mass separator must not be worse than $(2\div4) \times 10^{-4}$ Pa. One ion channel (2 in Fig. 2) will be passed through the turning magnet for deflecting the ion beam to PITRAP. The nuclide generation rates in the target in comparison to other planned installations for unknown or poorly studied masses are presented in Fig. 3. The productivity of the two types of targets (cold and hot) is approximately the same.

4. PITRAP-Setup

The proposed variant of a PITRAP system is shown in Fig. 4. The magnet (1) turns the incoming ion beam by 90°. The ions are directed into the deceleration section in front of a gas-filled radio frequency quadrupole (2) where they are moderated and focused. The radio-frequency quadrupole is filled with a buffer gas (helium) at pressure ~1 Pa. It consists of four segmented rods to which a radio-frequency voltage is applied in order to keep the ions in the radial plane. A constant voltage can also be applied to the segments of rods in order to form an

Fig. 4. Model of the measuring facility with a Penning ion-trap (4). (1) is a turning magnet, (2) indicates the gas-filled RFQ, (3) is the multi-ToF system.

electric field along the axis of the system, having the minimum potential in the region of the penultimate segments.

Colliding with the molecules of the buffer gas the ions lose kinetic energy and accumulate at the minimum of the potential. Thus, a group of ions is formed, which after reduction of the potential on the last segments of the rods can be extracted from the radio-frequency quadrupole. Most of the elements of the radio-frequency quadrupole are supposed to be arranged on a high-voltage platform surrounded by a mesh (2 in Fig. 4). The longitudinal emittance of a 2.5 keV beam after the quadrupole will be 10 eV·μsec and the transverse emittance ~10 π·mm·mrad. The cooling time in the gas-filled radio-frequency quadrupole is about 10 msec. The potential of the effluent group of ions is lowered by means of a pulsed change of their potential energy in the pulse cavity, comprising a tube where the voltage is switched as ions fly through it.

A group of ions enters a pulse cavity, where their potential is reduced to the required value. Next, ions are guided towards a time-of-flight mass-analyzer (3), which can operate as a mass-spectrometer for fast measurement of the ion mass or as a mass-separator for addition purification of the ion beam [12]. A group of ions in the operating regime of a time-of-flight mass-separator is transported toward the system of Penning traps, located in the superconducting magnet (4). In addition, an ion source for test measurements or experiments in the off-line regime, including calibrations, can be located in front of the turning magnet (1 in Fig. 4).

5. Penning Traps System

The expected rates for some exotic nuclides at the entrance of the traps are shown in Fig. 5. They are high enough for PTMS to perform the measurements.

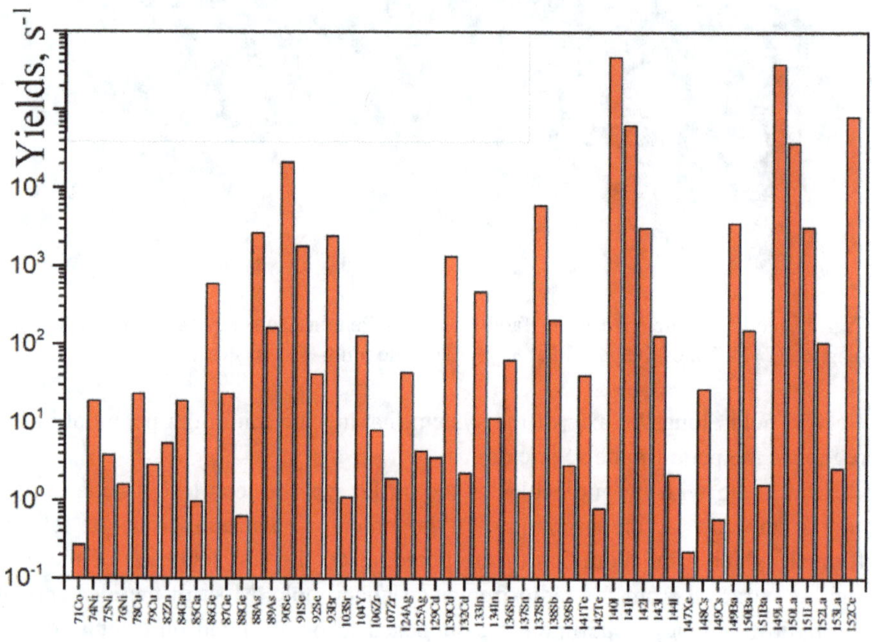

Fig. 5. The rates of ions in the entrance of traps. The target of ^{235}U with the weight of 1.5 g has been used for estimations. The efficiency of He-jet transportation and cooling are assumed to 50%, the ionization efficiency is of 0.1%.

It is proposed that a variant with two cylindrical traps arranged in one tube of a superconducting magnet (B=7 T) (analog of the SHIPTRAP setup [13]) be used. Two centers of the highest homogeneity of the magnetic field, should correspond to the centers of the traps. The first preparatory trap, intended for mass-selective cooling of the ions in the buffer gas, is filled with helium at working pressure 10^{-4} Pa. It permits separating ions with a certain charge-to-mass ratio. The second trap under high vacuum $10^{-6} \div 10^{-7}$ Pa is intended for high-precision measurements of masses. The traps are separated by a 1.5 mm (in diameter) diaphragm. In order to determine the masses (total binding energies) the cyclotron frequencies of ions which directly depends on the mass value in the traps ($\upsilon_c = Bm/q$) have to be measured. It can be done by the ToF ion cyclotron resonance method [14] as well as by the novel phase-imaging method [6].

6. Conclusion

The high intensity of the neutron flux of the PIK research reactor at the PNPI in combination with the high sensitivity and precision of the ion trap opens up prospects for multifunctional studies in different areas of science. The proposed Penning ion traps in tandem can be used on-line with secondary beams as well as off-line using fission products and products from neutron capture reactions. The on-line variant is of greatest interest for studying exotic neutron-rich nuclides for determining the true path of a fast neutron capture process in astrophysics. This r-process is assumed to be responsible for the creation of the chemical elements in nature. Its 'nuclide' path can be determined uniquely only by precise and reliable measurement of the masses of vast regions of exotic nuclides in a ground-based laboratory. Possibilities for using ion traps in the interests of atomic and neutrino physics by means of high-precision measurements of the mass difference of long-lived nuclides open up in the off-line measurement regime. The technical work up of the design takes account of these multifaceted problems. The design of the first Penning trap in Russia was developed using the many years of experience gained in working with different ion traps in foreign laboratories. The technical design project is presented in [15].

We are grateful to A. A. Vorobyov for stimulating the work on the Penning ion trap design at the PNPI and valuable discussions. We are also grateful to K. Blaum and M. Block for their interest in the project and for fruitful discussions.

References

1. K.A. Konoplev, Main scientific results of the PNPI in 2000–2004, 18–23, Gatchina, 2005.
2. V.L. Aksenov and M.V. Kovalchuk, www.NIKIET.ru (2015).
3. K. Blaum, Yu. Novikov, and G. Werth, *Contemp. Phys.*, **51**, 149 (2010).
4. E. Burbidge, M. Burbidge *et al.*, *Rev. Mod. Phys.*, **29**, 547–650 (1957).
5. S. Eliseev, C. Roux, K. Blaum *et al.*, *Phys. Rev. Lett.*, **107**, 152501 (2011).
6. S. Eliseev, K. Blaum M. Block *et al.*, *Phys. Rev. Lett.*, **110**, 082501 (2013).
7. S. Eliseev, Yu. Novikov, and K. Blaum, *J. Phys. G: Nucl. Part. Phys.*, **39**, 124003 (2012).
8. H.-J. Kluge and Yu. Novikov, Nucl. *Phys. News*, **17**, No. 4, 48–50 (2007).

9. A.G. Dernjatin, K.A. Mezilev, Yu.N. Novikov and A.W. Popov, Projects planned at the reactor PIK on 1991-1995. Abstracts, PNPI (1991).
10. V. Panteleev, A. Barzakh, D. Fedorov *et al.*, PNPI, High Energy Physics Division, Main Scientific Activities 2007–2012, 274–277 (2013).
11. D. Simonovski, Yu.I. Gusev *et al.*, Atom. Energya (accepted).
12. R. Wolf, D. Beck, K. Blaum *et al.*, *Nucl. Instrum. Meth. A* **686**, 82 (2012).
13. M. Block, D. Ackermann, K. Blaum *et al.*, *Eur. Phys. J. D* **45**, 39 (2007).
14. M. König *et al.*, *Int. J. Mass Spectrom. Ion Proc.*, **142**, 95 (1995).
15. Yu.N. Novikov, Yu.I. Gusev, S.A. Eliseev *et al.*, Report No. Φ-310. B.P. Konstantinov PNPI NRC "Kurchatov Institute", Gatchina, 2016.

Control and Data Acquisition Systems of the MASHA Setup

A. S. Novoselov*, A. M. Rodin, E. V. Chernysheva, S. N. Dmitriev, A. V. Gulyaev, A. B. Komarov, Yu. Ts. Oganessian, A. V. Podshibyakin, V. S. Salamatin, S. V. Stepantsov, V. Yu. Vedeneev and S. A. Yukhimchuk

Flerov Laboratory of Nuclear Reactions, Joint Institute for Nuclear Research, Dubna, 141980, Russia
**E-mail: noval@jinr.ru*

L. Krupa and M. Holik

Institute of Experimental and Applied Physics, Czech Technical University in Prague, Horska 3a/22, 128 00 Praha 2, Czech Republic

J. Kliman and D. Kamas

Institute of Physics SASc, Dubravska cesta 9, 84228 Bratislava, Slovakia

A. Opíchal and J. Pechousek

Faculty of Science of Palacký University in Olomouc, 17. listopadu 1192/12, 779 00 Olomouc, Czech Republic

The mass separator of super heavy elements also known as MASHA setup from the automation systems point of view is described. The setup uses ISOL (Isotope Separation On-Line) method and represents the modular online mass spectrometer where masses of new elements from helium up to mass A=450 are measured directly in the course of their synthesis with accelerated heavy-ion beams. During the last years the control, the data acquisition and the beam diagnostics systems were modified. The modernization was in changing the platforms to the modern one from the CAMAC to the NI PXI and from the SmartBOX to the Wago PLC and the NI CompactRIO. The new systems were integrated and tested on the beam experiments with heavy ions at the U-400M cyclotron. The future experimental program of the MASHA setup for study of the physical and chemical properties of superheavy elements at the SHE factory is discussed.

Keywords: Control systems; data acquisition and sorting; charged-particle spectrometers.

1. Introduction

Nowadays the successful experimental research is not possible to imagine without the complex automation. A great number of the different parameters should be under control to obtain the clear result of the fine process. The search of the new chemical elements, isotopes and its investigation is the field where modern technologies of the automation open the new opportunities for the nuclear physics.

This paper gives a short overview of the different automation systems at the MASHA setup. MASHA represents the modular online mass spectrometer using ISOL (Isotope Separation On-Line) method where masses of new elements from helium up to mass A=450 are measured directly in the course of their synthesis with accelerated heavy-ion beams. It was designed and built at the Flerov laboratory of the Joint Institute for Nuclear Research in Dubna (Russia). More detailed information about the setup was published at works [1,2].

Some years ago the modernization of the control, the beam diagnostics and the data acquisition systems was started. Now the systems were integrated and tested on the beam experiments with heavy ions at the U-400M cyclotron. All systems during the experimental work are remote controlled from the two multimonitor personal computers at the control room. The modernization is not stopped because the different new experiments usually need some changes of the MASHA setup and the automation systems is not an exception.

2. Control System

The control system of the MASHA setup is based on the Wago PLC and the NI CompactRIO platforms. Also there are some number of the SmartBOX logical controllers, but mostly they were replaced by the Wago PLC because of the instable work. The software for the control system is developed with the help of the LabView package from National Instruments.

The communication between the control room and the experimental hall is done using the network with the link speed up to the 1 Gbps. The serial devices which are using RS-232/422/485 are connected to the network through the MOXA device servers. Also the CAN bus is connected with the help of the gateway.

The logical structure of the MASHA control system:

- magnet system
- vacuum system
- ion source
- high voltage
- hot catcher
- rotating target

The magnet system control includes the control of the magnet elements and the magnetic field measurement system. The setup has 4 dipole magnets with Hall sensors. All other magnets are focusing and there are controlled only by the current and voltage from the remote controlled power supply. Two main dipole magnets additionally supplied with the NMR sensors for more precise measurement of the magnetic field. The software for this system automatically calculates and sets the magnetic field for the required mass of the isotope at the focal plane of the separator.

The vacuum system control includes the control of more than 10 of the turbomolecular pumps, some number of the backing pumps, more than 30 of the vacuum valves and the large number of the vacuum sensors. The software for this system helps to the operator to reach and control the vacuum level at the different parts of the MASHA setup.

The ion source for the setup is electron cyclotron resonance based. To make it work is needed to control the microwave source (2,45GHz), valve of the volume with buffer gas and also mechanical based adjustment of the resonance.

The high voltage system is used for extraction of the ions from the ion source direct to the mass separator. From the automation point of view it contains the control of the two high voltage power supplies.

The hot catcher automation includes the control of the high current power supply and obtaining the temperature of the carbon heater.

The rotating target is controlled by the commands and information from power supply of its motor. It is needed to mention about the automatic blocking system of this part of MASHA setup. If in some case during the operation the target wheel will be locked it automatically will be closed from the beam by the faraday cup.

There are also different types of the actuators such as pneumatic or with the step motor for beam absorbers, scattering foil, Faraday cups and etc.

3. Beam Diagnostics System

The beam diagnostics system contain the on-line and the off-line parts for the beam energy and current measurement. The term “on-line” here means that the beam is not interrupted and it irradiates the target material. The “off-line” methods are opposite and there are based on the beam stopping. These methods complement each other.

The on-line system for beam energy measurement contains 2 pick-up sensors, digitizers and TDC for precise time-of-flight measurement. For the on-line beam current measurement with digital ammeter there is the isolated ring after the target for the collection of the secondary electrons after the target.

The off-line system for beam energy measurement contains 2 multichannel plates (TOF scheme) and silicon detector. The off-line beam current measurement is classical and based on the Faraday cup with digital multimeter.

More detailed information is available at work [3].

4. Data Acquisition System

The data acquisition system is the National Instruments PXI based using the high speed multichannel digitizers from XIA (250 MHz,12 bit, 16 channels). The Si stripe detector of the well type is placed at the focal plane of the setup. The stripe number at the focal plane is 192 and for the side walls is 160. The total number of the spectrometric channels is 704. The software is written using the C++ language. For some experiments system is expanded with the TimePix pixel detector from the IEAP CTU in Prague.

More detailed information is available at work [4].

5. Conclusion

The control, the beam diagnostics and the data acquisition systems for the MASHA setup are created on the base of the modern technologies. The future experimental program for study of the physical and chemical properties of superheavy elements at the SHE factory define the future development of the MASHA setup. Nowadays we plan to develop gas catcher, pre-separator and this things also couldn't be without modern automation technologies.

References

1. L. Krupa et al., Mass spectrometer MASHA: Current status and perspectives for cyclotron DC280. 3rd International Symposium on Super-Heavy Elements "Challenges in the studies of super-heavy nuclei and atoms", SHE 2017, September 10–14, 2017, Kazimierz Dolny, Poland. Available from http://she2017.ncbj.gov.pl/talks/.
2. V.Yu Vedeneev et al., The current status of the MASHA setup. Hyperfine Interaction (2017), V. 238:19, Springer Netherlands.
3. S. Motycak et al., A new beam diagnostic system for the MASHA setup. *Physics of Particles and Nuclei Letters*, **13**(5), 586–590 (2016).
4. A.S. Novoselov et al., Data Acquisition System for the Focal Plane Detector of the Mass Separator MASHA. *Physics of Particles and Nuclei Letters*, **13**(5), 595–597 (2016).

Project of Priority Experiments at SHE Factory

Yu. Ts. Oganessian, S. N. Dmitriev and V. K. Utyonkov*
Flerov Laboratory of Nuclear Reactions, Joint Institute for Nuclear Research, Dubna, RU-141980, Russian Federation
**E-mail: utyonkov@jinr.ru*

A new experimental complex, the SHE Factory at FLNR, including a high-current cyclotron DC280 and a gas-filled separator DGFRS-2, is planned to be commissioned by the end of 2018. The project of priority experiments which are planned to be performed for testing the capabilities of DGFRS-2 coupled with DC280 is presented. These include study of the ^{243}Am+^{48}Ca reaction for measuring yields of Mc isotopes with large statistics at different beam energies. Investigation of chemical properties of elements Cn and Fl with use of the chemistry setup placed after DGFRS-2 is discussed as well. The first research experiments, including synthesis of new elements 119 and 120 in the reactions of ^{249}Bk and $^{249\text{-}251}$Cf with ^{50}Ti, are also considered.

Keywords: Superheavy elements; α decay; spontaneous fission.

1. Introduction

At present, the region of known superheavy nuclei (SHN) with $Z \leq 118$ and their α-decay descendants forms a relatively narrow "ridge" on the nuclear landscape [1]. In order to more fully understand the importance of shell stabilization in this region, it is essential to considerably extend the area of synthesized SHN. For example, isotopes of elements Ts and Lv that have a neutron more than the presently known ones could be produced in the 2*n*-evaporation channel of the ^{249}Bk(^{48}Ca,2*n*)^{295}Ts and ^{248}Cm(^{48}Ca,2*n*)^{294}Lv reactions together with their α-decay products. However, the cross sections of these reactions are predicted to be at the level of 0.3-0.4 pb by the model which reasonably reproduces available experimental data [2]. Exploration of the region of neutron-deficient SHN seems to be possible with use of the lighter target nuclei. But again, a considerable drop of the evaporation cross section of the ^{239,240}Pu+^{48}Ca reactions compared to that of the ^{244}Pu+^{48}Ca reaction was found [3]. Several attempts have been undertaken to synthesize new elements heavier than Og by using target nuclei ranging from ^{238}U to ^{249}Cf with projectiles heavier than ^{48}Ca, from ^{50}Ti to ^{64}Ni, but no indisputable decay chains of parent nuclei could be observed at the upper cross-section limits of 0.07-1.1 pb [4]. The experiments aimed at a more detailed investigation of the known nuclides resulted in the first experimental insights related to the decay schemes of ^{276}Mt and ^{272}Bh by measuring γ rays in coincidence with α particles and demonstrated the prospects for investigating the nuclear structure of SHN [5]. Further investigations of SHN can reveal interesting features of their structure predicted by theory, e.g., change or even coexistence of different shapes of nuclei in the transition region between shells at N=162 and N=184.

Obviously, further investigations of the heaviest nuclei and synthesis of the new elements call for considerable increase in the sensitivity of equipment used in experiments. The new laboratory for production of superheavy nuclei (SHE Factory) is scheduled to be commissioned by the end of 2018 at FLNR (JINR). SHE Factory includes a new high-current cyclotron DC280 with beam intensity of 3-6$\times10^{13}$ for ions like ^{48}Ca, ^{50}Ti, and ^{54}Cr [6] as well as several new separators one of which is the gas-filled separator DGFRS-2. Expected yield of SHN can be enhanced by almost two orders of magnitude compared to the current level at FLNR due to increased beam intensity of DC280, transmission of DGFS-2, and possible duration of experiments.

2. Dubna Gas-Filled Recoil Separator DGFRS-2

The separator DGFRS that is currently in operation at FLNR has a DQ_hQ_v magnetic configuration: a flat-field dipole magnet (D) followed by two horizontally (Q_h) and vertically (Q_v) focusing quadrupole lenses. The main loss of the complete-fusion reaction products (ERs) occurs in the dipole magnet because of a relatively narrow 60-mm gap between its poles. Besides, lack of vertical focusing of ERs, that undergo multiple-scattering in the target, does not allow use of the targets thicker than 0.4 mg/cm^2 that additionally reduces the overall yield of synthesized ERs. The transmission of the new separator DGFRS-2 will be increased by installing a quadrupole before the dipole magnet and by increasing the dipole gap. The DGFRS-2 has a $Q_vD_hQ_vQ_hD$ ion optical scheme [7]. The quadrupole Q_1 with a maximum field gradient G=13.3 T/m and an effective length L=0.46 m focuses ERs in the vertical direction (Figure 1). The dipole magnet D_1 with a deflection angle φ=31.5°, a bending radius ρ=1.8 m, a maximum field strength B=1.8 T, and a gap h=120 mm focuses ERs by the inclined pole edge in the horizontal direction. The Q_2 and Q_3 quadrupoles (G=5.4 T/m, L=0.60 m) focus ERs on the DGFRS-2 focal plane. The dipole D_2 (φ=10°, ρ=1.8 m, B=1.8 T, h=120 mm) is applied for additional reduce of the background, e.g., that caused by the scattered atoms of filling gas (H_2 or He).

With such configuration of DGFRS-2, the transmission should be larger and the thickness of the target layer can be increased by a factor of about 2 in comparison with those for DGFRS. The ion-optical calculations and modeling showed [7] that the overall product yield of DGFRS-2 is expected to be higher by a factor of 3 in comparison to that for DGFRS.

Fig. 1. Photo of the Dubna gas-filled recoil separator (DGFRS-2). From left to right: quadrupole lens Q_1, dipole magnet D_1, two quadrupole lenses Q_2 and Q_3, and dipole magnet D_2.

The new detection system will be similar to that used at DGFRS, and will include focal-plane double-sided strip detectors with 1-mm-wide strips, 48 horizontal and 128 vertical strips. This detector will be surrounded by an array of six detectors each with an active area of 65×120 mm^2 and 8 strips.

3. First-Day Experiments at DGFRS-2

For initial setting of the DGFRS-2 parameters, the computer code will be used which simulates the trajectories of ERs in the separator [7]. For tests of the code and choosing optimal parameters of DGFRS-2, we first plan to start with reactions, e.g., ^{nat}Yb+^{48}Ca and ^{170}Er+^{50}Ti resulting in production of Th isotopes with large cross section of about 10^8 pb. At this step, the image size of ERs, transmission and dispersion of DGFRS-2 could be measured.

The ^{243}Am(^{48}Ca,xn)$^{291-x}$Mc reaction seems to be important from several points of view. First, the optimal parameters of DGFRS-2 could be determined. The sensitivity of this experiment at the maximum reaction cross section σ=8.5 pb, with target thickness h_t=0.5mg/cm^2, separator transmission ε=60%, and beam intensity I_b=5 pμA could reach 17 decay chains per day. In addition, several issues may be clarified in these studies with large sensitivity:

- Some of the theoretical models predict the largest yield of SHN at lower excitation energies than those already tested in experiments by now (see, e.g., insert in Figure 2). Measurement of the excitation function at low energies is important for performing experiments on the synthesis of new elements.

- Isotope ^{289}Mc was identified by 16 decay chains of ^{293}Ts; α decay of ^{281}Rg was observed in two cases (Figure 1). Similar short chains were detected in the ^{243}Am+^{48}Ca reaction; in the 14 registered chains α decay of ^{281}Rg was not revealed. Observation of short lived ^{277}Mt, the descendant of ^{281}Rg, could additionally clarify the origin of these chains.
- For ^{277}Mt, product of the ^{243}Am(^{48}Ca,3n)^{288}Mc reaction, two different half-lives seems to be observed that could be caused by existence of the theoretically predicted isomers in the transition region between N=162 and N=184. Other possible example could be an isomeric state in ^{269}Sg [3].
- About 100 decay chains of ^{288}Mc were observed in [1,5] but no chains of ^{288}Fl or ^{284}Cn, products of the p2n channel or electron capture (EC) of ^{288}Mc or ^{284}Nh, were found. Increased statistics could clarify the cross section level of the pxn channels and EC branch for these nuclei.
- Only two chains of the 5n channel were observed in the reactions with ^{244}Pu [1] and ^{242}Pu [8]. Synthesis of the still unknown ^{286}Mc is important for further development of theory of formation of SHN.

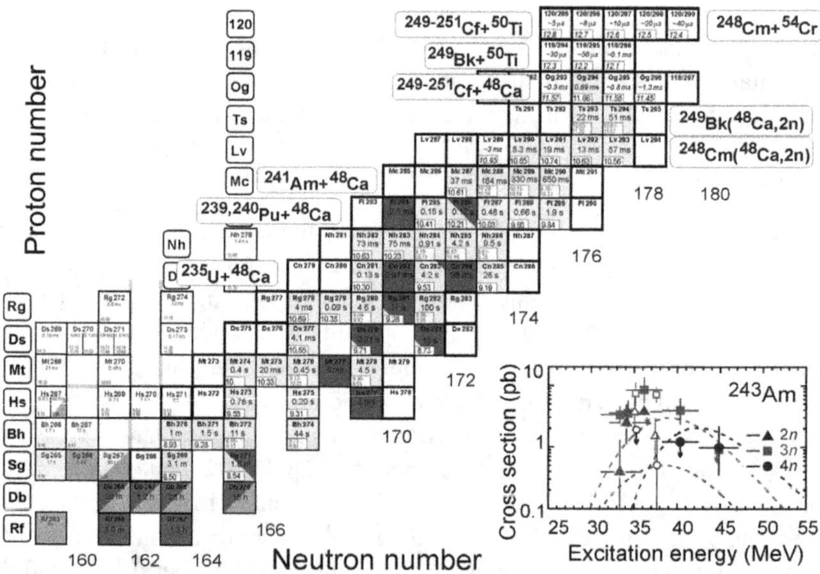

Fig. 2. Top part of Chart of Nuclei. New nuclides which can be synthesized in the given reactions are shown by open squares. Insert shows excitation functions of the 2n- to 4n-evaporation channels from the complete-fusion reaction ^{243}Am+^{48}Ca measured in [1] (solid symbols) and in [5] (open symbols). Lines show results of calculations [2].

The $^{242}Pu+^{48}Ca$ and $^{244}Pu+^{48}Ca$ reactions were used in study of the chemical properties of Fl [9]. However, only a few chains of Fl atoms were observed. In thermochromatography experiments performed at JINR, the deposition pattern of Fl on a Au surface indicates its high volatility. However, the measured at GSI lower limit for the adsorption enthalpy of Fl on a Au surface points to the formation of a metal-metal bond of Fl with Au. New chemistry experiments, which can be performed with use of DGFRS-2 as a pre-separator and with a yield of Fl of about 5 atoms per day, could resolve this issue.

The first series of experiments performed with use of the unique mixed-Cf target ($^{249-251}Cf$) and ^{48}Ca beam, aimed at the synthesis of the heaviest isotopes $^{295,296}Og$, ended in observation of the fifth decay chain of ^{294}Og [10]. The new target is under preparation at ORNL. The sensitivity of experiments at DC280 and DGFRS-2 could correspond to the yield of 4 events of Og isotopes per week at σ=0.5 pb, h_t=0.3 mg/cm^2, and I_b=5 pμA.

The same target can be used for synthesizing new element 120 in the reaction with ^{50}Ti. Unfortunately, the predicted level of cross sections of the reactions leading to production of elements with Z>118 is very uncertain. For better understanding the fusion-evaporation reaction mechanisms, the measurement of the cross section of the reactions of actinide nuclei with ^{50}Ti is vitally important. Of particular interest are the reactions leading to the same compound nucleus but with different beam particles, that is ^{48}Ca and ^{50}Ti. Among several candidates for such studies, the reactions $^{246}Cm+^{48}Ca\rightarrow^{294}Lv^*$ and $^{244}Pu+^{50}Ca\rightarrow^{294}Lv^*$ seem to be the most promising. The cross sections of the reactions with neighboring isotopes ^{245}Cm and ^{248}Cm are comparable, about 4-5 pb [1], and the same value may be expected for the reaction with ^{246}Cm. For the $^{244}Pu+^{50}Ti$ reaction, the cross section is predicted to be lower by about a factor of 20 [2]. The yield of Lv isotopes in this reaction could be about one event per week at σ=0.2 pb, h_t=0.3 mg/cm^2, and I_b=3 pμA. The decay properties of the 3n- and 4n-reaction channels, $^{290,291}Lv$, are known [1] (Figure 2). Note, localization of the excitation-function maximum is also extremely important for experiments aimed at the discovery of new elements.

This series of experiments allow to solve a significant part of both technical and scientific problems in the planning of future experiments on the synthesis of the new elements. One of the most important challenges in nuclear physics is the synthesis of elements 119 and 120. The most encouraging approach for the discovery of element 119 seems to be using of the complete fusion of ^{50}Ti with ^{249}Bk. The identification of the products of the 3n- and 4n-evaporation channels, $^{295,296}119$ can be based on the genetic correlation with known descendants

^{287,288}Mc (Figure 2). The production rate of the nuclides under study could reach one event per month at σ=50 fb, h_t=0.3 mg/cm^2, and I_b=3 pμA.

Due to larger expected cross section for synthesis of element 120, the reactions of ^{50}Ti with ^{249}Cf or mixed-Cf target nuclei look more promising compared to the ^{248}Cm+^{54}Cr reaction [2] although the latter leads to somewhat heavier isotopes. The decay properties of the descendants of the isotopes $^{295-299}$120, from the heaviest ^{294}Og and ^{291}Lv down to the lightest ^{263,265,267}Rf, have been determined in previous experiments; this allows application of the correlation method for identification. The discovery of element 120 would require about two months at σ=25 fb, h_t=0.3 mg/cm^2, and I_b=3 pμA at SHE Factory which opens new exciting opportunities for investigation of SHN.

Acknowledgments

We acknowledge the support by the JINR Directorate grant and the RFBR Grant No. 16-52-55002.

References

1. Yu.Ts. Oganessian and V.K. Utyonkov, *Rep. Prog. Phys.* **78**, 036301 (2015); *Nucl. Phys. A* **944**, 62 (2015).
2. V.I. Zagrebaev and W. Greiner, *Nucl. Phys. A* **944**, 257 (2015).
3. V.K. Utyonkov *et al.*, *Phys. Rev. C* **92**, 034609 (2015); *ibid.* **97**, 014320 (2018).
4. Yu.Ts. Oganessian *et al.*, *Phys. Rev. C* **79**, 024603 (2009); S. Hofmann *et al.*, GSI Scientific Report 2008, GSI Report No. 2009-1, 131 (2009); S. Hofmann *et al.*, *Eur. Phys. J. A* **52**, 180 (2016); J. Khuyagbaatar *et al.*, GSI Scientific Report 2012, GSI Report No. 2013-1, 131 (2013).
5. D. Rudolph *et al.*, *Phys. Rev. Lett.* **111**, 112502 (2013); J.M. Gates *et al.*, *Phys. Rev. C* **92**, 021301(R) (2015).
6. S.N. Dmitriev, this volume.
7. A.G. Popeko, *Nucl. Instrum. Methods Phys. Res. B* **376**, 144 (2016).
8. P.A. Ellison *et al.*, *Phys. Rev. Lett.* **105**, 182701 (2010).
9. R. Eichler *et al.*, *Radiochim. Acta* **98**, 133 (2010); A. Yakushev *et al.*, *Inorg. Chem.* **53**, 1624 (2014).
10. N.T. Brewer *et al.*, *Phys. Rev. C* **98**, 024317 (2018).

Features of the Solid-State ISOL Method for Fusion Evaporation Reactions Induced by Heavy Ions

A. M. Rodin[1], E. V. Chernysheva[1], S. N. Dmitriev[1], A. V. Gulyaev[1], D. Kamas[2], J. Kliman[2], L. Krupa[3], A. S. Novoselov[1], Yu. Ts. Oganessian[1], A. Opíchal[4], A. V. Podshibyakin[1], V. S. Salamatin[1], S. V. Stepantsov[1], V. Yu. Vedeneev[1] and S. A. Yukhimchuk[1]

[1]Flerov Laboratory of Nuclear Reactions, JINR, Dubna, 141980, Russia
[2]Institute of Physics SASc, Dubravska cesta 9, 84228 Bratislava, Slovakia
[3]Institute of Experimental and Applied Physics, Czech Technical University in Prague, Czech Republic, Horska 3a/22, 128 00 Praha 2
[4]Palacký University in Olomouc, Czech Republic

The solid state ISOL system used at the mass-separator MASHA is presented. The system separates products of complete fusion reaction induced by heavy ion beams. The poly-graphene foil is used as a hot absorber of the reaction products. The results of a total efficiency and a separation time measurement of short-lived mercury and radon isotopes are presented. The measurement of the total efficiency as a function of time was carried out at intense heavy ion beams (up to 0,5 pμA). It is shown that the decrease of the total separation efficiency is due to heating and radiation destruction of the hot absorber structure. A new design of the hot catcher is proposed based on the thin sheet made of carbon nanotubes and graphene. The test experiments showed a good potential of using such materials for the reactions studied. A new design of the separation system (consisting of the ECR ion source and the hot catcher) is proposed for experiments on the mass measurement of isotope ^{283}Cn. This isotope will be produced in the reaction ^{48}Ca+^{242}Pu as the decay product of the nucleus ^{287}Fl. The whole system will be heated up to 300°C in order to increase the separation efficiency for the system proposed. In addition, the inner surface of the vacuum pipelines and chambers will be covered with a special chemical inert glass-enamel coating.

Keywords: Mass spectrometer; ISOL-method.

1. Introduction. Motivation of the Solid-State ISOL System for the Mass-Separator MASHA

In the experiments on synthesis of superheavy nuclei the basic nuclide identification methods are radiochemistry and the ones based on the properties of the decay chains ended by well-known nuclei. Since the half-lives of the majority of superheavy nuclei are quite small (from 100 μs till 10 ms), the kinematic separators are widely used for the synthesis of superheavy nuclei, allowing to separate them with a high reliability and efficiency, but they do not measure the masses of nuclei. The mass-spectrometer MASHA [1] was developed at FLNR

JINR (Dubna) to measure the masses of superheavy nuclei and to investigate their α-decays or spontaneous fission. The MASHA set-up is a combination of the ISOL method of synthesis and separation of radioactive nuclei with the classical method of mass analysis, allowing the mass identification of the synthesized nuclides in a wide mass range (A = 1 – 450 u).

The experiments performed at FLNR JINR showed that some of the nuclei with Z=112 and 114 have long enough half-lives, e.g., 3.8 s for ^{283}Cn, 29 s for ^{285}Cn, and 2.6 s for ^{289}Fl [2]. These elements are volatile. It means that the ISOL method can be applied for investigation of these elements.

There are special requirements for the hot catcher in studies of short-lived superheavy nuclei that have low production cross-sections. The necessary separation time should be about ~ 1 s and the necessary efficiency must be higher than 10%. The analysis of different materials shows that a good choice for the hot catcher is porous graphite. The hot catcher is a very important component for the set-ups that use the ISOL method since the efficiencies of extraction of nuclides depend on its properties. It is well known that the use of porous graphite as the hot catcher improves the time characteristics and the efficiency of extraction. At ISOLDE [3] a liquid lead target is used as the hot catcher to provide a better diffusion of particles. At MASHA the extraction time was 1.8 s as compared with 30 s at ISOLDE. The catcher was made from the poly-graphene paper with a density of 0.5 g/cm^3 and a porosity of 75%. The catcher thickness was 0.3 mm and it was shaped as a 30 mm diameter disk. The operating temperature of the hot catcher was 1800-2000°C during the experiment. The ECR ion source has an ionization efficiency of about 90% for noble gases. It works at a pressure of 10^{-5} mbar. A separation time of 1.8 c of the MASHA setup seems to be a convenient value for the experiments on measuring the masses of superheavy elements. Since the production cross-sections of superheavy elements are of a few picobarns, the total efficiency of separation have to be more than 10%.

2. Test Experiment for Measurement of the Mass of a Superheavy Nucleus

The first attempt of determination of the mass of a Cn isotope produced as a daughter one in the reaction ^{48}Ca+^{242}Pu→^{290}Fl was made in 2015. The experiment was carried out at U-400M cyclotron of FLNR JINR. Two fission fragments and one alpha decay were detected at the decay of ^{283}Cn. The beam energy was 7.2 AMeV. Aluminum foils were used to decrease it down to the Coulomb barrier. The intensity of ^{48}Ca beam was about 5-10 μA. During the experiment a special attention was paid to the control of efficiency and reliability of all systems. A special construction of the target was developed for this purpose. The target of

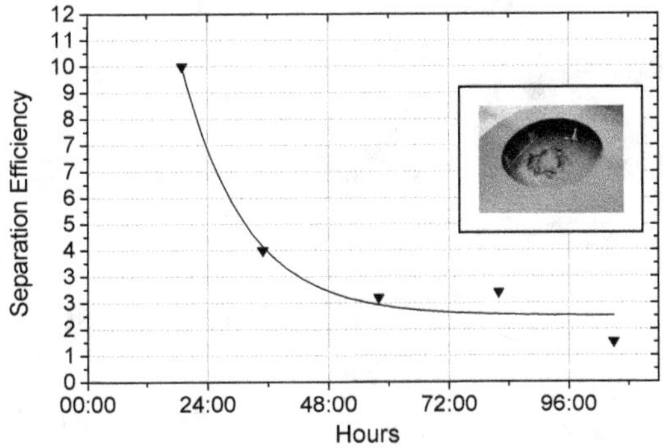

Fig. 1. The separation efficiency of the MASHA setup as a function of time. The poly-graphene paper is shown in frame after 300 h of irradiation. The ^{48}Ca beam current was 0.5 pμA. The electrical power density was 2 kW/cm^3 and the ion irradiation one was 8 kW/cm^3.

^{242}Pu (0.49 mg/cm^2) had an admixture of Nd isotopes (~1%). During the experiment the yields of mercury isotopes produced in the reaction $^{48}Ca+^{nat}Nd$ were measured. The yield of ^{185}Hg was tested each day (Fig. 1).

It turned out that the separation efficiency of the MASHA setup was not stable during the experiment in the case of a high beam intensity (typically 10 μA of ^{48}Ca) which corrupted the hot catcher shown in the frame in Fig. 1. As a result the structure of graphite was destroyed and the extraction efficiency of the hot catcher decreased by almost 6 times during a few days. It meant that a serious modernization of the hot catcher and the whole setup was necessary for the measurement of superheavy nucleus masses.

3. Modernization of the Mass-Spectrometer

The modernization of the main units of the mass-spectrometer MASHA was performed. It includes a new design of the hot catcher, the ECR ion source and vacuum chambers, the hot transport system, and installing an additional strip detector at the intermediate plane F1 for the permanent control of the separation efficiency.

The design of the hot catcher includes now a poly-graphene heater and a thin catcher made of 2 mg/cm^2 graphene foil or carbon nanotube paper. This construction has to prevent the catcher from heating and radiation destruction. The ECR ion source with a special ceramic coating was developed and tested. The possibility to work at a temperature of up to 300°C and the use of a special chemical inert glass-enamel coating for the inner surfaces of the vacuum pipelines

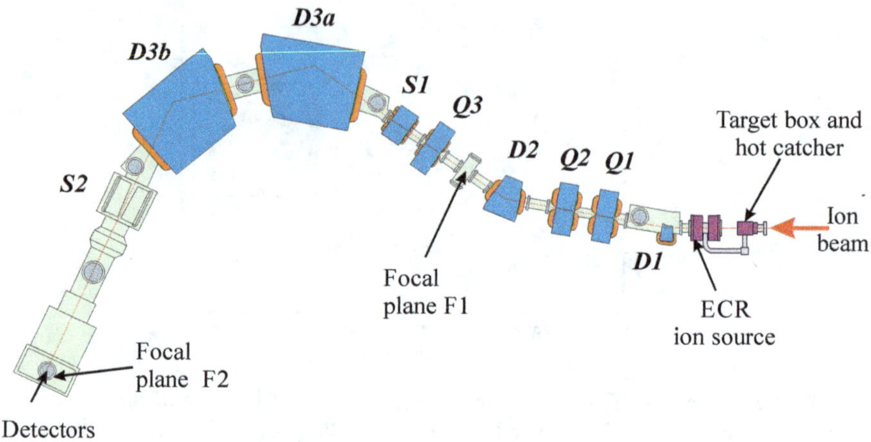

Fig. 2. Scheme of the MASHA setup.

and chambers are the particular features of this design. This is necessary for the MASHA setup to detect not only volatile elements (radon, mercury) but also others close to them (At, Fr, Ac).

In Fig. 2 the scheme of the MASHA setup is shown. It consists of the target box with the hot catcher, the ECR ion source, the ion-optical transport system with a 192-strip silicon detector placed in the focal plane F2. The additional strip detector at the intermediate focal plane F1 registers permanently the products of the complete fusion reaction $^{48}Ca+^{nat}Nd$ which takes place due to the admixture of Nd in the target, providing the Cn nuclei have to be detected in the focal plane detector.

In the test experiment we carried out simultaneously the reaction $^{40}Ar+^{144}Sm$ at the intermediate plane F1 and the reaction $^{40}Ar+^{166}Er$ at the focal plane F2. A special construction of a rotating target wheel was developed. The target wheel consists of the 6 identical sectors: 3 of ^{144}Sm and 3 of ^{166}Er. Fig. 3a presents the energy spectrum from the first strip of the intermediate plane 16-strip detector and Fig. 3b shows the (M, E_α) distribution for the reaction $^{40}Ar+^{166}Er$ measured by the focal plane 192-strip detector in the same experimental run.

The fusion-evaporation reaction $^{40}Ar+^{144}Sm$ was used for the determination of the separation efficiency of the mass-spectrometer MASHA. For measurement of the total one (including the extraction efficiency from the hot catcher, the efficiency of ionization in the ECR ion source and the efficiency of the ion-optic transport system) the experiments with a moving foil array were carried out and then the absolute production cross sections of Hg and Rn isotopes were obtained.

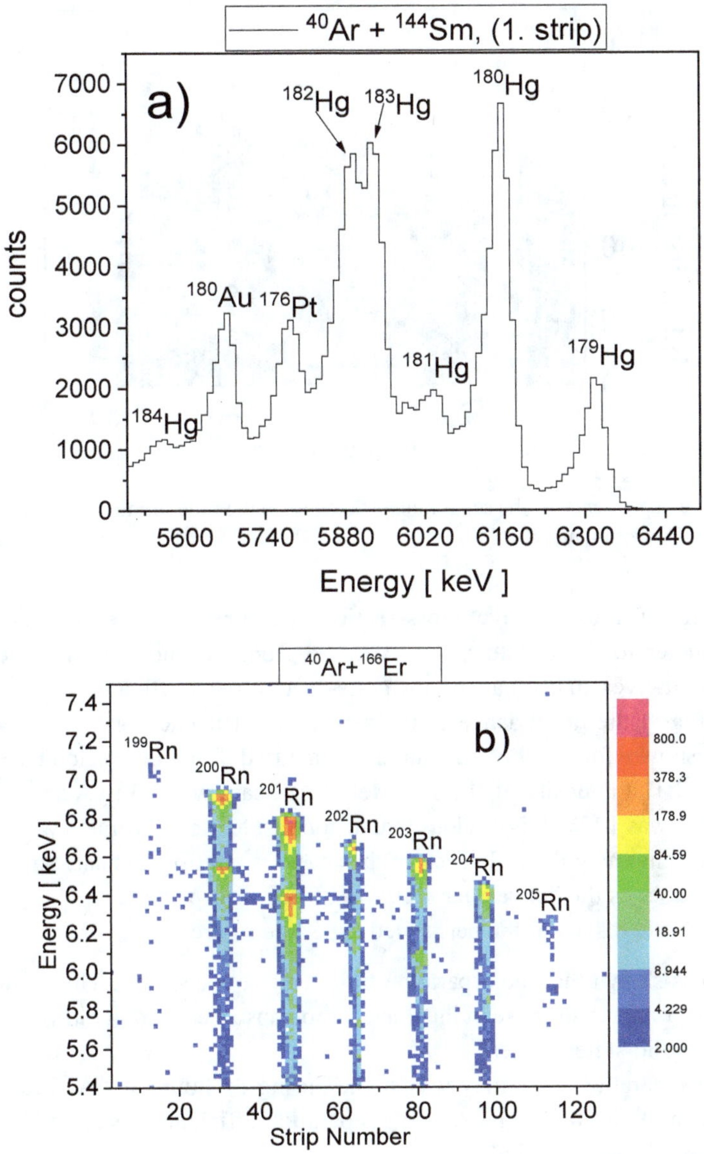

Fig. 3. a) The energy spectra of mercury isotopes (^{179}Hg–^{182}Hg) registered at the F1 intermediate plane of the MASHA setup in the complete fusion reaction ^{40}Ar+^{144}Sm. b) Radon isotopes measured at the focal plane F2 of the facility in the reaction ^{40}Ar+^{166}Er at E_{lab}=218 MeV in the middle of the target. Both reactions were carried out simultaneously with use of the rotating target wheel.

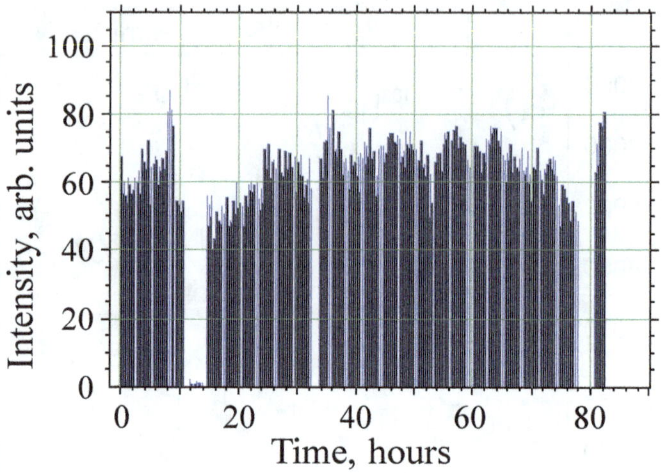

Fig. 4. The dependence of the separation efficiency on the time of measurement for the hot catcher made from nanotubes.

The ratio of the effective cross-section measured by the MASHA mass-spectrometer to the absolute one extracted from the independent catcher foil experiment gives an estimate of the mass-spectrometer efficiency.

In Fig. 4 the dependence of the separation efficiency on the measurement time is shown for the hot catcher manufactured from the nanotube paper of 2 mg/cm^2. The intensity of the 190 MeV ^{40}Ar beam was 0.3 pμA and the power deposition was 12 W. It is obvious that in contrast to the old version with the thick carbon catcher now the efficiency of the set-up is almost constant. So, the use of the nanotubes as the hot catcher seems to be very prospective.

The basic results of the performed experiments are:

(a) It was shown that the separation time did not exceed 1 s. This is two times shorter than in the case of the thick carbon absorber used in old version of the "hot solid-state catcher".

(b) The separation efficiency was stable during 85 hours at ^{40}Ar beam current $\leq$0.5 pμA. In the former case the separation efficiency decreased by 6 times during the same period.

4. Conclusions. Current Status of Modification of the Mass-Spectrometer MASHA

For the MASHA mass-spectrometer a series of improvements of its setup parameters was performed. The measurement of the total efficiency as a function of time was carried out at intense heavy ion beams (up to 0.5 pμA). It was shown that the decrease of the total efficiency was due to thermal and radiation destruction of the hot absorber structure.

The new versions of the hot catcher were designed based on a thin sheet made of carbon nanotubes and graphene. The test experiments showed a high stability of the separation efficiency in this case.

At the intermediate focal plane F1 the additional strip detector was installed for the permanent control of the system. In the test experiment we measured simultaneously the yields of ER's in the reactions ^{40}Ar+^{144}Sm in the intermediate plane F1 and ^{40}Ar+^{166}Er in the focal plane F2.

The new design of ECR ion source and its vacuum, catcher and transport line was developed. The chemical inert covering of the inner part of the vacuum chambers was realized.

The whole separation system will be heated in the future experiments up to 300°C in order to increase the efficiency separation for the proposed system.

References

1. A. M. Rodin *et al.*, *Instr. Exp. Tech.*, **57**(4), 386–393 (2014).
2. Yu. Ts. Oganessian, *J. Phys. G.*, **34**, R165–R242A (2007).
3. S. Lukic *et al.*, *Nucl. Instr. and Meth. in Phys. Res. A*, **565**, 784–800 (2006).

Upgrade Possibilities of the Spectrometer "MULTI"

I. Sivaček[1,2*], Yu. G. Sobolev[1] and S. S. Stukalov[1]

[1]*Flerov Laboratory of Nuclear Reactions, Joint Institute for Nuclear Research, 141980, Dubna, Russia*

[2]*Nuclear Physics Institute of the Czech Academy of Sciences, 250 68 Řež, Czech Republic*

**E-mail: sivacek@ujf.cas.cz*

The spectrometer for the measurement of the total reaction cross section "MULTI" is continuously upgraded to achieve better performance in the experimental measurements with radioactive beams. Characteristics of the gamma spectrometer with the new types of high precision detectors in various geometrical setups are investigated by Monte Carlo simulations in Geant4. Results of the simulations are discussed from the point of the registration efficiency, spectrometer resolution and geometrical conditions. Optimal geometry is suggested.

Keywords: Total reaction cross section; Monte Carlo simulations; gamma spectroscopy.

1. Introduction

The detector setup "MULTI" is a spectrometer designed for the measurement of the total reaction cross section of nuclear reactions with the beams of exotic nuclei. It consists of two parts. In-beam multilayer detector telescope is used for detection of the charged particles and a 4-π gamma spectrometer, situated around the target outside the vacuum volume, is used for spectroscopy of gamma and neutrons accompanying the nuclear reactions. Charged beam particles and reaction products are identified by the telescope consisting of a silicon dE_0 detector, two active collimators [1] used for the time-of-flight (TOF) measurement and a silicon dE_T detector as an active target [2]. Scheme of the setup can be found on the Figure 1.

Total reaction cross section is evaluated from the relative decrease of the beam intensity in the telescope elements. Reaction products are identified from the dE_0 – TOF matrix. For enhanced identification of the correct events in the target, prompt gamma and neutrons accompanying the nuclear reactions are detected by 4-π gamma spectrometer in coincidence with the TOF signal [3].

Identification matrix allows to measure cross sections for different secondary beam particles simultaneously.

Although the spectrometer MULTI is portable and can be mounted on any beamline, experiments are provided on the secondary beam of fragment

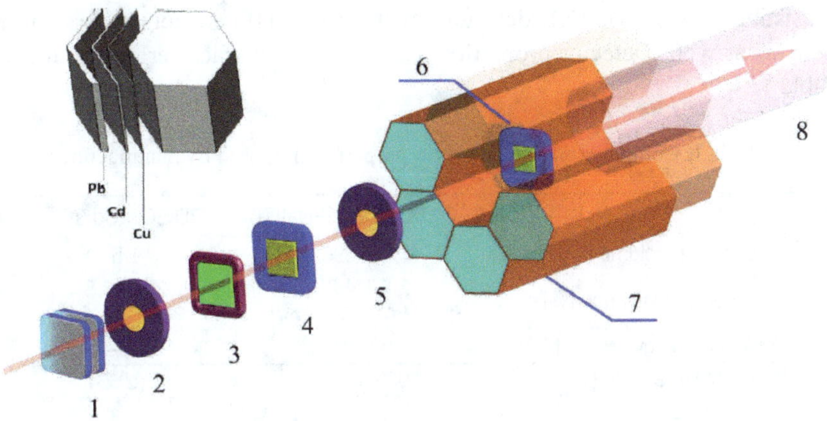

Fig. 1. Schematic view of the spectrometer MULTI. 1 — degrader, 2 — active collimator AC_1, 3 — removable pixel detector "16X×16Y", 4 — dE_0, 5 — active collimator AC_2, 6 — active target, 7 — 12×CsI(Tl) gamma spectrometer, 8 — beam dump.

separator ACULINA, in the Flerov Laboratory of Nuclear Reactions of the Joint Institute for Nuclear Research in Dubna [4]. Measurements with the low energy radioactive beams are planned on the MAVR spectrometer [5].

Registration efficiency of the gamma spectrometer is crucial, since it is an input parameter for the total reaction cross section evaluation. For this reason, the gamma spectrometer is being continuously upgraded to achieve better performance. Current upgrade is aimed for enhancement of the gamma spectrometer energy resolution and overall detection efficiency by employing new, high-resolution gamma detectors.

Previous upgrade reported in [6] was aimed to rise detection efficiency by employing additional 6 CsI(Tl) detectors into the setup, forming a spectrometer total of 12 scintillation detectors. This upgrade has risen detection efficiency by 6–12 per cent (depending on the gamma energy), to the 84% [7].

2. High Precision Gamma Detectors

Detection efficiency of the gamma spectrometer can be upgraded by rising the solid angle coverage. The subject of the Monte Carlo analysis in Geant4 [8] was a set of 10 RITEC® CdZnTe (CZT) portable semiconductor detectors, each with 1,5 cm^3 sensitive volume and a set of SCIONIX® $CeBr_3$ + NaI(Tl) phoswich scintillation detectors in 4 different geometrical setups, consisting of 4 and 6 detectors (the rectangular container contains 133 cm^3 $CeBr_3$ cube, following by 395 cm^3 NaI(Tl) rectangle and a photomultiplier). One additional geometry consisting of 8 SCIONIX® detectors, without CsI(Tl) detectors, was subject of analysis as well. Detector properties proposed by manufacturer can be found in Table 1.

Table 1. Comparison of the detector properties proposed by manufacturer.

	SCONIX® CeBr3+NaI(Tl)	RITEC® CdZnTe
sensitive volume [cm3]	133 + 395	1,5
detector resolution [%]	2,5 – 4,2 (2,5 MeV – 662 keV)	2,5 – 3,5 (-)
designed energy range [MeV]	> 0,1	0,02 – 3
power supply [V]	920 (HV)	5 (USB)

Two main parameters were evaluated – peak efficiency and total efficiency for gamma registration. The peak efficiency is recognized, when the particle is fully absorbed in one of the detectors. The total efficiency is the energy deposition over 150 keV threshold in any of the detectors in the setup [9]. For the future comparison of the selected geometry with the real data, an isotropic ^{60}Co point source was used in the simulation.

3. Results of the Simulations

Sketches of geometries with SCIONIX® $CeBr_3$ detectors are presented in Figure 2.

The geometry consisting of CsI(Tl) + 10 × RITEC® CZT detectors is analogous to Fig. 2A. Results of the Monte Carlo simulations are presented in

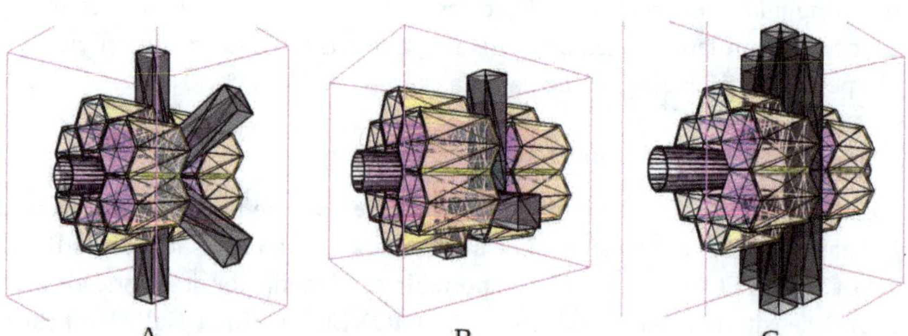

A B C

Fig. 2. The gamma spectrometer MULTI in the investigated geometries consisting of 2×6 CsI(Tl) detectors and A) 6×$CeBr_3$, B) 4×$CeBr_3$, C) 6×$CeBr_3$ detectors. Bright modules represents CsI(Tl) detectors, dark modules – $CeBr_3$ + NaI(Tl) detectors.

Table 2. Results of the Monte Carlo simulations for geometrical efficiency of the setup.

Geometry type	Total efficiency (CsI(Tl) + new detectors, %)	peak efficiency (upgrade, %)
12x CsI(Tl) (fig.1)	85 + 0	16
10x CZT (fig. 2A)	81 + 2	0.2
6x CeBr3 (fig. 2A)	62 + 26	7
4x CeBr3 (fig. 2B)	61 + 34	9
6x CeBr3 (fig. 2C)	60 + 35	9
8x CeBr3 (fig. 3)	0 + 63	17 *(see fig. 4)

the Table 2, which contains also simulation results for the actual 12 × CsI(Tl) setup. Values for total efficiency are divided between CsI(Tl) part of the setup plus new considered detectors in the geometry.

An option discussed in Chapter 4, consisting only of 8 × CeBr3 detectors, is included as well.

4. Discussion

Dimensions of the CZT detector together with its portability without need for high voltage power supply makes this detector ideal for low energy gamma and x-ray spectroscopy, or field applications. For the detection of high-energy gamma particles is necessary to employ a high-volume detector.

SCIONIX® $CeBr_3$ + NaI(Tl) solution is a good choice in this case. Together with 12 CsI(Tl) detectors, the detection system covers solid angle close to 4π. This represents almost 95% of the total detection efficiency (events over 150 keV as defined in the section 2). While the total efficiency is not significantly lower with the CZT detectors, it is due to the small dimensions of the detector, so the low resolution CsI(Tl) detectors covers bigger solid angle compared to the $CeBr_3$ detectors setup.

Peak efficiency represents the fraction of gamma particles, that loose all of their energy inside one of the detectors. It represents the crucial demand, how many gamma particles can contribute to the peak in the spectrum. The "Fig. 2C" case allows to fully employ the advantages of $CeBr_3$ crystal, compared to "Fig. 2B" (see Fig. 4).

Simulations had confirmed, that CZT detector have at least an order of magnitude higher peak-to-compton ratio, than the $CeBr_3$ detector. Comparison of the peak-to-total values showed, that $CeBr_3$ detector achieve 8–10 times better results in the energy interval 0-5 MeV due to bigger sensitive volume of the detector.

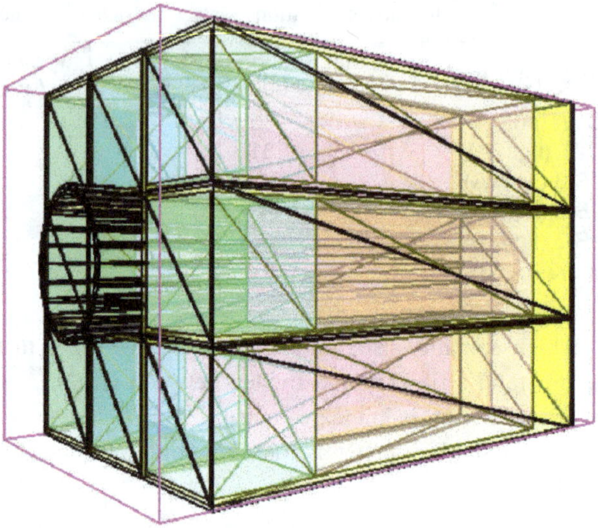

Fig. 3. Gamma spectrometer consisting of 8 $CeBr_3$ modules, where cubic part on the left side is $CeBr_3$ crystal and rectangular part on the right is NaI(Tl) crystal of the detector.

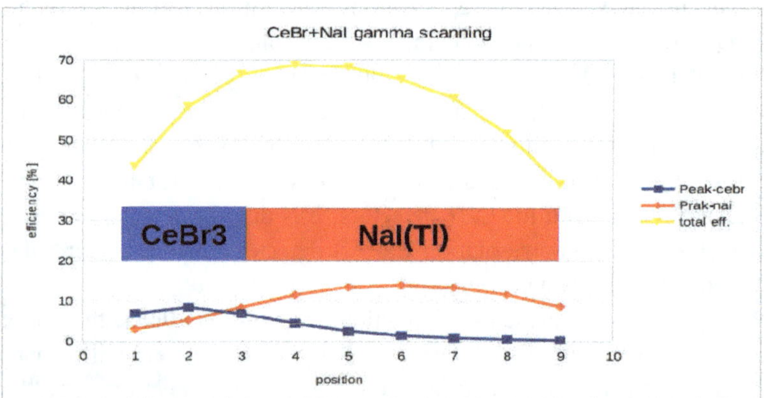

Fig. 4. Efficiency scanning of the SCIONIX® detector with the contributions of $CeBr_3$ and NaI(Tl) crystals to the detector response.

A geometry consisting only of 8 SCIONIX® detectors was investigated as well (Fig. 4). Simulation results showed the total efficiency of 64.4% and peak efficiencies 17.6% for 1174 keV and 16.0% for 1332 keV gamma. ^{60}Co source was in the center of the setup with respect to the sensitive detector volumes. Analysis of this setup brought a question about contribution of fast ($CeBr_3$)

and slow (NaI(Tl)) component of the detector response – where is the optimal position for the target. Figure 5 shows the dependency scanned in 9 points according to the detector.

5. Conclusions

Monte Carlo simulations in Geant4 showed the estimation of performance of the possible upgrade of the gamma spectrometer "MULTI" with new high-resolution detectors. Additional SCIONIX® $CeBr_3$ detectors allows to achieve total geometrical efficiency of the gamma spectrometer up to 94%. According to simulation results, expected peak efficiency for the prompt gamma is 9%. Geometry option consisting only of new **8** × CeBr3 detectors provides lower total efficiency of the spectrometer (63%), but allows to fully implement advantages of the high-resolution detectors for the spectroscopy of prompt gamma and neutrons accompanying nuclear reactions. Proposed upgrade allows to gather spectroscopic data during the total reaction cross section measurements, instead of only tagging the correct events.

References

1. Yu. G. Sobolev et al., *Instr. and Exp. Techniques* **54**(4), 449 (2011).
2. A. Khouaja et al., *Nuclear Physics A* **780**, 1 (2006).
3. Yu. G. Sobolev et al., *Instr. and Exp. Techniques* **55**(6), 618 (2012).
4. A. M. Rodin et al., *Nucl. Instr. and Meth. in Physics Research B* **204**, 114 (2003).
5. V. A. Maslov et al., Proceedings of the EXON 2014, p. 669, ISBN 9789814699457.
6. I. Sivacek et al., Proceedings of the EXON 2016, p. 367, ISBN 9813226536.
7. Yu. E. Peniozhkevich et al., *Phys. Rev. C* **99**, 014609 (2019).
8. S. Agostinelli et al., *Nucl. Instrum. Meth. A* **506**, 250 (2003).
9. M. G. Saint-Laurent et al., *Z. Phys. A* **332**, 457 (1989).

Data Acquisition Systems in VME Standard for the Researches with Radioactive Ions Beams

R. S. Slepnev[1], A. A. Bezbakh[1], S. G. Belogurov[1,2], V. Chudova[1,3], A. S. Fomichev[1], A. V. Gorshkov[1], S. A. Krupko[1], M. Mentel[1,4], E. V. Ovcharenko[1], P. Plucinski[1,4] and P. G. Sharov [1]

[1]*Flerov Laboratory of Nuclear Reaction Joint Institute for Nuclear Research, 141980, Dubna, Russia*

[2]*Federal State Autonomous Educational Institution of Higher Professional Education, National Research Nuclear University "MEPhI", Moscow Institute of Physics, Moscow, Russia*

[3]*Institute of Physics, Silesian University in Opava, Opava, Czech Republic*

[4]*AGH University of Science and Technology, Faculty of Physics and Applied Computer Science, Krakow, Poland*

Two data acquisition systems in VME standard based on RIO3/RIO4 processors (C.E.S.) and VM-USB controller (Wiener), which were tested in multiparameter nuclear physics experiments on ACCULINNA and ACCULINNA-2 facilities (Dubna, FLNR JINR), are described in the article. The advantages and disadvantages of both systems as well as the prospects for their further development are given.

Keywords: Data acquisition; nuclear physics experiments.

1. Introduction

In modern nuclear physics experiments, in order to obtain new information on the structure of nuclei and the mechanisms of nuclear reactions, as a rule, it is necessary to organize the correlation measurements of energy, the emission angle, and the time stamps for several interaction products of nuclear reaction. As a result, the number of parameters registered in one event can reach several hundred or even thousands, and the number of registered events (the counting rate) can be up to 10^4 triggers/s. In addition, during the experiment, the problem often arises of changing the conditions of observation and data analysis (taking into account the configuration of the trigger, new electronics blocks in different standards, time matching, etc.) Taking into account the specifics of carrying out experiments with radioactive ions (RI) beams [1-3], the following main criteria for a modern data acquisition system (DAQ) can be formulated:

(1) performance (stable operation with workload of detecting equipment up to 10^6 particles/s and the counting rate up to 10^4 triggers/s);

(2) flexibility and scalability, in other words possibility to connect several VME (as well as different standards: CAMAC, FASTBUS, VXI) and different electronics units (ADC, QDC, TDC, Flash-ADC, etc.)
(3) visualization of the collected data for the operational tuning of the correct operation of detecting equipment (detectors, electronics) and DAQ.

The aim of this work is to make a comparative analysis of two DAQs in VME standard and to show the prospects for their use on some specific examples, namely from the point of view of the possibility of using them in the ACCULINNA and ACCULINNA-2 facilities.

2. VME Based DAQ with the RIO Family of Processors and Event Builder

Earlier, in [4], we gave an example of the effective use of DAQ in the VME standard based on the RIO family of processors with the LynxOS real-time system and the MBS software (Multi Branch System, [5]).

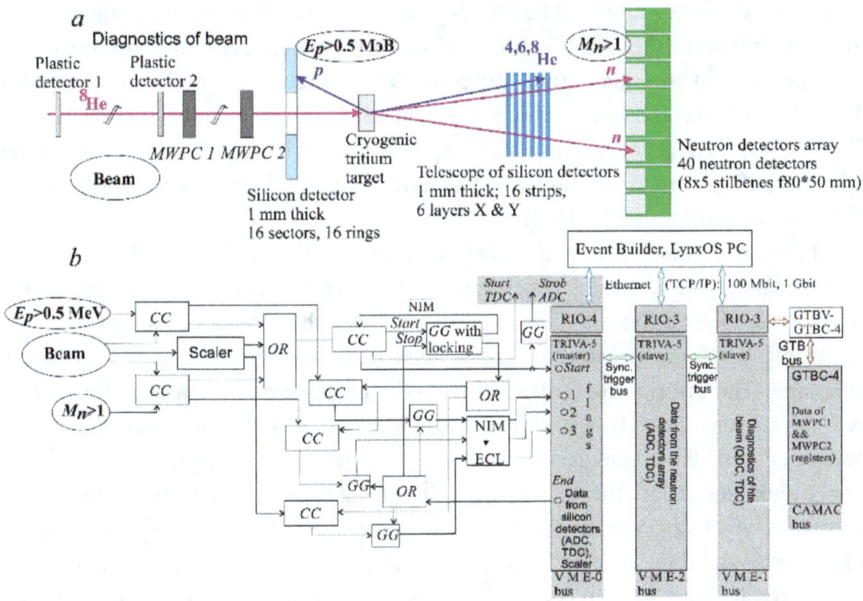

Fig. 1. Layout of planned ^{8}He+t experiment with 3 different triggers (flags); *panel a* – configuration of the detectors, *panel b* – DAQ with Event Builder and electronics. CC – coincidence circuit, OR – comparison circuit, GG – gate generator.

The merits of this system include the ability to read information only about the working channels of VME electronic blocks (zero suppression), which drastically reduces the workload of the DAQ. Further development of this approach is connected with the possibility of combining several VME crates.

For the organization of measurements with detectors (Figure 1), which are sufficiently far from each other (up to 100 m), a DAQ is needed, which can combine in one experimental event the signals coming from these detectors. It is desirable to minimize the number and length of cable connections of detectors with electronic units. For such a system, it is necessary to ensure the synchronization of the arrival of signals coming from different detectors. There are two ways to solve this problem: synchronization of data on the trigger bus and synchronization of data by time stamp. Example of the DAQ with the synchronization of the data by the trigger bus for experiments on the ACCULINNA-2 facility (FLNR JINR, Dubna) is shown in Figure 1. DAQ with the Event Builder consists of several VME crates with RIO-3 and RIO-4 processor modules running under LynxOS real-time system and an Event Builder based on a diskless computer under the Linux operating system. Processor VME modules RIO-3, RIO-4 and the Event Builder are connected via the Ethernet over the TCP/IP protocol using a switch. Synchronization of the experimental event data from different VME crates is carried out through TRIVA-5 trigger modules, which are connected by a special trigger bus. The CAMAC crate is connected via an expansion board GTBV, which is inserted into the RIO-3 processor module of one of the VME racks and connects to the GTBC-4 controller via the GTB data bus.

In each of the VME and CAMAC crates, electronic blocks (ADC, TDC, QDC, registers, etc.) are read using the MBS program running in LynxOS. The specific configuration of the experiment and the operating modes of the electronic units are prescribed in the user program (fuser.c) written in the C language. The information collected from the electronic units in the form of data buffers is sent over the Ethernet (100 Mbit or 1 Gbit per second) on the computer with the Event Builder, where it gathers into a single physical event. For quick setup of the experiment, configuration of detectors, modes of operation of electronics and various experimental characteristics, visualization of the received data is used. Data visualization is realized with the help of the Go4 software package, in which C ++ is used to decode data from electronic blocks and build histograms and drawings for online analysis.

It should be noted that this DAQ has a number of drawbacks: a limitation of the operation rate when working with a large number of triggers (~ 10^3-10^4 per second), while the dead time of the DAQ (i.e., the time when the DAQ is unavailable) should be less than 50%. The specific number of triggers and the dead time of the DAQ, depends on the experimental conditions, i.e. from the total number of the channels and the number of measured channels in the event. In addition, the disadvantages include a complex data structure (some places in the software are known to a limited number of people), cumbersome visualization mechanism, proprietary software.

In general, the DAQ partially satisfies the first condition described in the introduction, and completely satisfies the second and third. Thus, this DAQ can almost completely cover the experimental requirements of a large number of nuclear-physical facilities (including the ACCULINNA-2 facility), where a large number of parameters need to be measured from distant detectors.

3. VM-USB Controller Based DAQ

These shortages were largely overcome for the DAQ based on VM-USB controller of the Wiener [6]. This block has been mass-produced since 2005, but its wide application became possible only a few years ago thanks to a software package developed at the University of Michigan (East Lansing, USA) [7].

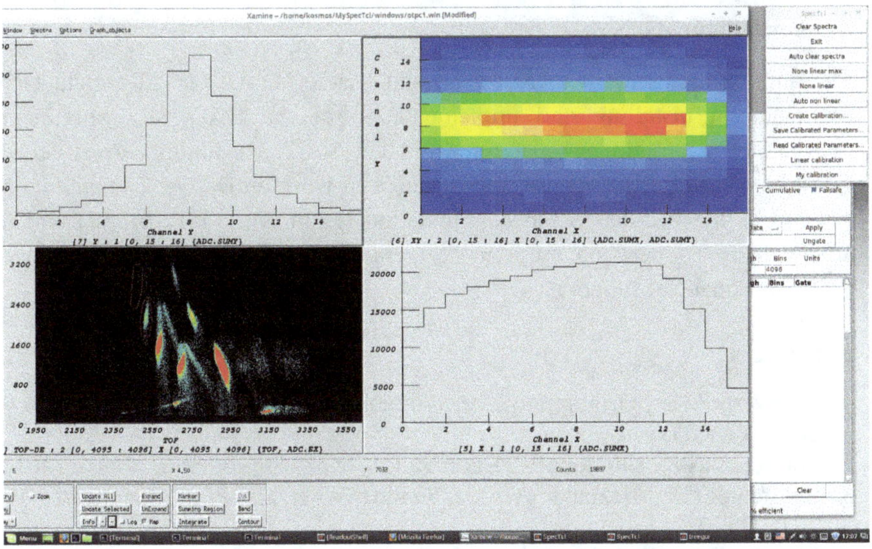

Fig. 2. Visualization for the experiment with using of the SpecTcl program.

DAQ based on the VM-USB controller runs under the Linux operating system distributions of the Debian family, consisting of free open source software. The ReadoutShell program of the data acquisition system works on a

Debian Linux computer, interrogates electronic VME blocks in the rack through a VM-USB VME controller from Wiener Company. In contrast to the previous DAQ, a specific configuration of the experiment and modes of operation of electronic units are prescribed in a special configuration file. In addition, as in the first example, for quick setup of the experiment, visualization of the received data is used. The SpecTcl visualization program decodes and displays data read by the ReadoutShell program in a graphical form on the computer display (Figure 2). The visualization program is presented as an executable file, configuration files written in Tcl and auxiliary files created by the user.

To the disadvantages of this DAQ one can attribute the lack of the possibility of scaling and working with electronics in other standards, for example, CAMAC. The advantage of this data collection system is that it is easier to customize for a specific experiment and, in general, the less complex structure of the entire DAQ. Thus, this DAQ satisfies the first and third of the conditions of the introduction, and does not correspond to the second one. Accordingly, this DAQ can be used in experiments, with a relatively small amount of electronics (for example one VME crate).

Acknowledgments

The authors are grateful to Kurtz N. and Kiselev O.A. (GSI, Darmstadt) for assistance in installing and implementation of the DAQ for experimental installation on a fragment-separator ACCULINNA-2. The authors are grateful to Golovkov M.S. for the general management of work on DAQs. The authors are grateful to the Russian Science Foundation (grant No. 17-12-01367) and the JINR-BMBF cooperation program "Superheavy Elements and Exotic Nuclei" for partial financial support.

References

1. Golovkov M.S., Grigorenko L.V., Fomichev A.S. et al., *Phys. Rev. Lett.* **93**, 262501 (2004).
2. Fomichev A.S., Chudoba V., Egorova I.A. et al., *Phys. Lett. B*, **708**, 6 (2012).
3. Sidorchuk S.I., Bezbakh A.A., Chudoba V. et al., *Phys. Rev. Lett.*, 108, 202502 (2012).
4. Slepnev R.S., Bezbakh A.A., Golovkov M.S. et al., *Instr. Exp. Tech.*, **55**(6), 645 (2012).
5. http://www-win.gsi.de/daq/
6. http://www.wiener-d.com/sc/modules/vme--modules/vm-usb.html
7. http://docs.nscl.msu.edu/daq/

Manifestation of Clustering in Fission Fragments at Crossing of Metal Foils

A. O. Strekalovsky*, D. V. Kamanin, A. A. Alexandrov,
I. A. Alexandrova, Z. I. Goryainova, E. A. Kuznetsova, A. N. Solodov,
O. V. Strekalovsky and V. E. Zhuchko

Joint Institute for Nuclear Research, 141980 Dubna, Russia
**E-mail: alex.strek@bk.ru*

Yu. V. Pyatkov

National Nuclear Research University MEPhI (Moscow Engineering Physics Institute), 115 409 Moscow, Russia, Joint Institute for Nuclear Research, 141980 Dubna, Russia

V. Malaza

University of Stellenbosch, Faculty of Military Science, Military Academy, Saldanha 7395, South Africa

The manifestation of a new original effect appeared at crossing of the metal foils by fission fragments is reported. The effect takes place predominantly in front impacts. The results obtained suggest the fragment from the conventional binary fission to be borne in a shape-isomer state which looks like di-nuclear system with magic core.

Keywords: Shape-isomer states; magic cluster.

1. Introduction

In our previous experiments [1, 2] at the COMETA setup we observed essential mass deficit in the total mass of the fission fragments detected in coincidence with the ions knocked out from the metallic foil. The brake-up takes place predominantly at frontal impacts. Such effect could be expected if the scattered fragment looks like a di-nuclear system which is destroyed due to inelastic scattering on the nucleus of the foil. Further experiments discussed here are dedicated to event by event "weighing" of the fragment before and after the foil.

2. Experiments and Results

The experiments were performed at LIS (Light Ions Spectrometer) spectrometer in the FLNR of JINR. The layout of the setup is shown in Figure 1. LIS is a double-armed time-of-flight spectrometer which includes three micro-channel based timing detectors and two PIN diodes. Each PIN diode provides information for estimating both FF energy and time-of-flight. Metal foils (degraders) of different thicknesses can be placed in the detector TD1. The aperture for fission fragments detected in coincidence in the opposite PIN diodes does not exceed 3^0. The data acquisition system consists of the fast digitizer CAEN DT5742 and a personal computer. The digital images of all the signals are obtained for further off-line processing. The construction of the spectrometer allows measuring of the FF mass using "two-velocities" (time-of-flights at the bases L3 and L4) and "velocity-energy" (time-of-flight at the base L5 and energy measured by PIN1) approaches simultaneously. Corresponding mass values will be marked below with M_{tt} and M_{te}. Thus, we know the mass of each FF before and after it crosses the degrader-foil in TD1.

Here we compare the results of two experiments carried out at the LIS setup. They differ by applied time-pickoff algorithms. Well known CFD (Constant Fraction Discriminator) method gives a time-pickoff point which is independent from the amplitude of the signal but is biased compared to a "true" moment when the ion hits the detector. The delay is due to the "plasma delay" effect in the PIN diode and should be taken into account using special parametrization. Mass reconstruction procedure based on CFD approach [3] was used in the experiment Ex1. In order to have a "true" time pickoff point, we proposed "sewing parabola time-pickoff" method [4] which was applied in Ex2.

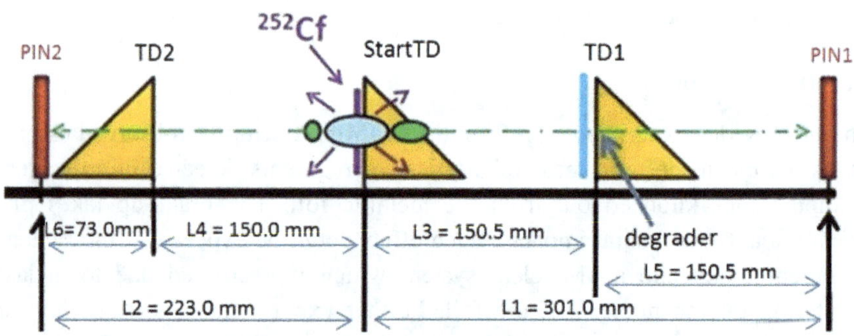

Fig. 1. Layout of the LIS spectrometer. It includes three timing detectors (Start TD, TD1, TD2), two PIN diodes and ^{252}Cf (sf) source. Additional metal foil (degrader) can be placed in TD1 detector.

The mass distributions in Figure 2 clearly show event-by-event how the initial mass M_{tt} of the heavy fragment changes after passing the foil.

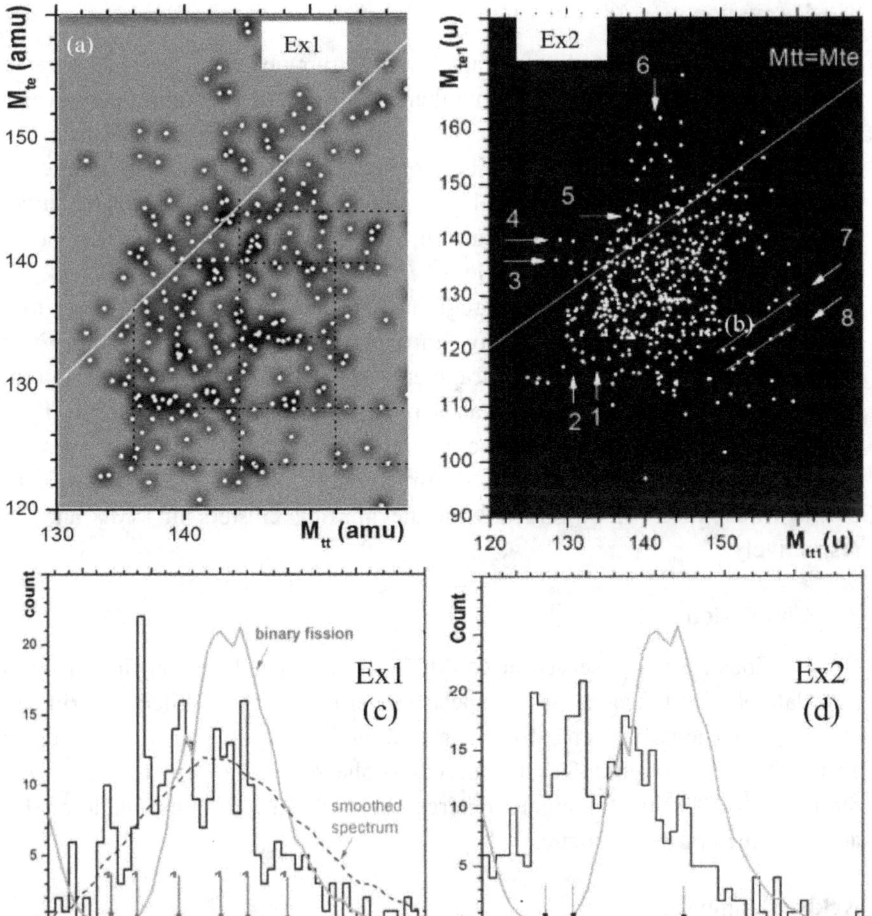

Fig. 2. Mass correlation distributions M_{tt}-M_{te} obtained for the heavy FFs mass peak are presented in Figures 2a, b. They were obtained in the experiments Ex1 and Ex2 respectively. Copper foil 4.11 microns thick was placed in the TD1 detector. Projections of the plots onto M_{te} axis are shown in Figures 2c, d. Masses of known magic nuclei are marked by arrows. Mass spectrum of the FFs from conventional binary fission is shown in gray. See text for details.

Projections of the plots onto M_{te} axis (Figures 2c, d) differ significantly from the mass spectrum of heavy fragments in conventional binary fission. The strongest difference is observed for the masses associated with magic isotopes of ^{128}Sn and ^{124}Cd.

3. Discussion

The most pronounced peculiarity of the M_{te} spectrum observed in Ex1 (Figure 2c), namely, strong peaks at the mass numbers 124 and 128 was reproduced well in Ex2 (Figure 2d). At the same time the mass correlation distribution from Ex2 shows some new interesting details. Corresponding structures are marked by numbered arrows (1–8). The vertical lines M_{tt} = (134, 136, 130) u attract attention. Currently we do not know the nature of such structures. Horizontal lines (3–5) are associated with the magic nuclei of $^{136,\ 140}$Xe and ^{144}Ba. The lines intersect the diagonal $M_{tt} = M_{te}$ and in the region to the left of the diagonal seem to be methodic artefact. As an alternative interpretation these parts are due to the effect of completion of the heavy fragments to the heavier magic cores. In other words likely a nucleon transfer takes place from the degrader to the flying fission fragments until they become magic nuclei. Similar effect may cause the locus (6). Loss of fixed mass by the fragments passing the foil serves cause of the tilted lines (7, 8) linked with the missing clusters of ^{32}Mg and ^{38}Si respectively.

4. Conclusion

Our previous results observed at COMETA setup have been confirmed by the new data obtained recently at LIS spectrometer using the modified experimental methodic: fission fragments from conventional binary fission at least during 15 ns after a fission manifest themselves as shape-isomers. By passing through the metal foil, fission fragments undergo brake-up with forming magic nucleus as one of the brake-up product.

Acknowledgments

This work was supported by the research project № 18-32-00538 of the Russian Foundation for Basic Research (RFBR), and partially by the MEPhI Academic Excellence Project (Contract No. 02.a03.21.0005, 27.08.2013) of the Russian Science Foundation.

References

1. Yu.V. Pyatkov et al., Proceedings of the 21th International Seminar on Interaction of Neutrons with Nuclei, Alushta, Ukraine, 20–25 May 2013. Dubna 2014, p. 127.
2. Yu.V. Pyatkov et al., Int. Symposium on Exotic Nuclei “EXON-2014”, Kalaningrad, Russia, 08–13 September 2014. Conf. proc., Editors: Yu.E. Penionzhkevich, and Yu.G. Sobolev, World Scientific , 2015. p. 383–389.

3. D.V. Kamanin et al., Proceedings of the 21th International Seminar on Interaction of Neutrons with Nuclei, Alushta, Ukraine, 20–25 May 2013. Dubna 2014, p. 107.
4. D.V. Kamanin et al., *Izvestiya Rossiiskoi Akademii Nauk. Seriya Fizicheskaya*, **82**(6), 804 (2018).

Radioactive Ion Beams for the Fission Study of Heavy Neutron-Rich Nuclei*

G. M. Ter-Akopian†, A. S. Fomichev, M. S. Golovkov and Yu. Ts. Oganessian
Flerov Laboratory of Nuclear Reactions, JINR, Dubna, 141980, Russia
Dubna State University, Dubna, 141982 Russia

A. Bezbakh, A. V. Gorshkov, S. A. Krupko and S. I. Sidorchuk
Flerov Laboratory of Nuclear Reactions, JINR, Dubna, 141980, Russia

E. Yu. Nikolskii
Flerov Laboratory of Nuclear Reactions, JINR, Dubna, 141980, Russia
NRC "Kurchatov Institute" Kurchatov sq. 1, 123182, Russia

R. Wolski
Flerov Laboratory of Nuclear Reactions, JINR, Dubna, 141980, Russia
Institute of Nuclear Physics PAN Radzikowskiego 152, Kraków, 31342, Poland

Radioactive ion beams (RIBs) offer a way to obtain moderately excited nuclei with atomic numbers Z≥100 produced in the multi-nucleon transfer reactions. The choice of suitable RIBs available from the ACCULINNA-2 fragment separator, coupled with the U-400 cyclotron, and possible results of relevant experiments are discussed.

Keywords: Radioactive ion beam; fragment separator; transfer reactions; super-heavy nuclei.

1. Introduction

Nowadays, when the existence of super-heavy nuclei is well established [1,2] accumulation of knowledge about the fission barriers of heavy nuclei having neutron numbers ~160 and more remains topical. The use of quasi-elastic multi-nucleon transfer (MNT) reactions for the production of excited compound nuclei with atomic numbers $Z \geq 100$ appears to be a good approach to the problem.

Transfer reactions (d,p), (t,p), (t,^{4}He), (^{3}He,t), (^{3}He,^{4}He), studied for a number of accessible heavy target nuclei were used in 1970th in Los-Alamos in experiments where the fission probabilities were measured for the compound

* Work partially supported by the Russian Science Foundation, Grant No. 17-12-01367.

nuclei formed with the quite well defined excitation energies [3-5]. Statistical-model analysis allowed the authors to obtain the barrier values for more than 30 isotopes of actinide elements ranking from thorium up to fermium. The lack of more heavy targets terminated progression towards the heavier and more neutron-rich nuclei.

Production of heavy actinide nuclei in MNT reactions was the subject of radiochemical study [6-12] made when ^{248}Cm [6-9], ^{249}Cf [7] and ^{254}Es [10-12] targets were bombarded with ^{13}C, ^{16}O, ^{18}O, ^{20}Ne, ^{22}Ne, and ^{40,48}Ca beams. The data of Refs. [6-12] show that the yields of neutron-excess actinide nuclei produced in the MNT reactions reach their maximum values at center-of-mass energy slightly exceeding the fusion barrier ($E_{cm} \approx 1.05 \times E_B$).

2. Radioactive Ion Beams (RIBs) Achievable from the U400M – ACCULINNA-2 – U400R Assembly

In 2017 the fragment separator ACCULINNA-2 came into operation [13,14], and the first experiments were carried out using RIBs provided by this setup [15]. ACCULINNA-2 is installed on a primary beam line of the U400M cyclotron. In 2019 this cyclotron will be upgraded [16,17]. Parameters expected for the primary beams obtained after this upgrade are given in Table 1.

Table 1. U400M→U400MM. Beam energies and intensities.

	At present		In 2020	
Beam ions	E, MeV/u	I, pμA	E, MeV/u	I, pμA
$^{11}B^{3+}$	33	3	38	12
$^{15}N^{5+}$	49	1	54	10
$^{18}O^{6+}$	-	-	54	10
$^{22}Ne^{7+}$	45	0.5	50	5.5
$^{26}Mg^{8+}$	42	0.5	48	5.5
$^{36}S^{11+}$	38	0.3	45	2
$^{36}S^{12+}$	–	–	54	0.3

Results of calculations made with the use of the LISE++ code [18] for some RIBs which will be obtained from ACCULINNA-2 are given in Table 2.

It is supposed in the DRIBs III project that the RIBs obtained at the final focal plane of ACCULINNA-2 will be injected in the U400R cyclotron (U400 converted into the smoothly-variable energy mode) to be accelerated to energy smoothly varied in a range of 1.6 – 27 MeV/u [17,18]. Procedures of RIB capture and extraction made with gas catcher, followed by charge breeding and injection in U400R, are associated inevitably with the loss of beam intensity. We

Table 2. RIBs available from the ACCULINNA-2 separator.

RIB	Half life time	Energy, MeV/u	Intensity, pps	Purity, %	Primary beam	$E_{primary}$, MeV/u
^{9}Li	178 ms	40	4×10^{7}	90	^{15}N	54
^{10}Be	1.5×10^{6} y	30	5×10^{8}	98	^{11}B	42
^{14}C	5700 y	35	3.7×10^{9}	98	^{15}N	54
^{16}C	192 ms	39	2.4×10^{8}	75	^{18}O	54
^{17}N	4.2 s	35	3.6×10^{9}	90	^{18}O	54
^{19}N	336 ms	35	1.1×10^{7}	50	^{22}Ne	50
^{20}O	13.5 s	35	1.5×10^{8}	90	^{22}Ne	50
^{22}O	2.25 s	36	1.0×10^{6}	70	^{26}Mg	48
^{21}F	4.2 s	33	1.5×10^{9}	98	^{22}Ne	50
^{23}F	2.2 s	33	2.6×10^{7}	40	^{26}Mg	48
^{24}Ne	3.4 min	33	1.1×10^{8}	80	^{26}Mg	48
^{30}Mg	335 ms	31	5×10^{6}	40	^{36}S	45
^{34}Si	2.8 s	26	2×10^{7}	40	^{36}S	45

assume that a drop factor of 10 will be typical for the intensities of RIBs extracted from U400R.

Experiments discussed here need RIBs having energy in a range of 4 – 6 MeV/u. Energy precision making 0.1% will be provided by U400R. This choice of the beam energy corresponds to experiments carried out when a heavy target (e.g. ^{244}Pu, ^{248}Cm, ^{249}Bk, $^{249\text{-}251}Cf$, ^{254}Es) is bombarded by RIBs with energies being close to fusion barrier (B_{fus}) or exceeding it by ~10 – 30%.

3. Prospective MNT Reactions

Sizable cross sections reported for the heavy nuclei produced in bombardments of heavy targets [6-12] can be understood taking into account that: (i) about 50% of the total reaction cross section appears in transfer channels [19], (ii) transfer products heavier than the target are formed predominantly in these reactions, (iii) because of the modest energy cost Q_{gg}, inherent to these reactions, primary reaction products receive low excitation energies (≤10 MeV), so that the observed final nuclei are directly formed or they arise from the one-neutron evaporation channel. It is noteworthy that these transfer reactions are characterized by not too large values of energy excess $E_{ex}=Q_{gg}+E_{c.m.}-E_{Coul.exit}$ where $E_{c.m.}$ and $E_{Coul.exit}$ denote the beam center-of-mass energy and the exit-channel separation energy, respectively. E_{ex} exhibits the maximum excitation energy which can be imparted to the binary products formed in a transfer reaction. The Coulomb energy of two touching spheres coming about in the reaction exit channel is taken usually for the value of $E_{Coul.exit}$.

A sample of MNT reactions is given in Table 3. These reactions seem to be prospective for the fission study of neutron-rich heavy compound nuclei having excitation energy (E*) close to their fission barriers and to the neutron binding energy. A new region of nuclei will be accessible for this study made with RIBs bombarding such target nuclei as ^{244}Pu, ^{248}Cm, ^{249}Bk, $^{249-251}$Cf, ^{254}Es.

Table 3. Critical parameters of MNT reactions.

Projectile	Relocated complex	Ejectile	Q_{gg}, MeV	$E_{ex}{}^{a}$, MeV	Q_{eff}, MeV
^{9}Li	1p4n	^{4}He	+10.8	+22.6	+21.0
^{17}N	1p4n	^{12}C	−3.9	+8.9	+4.7
^{19}N	1p6n	^{12}C	−2	+10	+7
^{10}Be	2p4n	^{4}He	−4.4	+19.5	+16.9
^{14}C	2p4n	^{8}Be	−22.4	+1.1	−1.8
^{16}C	2p6n	^{8}Be	−10.8	+12.0	+9.8
^{20}O	2p6n	^{12}C	−15.8	+7.4	−1.6
^{17}N	3p6n	^{8}Be	−15.8	+17.6	+13.6
^{21}F	3p6n	^{12}C	−20	+13.0	+7.0
^{25}Na	3p6n	^{16}O	−31	+19.7	−6.0
^{19}N	3p8n	^{8}Be	−16	+17	+14
^{23}F	3p8n	^{12}C	−24	+9	+3
^{14}C	4p6n	^{4}He	−29.8	+16.5	+12.4
^{16}C	4p8n	^{4}He	−18.4	+27.2	+23.0
^{20}O	4p8n	^{8}Be	−30.7	+12.5	+8.0
^{24}Ne	4p8n	^{12}C	−35.6	+8.3	+2.3
^{22}O	4p10n	^{8}Be	−51	−8	−13
^{30}Mg	4p10n	^{16}O	−41	0	−7
^{22}O	6p12n	^{4}He	−38	+28	+23
^{34}Si	6p12n	^{16}O	−60	0	−9
^{18}O	2p4n	^{12}C	−15.5	+7.5	+3.1
^{22}Ne	2p4n	^{16}O	−17.8	+7.3	+0.7

Note: Numbers in columns 4-6 are given for the case of ^{254}Es target.

[a] Projectile energies are assumed to be by 10% above the fusion barrier.

Radioactive nuclei appropriate as projectiles used in this study are listed in the first column. Nucleon complexes (XpYn) assumed to be transferred to the target nuclei are indicated in the second column, and the respective light products (ejectiles) emitted from the target are given in column 3. Among the few suggested ejectiles ^{4}He is marked out as outstanding [20] for its appearance as a cluster in nuclear structures. Occurrence of such clusters as ^{8}Be, ^{12}C, ^{16}O in the structures of light, neutron-rich nuclei seems to be quite probable.

Numerical values given in Table 3 characterize the transfer reactions observable in the case of the ^{254}Es target. In addition to the values Q_{gg} and E_{ex} presented in column 6 for each individual case is the effective reaction Q value $Q_{eff} = Q_{gg} - \Delta V_{Coul}$ with ΔV_{Coul} marking the Coulomb energy reduction obtained in the exit channel due to the occurred charge transfer. The ΔV_{Coul} values are calculated assuming for the entrance and exit channels configurations of two spherical nuclei located at the distance of closest approach. Study made for the ^{16,18}O, ^{19}F + ^{208}Pb reactions at the near-barrier energies revealed [21] that the multi-nucleon transfers resulting in a fixed charge exchange occur with probability increasing exponentially with the Q_{eff} value.

Data reported in Ref. [10] give reason to assume a value of 0.1 mb/sr for the cross section expected for the reactions ^{254}Es(^{18}O,^{12}C)^{260}Md* and ^{254}Es(^{22}Ne,^{16}O)^{260}Md* leading to the compound nucleus ^{260}Md* formed with excitation energy $E^* \approx 4 - 7$ MeV. Judging by the Q_{eff} values given in Table 3 one should expect that the cross sections will come to a level of 1 mb/sr or more in the case of the reactions ^{254}Es(^{10}Be,^{4}He)^{260}Md* and ^{254}Es(^{16}C,^{8}Be)^{262}Md*. Valuable clustering conditions and quite high Q_{eff} values inherent to many examples presented in Table 3 allow one to expect that the neutron-rich compound nuclei of Fm, Md, No, Lr might be obtained as a result of the quasi-elastic MNT reactions.

Having the known beam energy and measuring the emission angle and energy for the light ejectile nucleus one can define the excitation energy of the reaction products. Dealing with the ejectiles presented in Table 3 one can be sure that this excitation will be deposited entirely in the heavy compound nucleus emerging in the reaction exit channel. RIB energy variation allows to adjust the available excitation-energy range and test the influence of the compound-nucleus angular momentum on the measured fission-barrier height.

Ensuring good resolution obtainable for the ejectile energy imposes limitation on the target thickness. Having thickness equal to 0.1 mg/cm^2 one can achieve 300-keV resolution in the produced compound-nucleus excitation E*. The neutron binding energy (B_n) and the height of the first peak of the fission barrier (V_A) will be evaluated as a result of analysis, made in the framework of statistical model for the fission probability measured in a range of $E^* \approx 4 - 7$ MeV. Approximate mass defects of the studied nuclei are known, however, only from the present-day atomic mass tables. Therefore the intrinsic errors of the found values B_n and V_A might be quite large. But perhaps the continuation of the measured excitation spectra down to $E^* < 4$ MeV combined with extrapolation to the zero point region will result in independent estimation of unknown mass defects.

References

1. Yu. Oganessian, Proc. IPAC2017, Copenhagen, Denmark, https://doi.org/10.18429/JACoW-IPAC2017-FRYAA1
2. V. Utyonkov, Yu. Oganessian, S. Dmitriev, M. Itkis, K. Moody, M. Stoyer, D. Shaughnessy, J. Roberto, K. Rykaczewski, and J. Hamilton, EPJ Web of Conferences **131**, 06003 (2016).
3. B. B. Back, Ole Hansen, H. C. Britt, and J. D. Garrett, *Phys. Rev. C* **9**, 1924 (1976).
4. B. B. Back, H. C. Britt, Ole Hansen, B. Leroux, and J. D. Garrett, *Phys. Rev. C* **10**, 1948 (1974).
5. H. C. Britt, E. Cheifetz, D. C. Hoffman, J. B. Wilhelmy, R. J. Dupzyk, and R. W. Lougheed, *Phys. Rev. C* **21**, 761 (1980).
6. D. Lee et al., *Phys. Rev. C* **25**, 286 (1982).
7. D. Lee et al., *Phys. Rev. C* **27**, 2656 (1983).
8. D. C. Hoffman et al., *Phys. Rev. C* **31**, 1763 (1985).
9. J. Moody et al., *Phys. Rev. C* **33**, 1315 (1986).
10. M. Schädel et al., *Phys. Rev. C* **33**, 1547 (1986).
11. K. J. Moody et al., *Phys. Rev. C* **41**, 152 (1990).
12. K. J. Moody et al., *Nucl. Phys. A* **563**, 21 (1993).
13. A. S. Fomichev, L. V. Grigorenko, S. A. Krupko, S. V. Stepantsov, and G. M. Ter-Akopian, *Eur. Phys. J. A* **54**, 97 (2018).
14. A. V. Gorshkov et al., see talk presented at Symposium EXON 2018.
15. B. Zalewski et al., see talk presented at Symposium EXON 2018.
16. G. G. Gulbekian et al., 21st Int. Conf. Cyclotrons and their Applications, Zurich, Switzerland, 2016, doi:10.18429/JACoW-Cyclotrons 2016-THA03.
17. G. Gulbekian, B. Gikal, I. Kalagin, N. Kazarinov, PEPAN Letters, #7(163), 827 (2010).
18. O. B. Tarasov and D. Bazin, *Nucl. Instr. Meth. B* **266**, 4657 (2008).
19. F. Videbaek, R. B. Goldstein, L. Grodzins, S. G. Steadman, T. A. Belote, J. D. Garrett, *Phys. Rev. C* **15**, 954 (1977).
20. M. Freer, The Euroschool on Exotic Beams, V. 4, p. 1; Ch. Scheidenberger, M. Pfützner, Eds, Springer 2014; DOI 10.1007/978-3-642-45141-6.
21. D. C. Rafferty, M. Dasgupta, D. J. Hinde et al., *Phys. Rev. C* **94** 024607 (2016).

Calibration of Long Neutron Counter TETRA Using Reactions of Photo-Dissociation

D. Testov[1,2,*], D. Verney[3], V. Smirnov[1], M. Babo[3], Yu. Penionzhkevich[1,4], F. Ibrahim[3], E. Kuznetsova[1], S. Lukyanov[1], K. Mendibayev[1], Yu. Sobolev[1] and E. Sokol[1]

[1]*Joint Institute for Nuclear Research, Joliot-Curie 6, 141980 Dubna, Moscow Region, Russia*
[2]*Dipartimento di Fisica e Astronomia and INFN, Sezione di Padova, Padova, Italy*
[3]*Institut de Physique Nucleaire CNRS-IN2P4, Univ. Paris-Sud, Universite Paris-Saclay, 91507 Orsay Cedex, France*
[4]*National Research Nuclear University, Kashirskoye Shosse 31, Moscow 115409, Russia*
[*]*E-mail: dumon@jinr.ru*

TETRA long neutron counter has proven to be an unique tool to identify and to measure delayed neutron emission from photo-fission fragments produced at ALTO ISOL facility. TETRA uses robust technology based on ^{3}He filled-counters operating in the proportional mode. The cross-section of the reaction (n, ^{3}He) is very larger for the thermal neutron energy range. Thus, β-delayed neutrons are moderated and can be detected with a high efficiency. However, due to the moderation process, information on the initial neutron energy is typically lost. Energy of emitted neutrons is needed for detailed description of the process of delayed (multi) neutron emission from neutron rich nuclei. In the present paper we discuss one of a method which can be used to recover information on the initial neutron energy.

Keywords: ^{3}He counters; β-decay; neutron emission; neutron spectrum; radioactive beams.

1. Delayed Neutron Emission

Nowadays much effort is devoted to understand the role of neutron excess and its influence on the nuclear structure, specifically, in the vicinities of closed neutron shells. In contrast to the nuclear shell structure along the β-stability line, which has been studied theoretically and experimentally quite well, the yet unknown shell structure in drip-line nuclei is currently of great interest. Therefore, the recent studies at IPN Orsay in the frame of JINR-IN2P3 collaboration were aimed at systematic investigation of the gross-properties of β-decay and trends in nuclear structure of neutron-rich photo-fission fragments produced at ALTO ISOL facility [1, 2]. One of the work-horse of such investigations is the neutron detector TETRA: 4π high effective array which consists of 80 ^{3}He counters complimented by $4\pi\beta$ plastic scintillator and a High Purity Germanium Detector as is described in Ref. [4]. The efficiency 52(2)%

for a single neutron was measured using ^{252}Cf source by two methods: based on the know activity and based on the (prompt) neutron multiplicity distribution. This value is in good agreement with the MCNP simulations [7]. The simulations of TETRA response to a mono energetic isotropic neutron source revealed a flatness up to 1 MeV followed gradually decrease towards 10 MeV. Therefore, measuring $P_{1n,2n,3n,4n...}$ values (probability of 1n, 2n, ..., β-delayed neutron emission) one has to assume a certain β-delayed neutron energy spectrum which increases the systematic errors to obtained values. One of the way to improve the errors, using so-called "effective" efficiency, was introduced in Ref. [3, 4]. The "effective" neutron efficiency can be determined relying on the coincidence data between β, γ and neutron detectors composing the system uniquely for each radioactive source created [4].

However, it is generally true that despite of the high efficiency one of the major drawbacks of such ^{3}He detectors systems is loss of information on the initial neutron energy due to the neutron moderation process. In β-decay of neutron rich nuclei information on neutron energy can help to understand which states about the neutron separation energy in the emitter are populated leading to neutron emission.

2. Reactions of Photo-Dissociation

East and Walton [5] shown that for a ^{3}He counter it is still possible to exploit the energy dependence of the thermalization process to partially recover the neutron energy information, namely the average energy of the neutron spectrum E_n. They suggested that the information on E_n can obtained from the front-to-back ratio for the counter banks in the "slab" neutron detector. The method is adopted to TETRA considering the similar ratio for the inner-to-outer rings of counters, see Fig. 1. More technical aspects of the setup are given in Ref. [4]. Thus, TETRA response, the inner-to-outer ring ratio $R(E_n)$, has to be calibrated using a mono energetic isotropic neutron source. Indeed, neutron sources of different mean energy lead to different patterns of how (normalised) count rate is distributed. We illustrated it previously by comparing the patterns recorded for prompt neutrons from ^{252}Cf source and β-delayed neutrons from the radioactive ^{84}Ga source, see Fig. 6.3 in Ref. [6]. Thus calibration of $R(E_n)$ using set of mono-energetic neutron sources can drive us to estimation of the mean energy of a neutron spectrum from an unknown sample.

Nine Calibration sources can be produced using reactions of photo-dissociation. In nuclear physics (α, n)-sources (like Ra-Be, Pu-Be, Am-Be) are commonly used as standard neutron sources. The neutron emission rate (source strength) is

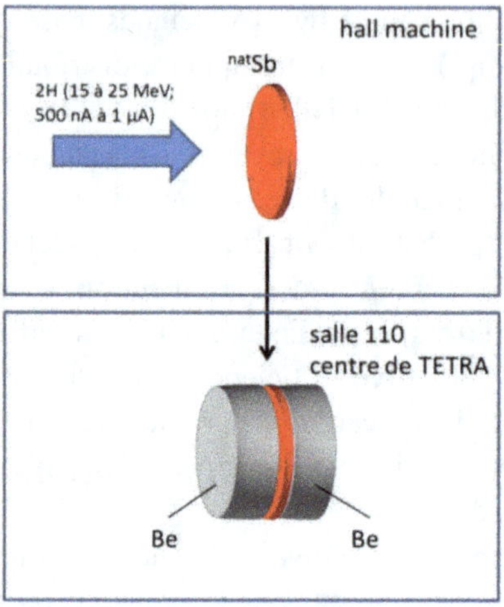

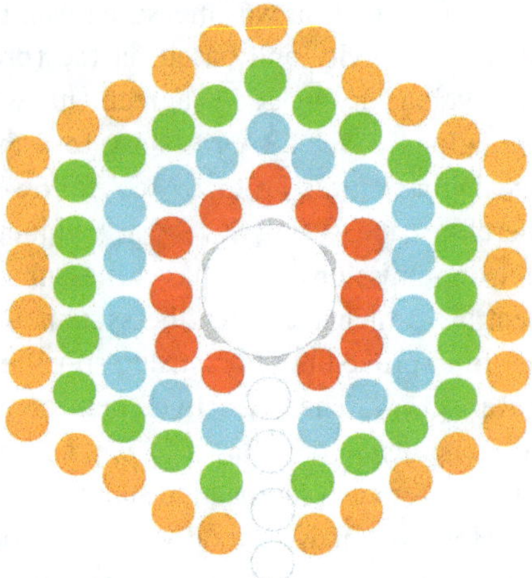

Fig. 1. *Left*: Heavy water (or beryllium) were exposed to intense flux from an artificially radioactive source were produced at the Tandem of IPN Orsay, schematic view. See text for explanations. *Right:* Counters inside TETRA are arranged in 4 rings, schematic view is shown. The response of the inner-most one (ring 1, marked by red color) to the most outer-one (ring 4, marked by orange color) was investigated as a function of the incident neutron energy.

nearly constant in time due to the long half-lives, but the emission spectrum is complex and insufficiently known. However, several photo-neutron sources emit nearly mono-energetic neutrons of the intermediate energy range $0.02 < E_n < 1$ MeV. In fact, these sources, widely used in the 1960s until the mid-1970s, have sadly been forgotten, mainly because of the technical difficulties in production and implementing them. The basic principle of a photo source neutrons is to expose heavy water or beryllium to an intense gamma flux. Thus, high-energy γ-rays falling on beryllium or deuterium will cause the emission of neutrons. In a photo-dissociation reaction a group of neutrons having a maximum energy (E_n-S_n) can be obtained. The gamma source is chosen so that in its radioactive decay only one gamma line is predominantly populated and has an energy higher than threshold for a (γ, n) reaction (Sn), for example, for Be S_n = 1.67 MeV; for ^{2}H S_n = 2.22 MeV. The energy of emitted neutrons is derived from the conservation of energy and momentum [5, 8]:

$$E_n = \frac{A-1}{A}\left[E_\gamma - Q - \frac{E_\gamma^2}{1862\cdot(A-1)}\right] + \delta,$$

where E_n is the energy of the neutrons (MeV); A - the atomic weight of the target nucleus; $E\gamma$ is the energy of the γ-ray (MeV); Q is the threshold for (γ, n) reaction in the nucleus on mass A, δ - is a function describing a small spread in the energy as a function of angle between the direction of the γ-ray and the direction in which neutron is emitted. Six sources of this type have been recommended as standard ones [8], covering the neutron spectrum from 25 keV to ~ 900 keV (exactly the β delayed neutron emission energy regime expected in the fission products studied at ALTO).

In the reported experiment γ-rays from artificially radioactive sodium and antimony, pre-reproduced at the TANDEM accelerator of IPN Orsay were used to make photo-neutron sources, see left part of Fig. 1. We characterized the energy response function R(E_n) of TETRA using these sources placed at the geometrical center of the array. We investigated the response of the inner-most and the outer-most ring of as a function of the incident neutron energy, see right part of Fig. 1.

3. Conclusion

We have investigated TETRA response as a function of incident neutron energy. Mono energetic neutron sources were produced in reactions of photo-dissociation. The results will be reported in a separate publication. The performed calibration will allow to estimate the mean energy of neutron spectra from radioactive sources produced at ALTO.

References

1. F. Azaiez, S. Essabaa, F. Ibrahim, and D. Verney, *Nucl. Phys. News.* **23**, (2013).
2. Yu. E. Penionzhkevich, *JINR NEWS* **4**, 17 (2017).
3. D. Verney, D. Testov, F. Ibrahim, *Phys. Rev. C* **95**, 054320 (2017).
4. D. Testov, D. Verney, F. Ibrahim et al., *Nucl. Inst. Meth. A* **815**, 96 (2016).
5. L. V. East and R. B. Walton, *Nucl. Inst.* **72**, 161 (1969).

High Resolution γ-Ray Spectroscopy Using GALILEO Array

D. Testov[1,a,*], J. J. Valiente-Dobón[2], D. Mengoni[1], F. Recchia[1], A. Goasduff[1], A. Boso[1,4], G. de Angelis[2], S. Lenzi[1], S. Bakes[1,2,b], C. Boiano[3], B. Cederwall[5], G. Colucci[2], M. Cicerchia[3], P. Čolović[6], F. Didierjean[7], M. Doncel[8], J. A. Dueñas[9], F. Galtarossa[2], A. Gozzelino[2], K. Hadyńska-Klęk[4], R. Isocrate[2], G. Jaworski[2], P. R. John[10], H. Liu[5], S. Lunardi[1], R. Menegazzo[1], A. Mentana[3], V. Modamio[11], A. Nannini[12], D. R. Napoli[2], M. Palacz[13], G. Pasqualato[1], M. Rocchini[12], S. Riccetto[3], B. Saygi[14], E. Sahin[15], M. Siciliano[3], Yu. G. Sobolev[16] and S. Szilner[6]

[1]Dipartimento di Fisica e Astronomia and INFN, Sezione di Padova, Padova, Italy
[2]INFN, Laboratori Nazionali di Legnaro, Legnaro, Italy
[3]Dipartimento di Fisica dellUniversità di Milano e Sezione INFN, Milano, Italy
[4]National Physical Laboratory, Teddington, Middlesex, TW11 0LW, UK
[5]KTH Royal Institute of Technology, Stockholm, Sweden
[6]Ruder Boskovic Institute, Zagreb, Croatia
[7]Institut Pluridisciplinaire Hubert CURIEN (IPHC), Strasbourg, France
[8]The University of Liverpool, UK
[9]Departamento de Ingenieria Elctrica, Escuela Técnica Superior de Ingeniería, Universidad de Huelva, Spain
[10]Institut für Kernphysik, Technische Universität Darmstadt, Germany
[11]University of Oslo, Norway
[12]Universit_a degli Studi and INFN Sezione di Firenze, Florence, Italy
[13]HIL, Poland
[14]Ege University, Faculty of Science, Department of Physics, Izmir, Turkey
[15]University of Surrey, UK
[16]Joint Institute for Nuclear Research, Dubna, Moscow Region, Russia
[a]also at Joint Institute for Nuclear Research, Dubna, Moscow Region, Russia
[b]also at University of Surrey, Department of Physics, UK
[*]E-mail: testov@lnl.infn.it

The GALILEO γ-ray spectrometer has been constructed at the Legnaro National Laboratory of INFN (LNL-INFN). It can be coupled to advanced ancillary devices which allows nuclear structure studies employing the variety of in-beam γ-ray spectroscopy methods. Such studies benefit from reactions induced by the intense stable beams delivered by the Tandem-ALPI-PIAVE accelerator complex and by the radioactive beams which will be provided by the SPES facility. In this paper we outline two experiments performed within the experimental campaign at GALILEO coupled to the EUCLIDES Si-ball and the Neutron Wall array. The first one was aimed at spectroscopic studies in A = 31 mirror nuclei and the second one at measurements of lifetimes of excited states in nuclei in the vicinity of ^{100}Sn.

Keywords: γ-Ray spectroscopy; γ-ray spectrometer; nuclear structure; nuclear state lifetime.

1. Introduction

GALILEO γ-ray array spectrometer is the resident array at National Legnaro Laboratories. In Phase I [1] it consists of 25 Compton-suppressed HPGe tapered detectors, originally from the GASP array [2].It is organized in 4 rings. Three backward rings made of 5 detectors each at 152°, 129° and 119°. The last ring at 90° comprises 10 detectors, see Fig. 1.

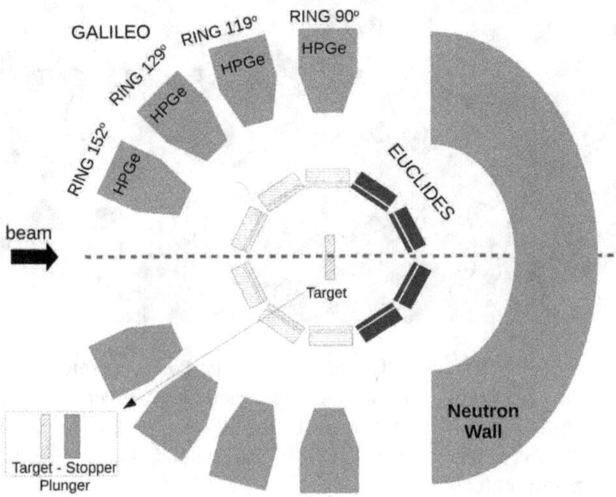

Fig. 1. Schematical view of GALILEO γ-ray array coupled to EUCLIDES, Neutron Wall. The target is installed at the center of the reaction chamber. The hatched part of ECULIDES is dismounted to allow plunger installation. See text for more details.

The measured absolute efficiency measured is ~2.3% for 1.3 MeV γ-ray and the average resolution around 2.5 keV [3]. The Peak to Total ratio is ~50%. GALILEO can be run in a stand-alone mode or coupled to ancillary devices. During the first experimental campaigns we used one or more ancillary detectors such as the light-charged-particle detector array EUCLIDES [4]; pixel-type silicon detector of TRACE [5, 6]; a heavy ion detector for Coulomb excitation measurements SPIDER [7]; the plunger device [8] for electromagnetic-moment measurements, the NEUTRON WALL [12, 13]. To increase the γ-ray efficiency for high-energy transitions a $LaBr_3$ array [14, 15] can be used complementary to other ancillary detectors. The GALILEO electronic system is fully digital and it is synchronized by a distributed clock delivered by the GTS (Global Trigger and Synchronization) system [16], which enables the time synchronization between the GALILEO spectrometer and all other ancillary detectors. Some of the results

Fig. 2. GALILEO γ-ray spectrometer. On the left part it is visible 10 HPGe detectors inside BGO shield placed at 90° around the reaction chamber. Neutron Wall array, positioned on forward angles with respect to the beam direction is visible on the right side. The EUCLIDES Si-ball is inserted in the reaction chamber of GALILEO which can be seen opened at the center.

of this first experimental campaign were already reported in the publications [17-19] and in the LNL Annual reports [20].

The EUCLIDES Si-ball array can be installed inside the chamber in the full or in the plunger configuration. In the full configuration EUCLIDES includes 40 ΔE-E telescopes covering ~80% of the solid angle. The light charged particle identification of EUCLIDES relies on the ΔE-E method. Almost 4π coverage of the solid angle and high granularity of EUCLIDES ensures its high particle detection efficiency and the possibility to reduce the Doppler broadening of peaks in the recorded γ-ray spectra by an event-by-event kinematic reconstruction of the trajectory of recoiling nuclei [4]. An EUCLIDES configuration with only forward positioned ΔE-E telescopes allows plunger installation as is described in Sec. 3.

The Neutron Wall, composed of 15 hexagonal detectors arranged in 2 rings around the central pentagonal unit, is installed on the forwards angles with respect to the beam direction. Each hexagonal unit is divided in 3 hermetically separated segments. Considering the solid angle coverage of ~1π the efficiency reached for a single neutron detection is 20-25%. Performance of Neutron Wall coupled to GALILEO as well as details of pulse-shape-analysis are reported in Ref. [21]. The picture of the setup is given in Fig. 2.

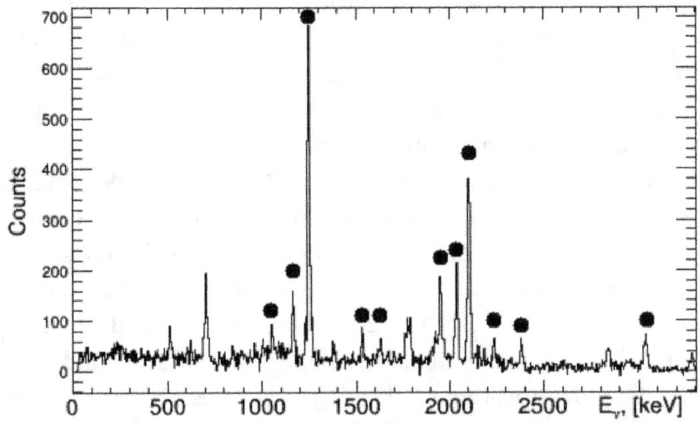

Fig. 3. Sum of the projections of γ-γ matrix gated on 1249-keV, 2102-keV and 1166-keV transitions in ^{31}S. 1α and 1 neutron conditions were requested. The marked peaks correspond to previously known transitions in ^{31}S.

To highlight the performance of GALILEO we have selected two experiments. Thus, in Sec. 2 it is reported an experiment to investigate excited levels of the ^{31}S populated in a fusion evaporation reaction. In Sec. 3 we describe the measurements of lifetimes of nuclear states in nuclei located in the vicinity of ^{100}Sn.

2. High Spin States in Mirror Nuclei ^{31}S and ^{31}P

One of the first experiments performed using GALILEO was aimed at identification of high spin states in mirror nuclei ^{31}S and ^{31}P. ^{12}C beam at 45 MeV impinged on the self-supported ^{24}Mg target of 400 μg/cm^2. The most recent studies of ^{31}S performed [22] revealed large oscillation behaviour of MED values for the negative-parity sequence as a function of spin, see Ref. [22]. These oscillation may be explained including in the wave function excitations to the *fp* shell considering thus the electromagnetic spin-orbit effect. Description of the MED in *sd* shell nuclei for negative parity and high spin states involving the electromagnetic spin orbit term is up to now only qualitative (because it involves interactions in two main shells). Additionally, shell-model calculations performed using the USD residual interaction and the Monte Carlo shell model with the SDPF-M interaction reproduce well the excitation energies and the reduced transition probabilities for positive-parity states up to the spin $13/2^-$, see Ref. [23]. An interesting feature revealed by these calculations is that the yrast negative-parity states show an alternating structure: the $7/2^-$, $11/2^-$, and $15/2^-$ states are described by almost equal contributions of the proton and neutron excitation to the *fp* shell, whereas the $9/2^-$ and $13/2^-$ states have only a neutron excitation to the $f_{7/2}$ shell. On the experimental side, MED values are available up to spin $J = 13/2$ for both negative and positive parity [22] which is not sufficient to disentangle the theoretical puzzle. Therefore, data on MED values for higher spin states are needed. Excited levels of ^{31}P was were previously studied in ^{12}C(^{20}Ne,p(n)), see Ref. [22] and ^{24}Mg(^{16}O,2αp(n)) reactions, Ref. [23, 24] up to high spins. In contrast, the more exotic ^{31}S was observed up to only $13/2^+$ and $13/2^-$ spin Ref. [22]. Therefore, the goal of reported experiment was to extend the level schemes and study the mirror energy differences in the A=31, T=1/2 mirror nuclei. ^{31}P and ^{31}S were produced in the same fusion evaporation reaction ^{24}Mg(^{12}C,αp) and ^{24}Mg(^{12}C,αn) respectively. The identification of ^{31}P (1α1*p*) and ^{31}S (1α1*n*) was performed in the off-line analysis using EUCLIDES and Neutron Wall. In the preliminary data analysis the sum of projections of γ-γ matrix recorded requesting coincidence with 1α and 1*n* neutron is shown in Fig. 3. The γ-ray transitions in ^{31}S known from experiments cited in the literature are marked. A further analysis is ongoing [11].

3. Probing the Strength of the 100Sn Shell Closure via Life-Time Measurements in Light Sb and Te

The structure of nuclei far from β stability has been a key subject for both experimental research and theoretically investigations. The information on the electromagnetic properties of the neutron-deficient nuclei in the vicinity of ^{100}Sn

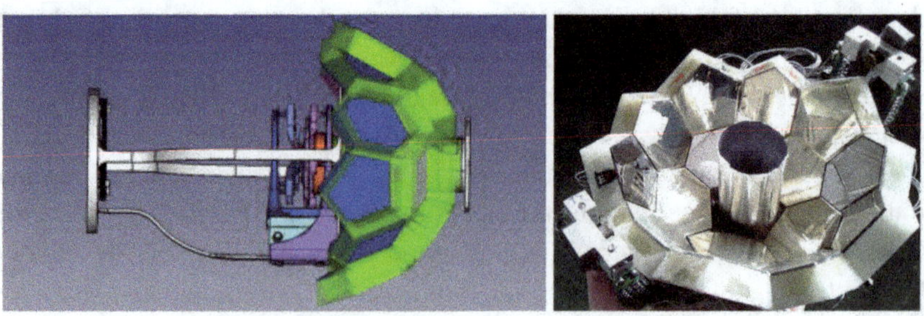

Fig. 4. Plunger device at LNL coupled to EUCLIDES (left); EUCLIDES plunger configuration consisting of 5 segmented and 10 single-plate ΔE-E telescopes (right).

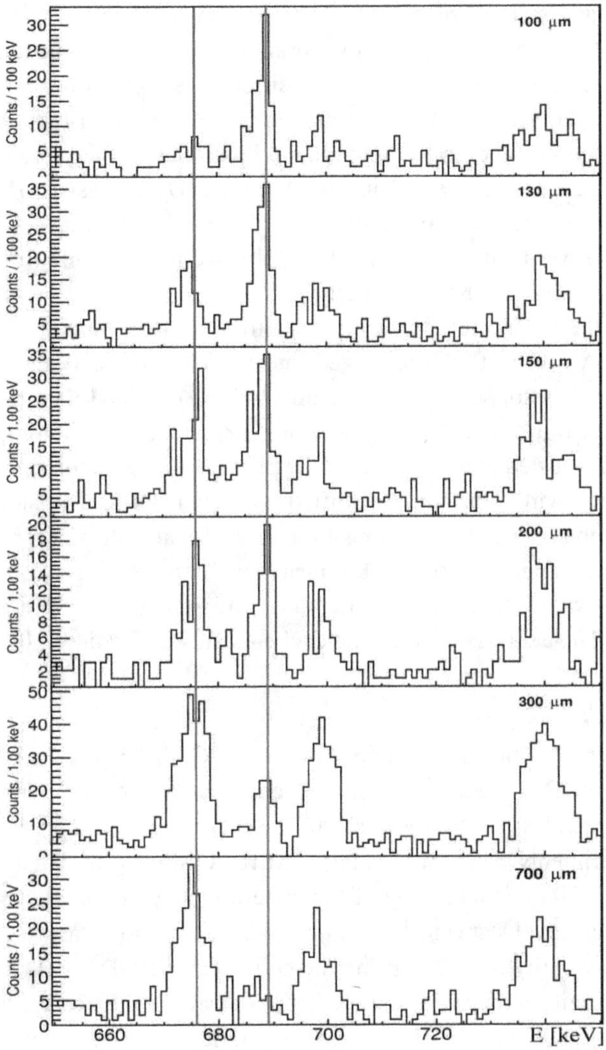

Fig. 5. Coincidence spectra resulting from gating on the $4^+ \rightarrow 2^+$ shifted component in ^{112}Te observed in detector-ring 1 (Θ_1=129°). 3p conditioned was requested. Indicated is the $2^+ \rightarrow 0^+$ shifted and stopped component in ^{112}Te in detector-ring 2 (Θ_1=119°).

represent unique testing ground for many different theoretical approaches. Recoil Distance Doppler-Shift (RDDS) method [25] is extensively used in the nuclear structure experiments to determine excited states lifetime in the range of the ps to the ns. Thus a dedicated plunger device [8] to fit in the GALILEO reaction chamber with part of the EUCLIDES array was constructed. This allows the application of the RDDS method relying on γ-γ coincidence measurements. In order to select evaporation channels in a fusion-evaporation reaction we removed the backward positioned Si telescopes of EUCLIDES Si-ball array [4], see Fig. 1. The removed telescopes contribute weakly to the overall detection efficiency of light charged particles in the present experiment. In this configuration, presented in Fig. 4, EUCLIDES consists of 5 segmented ΔE-E telescopes placed at the forward angle and 10 single-plate telescopes in the second forward ring. Thus, the channel selection capability of EUCLIDES can be exploited also in RDDS experiments.

During the performed experiment neutron-deficient nuclei were populated using a 2 pnA beam of ^{58}Ni impinged into a 1 mg/cm^2 ^{58}Ni target followed by a 15 mg/cm^2 Au-stopper foil. The quality of the obtained data can be seen in Fig. 5. In this figure it is shown the coincidence spectra resulting from gating on the $4^+\rightarrow 2^+$ shifted component in ^{112}Te observed in detector-ring 1 (Θ_1=129°) in coincidence with 3 protons identified by EUCLIDES. The stopped and the shifted component of $2^+\rightarrow 0^+$ transition in ^{112}Te are shown for the a set of the target-to-stopper distances. The measured lifetime of the 2^+ coincide with the previous values cited in the literature and will be reported in a separate publication. The analysis to study the excited states is ongoing [10].

4. Conclusion

In the paper we briefly describe the new GALILEO γ-ray spectrometer constructed at LNL Legnaro. It can be coupled to ancillary devices to allow variety of nuclear structure research. In the paper we report preliminary results of two experiments performed using GALILEO during the first experimental campaign at LNL Legnaro. One of the experiments was aimed at spectroscopic studies of high-level states in ^{31}S, and the second one, to measure lifetime in the nuclei in the vicinity of 100Sn using the coincidence RDDS method conditioned on charged particles detected by EUCLIDES array. In the nearest future nuclear structure research at GALILEO will benefit from accelerated beams of radioactive isotopes to be delivered by SPES facility [26].

References

1. J. J. Valiente-Dobón et al., LNL Annual report p. 79 (2014).
2. D. Bazzacco, Proc. Workshop on Large γ-ray Detector Arrays (Chalk River Canada) AECL10613, 376 (1992).
3. I. Zanon, Master thesis (2018), http://tesi.cab.unipd.it/61172/1/Zanon Irene tesi.pdf.

4. D. Testov et al., LNL Annual Report p. 105 (2015); accepted for publication at *Eur. Phys. J. A*.
5. D. Mengoni, J. A. Dueñas, M. Assié et al., *Nucl. Instr. and Meth. A* **764**, 241 (2014).
6. N. Cieplicka-Oryńczak, D. Mengoni, M. Ciemala et al., *Eur. Phys. J. A* **54**, 209 (2018).
7. M. Rocchini, K. Hadyńska-Klęk, A. Nannini et al., *Phys. Scr.* **92**, 074001 (2017).
8. C. Muller-Gatermann, F. von Spee, A. Goasduff et al., *Nucl. Inst. and Meth. A* **920**, 95 (2019).
9. D. Testov et al., LNL Annual Report p. 59 (2016).
10. S. Bakes et al., LNL Annual Report p. 22 (2017).
11. D. Testov et al., LNL Annual Report p. 61 (2017).
12. O. Skeppstedt, H. A. Roth, L. Lindstrom et al., *Nucl. Instr. and Meth. A* **421**, 531 (1999).
13. J. Ljungvall, M. Palacz, J. Nyberg, *Nucl. Instr. and Meth. A* **528**, 741 (2004).
14. A. Giaz et al., *Nucl. Instr. and Meth. A* **729**, 910 (2013).
15. G. Benzoni et al., LNL Annual Report (2015) p. 84, and S. Ceruti et al., LNL Annual Report p. 31 (2016).
16. M. Bellato et al., *Journal of Instrumentation* **8**, P07003 (2013).
17. S. P. Hua, L. Zhang, X. Zhang et al., *Nucl. Inst. and Meth. A* **815**, 96 (2016).
18. G. L. Zhang, G. X. Zhang, S. P. Hu, *Phys. Rev. C* **97**, 014611 (2018).
19. Y. H. Qiang et al., *Phys. Rev. C* **99**, 014307 (2019).
20. http://www.lnl.infn.it/_annrep/index.htm
21. A. Lonardi, Master thesis (2015), http://tesi.cab.unipd.it/52165/1/tesi L Lonardi Alberto.pdf.
22. D. Jenkins et al., *Phys. Rev. C* **72**, 031303(R) (2005).
23. M. Ionescu-Bujor et al., *Phys. Rev. C* **73**, 024310 (2006).
24. F. Della Vedova et al., *AIP Conf. Proc.* **764**, 205 (2005).
25. A. Dewald et al., *Prog. Part. Nucl. Phys.* **679**, 786 (2012).
26. G. Bison et al., *Nucl. Instr. and Meth. B* **376**, 240 (2016).

Latest Updates on Developments of the Underground Neutrino Facility in South Africa

Z. Z. Vilakazi[1], S. M. Wyngaardt[2], R. T. Newman[2,5], R. Lindsay[3], A. Buffler[4], R. de Meijer[3], P. Maleka[5], J. Bezuidenhout[6], R. Nchodu[5], M. van Rooyen[2], Z. Ndlovu[2]

[1]*School of Physics, University of the Witwatersrand, Johannesburg, Private Bag Wits 2050, South Africa*

[2]*Department of Physics, Stellenbosch University, Private Bag X1, Matieland, Stellenbosch, 7602*

[3]*Department of Physics, University of the Western Cape, Private Bag X17, Bellville, 7530*

[4]*Department of Physics, University of Cape Town, Rondebosch, 7700, South Africa*

[5]*iThemba LABS, P.O. Box 722, Somerset West, 7129, South Africa*

[6]*Military Academy, Private Bag X2, Saldanha, 7395*

Experiments in physics, Astro-particle physics and cosmology that require careful shielding against cosmic rays includes dark matter searches, studies of radioactive decays, and neutrino detection experiments. The need for such shielding has motivated the construction of laboratory caverns in mines and adjacent to tunnels under mountains. There are currently about a dozen such laboratories, in existence or under construction, all in the Northern Hemisphere. There is a well established case put forward for a Southern Hemisphere facility. Presented in this paper is a feasibility study of measurements of radon in air (using electret ion chambers and alpha spectroscopy), background gamma ray measurements (inside/outside) the tunnel using scintillator (inorganic) detectors, cosmic ray measurements using organic scintillators and radiometric analyses of representative rock samples for the establishment of such a facility in the South Africa.

Keywords: Underground laboratory; neutrinos; gamma ray; radon; dark matter; background.

1. Introduction

The observations of statistically robust annual modulation signals from DAMA/LIBRA experiment [1] [2] though statistically sound, have raised doubts as to whether this is a true effect, or an artifact resulting from seasonal variations in temperature, humidity, radioactive backgrounds, etc. Having a Southern Hemisphere analogue of the DAMA/LIBRA experiment would therefore give a control against many such artifacts, since seasonal fluctuations would be 6 months out of phase with those in the Northern Hemisphere. Another place where a Southern Hemisphere experiment would be valuable would be in the study of possible annual modulation effects in certain radioactive decays, as surveyed by Fischbach [3] [4]. These possible effects are culled from experiments in Northern Hemisphere laboratories, and may be due to seasonal variations. Again, having a Southern Hemisphere facility in which to repeat the experiments, with good shielding, would be important in seeing whether these effects are indications of new physics [5]. If radioactive decay rates did show an annual modulation, they could also be an additional potential background for dark matter searches that look for annual modulation effects.

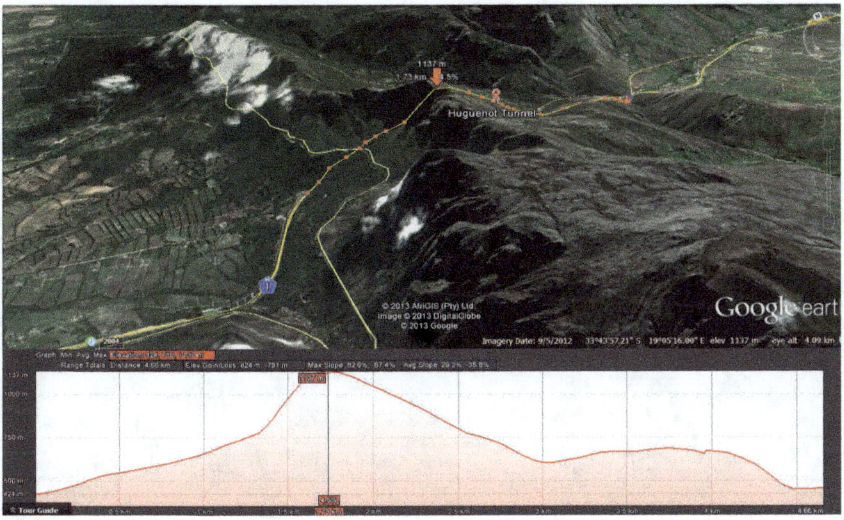

Fig. 1. The height profile of du Toits kloof mountain range, through which the Huguenot tunnel passes.

Discussions about an underground research facility in SA started in 2011. As one of the world's largest gold producers, South Africa has a number of the world's deepest gold mines (TauTona Gold Mine 3.9 km). Furthermore, it has to be noted that South Africa was active in this area of research in the 1960's through pioneering experiments done at a gold mine east of Johannesburg, the where the late Nobel Laureate F. Reines [6] [7] [8] [9] and colleagues (including the JPF Sellschop of the University of the Witwatersrand) observed the first natural neutrinos along with the Indian team led by Goku Menon and colleagues in Kolar Gold fields in India, thereby setting the first astrophysical limits!

The primary focus of the South African nuclear physics community is on establishing an underground facility in one of South Africa's deep gold mines. Running parallel to these developments is the alternative of developing such an underground laboratory inside the Huguenot Tunnel located between the towns of Paarl and Worcerster in the Western Cape Province of South Africa. This would also act as a prototype for the 3 – 4 km deep mine facility in the

Johannesburg area. Presented in this paper is a progress report on these on-going developments [11].

2. Feasibility Investigations

In order to decide on the best course to take in order to develop sensible research programs around the Huguenot tunnel facility a number of measurable parameters will have to be known.

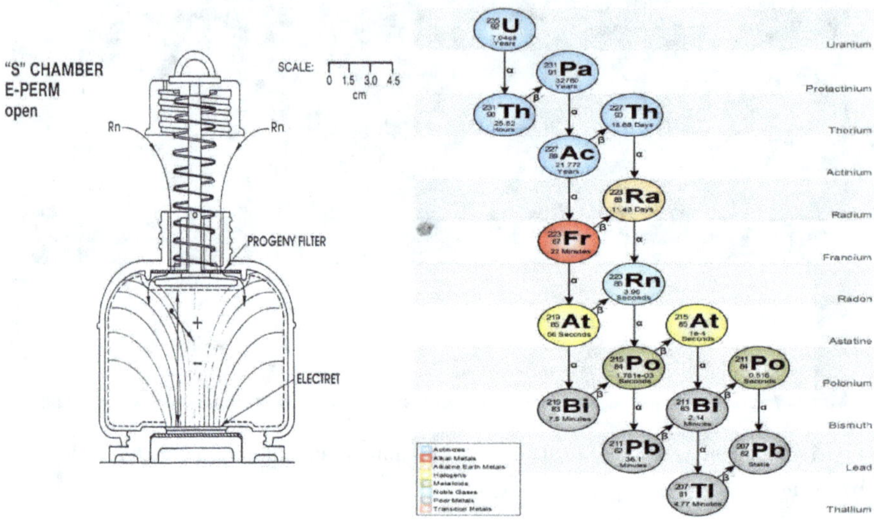

Fig. 2. Electret Passive Environmental Radon Monitor (E-PERM) (left) and radon decay chain (right panel).

According to the geological survey which was done during the excavation period (1973 - 1984) the composition of the mountain range mainly composes of quarzitic sandstone (also refered to as Table mountain sandstone). It is therefore expected that radon concentration levels within the tunnel should be insignificant, and should therefore not have any considerable contribution on the background radiation signature. Figure 1 shows the height profile of the du Toitskloof mountain range above the huguenot tunnel. The peak height is 1,995 m (6,545 ft).

Characterization of radioactivity within the tunnel

Paramount among the considerations of establishing an underground neutrino observatory is to ensure control and understanding of background due to the

Fig. 3. Electrets mounted in groups of 3 on the walls of the tunnel (left) close to the vehicle cross-cut area (right).

presence of radon from the Uranium decay chain (shown in figure 2) that can influence our measurements.

These studies began in 2013 when the environmental radiation groups at Stellenbosch University and iThemba LABS performed a preliminary radon measurements by placing 3 electrets [11] at the three vehicle cross-cut (VCC) areas to monitor radon in the undeveloped Northern tunnel seem in figure 3.

Table 1. Radon levels measured at vehicle cross-cuts.

	Mean	Standard Deviation
location	Air Radon Concentration	Air Radon Concentration
	(Bq.m^{-3})	(Bq.m^{-3})
VCC1	45.4	0.1
VCC2	52.7	7.1
VCC3	64.9	5.2

The measurements in the northern bore and connecting sections using Electret Ion Chambers (EIC) and a RAD7 continuous radon monitor were undertaken. The EIC measurements involved the deployment of EIC for a period of about a week when using SST electrets while RAD7 measurements involved sniff measurements over a period of about 1 hour. The measurements of radon concentrations at the three measured sites shown in Table 1 confirm that the level of radon is well below any considerable levels. Comparison between the radon levels at the three sites also shows a slightly higher level at VCC2 as compared to VCC1 and VCC3.

The radon background measurements have been followed up by an accumilated gamma ray measurement with the multi-element detector using a scintillator array (MEDUSA) detection system whose gamma ray spectrum was then compared to the known gamma energy spectral of 40*K*, ^{238}U, ^{232}Th, and ^{137}Cs shown in figure 4.

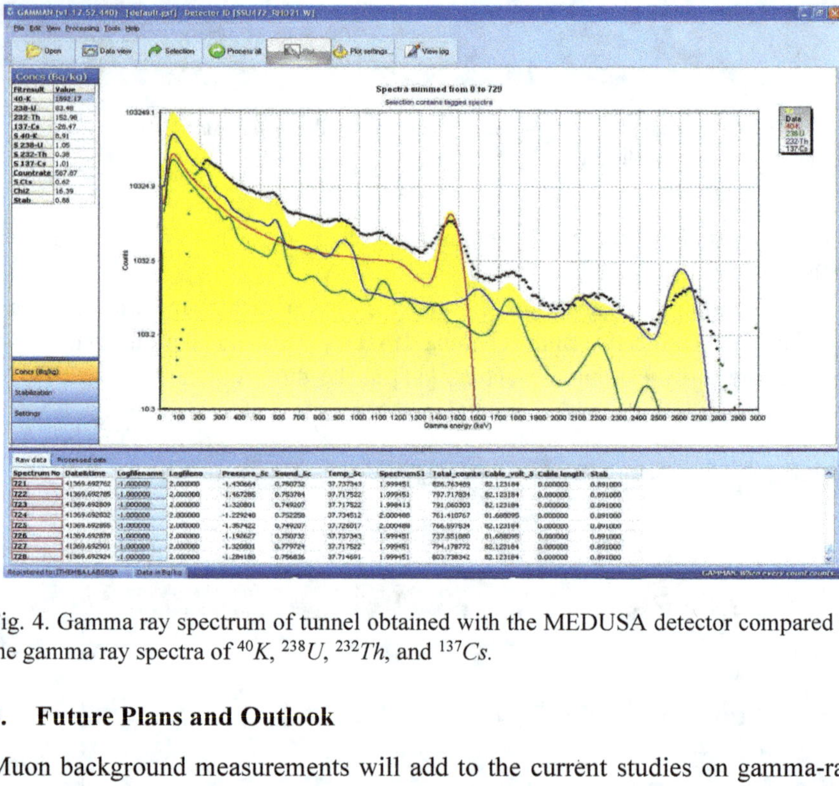

Fig. 4. Gamma ray spectrum of tunnel obtained with the MEDUSA detector compared to the gamma ray spectra of ^{40}K, ^{238}U, ^{232}Th, and ^{137}Cs.

3. Future Plans and Outlook

Muon background measurements will add to the current studies on gamma-ray spectra and radon background measurements within the tunnel. In addition to the background measurements, there were bio-monitors within the tunnel in place for the measurement of the air pollution levels. Following the current feasibility study, plans are in place begin extensive discussion with the South African Department of Science and Technology, South African Roads Agency Limited (SANREL), potential role players (SA Universities, iThemba LABS and International community in order to establish a permanent facility in place within the tunnel. One of the first steps in setting up the facility would be to develop an established dedicated low-level research laboratory and programs

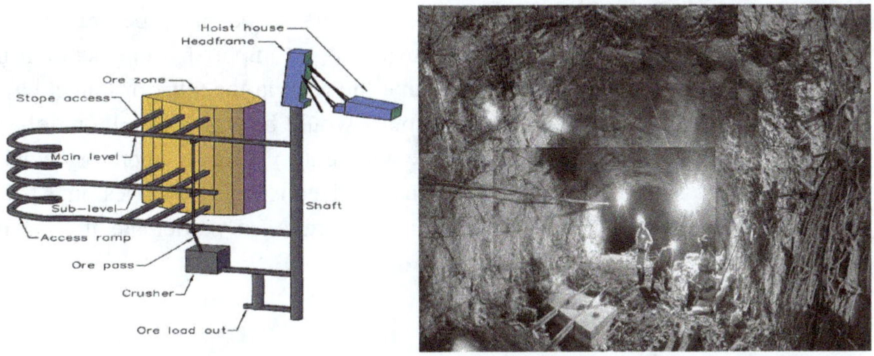

Fig. 5. A 3D image of an underground mining complex (left) and; (right) near the bottom of the Moab-Khotsong gold mine in South Africa, scientists aim to drill to the site of a recent earthquake where geo-seismic tests are conducted and a potential site for a makeshift detector.

which will detect geo-neutrinos, look for a possible seasonal signature of dark matter and search for opportunities which could uniquely be studied in the southern hemisphere.

In the not to distant future, a conversation could be entered into with mining companies to allow research to use of one of the side shafts of an active mine (as seen in a sample depiction in figure 5) as a dedicated cavern which could be used for deep underground neutrino studies. Furthermore, there have been extensive studies to investigate geo-seismic effects in order to assure appropriate calibration and handling of such effects. These studies were carriedout by the **J**apanese-**G**erman **U**nderground **A**coustic emission **R**esearch in **S**outh Africa (JAGUARS) where Semi-controlled earthquake-generation experiments at deep gold mines, 1995-2008 [12] - using a network of eight acoustic emission (AE) sensors 3.3 km below surface in the Mponeng gold mine to measure seismic events.

4. Conclusions

Having a Southern Hemisphere facility in which to repeat the experiments, with good shielding, would be important in seeing whether these effects are indications of new physics. If radioactive decay rates did show an annual modulation, they could also be an additional potential background for dark matter searches that look for annual modulation effects. In addition to the above experiments, an underground laboratory in the Southern Hemisphere could mount all the other types of experiments currently underway in the Northern

Hemisphere, such as solar and cosmic neutrino detection experiments and searches for rare beta decays. Aside from depth, an important consideration in setting up an underground laboratory is that the rock in the mine or tunnel have low radioactivity. A very rough cost estimate would be from $8 million dollars up, depending on how large and elaborate a laboratory. Such a laboratory would be a unique way for a consortium of Southern Hemisphere countries, in Africa or South America, and perhaps including Australia, to participate in current developments in physics research.

Acknowledgements

This study has received support from the South African National Research Foundation (NRF) and the Department of Science and Technology through the SA-JINR collaboration. The generosity of the SAUL collaboration is highly appreciated.

References

1. R. Bernabei et al., *Nuclear Instruments and Methods in Physics Research A* **592**, 3 (2008).
2. R. Bernabei et al., *European Physical Journal C* **67**, 39 (2010). (and references therein)
3. D. Javorsek et al., Power Spectrum Analyses of Nuclear Decay Rates, *Astropart. Phys.* **34**, 173 (2010).
4. E. Fischbach et al., Evidence for Solar Influences on Nuclear Decay Rates in Proceedings of the Fifth Meeting on CPT and Lorentz Symmetry, A. Kostelecky, Ed. 2011, World Scientific: Singapore. 168-172 e-Print: arXiv: 1007. 3318 [hep-ph].
5. S. Adler, priv comm. (2011).
6. F. Reines, M.F. Crouch, T.L. Jenkins, W.R. Kropp, H.S. Gurr, G.R. Smith, J.P.F. Sellschop, Evidence for high-energy cosmic ray neutrino interactions. *Phys. Rev. Lett.* **15**, 429 (1965).
7. B.S. Meyer, J.P.F. Sellschop, M.F. Crouch, W.R. Kropp, H.W. Sobel, H.S. Gurr, J. Lathrop, F. Reines, (UC, Irvine), Cosmic ray muon intensity deep underground versus depth, *Phys. Rev. D* **1**, 2229 (1970).
8. F. Reines et al., Upper limit on high-energy neutrinos from Weber pulses, *Phys. Rev. Lett.* **26**, 1451 (1971).
9. F. Reines, W.R. Kropp, H.W. Sobel, H.S. Gurr, J. Lathrop (UC, Irvine), M.F. Crouch, J.P.F. Sellschop, B.S. Meyer, Muons produced by atmospheric neutrinos: Experiment, *Phys. Rev. D* **4**, 80 (1971).

10. Z.Z. Vilakazi, S.M. Wyngaardy, R.T. Newman et al., Proceedings of New Horizons in Fundamental Physics, S. Schramm and M. Schaefer (eds), FIAS Interdisciplinary Series, DOI: 10.1007/978/3-319-44165-8_22, p. 307 (2017).
11. Kotrappa et al., *Health Phys.* **58**(4), 461 (1987).
12. Ogasawara et al., *Earth Planets Space* **61**, 49 (2009).

Production of Neutron-Rich Nuclei by Multinucleon Transfer Reactions at KISS Project*

Y. X. Watanabe†, Y. Hirayama, P. Schury, Y. Kakiguchi, M. Oyaizu, M. Wada and M. Miyatake

Wako Nuclear Science Center, Institute of Particle and Nuclear Studies, High Energy Accelerator Research Organization (KEK), Wako, Saitama 351-0198, Japan
†E-mail: yutaka.watanabe@kek.jp
www.kek.jp

M. Mukai, M. Ahmed and S. Kimura

Graduate School of Pure and Applied Science, University of Tsukuba, Tsukuba, Ibaraki 305-0006, Japan

H. Ishiyama, S. C. Jeong, J. Y. Moon and J. H. Park

Rare Isotope Science Project, Institute for Basic Science (IBS) Daejeon 305-811, Republic of Korea

We have been developing the KEK Isotope Separation System (KISS) to perform the β–γ spectroscopy of the neutron-rich nuclei around the neutron magic number of 126 to investigate the astrophysical environments of the r-process nucleosynthesis. The nuclei of interest are produced by the multinucleon transfer reactions between ^{136}Xe and ^{198}Pt. The experimental study demonstrated its promising potential to produce them. We have successfully performed the β–γ spectroscopy and the laser ionization spectroscopy at the KISS using the nuclear production of the reaction system.

Keywords: Nuclear production; multinucleon transfer reaction; gas cell; laser resonance ionization; β–γ spectroscopy; laser ionization spectroscopy.

1. Introduction

Properties such as half-lives and masses of the nuclei at the neutron closed shell of 126 are essential to identify the environment of the r-process in the astrophysical nucleosynthesis[1], because those nuclei are considered to be waiting point nuclei which are progenitors of the peak around the mass number of 195 in the solar r-abundance distribution. The experimental studies of the nuclear β-decays and masses as well as half-lives of neutron-rich nuclei around the neutron number (N) of 126 are critical. However most of those nuclei remain unexplored,

* This work is supported by JSPS KAKENHI Grant Nos. JP23244060, JP24740180, JP26247044, JP15H02096, 17H06090, JP17H01132 and JP18H03711.

because they are difficult to access by fusion, fission and fragmentation with stable projectiles.

We are developing KEK Isotope Separation System (KISS)[2] to perform the nuclear spectroscopy of those neutron-rich nuclei, which are produced in the multinucleon transfer (MNT) reactions between the ^{136}Xe projectile and the ^{198}Pt target. Recently, promising expectations have been focused on for their nuclear production by MNT reactions both theoretically and experimentally. Using a macroscopic approach based on Langevin-type dynamical equation of motion, Zagrebaev *et al.* demonstrated that the MNT reaction between the Xe projectile and the Pb target gave large cross sections enough to produce those neutron-rich nuclei.[3] The MNT reaction measurements performed at ANL indicated a general agreement with the macroscopic calculations.[4] However, they did not reach $N =$ 126 for the elements with the atomic number less than 80. Thus the experimental data were scarce to evaluate the nuclear production at $N = 126$.

2. Nuclear Production Around $N = 126$ by MNT Reactions

To investigate the feasibility of the reaction system, ^{136}Xe and ^{198}Pt, for the production of the neutron-rich nuclei around $N = 126$, we performed an experimental study using the large acceptance magnetic spectrometer VAMOS++[5] coupled to the EXOGAM gamma-ray array[6] at GANIL.[7,8] The projectile-like fragments produced in the MNT reactions were detected by the spectrometer and the target-like fragment (TLF) distributions were deduced from those measurements. Relatively large cross sections measured for the production of the neutron-rich nuclei around $N = 126$ as compared with the fragmentation suggested the advantage of the MNT reactions. It was observed that the neutron-rich side around $N = 126$ in the isotopic distribution mainly arises from the lowest total kinetic energy loss (TKEL) values less than 75 MeV and the high-TKEL components contribute to the production of neutron-deficient isotopes. Recently, Karpov and Saiko demonstrated good agreements between our experimental data and the macroscopic calculations for the total isotopic distributions and decomposed ones in different TKEL values.[9]

3. KISS Project and Recent Experimental Results

The KISS has been developed at RIKEN RIBF facility to separate a nuclear species from products in the MNT reactions for the nuclear spectroscopy of the neutron-rich nuclei around $N = 126$. Figure 1 shows a schematic view of the KISS. It consists of a gas cell system, a laser system, an extraction chamber, an Isotope Separator On-Line (ISOL) and a detection system for the β–γ spectroscopy. The ^{198}Pt target (12.5 mg/cm^2) is bombarded by the ^{136}Xe beam (10.75 MeV/nucleon) from the RIKEN Ring Cyclotron after energy-degradation by three layers of titanium foils (3 μm × 3) to optimize the reaction energies in the target. The TLFs produced by MNT reactions are ejected from the target and go into a doughnut-shaped gas cell[10] through a window made of a polyimide film (5 μm). They are thermalized and neutralized by the argon gas filled in the gas cell at the pressure of 88 kPa. Those neutral atoms are transported by the laminar gas flow, and are element-selectively ionized by two-color two-step laser resonance ionization technique just before the exit of the gas cell. The primary beam passes through the aperture with the diameter of 15 mm in the doughnut-shaped gas cell, and is stopped in the water cooled beam dump. The doughnut-shaped gas cell prevents the primary beam from entering the gas cell to suppress the plasma effect induced in the gas by the primary beam, which would neutralize the laser-ionized atoms of interest, resulting in the reduction of their extraction yields. The extracted singly charged ions are transported by RF ion guides and are accelerated by application of a high voltage of 20 kV. They are mass-separated with the ISOL method, and finally they are transported to the detection system for the β–γ spectroscopy.

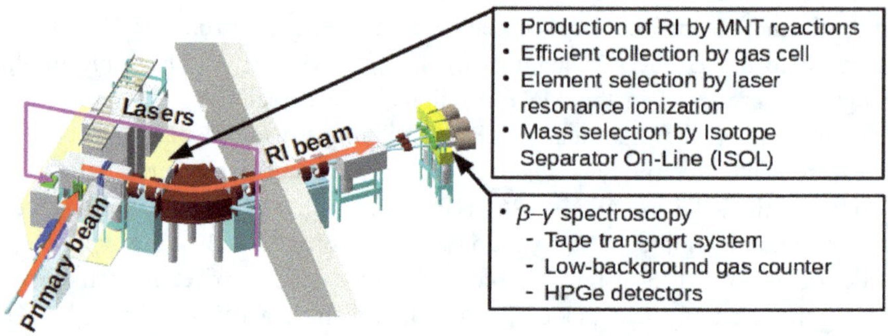

Fig. 1. Schematic view of the KEK Isotope Separation System.

The detection system consists of a tape transport system, a low-background gas counter called Multi-Segmented Proportional Gas Counter (MSPGC)[11] and four Super Clover High-purity Germanium (SC HpGe) detectors. The mass-separated ions are implanted in the aluminized Mylar tape stretched vertically. The KISS beam is pulsed with certain beam-on and off periods specific for the expected half-lives for β-decay measurements, and the tape is moved in synchronization of the beam pulsing to remove the background from the activity of the accumulated daughter nuclei. The tape is surrounded by the MSPGC to detect β-rays, which is a two-layered proportional gas counter. Each layer is placed concentrically with the vertical length of 200 mm, and has 16 angularly divided segments. To measure the lifetimes of rare reaction products, high efficiency for the β-ray detection and the low background rate are essential. The MSPGC achieves 40% detection efficiency for the β-rays with Q_β of 1 MeV including the geometrical acceptable solid angle of 80% of 4π. The achieved background rate is as low as 0.1 cps. The MSPGC is surrounded by four SC HpGe detectors for the γ-ray detection with close geometry of around 5 cm spacing from the tape. The efficiency for the photo peak detection is 10% for the γ-ray of 1 MeV.

We have extracted twelve kinds of neutron-rich isotopes, $^{199-201}$Pt, $^{196-200}$Ir and $^{195-208}$Os, using the ^{198}Pt target, and three kinds of neutron-rich isotopes, $^{185-187}$Ta, using the $^{nat.}$W target so far. The extracted isotopes were identified by measuring the half-lives of their β-decays, and their β–γ spectroscopy were performed. Figure 2(a) indicates the growth and decay curves of the β-ray detection for ^{198}Os in the beam-on and off periods of 252 s and 672 s, respectively, in coincidence with the γ-rays of 230.6 keV, which was newly found in the β-delayed γ-ray events. We have determined the half-life of the β-decay of ^{198}Os ground state as 125 ± 28 s for the first time.[12]

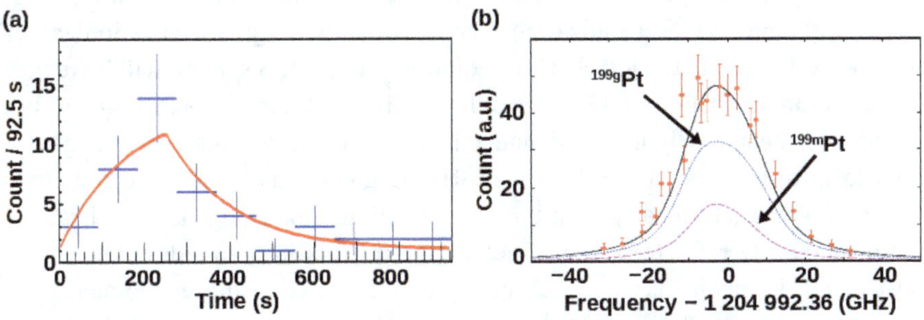

Fig. 2. (a) Growth and decay curves of the β-ray detection for ^{198}Os in coincidence with the γ-rays of 230.6 keV[12] and (b) HFS spectrum by detecting β-rays for ^{199}Pt.[13]

We have performed the laser ionization spectroscopy for ^{199}Pt and $^{196-198}$Ir to measure the isotope shift and the magnetic dipole moment in the hyperfine structure (HFS) spectrum. Figure 2(b) shows the HFS spectrum measured for ^{199}Pt by detecting β-rays. The ^{199}Pt isotopes were extracted as the mixture of the ground state ^{199g}Pt and the isomeric state ^{199m}Pt. Gating on the γ-rays of the isomeric transitions, the component of the isomeric state in the HFS spectrum was decomposed. The figure indicates the fitting results with both components of the ground state and the isomeric state to evaluate their isotope shifts and magnetic dipole moments. The detailed discussion is found in Ref. 13.

4. R&D and Future Plan of KISS

In order to go further to the lifetime measurements of more neutron-rich nuclei in the vicinity of the neutron magic number of 126, still lower background rate for the β-ray detection is necessary, because the extraction yields become smaller. The present MSPGC achieved the background rate of 0.1 cps by enabling two-dimensional tracking on the horizontal plane. We are now adapting the MSPGC to enable three-dimensional tracking using resistive (3 kΩ/cm) carbon wires for anode wires stretched vertically. The resolution of 2 mm in the full-width at the half-maximum (FWHM) was achieved for the vertical position determination by measuring the charges at both ends of the anode wires. It is good enough to reduce the background rate to 0.01 cps.

In addition to the β–γ spectroscopy, a new branch was constructed in the KISS beam line for the mass measurements. We will use a multi-reflection time-of-flight mass spectrograph (MRTOF-MS) for the mass measurements, which achieved the mass resolving power of 150 000 in the atomic mass region of 200.[14] An MRTOF-MS with a helium gas cell cooler buncher, which accumulates and thermalizes the ions extracted from the KISS and bunch them for the MRTOF-MS, were installed already. They are now under commissioning.

For the precise laser ionization spectroscopy, the in-gas-jet laser ionization[15] will be performed at the KISS. The resolution for the HFS spectra will be reduced from the present value of 12 GHz to 0.35 GHz in FWHM. A narrow-band laser system has been newly installed, and a de Laval nozzle was placed at the gas-cell exit to form the gas jet. An S-shaped RF ion guide was installed to radiate the excitation laser collinearly with the gas jet, and the ionization laser is irradiated perpendicularly reflected many times by the mirrors in both sides. The off-line test of the in-gas-jet laser ionization spectroscopy using the evaporated atoms from the platinum filament resulted in the resolution of 0.6 ± 0.1 GHz in the HFS spectrum for ^{194}Pt.

5. Summary

The production of neutron-rich nuclei around the neutron magic number of 126 by MNT reactions has been recently focused on both theoretically and experimentally. The study of those nuclei are important for the investigation of astrophysical environments of the r-process. However the experimental data had been scared. We have performed the MNT measurements of ^{136}Xe and ^{198}Pt. The results indicated the advantage of the MNT reactions over fragmentation for production of the neutron-rich nuclei around $N = 126$. Encouraged by this results, we have been developing the KISS, where the MNT reactions of ^{136}Xe and ^{198}Pt are used to produce the neutron-rich nuclei around $N = 126$. The gas cell, the laser resonant ionization and the ISOL are used for efficient collection and separation of those MNT reaction products. We have performed the lifetime measurements and the β–γ spectroscopy for $^{199-201}$Pt, $^{196-200}$Ir, $^{195-208}$Os and $^{185-187}$Ta. The laser ionization spectroscopy was performed for ^{199}Pt and $^{196-198}$Ir. The three-dimensional tracking gas counter is under development for lower background rate of β-ray detection. The mass measurements with the MRTOF-MS is planned. The in-gas-jet laser ionization is ready for the precise laser ionization spectroscopy.

Acknowledgments

The experiments at the KISS were performed at the RI Beam Factory operated by RIKEN Nishina Center and CNS, University of Tokyo. The authors acknowledge the staff of the RIKEN accelerator for their support.

References

1. E.M. Burbidge *et al.*, *Rev. Mod. Phys.* **29**, 547 (1957).
2. Y. Hirayama *et al.*, *Nucl. Instrum. and Methods B* **353**, 4 (2015).
3. V. Zagrebaev and W. Greiner, *Phys. Rev. Lett.* **101**, 122701 (2008).
4. J.S. Barrett *et al.*, *Phys. Rev. C* **91**, 064615 (2015).
5. M. Rejmund *et al.*, *Nucl. Instrum. and Methods A* **646**, 184 (2011)
6. J. Simson *et al.*, *Acta Phys. Hung. New Ser.: Heavy Ion Phys.* **11**, 159 (2000).
7. Y.H. Kim *et al.*, *EPJ Web of Conferences* **66**, 03044 (2014).
8. Y.X. Watanabe *et al.*, *Phys. Rev. Lett.* **115**, 172503 (2015).
9. A.V. Karpov and V.V. Saiko, *Phys. Rev. C* **96**, 024618 (2017).
10. Y. Hirayama *et al.*, *Nucl. Instrum. and Methods B* **412**, 11 (2017).
11. M. Mukai *et al.*, *Nucl. Instrum. and Methods A* **884**, 1 (2018).
12. Y. Hirayama *et al.*, *Phys. Rev. C* **98**, 014321 (2018).
13. Y. Hirayama *et al.*, *Phys. Rev. C* **96**, 014307 (2017).
14. P. Schury *et al.*, *Phys. Rev. C* **95**, 011305(R) (2017).
15. Yu. Kudryavtsev *et al.*, *Nucl. Instrum. and Methods B* **297**, 7 (2013).

Production and Study of Neutron Rich Heavy Nuclei, GALS Setup

S. G. Zemlyanoy[1,*], V. I. Zagrebaev[1], K. A. Avvakumov[1], N. Yu. Kazarinov[1], Yu. Kudryavtsev[2], V. Fedosseev[3], R. Bark[4] and Z. Janas[5]

[1]*Joint Institute for Nuclear Research, Dubna 141980, Russia*
[2]*KU Leuven, Instituut voor Kern- en Stralingsfysica, B-3001 Leuven, Belgium*
[3]*CERN, 1211, Geneva 23, Switzerland*
[4]*iThemba LABS, Nat. Research Foundation, South Africa*
[5]*Faculty of Physics, University of Warsaw, Hoza 69, PL-00681 Warsaw, Poland*
[*]*E-mail: zemlya@jinr.ru*

To produce and investigate exotic heavy neutron rich nuclei located in the “north-east” region of nuclear map the new setup is under construction at Flerov Laboratory for Nuclear Reactions (FLNR) of JINR, Dubna. In this setup available U-400M cyclotron beams will be used in low energy multi-nucleon transfer reactions. Products of 4.5 to 9 MeV per nucleon heavy-ion collisions, such as ^{136}Xe beam on ^{208}Pb target, are to be captured in a gas cell, selectively laser-ionized and then transported towards mass separator by a sextupole (or quadrupole) ion guide extraction system.

Keywords: Exotic heavy nuclei; neutron rich nuclei; laser ionization.

1. Introduction

In recent years the production and study of heavy neutron-rich nuclides are of great interest for nuclear physics investigations [1-5]. It is very important for nuclear astrophysics, especially for understanding the r-process of the astrophysical nucleosynthesis. The study of the nuclear structure of extremely neutron rich nuclei is also important for the basic nuclear spectroscopy. In order to produce and study such neutron-rich products of multinucleon transfer reactions the GAs cell based Laser ionization and Separation setup (GALS) is constructed at Flerov Laboratory for Nuclear Reactions of JINR. This setup is intended for production of heavy nuclei from the north-east part of nuclear map, which are near to or above the magic neutron number N = 126. New information about the changes of nuclear ground state properties can be obtained, e.g. appearance of new magic numbers or disappearance of the classical ones (as can be observed for light neutron-rich nuclei) as well as the occurrence of rapid nuclear structural changes, due to sudden collectivity onset.

The region of heavy neutron rich nuclei along the neutron line N = 126 is still a blank field of the nuclide chart. Recently some new isotopes were identified in reactions with a ^{238}U beam impinging on a Be target at

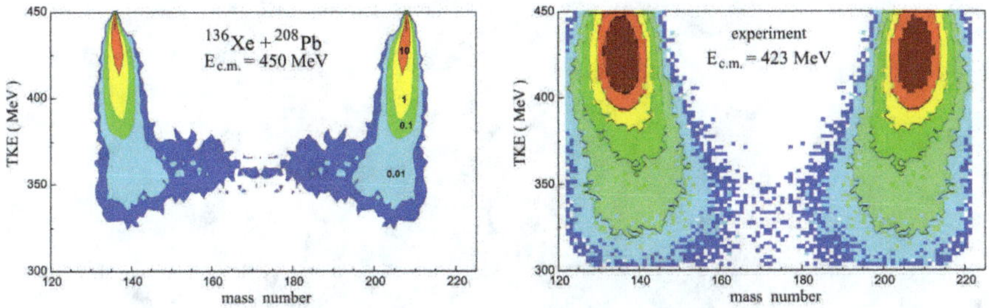

Fig. 1. Dependence of the transfer reactions products yield on the total kinetic energy and mass number for ^{136}Xe + ^{208}Pb reaction. The figure compares the theoretical predictions of [4] (on the left) and the experimental results (on the right) from the JINR cyclotron U-400M (courtesy of V. Zagrebaev and E. Kozulin). The numbers in the right-hand side of the theoretical predictions present the cross-section in mb.

1 GeV/nucleon [6]. The production cross-section for the new isotopes has been measured down to the pico-barn level and compared with predictions of different model calculations. However the magic neutron number N = 126 for atoms with Z = 75 – 78 is as still not achieved, because production and study of heavier neutron rich nuclei with A > 160 encounters both physical and technical problems (see [1, 4] for details). Using low-energy multinucleon transfer reactions for the production of neutron-rich heavy and superheavy nuclei was discussed by Zagrebaev and Greiner [4]. Figure 1 demonstrates high enough predicted yield for such reactions, which opens new opportunities in the studies at heavy ion facilities.

2. Experimental Setup

The experimental technique is already described in detail in [1-3, 7, 8]. Here we present only a brief overview of the experimental method that the GALS setup will use. This method, as proposed and applied in [7, 8], combines simultaneous Z and A separation based on stopping of the products of nuclear reactions in a gas cell, their subsequent resonance laser ionization and mass-separation. The scheme of the GALS facility can be found in [3].

The operation principle was described in a number of papers (see e.g. [9] and the references therein). The first stage of laser system based on three Sirah dye lasers pumped by two Nd:YAG EdgeWave lasers and the laser laboratory preparation are almost finished. Now we concentrate on the GALS subsystems in the experimental room within the cyclotron U-400M hall. Figure 2 presents the view of GALS facility with its subsystems (Front-End, Einzel lens, mass-separator, focal chamber and registration system). Work on these components

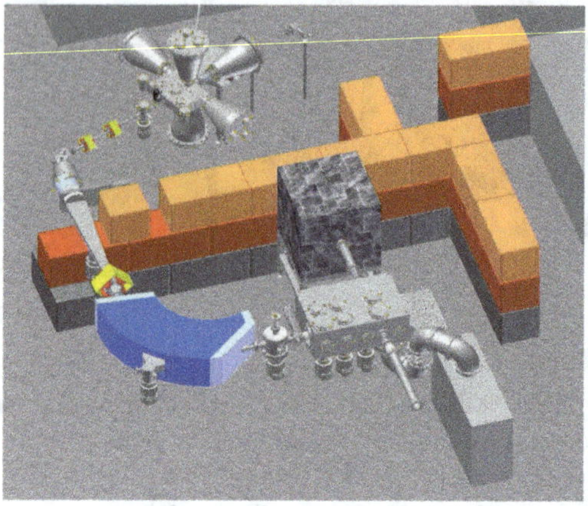

Fig. 2. The view of the whole GALS facility in experimental room within the FLNR U-400M cyclotron hall.

are now in the advanced stage of implementation. The mass separator could be a standard magnet separator similar to GPS at ISOLDE II [2, 10]), but it still requires considerable effort for its production. The same refers to the ion extraction and gas purification systems.

2.1. *The analyzing magnet simulation*

3D simulations of the GALS mass separator and beam dynamics within it were carried out in [11]. This simulations used 3D map of the AM magnetic field calculated using OPERA 3D code [12]. The number of the ions used in this simulation was equal to $2 \cdot 10^4$.

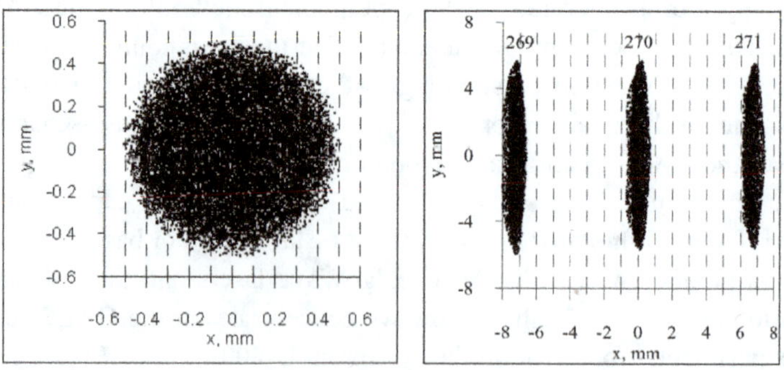

Fig. 3. Initial ion distribution in {x,y} plane (on the left side), and the ion distribution in the magnet focal plane (right side).

2.2. *The ion guide simulation*

In the previous publications it was shown that a sextupole ion guide (SPIG) can be used as an ion extraction and guiding system [13]. In order to estimate possible losses of the ions of interest, coming out of the gas cell, their interactions with the gas jet and background gas were taken into account in simulations (see Fig. 4). The simulations were performed in a SIMION software package [14] using the hard sphere collision model and the additional code describing the gas jet [15, 16]. In the model, the thermal ions propagate in a filled-cone direction distribution of half angle 30° out of the gas cell orifice with a diameter of 0.5 mm. The SPIG consists of six rods (each of them has a diameter of 1.5 mm and a length of 630 mm) cylindrically mounted on a sextupole structure. The SPIG inner diameter is 3 mm and its distance from the orifice was set to 2 mm. The radiofrequency (4.7 MHz) voltage with peak-to-peak amplitude up to 300 V is applied to the neighboring rods in the opposite phase. Also the whole SPIG has a voltage of −210 V compared to the gas cell [17]. The pressure of gas (argon) within the gas cell was set to 500 mbar, and the background pressure in the vacuum chamber was set to $2 \cdot 10^{-2}$ mbar. The simulations predict transport efficiency close to 100% on the first 65 mm of ions' path within the SPIG. When the ions proceed from the gas jet to the background, they quickly lose most of the kinetic energy, which lead to high losses or at least very high time of flight. To improve the expected transport efficiency and lower ions' time of flight, a segmented quadrupole ion guide is being developed.

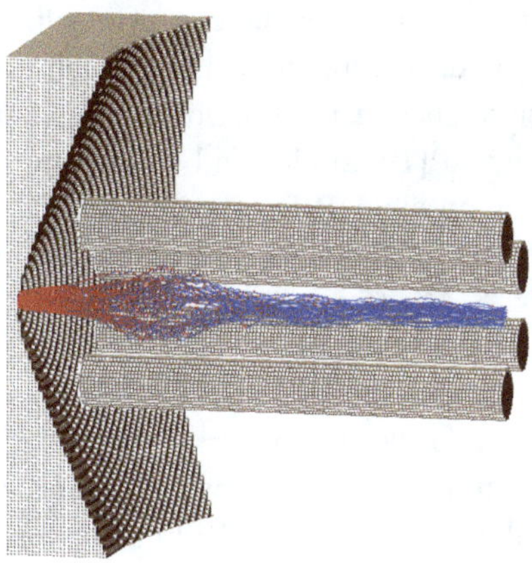

Fig. 4. Simulated trajectories of the ions at SPIG entrance. The collisions with the buffer gas atoms are marked with red dots.

3. Planned Experiments

The first online tests are planned to be performed on Os I. As already shown in [6], Os is very perspective candidate for reaching the neutron magic line N = 126. One of the recent calculations performed by A.V. Karpov [18] predicted sufficiently high reaction cross section for Os isotopes even above this magic line.

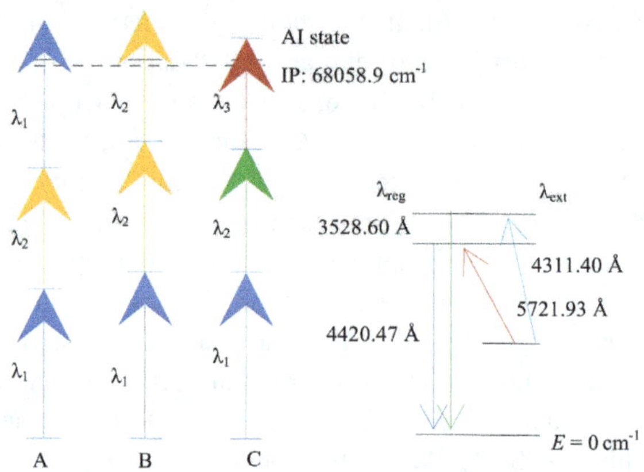

Fig. 5. *On the left side*: possible ionization schemes of Os I. The wavelengths (in air) are the following. A: $\lambda_1 = 3267.94$Å, $\lambda_2 = 5509.33$Å; B: $\lambda_1 = 3018.04$Å, $\lambda_2 = 5580.66$Å; C [20]: $\lambda_1 = 2909.06$Å, $\lambda_2 = 4752.16$Å, $\lambda_3 = 6043.15$Å. *On the right side*: transitions for the off-line test experiments.

In order to find suitable two- or three-step transitions for laser ionization, many Os I spectroscopic data were studied. Although there are many observed spectral lines, the atomic spectra information is not complete yet: most of the transitions are not assigned [19, 22-24] and the atomic strengths are unknown. However, some of the possible RIS schemes for Os have been previously studied [20, 21], and three of them which seem to be more effective (see e.g. [21]) are shown in Fig. 5. Most ionization schemes are 3-step ionization with a non-resonant transition to continuum.

The three-step resonant ionization scheme, with one step via an auto-ionizing state, is expected be the most favorable. Currently, there is no information in the literature about such states, except for the one at 71032 cm^{-1} corresponding to the third step of case C [20]. For this reason, at the second stage of the GALS activity, the search for auto-ionizing states is planned. At the next phase of experiments an off-line study of the optimal ionization scheme will be carried out. It supposes a detailed experimental study of the efficiency of the different ionization schemes.

4. Conclusion

A new GALS facility at FLNR U-400M cyclotron is in a stage of build-up. In this setup, the highly selective and efficient technique of stepwise resonant ionization in a gas cell (of the element of interest) with subsequent mass separation will be applied. The most efficient method to produce heavy neutron rich nuclei, as motivated in [4], is the multi-nucleon transfer reactions. Our estimations show that using primary ion beam of 0.1 pμA, target thickness of 0.3 mg/cm and setup efficiency of 10%, we will be able to measure decay properties of 1 new isotope per day.

References

1. Zagrebaev V. I., Zemlyanoy S. G., Kozulin E. M., Kudryavtsev Yu., Fedosseev V., Bark R., Othman H. A., Production and study of heavy neutron rich nuclei formed in multi-nucleon transfer reactions, *Hyperfine Interact,* **216**, 109 (2013).
2. Zagrebaev V. I., Zemlyanoy S. G., Kozulin E. M. *et al.*, Gas-cell-based setup for the production and study of neutron rich heavy nuclei, *Hyperfine Interact* **227**, 181 (2014).
3. Zagrebaev V. I., Zemlyanoy S. G., Kozulin E. M. *et al.*, GALS – setup for production and study of multinucleon transfer reaction products: present status, *J. Phys.: Conf. Ser.*, **724**, 012057 (2016).
4. Zagrebaev V. and Greiner W., Production of New Heavy Isotopes in Low-Energy Multinucleon Transfer Reactions, *Phys. Rev. Lett.*, **101**, 122701 (2008).
5. Long Zh., Zhao-Qing Feng, Feng-Shou Zhang, Production of heavy neutron-rich nuclei in transfer reactions within the dinuclear system model, *J. Phys. G: Nucl. Part. Phys.*, **42**, 085102 (2015).
6. Kurcewicz J., Farinon F., Geissel H. *et al.*, Discovery and cross-section measurement of neutron-rich isotopes in the element range from neodymium to platinum with the FRS, *Phys. Lett. B*, **717**, 371 (2012).
7. Kudryavstev Y. *et al.*, Beams of isotopes produced at LISOL by laser ionization after thermalization of energetic ions in a gas cell, *Nucl. Instrum. Methods Phys. Res. B*, **204**, 336 (2003).
8. Moore I. D. *et al.*, Development of a laser ion source at IGISOL, *J. Phys. G: Nucl. Part. Phys.*, **31**, S1499 (2005).

9. Campbell P., Moore I. D. and Pearson M. R., Laser spectroscopy for nuclear structure physics, *Prog. Particle Nucl. Phys.*, **86**, 127 (2016).
10. Ravn H. L. and Allardyce B. W., On-Line Mass Separators, *Treatise on Heavy-Ion Science*, **8**, 363 (1989).
11. Kazarinov N. Yu., 3D simulation of the ion-beam transport in bending magnets and electrostatic deflectors, *Physics of Particles and Nuclei Letters*, **13**, No. 7, 868 (2016).
12. Opera3D 2012 Oxford OX5 1JE, England.
13. Zemlyanoy S. *et al.*, Heavy neutron rich nuclei: Production and investigation, *J. Phys.: Conf. Ser.*, **1023**, 012004 (2018).
14. http://simion.com/
15. Traykov E. (GANIL, CEA/DSM-CNRS/IN2P3, Caen, France) private communication.
16. Kudryavtsev Y. *et al.*, The in-gas-jet laser ion source: Resonance ionization spectroscopy of radioactive atoms in supersonic gas jets, *Nucl. Instrum. Methods Phys. Res. B*, **297**, 7 (2013).
17. Zemlyanoy S., Avvakumov A., Fedosseev V. *et al.*, Current status of GALS setup in JINR, *Hyperfine Interact.*, **238**, 31 (2017).
18. Karpov A. V., Saiko V. V., Production of Neutron-Rich Nuclei in ^{136}Xe + ^{208}Pb Multinucleon Transfer Reaction, *Singapore: World Scientific*, ISBN 978-981-3226-53-1, 249 (2017).
19. Corliss, Charles H. and Bozman, William R., Experimental transition probabilities for spectral lines of seventy elements, *NBS Monograph*, **53** (1962).
20. Meggers, William F., Corliss, Charles H., Bourdon F. Scribner, Tables of spectral lines intensities (second edition), *NBS Monograph*, **145** (1975).
21. https://www.nist.gov/pml/atomic-spectra-database (version 5, updated October 2015).
22. Sansonetti J. E., Martin W. C., Handbook of Basic Atomic Spectroscopic Data, *J. Phys. Chem. Reference Data*, **34**, 1559 (2005).
23. Blum Joel D., Pellin M. J., Calawa W. F. *et al.*, Resonance ionization mass spectrometry of sputtered osmium and rhenium atoms, *Anal. Chem.*, **62**, 209 (1990).
24. Calaway W. F., Wien R. C., Burnett D. S. *et al.*, Simultaneous dual-element analyses of refractory metals in naturally occurring matrices using resonance ionization of sputtered atoms, *J. Vac. Sci. Technol. A*, **13**, 1310 (1995).

AUTHOR INDEX

LIST OF PARTICIPANTS

Md. Nure Alam **ABDULLAH**
Department of Physics, Jagannath
University, Dhaka-1100
BANGLADESH

Dieter **ACKERMANN**
Grand Accélérateur National d'Ions
Lourds — GANIL
Bd. Becquerel, BP 55027
14076 CAEN Cedex 05
FRANCE

Nikolay **AKSENOV**
Flerov Laboratory of Nuclear
Reactions
Joint Institute for Nuclear Research
141980 Dubna, Moscow region
RUSSIA

Yoshihiro **ARITOMO**
Kindai University
3-4-1 Kowakae
Higashi - Osaka City, Osaka
JAPAN

Dimiter **BALABANSKI**
ELI-NP, IFIN-HH
Horia Hulubei National Institute for
Physics and Nuclear Engineering
Str. Reactorului -30 P.O.BOX
MG-6, Bucharest-Magurele
ROMANIA

Christian **BECK**
IPNO CNRS
15 rue Georges Clemenceau
91406 Orsay Cedex
FRANCE

Piotr **BEDNARCZYK**
IFJ PAN, ul. Radzikowskiego 152
31-342 Kraków
POLAND

Olga **BEREZINA**
Petrozavodsk State University
Petrozavodsk, 185910
RUSSIA

Andrey **BEZBAKH**
Flerov Laboratory of Nuclear
Reactions
Joint Institute for Nuclear Research
141980 Dubna, Moscow region
RUSSIA

Catalin **BORCEA**
Horia Hulubei National Institute for
Physics and Nuclear Engineering
Str. Reactorului -30 P.O.BOX
MG-6, Bucharest-Magurele
ROMANIA

Ecaterina **BORCEA**
Horia Hulubei National Institute for Physics and Nuclear Engineering
Str. Reactorului -30 P.O.BOX
MG-6, Bucharest-Magurele
ROMANIA

Ivan **BORZOV**
NRC Kurchatov Institute
Moscow, 123182
RUSSIA

Nicolae **CARJAN**
Flerov Laboratory of Nuclear Reactions
Joint Institute for Nuclear Research
141980 Dubna, Moscow region
RUSSIA
Horia Hulubei - National Institute for Nuclear Physics and Engineering, P.O.Box MG-6
RO-76900, Bucharest
ROMANIA

Elena **CHERNYSHEVA**
Flerov Laboratory of Nuclear Reactions
Joint Institute for Nuclear Research
141980 Dubna, Moscow region
RUSSIA

Maxim **CHUYASHKIN**
Institute for Information Transmission Problems RAN
Bolshoy Karetny per. 19, build.1
Moscow 127051
RUSSIA

Andrey **DANILOV**
National Research Centre
Kurchatov Institute
Moscow 123182
RUSSIA

Alla **DEMYANOVA**
National Research Centre
Kurchatov Institute
Moscow 123182
RUSSIA

Sergey **DMITRIEV**
Flerov Laboratory of Nuclear Reactions
Joint Institute for Nuclear Research
141980 Dubna, Moscow region
RUSSIA

Natalia **DOKALENKO**
Joint Institute for Nuclear Research
141980 Dubna, Moscow region
RUSSIA

Alexander **D'YACHENKO**
Emperor Alexander I Petersburg State Transport University
St. Petersburg
RUSSIA

Vyacheslav **DYACHKOV**
Scientific Research Institute of Experimental and Theoretical Physics, Almaty
KAZAKHSTAN

Sergey **ELISEEV**
Max-Planck Institute for Nuclear Physics
Saupfercheckweg 1
D 69117 Heidelberg
GERMANY

Hideto **EN'YO**
RIKEN Nishina Center for Accelerator-Based Science
2-1 Hirosawa
Wako, Saitama 3510198
JAPAN

Batchuluun **ERDEMCHIMEG**
Flerov Laboratory of Nuclear Reactions
Joint Institute for Nuclear Research
141980 Dubna, Moscow region
RUSSIA

Tommi **ERONEN**
Department of Physics
University of Jyväskylä
P.O. Box 35 (Y5) FIN-40351 Jyväskylä
FINLAND

Naoki **FUKUDA**
RIKEN Nishina Center for Accelerator-Based Science
2-1 Hirosawa,
Wako, Saitama 3510198
JAPAN

Sydney **GALES**
IPNO CNRS
91406 Orsay Cedex
FRANCE

Ildar **GAZIZOV**
JSC "Institute in Physical - Technical Problems"
141980 Dubna, Moscow region
RUSSIA

Alexander **GORSHKOV**
Joint Institute for Nuclear Research
141980 Dubna, Moscow region
RUSSIA

Leonid **GRIGORENKO**
Flerov Laboratory of Nuclear Reactions
Joint Institute for Nuclear Research
141980 Dubna, Moscow region
RUSSIA

Hiromitsu **HABA**
Nishina Center for Accelerator-Based Science, RIKEN
Wako, Saitama 351-0198
JAPAN

Joseph **HAMILTON**
Department of Physics and Astronomy, Vanderbilt University
6301 Stevenson Center
VU Station B #351807
Nashville, TN 37235
USA

Karl **HAUSCHILD**
Institut de Physique Nucléaire d'Orsay UMR 8608 – CNRS-IN2P3
Université Paris Sud 91406 Orsay
FRANCE

Nobuo **HINOHARA**
Center for Computational Sciences
University of Tsukuba
Tsukuba, 305-8577
JAPAN

Fadi **IBRAHIM**
Institut de Physique Nucléaire
CNRS IN2P3 Univ Paris Sud
91406 Orsay Cedex
FRANCE

Andrey **ISAEV**
Flerov Laboratory of Nuclear Reactions
Joint Institute for Nuclear Research
141980 Dubna, Moscow region
RUSSIA

Talgat **ISSATAYEV**
Flerov Laboratory of Nuclear Reactions
Joint Institute for Nuclear Research
141980 Dubna, Moscow region
RUSSIA

Yulia **ITKIS**
Flerov Laboratory of Nuclear Reactions
Joint Institute for Nuclear Research
141980 Dubna, Moscow region
RUSSIA

Mikhail **ITKIS**
Joint Institute for Nuclear Research
141980 Dubna, Moscow region
RUSSIA

Makoto **ITO**
Department of Pure and Applied Physics, Kansai University
3-3-35 Yamate-cho
Suita, 564-8640
JAPAN

Vladimir **IVANSHIN**
Flerov Laboratory of Nuclear Reactions
Joint Institute for Nuclear Research
141980 Dubna, Moscow region
RUSSIA

Yoritaka **IWATA**
Tokyo Institute of Technology
Tokyo 152-8550
JAPAN

Igor **IZOSIMOV**
Flerov Laboratory of Nuclear Reactions
Joint Institute for Nuclear Research
141980 Dubna, Moscow region
RUSSIA

Daniyar **JANSEITOV**
Joint Institute for Nuclear Research
141980 Dubna, Moscow region
RUSSIA

Rostislav **JOLOS**
Bogoliubov Laboratory of Theoretical Physics, Joint Institute for Nuclear Research
141980 Dubna, Moscow region
RUSSIA

Asset **KABYSHEV**
L.N.Gumilyov Eurasian National University 5 Munaitpasov Street
010008, Astana
KAZAKHSTAN

Dmitry **KAMANIN**
Flerov Laboratory of Nuclear Reactions
Joint Institute for Nuclear Research
141980 Dubna, Moscow region
RUSSIA

M.T. Senthil **KANNAN**
Department of Physics
Bharathiar University
Coimbatore - 641046
INDIA

Alexander **KARPOV**
Flerov Laboratory of Nuclear Reactions
Joint Institute for Nuclear Research
141980 Dubna, Moscow region
RUSSIA

Evgeny **KONOBEEVSKI**
Institute for Nuclear Research of the Russian Academy of Sciences
Moscow, 117312
RUSSIA

Daria **KOSTYLEVA**
Justus-Liebig-Universität Gießen
II. Physikalisches Institut
Heinrich-Buff-Ring 16
35392 Gießen
GERMANY

Kairat **KUTERBEKOV**
L.N.Gumilyov Eurasian National University 5 Munaitpasov Street
010008, Astana
KAZAKHSTAN

Elena **KUZNETSOVA**
Flerov Laboratory of Nuclear Reactions
Joint Institute for Nuclear Research
141980 Dubna, Moscow region
RUSSIA

Kobus **LAWRIE**
iThemba LABS
PO Box 722
Somerset West 7129
SOUTH AFRICA

Zhong **LIU**
Key Laboratory of High Precision Nuclear Spectroscopy
Institute of Modern Physics
Chinese Academy of Sciences
Lanzhou 730000
PEOPLE'S REPUBLIC OF CHINA

Radomira **LOZEVA**
CSNSM, CNRS/IN2P3, Université Paris-Saclay, F-91405 Orsay
FRANCE

Sergey **LUKYANOV**
Flerov Laboratory of Nuclear Reactions
Joint Institute for Nuclear Research
141980 Dubna, Moscow region
RUSSIA

Adam **MAJ**
IFJ PAN, ul. Radzikowskiego 152
31-342 Kraków
POLAND

Tatiana **MAMONOVA**
Flerov Laboratory of Nuclear Reactions
Joint Institute for Nuclear Research
141980 Dubna, Moscow region
RUSSIA

Vladimir **MASLOV**
Flerov Laboratory of Nuclear Reactions
Joint Institute for Nuclear Research
141980 Dubna, Moscow region
RUSSIA

Nur Liyana Binti **MOHD ANUAR**
Kindai University
3-4-1 Kowakae
Higashi - Osaka City, Osaka
JAPAN

Evgeny **MOLCHANOV**
Joint Institute for Nuclear Research
141980 Dubna, Moscow region
RUSSIA

Daisuke **NAGAE**
RIKEN Nishina Center for Accelerator-Based Science
Wako Nuclear Science Center
RBF-415/416, 2-1 Hirosawa
Wako, Saitama, 351-0198
JAPAN

Mikhail **NAUMENKO**
Flerov Laboratory of Nuclear Reactions
Joint Institute for Nuclear Research
141980 Dubna, Moscow region
RUSSIA

Ivan **NIKONOROV**
JSC "Institute in Physical - Technical Problems"
141980 Dubna, Moscow region
RUSSIA

Hiroki **NISHIBATA**
RIKEN Nishina Center for Accelerator-Based Science
2-1 Hirosawa, Wako
Saitama 3510198
JAPAN

Katsuhisa **NISHIO**
Advanced Science Research Center, Japan Atomic Energy Agency
Tokai, Ibaraki, 319-1195
JAPAN

Yuri **NOVIKOV**
Petersburg Nuclear Physics Institute, Research National Center "Kurchatov Institute", Gatchina
188300, RUSSIA

Aleksey **NOVOSELOV**
Flerov Laboratory of Nuclear Reactions
Joint Institute for Nuclear Research
141980 Dubna, Moscow region
RUSSIA

Yuri **OGANESSIAN**
Flerov Laboratory of Nuclear Reactions
Joint Institute for Nuclear Research
141980 Dubna, Moscow region
RUSSIA

Igor **PANOV**
National Research Center "Kurchatov Institute"
123182, Moscow,
RUSSIA
NRC "Kurchatov Institute-ITEP"
117218, Moscow
RUSSIA

Yulia **PARFENOVA**
Flerov Laboratory of Nuclear Reactions
Joint Institute for Nuclear Research
141980 Dubna, Moscow region
RUSSIA

Yuri **PENIONZHKEVICH**
Flerov Laboratory of Nuclear Reactions
Joint Institute for Nuclear Research
141980 Dubna, Moscow region
RUSSIA

Valery **PERSHINA**
Helmholtzzentrum für Schwerionenforschung GmbH
Darmstadt, Planckstr. 1, D-64291 Darmstadt,
GERMANY

Andrei **POPEKO**
Flerov Laboratory of Nuclear Reactions
Joint Institute for Nuclear Research
141980 Dubna, Moscow region
RUSSIA

Alessandro **PULCINI**
Dipartimento di Fisica "Ettore Pancini", Università degli Studi di Napoli Federico II
Napoli, 80126
ITALY
Istituto Nazionale Fisica Nucleare
Sezione di Napoli, Napoli
ITALY

Yuri **PYATKOV**
Flerov Laboratory of Nuclear Reactions
Joint Institute for Nuclear Research
141980 Dubna, Moscow region
RUSSIA

Alexander **RODIN**
Flerov Laboratory of Nuclear Reactions
Joint Institute for Nuclear Research
141980 Dubna, Moscow region
RUSSIA

Vyacheslav **SAIKO**
Flerov Laboratory of Nuclear Reactions
Joint Institute for Nuclear Research
141980 Dubna, Moscow region
RUSSIA

Satoshi **SAKAGUCHI**
Department of Physics, Kyushu University, 10, Higashi Ward
Fukuoka, 812-0053
JAPAN

Vyacheslav **SAMARIN**
Flerov Laboratory of Nuclear Reactions
Joint Institute for Nuclear Research
141980 Dubna, Moscow region
RUSSIA

Tetsuya K. **SATO**
Japan Atomic Energy Agency
Tokai, Ibaraki 319-1195
JAPAN

Hervé **SAVAJOLS**
Grand Accélérateur National d'Ions Lourds — GANIL
Bd. Becquerel, BP 55027, 14076
CAEN Cedex 05
FRANCE

Vasili **SEMIN**
Flerov Laboratory of Nuclear Reactions
Joint Institute for Nuclear Research
141980 Dubna, Moscow region
RUSSIA

Yuri **SEREDA**
Flerov Laboratory of Nuclear Reactions
Joint Institute for Nuclear Research
141980 Dubna, Moscow region
RUSSIA

Alexey **SEVERYUKHIN**
Bogoliubov Laboratory of Theoretical Physics Joint Institute for Nuclear Research 141980
Dubna, Moscow reg.
RUSSIA

Boris **SHARKOV**
Joint Institute for Nuclear Research
141980 Dubna, Moscow region
RUSSIA

Madan M. **SHARMA**
Physics Department
Kuwait University
Jamal Abdul Nasser St. 13060
KUWAIT

Andrey **SHIROKOV**
Skobeltsyn Institute of Nuclear Physics
Lomonosov Moscow State University
Moscow 119991
RUSSIA

Maxim **SHUMEIKO**
Flerov Laboratory of Nuclear Reactions
Joint Institute for Nuclear Research
141980 Dubna, Moscow region
RUSSIA

Roman **SLEPNEV**
Flerov Laboratory of Nuclear Reactions
Joint Institute for Nuclear Research
141980 Dubna, Moscow region
RUSSIA

Yuri **SOBOLEV**
Flerov Laboratory of Nuclear Reactions
Joint Institute for Nuclear Research
141980 Dubna, Moscow region
RUSSIA

Viktar **STARASTIN**
National Research Centre
Kurchatov Institute
Moscow 123182
RUSSIA

Boris **STARCHENKO**
Joint Institute for Nuclear Research
141980 Dubna, Moscow region
RUSSIA

Patrick **STEINEGGER**
Flerov Laboratory of Nuclear Reactions
Joint Institute for Nuclear Research
141980 Dubna, Moscow region
RUSSIA

Alexander **STREKALOVSKIY**
Flerov Laboratory of Nuclear Reactions
Joint Institute for Nuclear Research
141980 Dubna, Moscow region
RUSSIA

Oleg **TARASOV**
National Superconducting Cyclotron Laboratory
Michigan State University
East Lansing, MI 48824
USA

Gurgen **TER-AKOPIAN**
Flerov Laboratory of Nuclear Reactions
Joint Institute for Nuclear Research
141980 Dubna, Moscow region
RUSSIA

Dmitry **TESTOV**
Flerov Laboratory of Nuclear Reactions
Joint Institute for Nuclear Research
141980 Dubna, Moscow region
RUSSIA

Yasuhiro **TOGANO**
Rikkyo University
3-34-1 Nishi-Ikebukuro
Toshima-ku, Tokyo, 171-8501
JAPAN

Smarajit **TRIAMBAK**
University of the Western Cape
Private Bag X17
Bellville 7535
REPUBLIC OF SOUTH AFRICA

Tomohiro **UESAKA**
RIKEN Nishina Center for Accelerator-Based Science
2-1 Hirosawa, Wako
Saitama 3510198
JAPAN

Anna **UNZHAKOVA**
Saint Petersburg State University
St. Petersburg, 199034
RUSSIA

Vladimir **UTYONKOV**
Flerov Laboratory of Nuclear Reactions
Joint Institute for Nuclear Research
141980 Dubna, Moscow region
RUSSIA

Emanuele **VARDACI**
Dipartimento di Fisica "Ettore Pancini", Università degli Studi di Napoli Federico II
Napoli, 80126
ITALY

David **VERNEY**
Institut de Physique Nucléaire d'Orsay UMR 8608 – CNRS-IN2P3 Université Paris Sud 91406 Orsay
FRANCE

Zeblon **VILAKAZI**
School of Physics
University of the Witwatersrand
Johannesburg
Private Bag Wits 2050
SOUTH AFRICA

Alexey **VOINOV**
Flerov Laboratory of Nuclear Reactions
Joint Institute for Nuclear Research
141980 Dubna, Moscow region
RUSSIA

Anatolij **VORONIN**
Petrozavodsk State University
Petrozavodsk, 185910
RUSSIA

Victor **VORONOV**
Bogoliubov Laboratory of Theoretical Physics
Joint Institute for Nuclear Research
141980 Dubna, Moscow region
RUSSIA

Yutaka **WATANABE**
Wako Nuclear Science Center
Institute of Particle and Nuclear Studies, High Energy Accelerator Research Organization (KEK)
Wako, Saitama 351-0198
JAPAN

Roman **WOLSKI**
Flerov Laboratory of Nuclear Reactions
Joint Institute for Nuclear Research
141980, Dubna, Moscow region
RUSSIA

Alexander **YUSHKOV**
Scientific Research Institute of Experimental and Theoretical Physics, Almaty
KAZAKHSTAN

Bogumił **ZALEWSKI**
Flerov Laboratory of Nuclear Reactions
Joint Institute for Nuclear Research
141980 Dubna, Moscow region
RUSSIA

Yuliya **ZARIPOVA**
Scientific Research Institute of Experimental and Theoretical Physics, Almaty
KAZAKHSTAN

Pavel **ZARUBIN**
Joint Institute for Nuclear Research
141980 Dubna, Moscow region
RUSSIA

Irina **ZARUBINA**
Joint Institute for Nuclear Research
141980 Dubna, Moscow region
RUSSIA

Elena **ZEMLYANAYA**
Joint Institute for Nuclear Research
141980 Dubna, Moscow region
RUSSIA

Sergey **ZEMLYANOY**
Flerov Laboratory of Nuclear Reactions
Joint Institute for Nuclear Research
141980 Dubna, Moscow region
RUSSIA

PITER INN
HOTEL & CONFERENCE CENTRE
SHOP

PETROZAVODSK

PETROZAVODSK

PETROZAVODSK

PETROZAVODSK

lenovo

PETROZAVODSK
EX ON

PETROZAVODSK

PETROZAVODSK
lenovo

lenovo

lenovo

lenovo

lenovo

EXPERIMENTAL STUDIES OF TOTAL REACTION CROSS SECTION
U-400M
ACCULINNA

Evaluation for Possibility of Synthesizing
New Superheavy Elements
近畿大学
EXON 201
10-15 Sept 2
Introduction
Method
Results

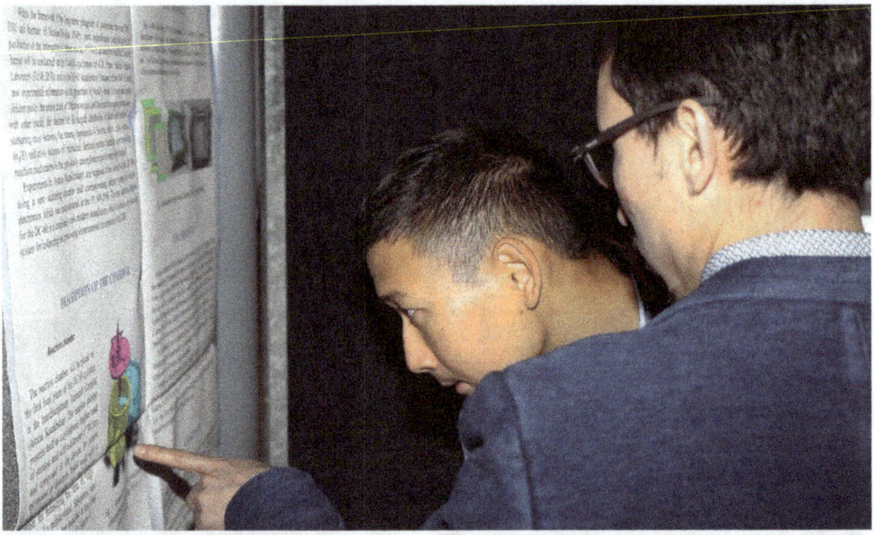

POSTER
SESSION
2
IX International Symposium on Exotic Nuclei
Russia, Petrozavodsk, September 10-16, 2018
Determination of separation efficiency of MASHA spectrometer by means of measurement of absolute cross-sections of evaporation residues.

of statically deformed heavy nuclei
transfer reactions
Model

MOTIVATION
EXPERIMENTAL SETUP
RESULTS
METHOD

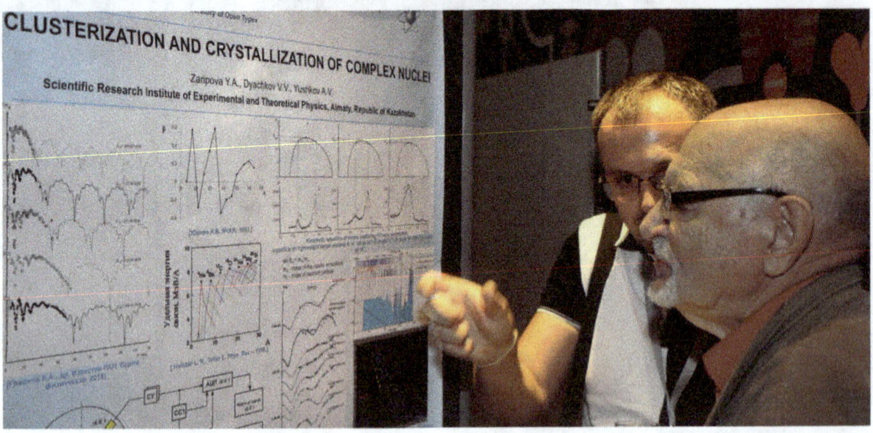
CLUSTERIZATION AND CRYSTALLIZATION OF COMPLEX NUCLEI
Zaripova Y.A., Dyachkov V.V., Yushkov A.V.
Scientific Research Institute of Experimental and Theoretical Physics, Almaty, Republic of Kazakhstan

SONOR

ВЫНОС МУСОРА
ЗА БОРТ
ЗАПРЕЩЁН

ТАИНСТВ
РЕСТАВРАЦ
SECRETS
OF RESTORATIO

www.ingramcontent.com/pod-product-compliance
Lightning Source LLC
LaVergne TN
LVHW010537300926
849162LV00002B/159